Sustainable Development and Biodiversity

Volume 11

Series editor

Kishan Gopal Ramawat
Botany Department, M.L. Sukhadia University, Udaipur, Rajasthan, India

This book series provides complete, comprehensive and broad subject based reviews about existing biodiversity of different habitats and conservation strategies in the framework of different technologies, ecosystem diversity, and genetic diversity. The ways by which these resources are used with sustainable management and replenishment are also dealt with. The topics of interest include but are not restricted only to sustainable development of various ecosystems and conservation of hotspots, traditional methods and role of local people, threatened and endangered species, global climate change and effect on biodiversity, invasive species, impact of various activities on biodiversity, biodiversity conservation in sustaining livelihoods and reducing poverty, and technologies available and required. The books in this series will be useful to botanists, environmentalists, marine biologists, policy makers, conservationists, and NGOs working for environment protection.

More information about this series at http://www.springer.com/series/11920

Vijay Rani Rajpal · S. Rama Rao
S.N. Raina
Editors

Molecular Breeding for Sustainable Crop Improvement

Volume 2

Springer

Editors
Vijay Rani Rajpal
Delhi University
Delhi
India

S.N. Raina
Amity University of Biotechnology
Noida, Uttar Pradesh
India

S. Rama Rao
Department of Biotechnology and Bioinformatics
North Eastern Hill University
Shillong, Megalaya
India

ISSN 2352-474X ISSN 2352-4758 (electronic)
Sustainable Development and Biodiversity
ISBN 978-3-319-27088-3 ISBN 978-3-319-27090-6 (eBook)
DOI 10.1007/978-3-319-27090-6

Library of Congress Control Number: 2015957782

Printed on acid-free paper

This Springer imprint is published by SpringerNature
The registered company is Springer International Publishing AG Switzerland

Preface

The advent of modern tools and techniques in genetics, cytogenetics, and molecular biology has revolutionized the plant breeding approaches to help overcome the limitations associated with the conventional plant breeding. The recent integration of advances in biotechnology, genomic research, and molecular marker applications with conventional plant breeding practices has created the foundation for molecular plant breeding, an interdisciplinary science that is revolutionizing twenty-first-century crop improvement. Molecular breeding offers opportunities to apply marker assisted selection (MAS) as compared to phenotypic selection in conventional breeding and is moving ahead at an unprecedented pace in all major crops and has become the standard practice in many.

Molecular breeding involves alien gene introgression to mobilize useful genetic variation to breeding programs to widen the genetic base of crop cultivars and relies on the use of molecular markers, and linkage, QTL, and association mapping of agronomic traits to identify candidate genes and to design functional markers for MAS. MAS coupled with marker assisted backcross breeding (MABB) that helps in gene pyramiding, next-generation sequencing (NGS) that generates genome-wide markers and screening of new alleles, and targeting induced local lesions in genomes (TILLING) or ecotype TILLING (EcoTILLING) for the screening of either mutant or natural germplasm collections are used to integrate genomic information into directional and selective breeding in crops to maximize genetic gains.

This book encompasses articles on the application of above-mentioned genomic approaches, tools, and resources in a precision breeding approach. Each chapter elucidates an authoritative account on the topic. We are sincerely grateful to all the authors for their valuable contributions. We would like to acknowledge cooperation, patience, and support of our contributors who have put in their serious efforts to ensure a high scientific quality of this book with up-to-date information. We sincerely thank Dr. K.G. Ramawat for motivating us to take up this assignment. Sincere thanks are due to Khushboo Arora for her help during the editing process. One of the editors, Vijay Rani Rajpal, is sincerely grateful to her daughter Navya and husband Susheel for their help, patience, and understanding. Without their unconditional support and encouragement, this journey would have been even more difficult. This work could not be completed without the active support of Springer team who took pains in streamlining the production process. We particularly appreciate Dr. Valeria for her continued support.

Plant breeders, taxonomists, geneticists, cytogeneticists, molecular biologists, and biotechnologists will greatly benefit from this book. We sincerely hope that this book will serve as a milestone in the precision breeding of crops to achieve meaningful plant genetic improvement.

Vijay Rani Rajpal
S. Rama Rao
S.N. Raina

Contents

Contributors

Sazda Abdi Department of Botany, University of Delhi, Delhi, India

Ramesh K. Aggarwal Centre for Cellular and Molecular Biology (CSIR-CCMB), Hyderabad, Telangana, India

Paramvir Singh Ahuja Biotechnology Division, CSIR-Institute of Himalayan Bioresource Technology, (Council of Scientific and Industrial Research), Palampur, Himachal Pradesh, India

Sapinder Bali Department of Botany, University of Delhi, New Delhi, India

Vishnu Bhat Department of Botany, University of Delhi, Delhi, India

David Bonnett Bayer Crop Science, Fargo, ND, USA

Atika Chandra Maitreyi College, University of Delhi, Delhi, India

Parveen Chhuneja School of Agricultural Biotechnology, Punjab Agricultural University, Ludhiana, India

Jose Crossa CIMMYT Int., Mexico, DF, Mexico

Jeena Devasia Central Coffee Research Institute, Chikmagalur, Karnataka, India

H.S. Dhaliwal Eternal University, Baru Sahib, Himachal Pradesh, India

Harcharan Singh Dhaliwal Eternal University, Sirmour, Himachal Pradesh, India

Susanne Dreisigacker CIMMYT Int., Mexico, DF, Mexico

Anuj Dwivedi Department of Botany, University of Delhi, New Delhi, India

Nicholas W. Ellison AgResearch Grasslands, Palmerston North, New Zealand

Shailendra Goel Department of Botany, University of Delhi, New Delhi, India

S. Gopala Krishnan Division of Genetics, ICAR-Indian Agricultural Research Institute, New Delhi, India

P.K. Gupta Department of Genetics and Plant Breeding, Chaudhary Charan Singh University, Meerut, UP, India

Carlos Guzmán CIMMYT Int., Mexico, DF, Mexico

Xinyao He CIMMYT Int., Mexico, DF, Mexico

Satinder Kaur School of Agricultural Biotechnology, Punjab Agricultural University, Ludhiana, India

Kuldip Biotechnology Division, CSIR-Institute of Himalayan Bioresource Technology, (Council of Scientific and Industrial Research), Palampur, Himachal Pradesh, India

Kishor Kumar Punjab Agricultural University, Ludhiana, Punjab, India

Rahul Kumar Biotechnology Division, CSIR-Institute of Himalayan Bioresource Technology, (Council of Scientific and Industrial Research), Palampur, Himachal Pradesh, India; DAV University, Jalandhar, India

Sundip Kumar Molecular Biology and Genetic Engineering, College of Basic Sciences and Humanities, G.B. Pant University of Agriculture and Technology, Pantnagar, India

Upendra Kumar Centre of Excellence for Mountain Biology, Uttarakhand Council for Biotechnology, Haldi, India

Caixa Lan CIMMYT Int., Mexico, DF, Mexico

Divya Mohanty Department of Botany, University of Delhi, Delhi, India

Yasuhiko Mukai Laboratory of Plant Molecular Genetics, Division of Natural Sciences, Osaka Kyoiku University, Kashiwara, Osaka, Japan

Kumari Neelam Punjab Agricultural University, Ludhiana, Punjab, India

B.M. Prasanna Global Maize Program, CIMMYT (International Maize and Wheat Improvement Center), Nairobi, Kenya

Priyanka Centre of Excellence for Mountain Biology, Uttarakhand Council for Biotechnology, Haldi, India

Y. Raghuramulu Central Coffee Research Institute, Chikmagalur, Karnataka, India

Soom Nath Raina Amity Institute of Biotechnology, Amity University, Noida, Uttar Pradesh, India

Vijay Rani Rajpal Department of Botany, HansRaj College, University of Delhi, Delhi, India

Deepmala Sehgal International Maize and Wheat Improvement Center (CIMMYT), Texcoco, Edo. de México, Mexico; Wheat Geneticist and Molecular Breeder, CIMMYT, Texcoco, CP, Mexico

Ram Kumar Sharma Biotechnology Division, CSIR-Institute of Himalayan Bioresource Technology, (Council of Scientific & Industrial Research), Palampur, Himachal Pradesh, India

Santosh Kumar Sharma Laboratory of Plant Molecular Genetics, Division of Natural Sciences, Osaka Kyoiku University, Kashiwara, Osaka, Japan

Shashi Department of Botany, University of Delhi, Delhi, India

A.K. Singh Division of Genetics, ICAR-Indian Agricultural Research Institute, New Delhi, India

Kuldeep Singh Punjab Agricultural University, Ludhiana, Punjab, India

Richa Singh Department of Molecular Biology and Biotechnology, Chaudhary Charan Singh Haryana Agricultural University (CCSHAU), Hisar, Haryana, India

Alan V. Stewart PGG Wrightson Seeds, Lincoln, New Zealand

Sivakumar Sukumaran CIMMYT Int., Mexico, DF, Mexico

N. Surya Prakash Central Coffee Research Institute, Chikmagalur, Karnataka, India

Rajesh Tandon Department of Botany, University of Delhi, Delhi, India

Maki Yamamoto Department of Rehabilitation Sciences, Kansai University of Welfare Sciences, Kashiwara, Osaka, Japan

Chapter 1
Use of Alien Genetic Variation for Wheat Improvement

P.K. Gupta

Abstract Wheat production and productivity at the global level has witnessed a remarkable improvement during the last five decades, thus helping in providing food security. However, the annual growth rate in wheat production has declined from ~3 % in earlier decades to 0.5–0.7 % in recent years causing concern. Therefore, major worldwide efforts are being made to improve the yield potential of bread wheat. In this connection, alien genetic variation has been found to be an important source of genetic variation both for qualitative and quantitative traits of agronomic importance. A number of alien species belonging to the tribe Triticeae of the family Poaceae have been utilized for this purpose. These alien species have been utilized through the production of amphiploids, whole chromosome alien addition and substitution lines, whole-arm Robertsonian translocations, and the translocations involving small segments of alien chromosomes. The transfer of small segments carrying desirable alien genes was achieved through several approaches including irradiation, use of mutants, and suppression of diploidizing gene (Ph1). These alien resources along with the details of their successful utilization for wheat improvement have been described in this chapter.

Keywords Bread wheat · Alien species/genes/additions/substitutions · Amphiploids

1.1 Introduction

Wheat yield and production, worldwide and in India, have shown dramatic improvement during the last five decades. According to recent estimates, the mean global wheat yield has increased from 1.2 tonnes per hectare (t/ha) in the year 1961

The erratum of this chapter can be found under DOI 10.1007/978-3-319-27090-6_17

P.K. Gupta (✉)
Department of Genetics and Plant Breeding, Chaudhary Charan Singh University,
Meerut, UP, India
e-mail: pkgupta36@gmail.com

V.R. Rajpal et al. (eds.), *Molecular Breeding for Sustainable Crop Improvement*, Sustainable Development and Biodiversity 11,
DOI 10.1007/978-3-319-27090-6_1

to $\sim$3 t/ha in the year 2014 (in UK, the yield increased from 2.8 t/ha in 1948 to 8 t/ha in 2014). The world wheat production during the same period improved from $\sim$225 million tones (mt) to $\sim$715 mt. However, the annual growth rate in production has gone down from $\sim$3 % during the 1970s and 1980s to 0.5–0.7 % in recent years, which has been a matter of concern. It has also been observed that major advances in wheat yield and production have been due to better management and improved agricultural practices rather than by an improvement in the genetic potential (however, in UK, according to some estimates, the increase has been mainly due to improved genetic potential). A flow of newer varieties that are resistant to emerging new races of pathogens, particularly the three rusts, has also contributed to an improvement in yield and production, although it has been difficult for wheat breeders to sustain this activity. These aspects are being discussed at the global level through a Wheat Initiative launched in the year 2011 and have been more recently discussed during a meeting held as a part of Norman Borlaug Centenary Celebrations. The meeting was appropriately called 'Wheat for Food Security' and was held at the International Institute for Wheat and Maize Improvement (CIMMYT) in Mexico. The new 'International Wheat Yield Partnership' (IWYP) program, as a part of Wheat Initiative, was also launched during this meeting.

The major limitations in wheat yield and production have been due to biotic and abiotic stresses, which have been regularly addressed in recent years with limited success. It has also been recognized that the genetic variability, being limited within the wheat germplasm (primary gene pool), secondary, and tertiary gene pools can be exploited for additional genetic variability. This useful variability has been described as alien genetic variation, since it is available in taxa, which though lie outside the species delimitations of cultivated wheat, but can be crossed with wheat, even though sometimes using embryo rescue. Considerable information on the use of alien genetic variation for wheat improvement is available in several reviews published in the past (Sears 1981; Sharma and Gill 1983; Knott 1987; Islam and Shepherd 1991, Jiang et al. 1994a, b; Gill et al. 1996; Mujeeb-Kazi and Rajaram 2002). In this chapter, we briefly describe these alien genetic resources and discuss their utility in wheat improvement, with emphasis on relatively recent work, since earlier work has been adequately covered in several reviews published in the past.

1.2 Alien Genetic Resources

The major sources of alien genetic variation for wheat improvement are the taxa belonging to the sub-tribe Triticinae of the tribe Triticeae. These taxa include species from the following old and classical genera: *Triticum*, *Secale*, *Aegilops*, *Agropyron*, and *Haynaldia* (now called *Dasypyrum*). The old classical genus *Agropyron* is now known to include several newly described genera including *Thinopyrum*, *Lophopyrum*, *Elymus*, and *Leymus*. These and other alien genera have been classified in primary, secondary, and tertiary gene pools. The primary gene pool includes species of genera *Triticum* and *Aegilops*, which represent the progenitors of hexaploid wheat; the chromosomes of these species can pair with wheat chromosomes and genes can be transferred by normal recombination process without any aid of induced recombination or irradiation. In other words, the species carrying genomes A, B, or D constitute the primary gene pool. This gene pool contains hexaploid wheat landraces, cultivated tetraploid wheats, the wild tetraploid wheats (*T. turgidum*; syn. *Triticum dicoccoides*), and diploid progenitors of the A, B, and D sub-genomes to durum and bread wheats. Some cross-combinations require embryo rescue, but no cytogenetic manipulation is needed in exploiting this primary gene pool. The secondary gene pool consists of species belonging to the genera *Triticum* and *Aegilops*, which share only one sub-genome with the three sub-genomes of wheat, the other one or more genomes being not shared. Several diploid species of the Sitopsis section are included in this pool, and the F_1 hybrids within this gene pool exhibit reduced chromosome pairing. In such cases, gene transfers have been achieved through direct crosses, breeding protocols, homologous exchange of chromosome segments between the related genomes and through the use of special manipulation strategies. Embryo rescue is a complementary aid for obtaining hybrids. The tertiary gene pool includes diploid and polyploid species of the genera belonging to the tribe Triticeae, which carry genomes other than those present in hexaploid wheat. Homoeologous relationship between the chromosomes of these species and wheat allow alien gene transfers through more complex chromosome manipulations. Special techniques involving irradiation, use of *ph1* mutants or callus culture-mediated induction of translocations are also used. The wild hexaploid wheat landraces and distant relatives of wheat [e.g., *Secale cereale* (rye), *Thinopyrum bessarabicum*] also provide a vast and largely untapped reservoir of genetic variation that is used for improvement of target traits in wheat. Some of the alien species listed in Table 1.1 have been successfully utilized for the development of whole chromosome alien addition/substitution lines and/or translocations carrying each a complete alien chromosome arm or a small alien chromosome segment.

In UK, a major program in the form of WISP (Wheat Improvement Strategic Program) consortium has recently been launched to utilize alien genetic variation for wheat improvement. It is a BBSRC-funded collaborative program, which brings together experts from the following five UK institutions, and will run from 2011 to 2017: (i) John Innes Centre, (ii) National Institute for Agricultural Botany (NIAB), (iii) University of Nottingham, (iv) University of Bristol, and (v) Rothamsted Research Institute. The program will collect and generate new and novel wheat germplasm characterized for traits relevant to academics and breeders. It will also identify gene-based or DNA-based molecular markers for selecting these traits. The program is structured around three complementary 'pillars' (landraces, synthetics, and ancestral gene introgression), each of which will broaden the pool of genetic variation in wheat by a different route. Among the three pillars, the last one will deal with new and useful alien genetic variation from related species.

Table 1.1 Some examples of alien species, which have been used for the production of alien–wheat chromosome additions, substitutions, and translocations

Species and genome	Reference
Aegilops tauschii (*D*)	Joppa (1987), Friebe et al. (1992a)
Aegilops speltoides (S)	Friebe et al. (2000)
Aegilops longissima (S^l)	Friebe et al. (1993)
Aegilops searsii (S^s)	Friebe et al. (1995a)
Aegilops umbellulata (U)	Friebe et al. (1995b)
Aegilops caudata (C)	Friebe et al. (1992b)
Aegilops biuncialis (Mb)	Farkas et al. (2014)
Aegilops peregrina (U^US^U)	Friebe et al. (1996b)
Secale cereale (R)	Gill and Kimber (1974), Mukai et al. (1992)
Thinopyrum intermedium (E1E2X)	Friebe et al. (1992c)
Leymus racemosus (JN)	Qi et al. (1997)
Elymus trachycaulus (SH)	Jiang et al. (1994a)
Hordeum chilense (H^{ch})	Cabrera et al. (1995)

1.3 Strategies for Using Alien Genetic Variation

The process for the transfer of target genes from alien species into wheat often involves prior development of amphiploids and the alien addition/substitution lines. This is followed either by irradiation or by induced recombination (facilitated by absence of chromosome 5B or presence of *ph1* mutant) to produce translocations. Different strategies for using alien genetic variation will be briefly discussed in this section.

1.3.1 Amphiploids for Transfer of Alien Chromosome Segments

During the 1950s and 1960s, production of amphiploids used to be the first step for utilization of alien genetic variation for wheat improvement. The production and utilization of amphiploids generally involved the following steps: (i) crossing tetraploid wheat (AABB) or hexaploid wheat (AABBDD) with a diploid or tetraploid alien species to produce F_1 hybrids (sometimes using embryo rescue); (ii) doubling the chromosome number in the F_1 hybrids through colchicine treatment to produce amphiploids; (iii) crossing and backcrossing the amphiploid (some times after irradiation treatment; see next paragraph) with hexaploid wheat to facilitate transfer of either an alien segment or the whole alien chromosome or only one arm of a chromosome to wheat [for details about transfer of a whole chromosome leading to production of alien addition and alien substitution lines, and an arm leading to Robertsonian translocations, see later in this chapter].

As an example of the above procedure, in a recent study, a 10*x* amphiploid ($AABBDDU^bU^bM^bM^b$) was produced by combining hexaploid wheat (AABBDD) and tetraploid *Ae. biuncialis* ($U^bU^bM^bM^b$). This amphiploid was used for irradiation leading to transfer of alien segments, which could be identified using multicolor FISH (McFISH) (Molnar et al. 2009).

In another study conducted in Pakistan, Mujeeb-Kazi and his coworkers utilized *Ae. variabilis* ($2n = 4x = 28$; UUSS) as an alien species for imparting resistance against *Cochliobolus sativus* (spot blotch) and *Tilletia indica* (Karnal bunt). The amphiploids with $2n = 8x = 56$ (AABBUUSS) and $2n = 10x = 70$ (AABBDDUUSS) were obtained using this alien species. The morphology and cytogenetics of these amphiploids were examined with a view to transfer alien chromosome segments carrying genes for resistance against spot blotch and Karnal bunt into the recipient durum and bread wheat germplasm.

Direct use of amphiploids for wheat improvement through the transfer of an alien segment from crested wheatgrass (*Agropyron cristatum* = PP) carrying genes for resistance against leaf rust was also reported recently (Ochoa et al. 2015). The transfer of the alien segment involved crossing and backcrossing of the amphiploid AABBDDPP with hexaploid wheat (AABBDD). The amphiploid (AABBDDPP) itself was earlier obtained by crossing tetraploid wheat (*Triticum turgidum* L. conv. Durum Desf. $2n = 4x = 28$; AABB) with a self-fertile allotetraploid ($2n = 4x = 28$; DDPP) obtained through a cross between diploid wheat (*Aegilops tauschii* Coss.) and crested wheatgrass (*A. cristatum*). After three backcrosses, a fertile stable line (named TH4) was obtained with 42 chromosomes. Fluorescence in situ hybridization (FISH), genomic in situ hybridization (GISH), and use of genome-specific molecular markers confirmed that TH4 carried a compensating Robertsonian translocation involving the long arm of wheat chromosome 1B and the short arm of an unidentified *A. cristatum* chromosome. This TH4 line and similar other lines produced using this approach will certainly be used in future wheat breeding programs, as sources of resistance against a number of biotic and abiotic stresses (see later for details).

1.3.2 Use of Synthetic Hexaploid Wheats (SHWs)

One of the most successful programs for utilization of alien genetic variation for wheat improvement has been the development of a large number of SHWs by combining the genomes of tetraploid wheat (AABB) and *Aegilops tauschii* (DD) (Fig. 1.1). Since late 1980s, these SHWs have been produced in thousands at CIMMYT and are being utilized all over the world for the introgression of alien genetic material from *Ae. tauschii* into modern bread wheat cultivars (Lage et al. 2004; Talbot 2011). It has been shown that the hybrids produced from a cross between SHW and an improved variety exhibit double the genetic diversity relative

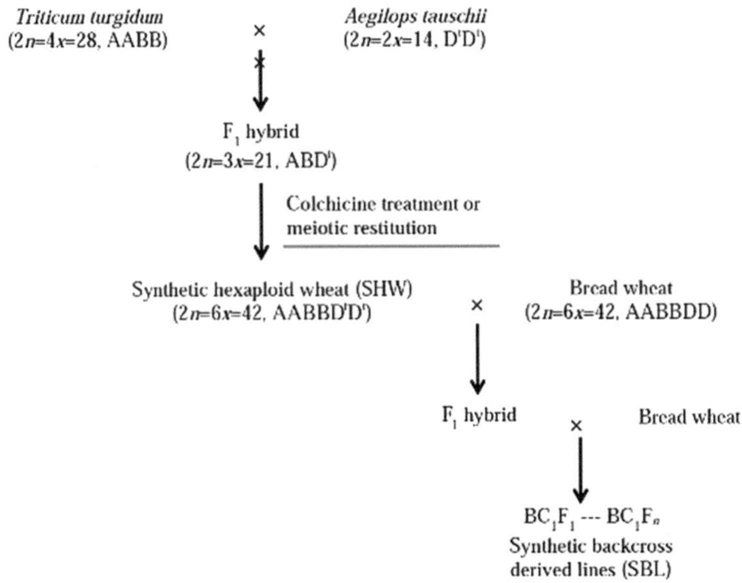

Fig. 1.1 Different steps involved in the transfer of desirable traits from synthetic hexaploid wheat (*SHW*) into elite cultivars in the form of synthetic backcross-derived lines (*SBLs*)

to its parents (Kazi and Van Ginkel 2004). Once created, a SHW (like any other amphiploid, as mentioned above) can be crossed with one or more bread wheat cultivars, followed by repeated backcrosses (Lange and Jochemsen 1992; Trethowan and Mujeeb-Kazi 2008) to produce synthetic backcross-derived lines (SBLs; Fig. 1.1). In actual practice, these SBLs have been subjected to selection, and the selected SBLs exhibited significant yield increases across a diverse range of environments, demonstrating their potential for improving wheat productivity worldwide.

The desirable traits, which could be improved using SHWs include the following: (i) resistance or tolerance to a number of biotic stresses including resistance to leaf blotch, glume blotch, crown rot, yellow leaf spot, leaf blight, powdery mildew, and karnal bunt; they also exhibited resistance to certain insect pests such as Green bugs and Hessian fly (Van Ginkel and Ogbonnaya 2007); (ii) tolerance to a number of abiotic stresses including drought, heat, and salinity; in some cases, the SBLs were shown to have deeper or thicker roots assisting wheat plant in water uptake, which helped the plant to survive under water stress (Schachtman et al. 1991, 1992); (iii) tolerance to preharvest sprouting; (iv) large kernels and heavy spikes; and (v) a higher concentration of both micro- and macronutrients (Fe, Mn, K, and P), relative to that in *T. aestivum* (Calderini and Ortiz-Monasterio 2003). A brief account of activities involving use of SHWs in different countries will be presented in this section.

1.3.2.1 SHWs in Europe, Asia, and Australia

In several countries including Australia, Argentina, Equador, China, India, and Pakistan, the SBLs have shown 5–40 % increase in yield over local modern check varieties (Ogbonnaya et al. 2007). In Australia (DPI, Victoria), under their 'Synthetic-Enriched Resources for Genetic Enhancement' (synERGE) program, 30 % increase was achieved through the use of SHWs. In China also, use of synthetic wheats started in 1995, and four varieties (Chuanmai 38, Chuanmai 42, Chuanmai 43 and Chuanmai 47) were released in Sichuan since 2003; Chuanmai 42—out-yielded commercial check variety by 23 %—and has been grown on >100,000 ha since 2006 (Yang et al. 2009). In Norway also, 448 synthetic wheats from CIMMYT were tested for powdery mildew, and synthetic wheats with partial resistance were crossed with local cultivars (unpublished results).

At the National Institute of Agricultural Botany (NIAB) of Cambridge University, SHWs were used for developing a 'super wheat', which had 30 % higher yield than existing wheat cultivars (in 2012) and also carried tolerance against a number of biotic and abiotic stresses. This will be the first commercialized form of synthetic wheat, which is likely to be released by 2019–2022 (http://www.niab.com/ news_and_events/article/281).

In Pakistan, during the two years of yield trials, two varieties derived from SHWs had 20 and 35 % higher yields than the commercial check variety (Kazi and Van Ginkel 2004). In 2003, a CIMMYT synthetic wheat derivative was also registered in Spain under the name Carmona. This is an early maturing variety that provides seed in a shorter period relative to most commercial cultivars and is valuable for those wheat growers, who often plant late in the year in southern Spain. Carmona also has a better grain quality and is suited to zero-tillage systems, where it resists foliar diseases and produces higher yields (Kazi and Van Ginkel 2004).

1.3.2.2 SHW-Derived 'VOROBEY' at CIMMYT

The biggest breakthrough in wheat breeding after 'Veery,' which carried the 1BL.1RS translocation, was the development of 'Vorobey' involving synthetic hexaploid wheats (SHW). Under the CIMMYT's program of Semiarid Wheat Yield Trial (SAWYT) that was started in 1991 for the development of wheats for the drought-prone areas, as many as 8 % of the lines in 5th SAWYT (1996) represented SHW-derived lines; this proportion increased to 46 % in 15th SAWYT (2006). However, the average coefficient of parentage of SHW in all synthetically derived crosses decreased from 75 to 19 %. Using yield across locations as an index, the average rank of the SHW-derived lines improved during the 5th to 12th SAWYT, so that in SAWYT 11 and 12, SHW-derived line Vorobey was a top-performing line. Vorobey performed well across all environments, giving yields up to 8 t/ha. The use of SHW in wheat breeding for rainfed environments at CIMMYT has increased significantly over the past 10–20 years and the performance and effect of SBLs improved with time. High grain yield also had a positive correlation with

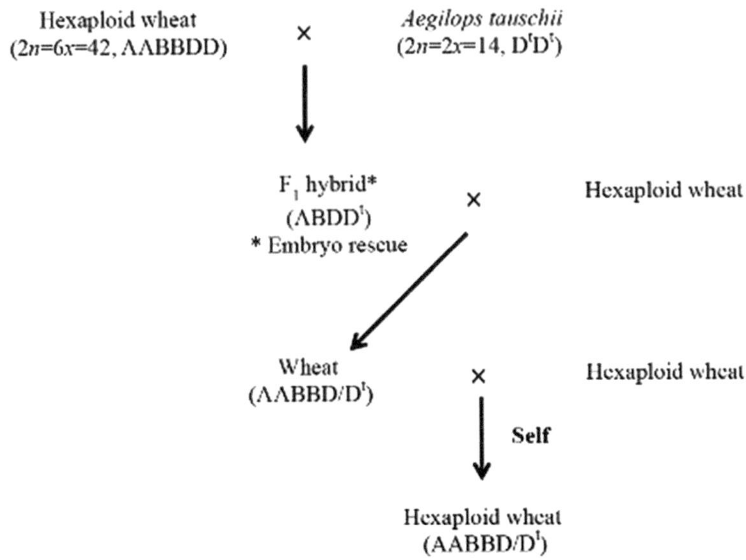

Fig. 1.2 Development of an advanced backcross population for AB-QTL analysis from the cross, wheat (6*x*) × *Ae. tauschii* (2*x*)

improved harvest index, improved grain weight and increased above ground biomass (Rattey et al. 2009, 2011; Shearman et al. 2005). The primary synthetic superior progenies also had larger seed size/weight, improved number of seeds/spike and number of spikes/plant (Cooper 2013; Cooper et al. 2012).

1.3.2.3 SHWs for QTL Mapping

SHWs have also been utilized for developing mapping populations that have been used for QTL interval mapping. A list of these mapping populations has been compiled by Ogbonnaya et al. (2013). SHWs have also been utilized for developing a number of AB-QTL populations, which have been utilized for QTL mapping (Huang et al. 2003, 2004; Moorthy et al. 2006; Naz et al. 2008). Different steps involved in the production of AB-QTL populations in the form of synthetic backcross-derived lines (SBLs) are depicted in Fig. 1.2.

1.3.3 Use of F_1 Hybrids (Wheat × Alien Species) for Alien Gene Transfer

An alternative to the above approach of using amphiploids (including SHWs) involved production and utilization of tetraploid/pentaploid F_1 hybrids ($2n$ = 28 or

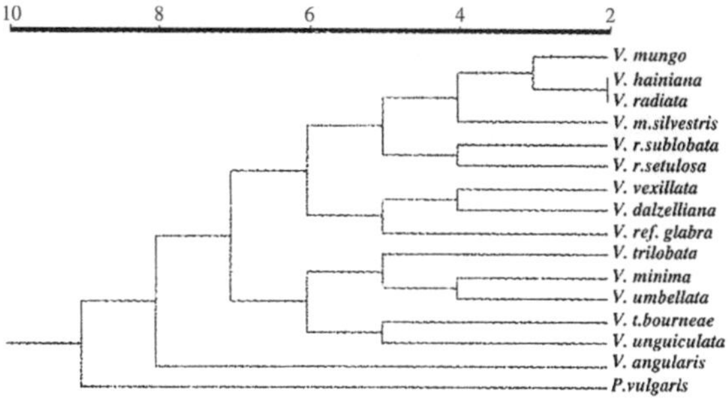

Fig. 1.3 Direct cross for transfer of one or more genes from tetraploid wheat into hexaploid wheat

$2n = 35$) from the cross, wheat ($6x$) × alien species ($2x$ or $4x$). Following steps are involved in this alternative approach: (i) hybridization of wheat with an alien species (Figs. 1.2 and 1.3); (ii) selfing or backcrossing the F_1 hybrids to hexaploid wheat to obtain stable hexaploids ($6x = 42 = 21^{II}$); alternatively, the seed of F_1 hybrids may be irradiated or allowed to undergo induced recombination between the chromosomes of wheat and those of the alien species via suppression of the pairing control gene *Ph1* (see later), and (iii) production and identification of wheat progeny, which contains a small alien chromosome segment carrying the target alien gene, but avoiding the simultaneous transfer of associated deleterious genes.

1.3.4 Alien Addition and Substitution Lines in Hexaploid Wheat

Starting in the 1950s and 1960s, additions and substitutions of whole individual alien chromosomes from a number of alien species to hexaploid wheat genome were successfully achieved (Evans and Jenkins 1960; Riley 1960; Islam et al. 1981; Friebe et al. 1998). Generally, disomic alien chromosome substitutions could be obtained only for the corresponding homoeologous wheat chromosomes. For instance, 1R rye chromosome could be substituted only for 1A, 1B, and 1D chromosomes, so that for any diploid alien species with $2n = 14$, only 21 disomic compensating substitutions were possible. This production of disomic alien additions/substitutions was initially attempted with the hope that these alien additions and substitutions may give birth to new cultivated species that may prove to be superior to cultivated hexaploid wheat (*T. aestivum*); this hope was never materialized, partly due to an instability of these lines and partly due to lack of their superiority over best wheat cultivars. Some of the alien species used for this purpose include the following (Gupta 1995): *Secale cereale*, *Aegilops comosa*, *Ae. geniculata*, *Ae. longissima*, *Ae. bicornis*, *Ae. biuncialis*,

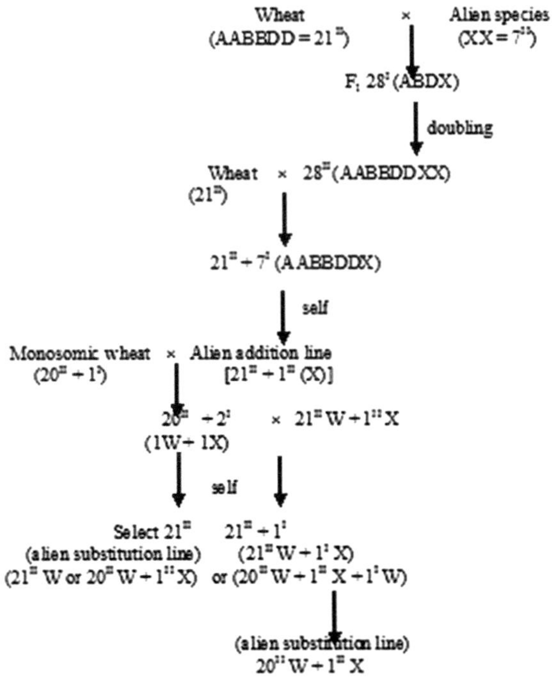

Fig. 1.4 Different steps involved in the production of whole chromosome alien addition and alien substitution lines in hexaploid wheat

Agropyron junceum, *Ag. intermedium*, *Ag. elongatum*, *Dasypyrum villosum,* and *Hordeum vulgare* (note that some of the *Agropyron* species listed here are now known to belong to new genera like *Thinopyrum*).

Alien addition lines have largely been produced and utilized for gene transfer using the following steps (Fig. 1.4): (i) hybridization between wheat and the alien species followed by colchicine doubling of chromosome number to produce an amphiploid (e.g., $8x$ = AABBDDUU in case of alien species *Aegilops umbellulata*); (ii) crossing of the amphiploid with hexaploid wheat to get a heptaploid hybrid ($7x$ = AABBDDU), which will form 21^{II}W + 7^{I}U at meiosis; (iii) selfing this hybrid followed by selection of monosomic/disomic addition lines (21^{II}W + 1^{I}U; 21^{II}W + 1^{II}U); (iv) crossing of these addition lines with wheat monosomics to get alien substitution lines (20^{II}W + 1^{II}U); (v) use of the alien addition/substitution lines for irradiation or induced recombination to produce translocations including those involving whole-arm substitutions (centric fusion).

The alien addition and substitution lines carried a large number of desirable traits including resistance against a variety of diseases. Therefore, these addition/substitution lines have been used for systematic transfer of these traits to elite and high yielding wheat cultivars. In a recent study, *Aegilops biuncialis* chromosome $3M^b$ addition and substitution [$3M^b$(4B)] or translocation ($3M^b$.4BS centric fusion) have been obtained with a view to improve the grain micronutrient (Fe, Zn, Mn, K) contents in the grain (Farkas et al. 2014).

1.3.5 Alien Substitution/Translocation Lines in Tetraploid Wheat

At USDA-ARS Cereal Crops Research Unit, Fargo, ND (USA), alien substitution/translocation stocks were also produced in tetraploid wheat cultivar 'Langdon.' As listed by Joppa (1993), these stocks included the following: (i) 39 Langdon-*T. dicoccoides* substitution lines; (ii) 14 Langdon D-genome substitution lines; (iii) 4 translocation lines representing 1AS-1AL.1DL translocations. The details of these stocks are also available at GrainGenes.

More recently, at the same station at Fargo (USA), the alien species *Thinopyrum bessarbicun* ($2n = 14$) was utilized to produce at least 14 alien addition lines in tetraploid wheat. These addition lines were studied using mcFISH and molecular markers (Jauhar and Peterson 2013).

1.3.6 Reconstitution of New Genomes Involving Alien Species

Efforts were also made to reconstitute entirely new genomes, by combining variable number of chromosomes from two different genomes (Evans 1964). In this study, an amphiploid (AABBEE), derived by combining tetraploid wheat and a diploid alien species (*Agropyron elongatum* = EE), was crossed with bread wheat. The F_1 hybrid plants (AABBDE) exhibiting $14^{II} + 14^{I}$ at meiosis were selfed, and in the progeny, plants exhibiting 21^{II} at meiosis were selected. The hexaploids were backcrossed again with wheat to identify the number of D and E chromosomes in the reconstituted genome, through a study of meiosis in the F_1 hybrids (Fig. 1.5).

1.3.7 Whole Alien Chromosome Arm Translocations

One of the most promising approach for alien gene transfer for wheat improvement has been to produce Robertsonian translocations, each involving a whole chromosome arm from an alien species (Fig. 1.6). The most important of these translocations is the 1BL.1RS translocation found in the 'Veery' series of wheat cultivars. The 1RS arm from rye (*Secale cereale*) carries a battery of resistance genes specifying resistance to leaf rust (*Lr26*), stem rust (*Sr31*), stripe rust (*Yr9*), and powdery mildew (*Pm8*) (Friebe et al. 1996a, b), and genes for adaptation to abiotic stresses, including a robust drought-tolerant root system (Sharma et al. 2011). Because of tight linkage of these genes on the chromosome arm 1RS, the genes are inherited as a single linkage block.

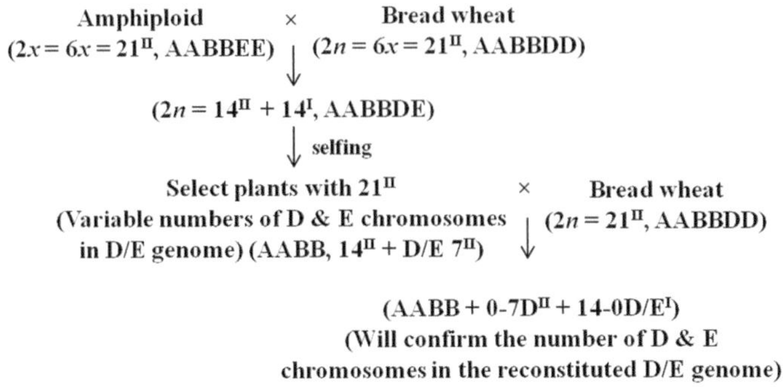

Fig. 1.5 Steps involved in the reconstitution of a new genome utilizing chromosomes from the D-genome of wheat and E genome of *Agropyron elongatum*

Fig. 1.6 Different steps involved in the production of alien introgression lines in wheat, where a complete arm of alien chromosome is substituted for a homoeologous wheat chromosome arm

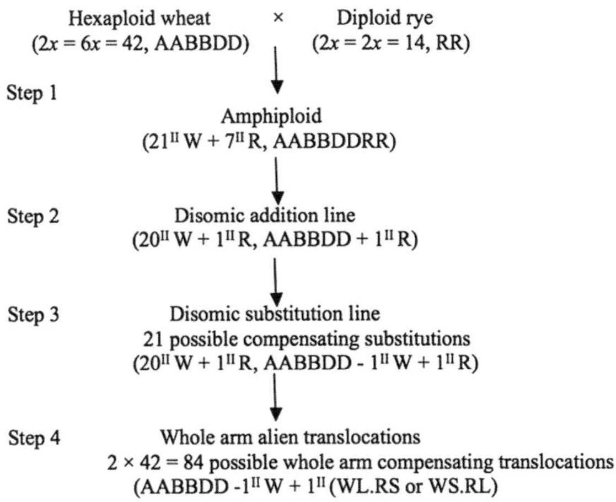

The second example of the transfer of a whole arm of an alien chromosome is T6AL.6VS, where the 6AS arm of wheat was replaced by the 6VS arm from *Dasypyrum villosum* (syn. *Haynalidia villosa*). The alien chromosome arm 6VS carries genes for resistance against (i) all known races of powdery mildew; (ii) wheat curl mite, (ii) stripe rust, (iv) Fusarium head scab, and (v) soilborne mosaic virus. In the plants heterozygous for *Pm21*, the chromosome T6AL·6VS does not pair with the normal 6AL·6AS in their short chromosome arms during meiosis. Thus, all of the genes on the 6VS arm are inherited as a single linkage block. In another recent study, *Ae. biuncialis* chromosome $3M^b$ addition and substitution [$3M^b$(4B)] and translocation ($3M^b$L.4BS, a centric fusion) were obtained with a view to improve the grain micronutrient (Fe, Zan, Mn, K) contents in the grain. Some examples of whole-arm substitutions (including the above two examples) are listed in Table 1.2.

Table 1.2 Some examples of the transfer of alien chromosome arms for wheat improvement

Donor species	Target gene	Translocation	Method of transfer
Ae. umbellulata	*Lr9*	T6BS.6BL-6UL	Irradiation
Ae. biuncialis	Genes for K, Zn, Mn	T4B.3Mb	Induced recombination
Secale cereale	*Pm8, Sr31, Lr26*	T1BL.1RS	Spontaneous
	Yr9, Pm17, Gb2	T1AL.1RS	Irradiation
Ag. elongatum	*Lr24, Sr24*	T3DS.3DL-3Ae#1L	Spontaneous
	Lr24, Sr24	T3DS.3DL-3Ae#1L	Induced recombination
	Sr26	T6AS.6AL-6Ae#1L	Irradiation
	Lr19, Sr25	T7DS.7DL-7Ae#1L	Irradiation

Many of the above whole-arm translocations could be easily identified with the help of FISH. For instance, 1BL/1RS translocation in Indian wheat cultivars were identified by Kumar et al. (2003), and those in Pakistani wheats were identified by Tahir et al. (2014).

1.3.8 Transfer of Alien Segments: Translocation Lines

Transfer of alien chromosome segments (each representing part of a chromosome arm) was achieved using either of the following two major approaches: (i) seed of an alien addition/substitution line carrying the gene of interest on the alien chromosome was irradiated, plants raised from irradiated seed backcrossed to a wheat cultivar, and selection exercised in segregating backcross population for desirable translocations; (ii) mutants for *Ph1* locus (*ph1* mutants) were used to facilitate meiotic pairing between the alien chromosome and its wheat homoeologue. Spontaneous translocations, following crosses made between wheat and an alien species, have also been reported.

1.3.8.1 Translocations Due to Irradiation

A large number of lines with translocations between wheat and alien chromosomes were produced through irradiation of seed for alien addition/substitution lines. These translocation lines carried each a small chromosome segment from an alien species and are largely listed in earlier reviews on the subject (for list of reviews, see above). Some of the examples of these translocations produced during the 1950s and 1960s included the following: (i) a translocation (called '*Transfer*') with an *Aegilops umbellulata* chromosome segment carrying a gene for leaf rust resistance (Sears 1956); this was the first example of useful transfer of an alien chromosome segment for improvement of wheat; (ii) a translocation (called '*Translocation-4*') with an *Agropyron elongatum* chromosome segment carrying a gene for leaf rust resistance (Sharma and Knot 1966); (iii) a 6A-6E translocation with a segment from *Ag. elongatum* chromosome 6E carrying stem rust resistance gene *Sr26*; this stock carrying alien segment was used extensively for the development of several Australian wheat cultivars (see Knott 1971, for a review); (iv) a 4A-2R translocation (called '*Transec*') carrying resistance against leaf rust and powdery mildew (Driscoll and Jensen 1964); this stock had low yields perhaps due to non-homoeologous nature of the translocation and due to loss of some essential genes from wheat.

One of the recent examples of translocations involving alien chromosome segments involved the use of an amphiploid, which combined the genomes of hexaploid wheat (AABBDD) and tetraploid alien species, *Ae. biuncialis* ($U^bU^bM^bM^b$). The 10*x* amphiploid (AABBDD$U^bU^bM^bM^b$) was irradiated, and translocations were identified in M_0 and M_1 generations using multicolor FISH (McFISH) (Molnar et al. 2009). Translocations obtained in this and similar other studies will prove useful for wheat improvement.

Majority of the above translocations, however, were long terminal or long proximal in nature, except 'Transfer,' which carried an intercalary translocation. Such long translocations would carry substantial linkage drag, if the gene of interest (GOI) in the alien chromosome is located farther away from the telomere/centromere, at an interstitial position. Sears (1981) suggested that interstitial transfers can be obtained by intercrossing either the two translocations each carrying the GOI (one carrying a proximal translocation and the other carrying a distal translocation), or between the translocation line (carrying the GOI in a terminal alien chromosome segment) and the recipient wheat cultivar. This would allow recombination to produce a translocation line carrying a small interstitial segment (for details, see Islam and Shepherd 1991).

1.3.8.2 Translocations Due to Induced Meiotic Recombination

In addition to the above translocations, which were introduced following irradiation, strategies were also developed for utilizing induced homoeologous pairing between one or more wheat chromosomes and their corresponding alien homoeologues available within an alien addition or an alien substitution line. (i) use of nullisomy for chromosome 5B carrying *Ph1*; (ii) use of 5B/5D nullisomic-tetrasomic line devoid of 5B chromosome; (iii) use of a *ph1* mutant, which does not suppress, but instead allows homoeologous pairing; (iv) suppression of the diploidizing effect of *Ph1* locus by the genome of a specific strain of *Ae. speltoides*/*Ae. mutica*.

(i) *Use of 5B nullisomy*. Nullisomy for 5B can be achieved through different strategies, involving either the use of nullisomic/monosomic for 5B, mono-5B wheat (carrying a disomic alien substitution of choice) or by using 'nullisomic-5B tetrasomic-5D.' Each of these cytogenetic stocks could be used as a female parent in a cross with either an alien species or with an alien addition/substitution line. Monosmics for 5B were used by Riley (1966) for transfer of alien genetic variation from *Ae. bicornis*, and by Joshi and Singh for transfer of alien genetic variation from rye (*Secale cereale*). Later, Sears (1972, 1973) utilized nulli-5B tetra-5D for crosses with alien substitution lines, each carrying an alien chromosome from *Ag. elongatum*, and successfully transferred two leaf rust resistance genes (*Lr 19*, *Lr24*) from this alien species. These lines carrying alien chromosome segments were later shown to be 3D-3Ag and 7D-7Ag translocations.

(ii) *Use of ph1 mutants*. In the second approach, one may use a homozygous *ph1* mutant line and cross it to either an alien species (Fig. 1.7) or to a mono-5B, alien substitution line (Fig. 1.8), so that in the F_1 hybrid, mutant *ph1* allele will be in hemizygous condition and will induce recombination between alien chromosomes and their wheat homoeologues to allow transfer of alien segments (Figs. 1.7 and 1.8). Once alien transfer is achieved in disomic condition, one may use this stock in the normal backcrossing program to transfer the alien segment to a high yielding elite genotype.

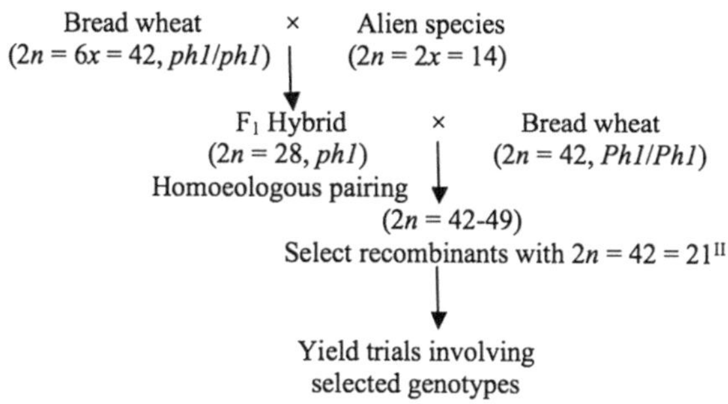

Fig. 1.7 Steps involved in the transfer of an alien segment through a cross between a homozygous *ph1/ph1* mutant and a diploid alien species

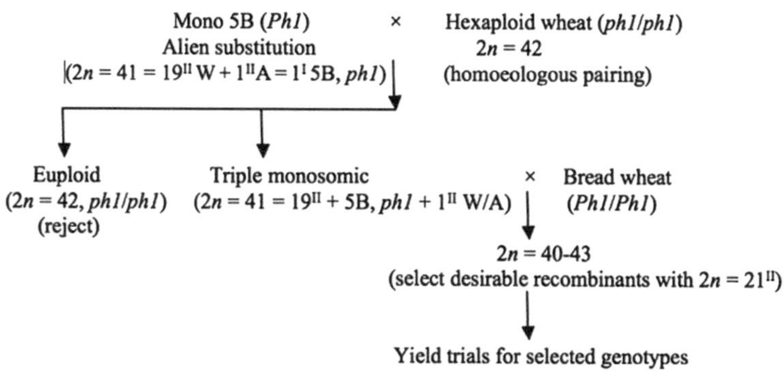

Fig. 1.8 Steps involved in the use of *ph1* mutant/deletion for transfer of alien segment to wheat genome through recombination

In a recent study, crossing of wheat—*Ae. biuncialis* disomic addition lines with *CSph1b* (Chinese Spring wheat mutant for *ph1b*) led to an increased frequency of pairing between wheat and *Aegilops* chromosomes in the F_1 hybrids (Schneider et al. 2005; Molnár and Molnár-Láng (2010).

(iii) *Suppression of Ph1 activity*. It is known that certain strains of *Ae. speltoides* or *Ae. mutica* suppress the effect of *Ph1* locus in F_1 hybrids obtained through a cross between wheat and one such strain. Although this feature can be directly utilized for transfer of desirable alien genes from these two alien species, the system can also be utilized for transfer of alien genes from other alien species. For instance, Riley et al. (1968) successfully utilized this system for transfer of stripe rust resistance (*Yr8*) associated with an alien segment from *Ae. comosa*. In this study, an alien addition line carrying relevant chromosome from *Ae. comosa* was crossed to an *Ae. speltoides* strain that was known for suppression of *Ph1* effect. The F_1 hybrids were crossed and backcrossed to a specific elite wheat cultivar, and desirable recombinants were selected in each segregating backcross generation.

1.3.8.3 Use of DNA-Based Molecular Markers for Alien Gene Transfer

Protein and DNA-based molecular markers have also been used for the detection of alien segments transferred to wheat using the approaches described above. These markers also facilitated selection of plants carrying small interstitial segments derived through recombination between two translocations that each carried a long terminal or proximal segment. For instance, Koebner and Shepherd (1986) and Sears (1981) induced recombination (using *ph1b* mutant, or nulli 5B-tetra 5D line) between wheat and rye chromosome segments involving the translocations 1DL.1RS and 1BL.1RS. Following markers were utilized for identification of recombinants, each carrying a desirable small alien segment: (i) protein markers, Tri-1 and Gli-D1 encoded by genes located on 1DS, and (ii) the gene encoding protein Sec-1 (a storage protein), and the gene *Sr31* for stem rust resistance, both located on rye chromosome arm 1RS. In another study, a number of isozymes (Aco-1, Est-2, Got-2, and Acph-3) encoded by genes carried by one or more *Agropyron* chromosomes were also used for tracking the presence of a small desirable alien chromosome segment in a translocation line (Nichols 1983; Jenkin et al. 1984).

DNA-based molecular markers were also utilized for following the transfer of alien chromosome segments in the translocation lines. These molecular markers included SCAR, PLUG, SSR, and SNP markers. FISH, mcFISH, and GISH were also used for ascertaining the identity of the transferred alien chromosome segments in improved wheat cultivars. A summary of some of the studies involving transfer of alien chromosome segments using DNA-based molecular markers and FISH/mcFISH/GISH is presented in Table 1.4. Some details of the studies involving improved wheat cultivars for ascertaining the presence/absence of alien chromosome segments are briefly described in the next section.

1.4 Molecular Dissection of Alien Segments in Improved Wheat Cultivars

The alien segments that are present in a large number of improved wheat cultivars have been identified following several approaches including physical mapping through FISH, McFISH, GISH, and genetic mapping using DNA-based molecular markers. These are briefly described.

1.4.1 Use of FISH and GISH for Detection and Physical Mapping of Alien Segments

Fluorescence in situ hybridization (FISH) and genomic in situ hybridization (GISH) have been extensively used to identify alien chromosome segments in wheat genome. The first such attempts were made during the 1990s, when segments of rye and *Aegilops tauschii* chromosomes could be identified using FISH, GISH (Friebe et al. 1995a, b, c), and multicolor FISH (McFISH; Mukai and Yamamoto 1998). As early as 1992, GISH was also used to identify alien chromatin from several alien species (*Leymus multicaulis*, *Thinopyrum bessarabicum*, *Hordeum chilense*, *H. vulgare* L., and *Secale cereale*) in wheat chromosomes (Schwarzacher et al. 1992). Research work on detection of *Thinopyrum* alien segments in wheat using GISH were also reviewed by Chen (2005). Recently, a wheat-*Ae. biuncialis* $3M^b(4B)$ disomic substitution and a $3M^b.4BS$ centric fusion in the F_3 progenies were identified using in situ hybridization with genomic DNA probes (FISH and GISH) and SSR markers (Farkas et al. 2014).

1.4.2 Use of Molecular Markers for Detection of Alien Segments

Molecular markers have also been utilized to locate and map alien segments on wheat chromosomes. In a recent study, Olson et al. (2013) transferred an *Ae. tauschii* alien segment carrying resistance against African stem rust fungus race TTKSK (Ug99) to an elite hard winter wheat line, KS05HW14. In BC_2 mapping populations, bulked segregant analysis (BSA) allowed identification of marker loci on 6DS and 7DS linked to stem rust resistance genes transferred from *Ae. tauschii* accessions TA10187 and TA10171, respectively. Linkage maps were developed for both genes and closely linked markers were identified, which can be used for selection and pyramiding with other Ug99-effective stem rust resistance genes. The *Ae. tauschii*-derived resistance genes were temporarily designated *SrTA10187* (associated with SSR loci *Xcfd49* and *Xbarc173*) and *SrTA10171* (associated with SSR loci *Xgdm88* and *Xwmc827*). These genes will serve as valuable resources for stem rust resistance breeding. In another recent study, alien segments carrying genes *Lr57*, *Yr40,* and *Sr53* earlier transferred from chromosome $5M^g$ of *Ae. geniculata* to chromosome 5D of wheat (Kuraparthy et al. 2007a, b, c; Liu et al. 2011) were recently mapped using SNPs (Tiwari et al. 2014). These examples illustrate that molecular markers can be effectively used for mapping and transfer of alien segments across wheat genotypes.

1.4.3 Radiation Hybrids for Identification of Alien Genes

Efforts have also been made to decipher the genetic architecture of alien segments available in wheat cultivars. Such an analysis was often difficult due to lack of pairing between the alien segment with the corresponding wheat segment, thus eliminating the possibility of obtaining recombinants for genetic analysis. An alternative approach was followed in a recent study, where Cao et al. (2011) localized the *Pm21* gene to a segment using radiation hybrid mapping and also used microarray analysis to identify candidate genes induced on infection of this cultivar with powdery mildew pathogen, *Bipolaris graminis* f. sp. *tritici* (*Bgt*).

1.5 Limitations in Alien Gene Transfer

In the past, there have been two major factors that have reduced the effectiveness of wheat/alien gene transfer. These are briefly discussed in this section.

1.5.1 Insufficient Number of Markers

The first limitation of alien gene transfer has been the availability of insufficient number of markers to screen large populations of wheat for the presence of alien chromosome segments. However, by combining comparative mapping with next generation sequencing (NGS) technology, it is now possible to develop an unlimited number of markers for any part of the genome. These markers are now being used to screen large populations for specific introgressions in monocot species (Tiwari et al. 2014).

1.5.2 Linkage Drag in Alien Gene Transfer

The second limitation in alien gene transfer is that an alien segment introgressed into wheat is often associated with some deleterious genes resulting in reduction in yield and/or fitness. Unfortunately, it is often very difficult to reduce the chromosome segment further, even by additional suppression of pairing control genes like *Ph1*. An alternative approach to reduce the size of a chromosome segment involved intercrossing of two lines with different but overlapping alien chromosome segments that carry the same target gene. As a result of recombination between the two overlapping alien segments, in the presence of pairing control genes, some of the progeny produced will carry a reduced alien chromosome segment that would carry the target gene but not the deleterious genes. However, this approach described by Sears for two *Aegilops umbellulata* chromosome segments carrying a gene for stem rust resistance (*Sr* genes) required identification of lines possessing overlapping alien chromosome segments in the first place (see Islam and Shepherd for a review). This strategy has rarely been followed due to lack of markers available to identify individuals carrying overlapping alien chromosome segments. However, the sequencing of the model genomes and the development of NGS technology provide means by which markers would be available for the whole genome of an alien species to allow selection of individuals with overlapping alien segments carrying the target gene(s).

1.6 Taxonomy of Alien Species and Wheat

The taxonomy of the grasses belonging to the tribe Triticeae has also been a subject of discussion for the last more than five decades. Two important and unfortunate revisions involved the following: (i) During the late 1950s and early 1960s, all species of the genus *Aegilops* were merged within the genus *Triticum* on the ground that two of the three sub-genomes of hexaploid wheat (A, B, and D sub-genomes), namely B and D sub-genomes are derived from the genus *Aegilops* (*Ae. speltoides* and *Ae. tauschii*). It was argued that progenitors of two of the three sub-genomes of bread wheat belonging to the genus *Triticum* cannot come from another genus, *Aegilops*. The revised classification with new names for all species of *Aegilops* within the genus *Triticum* proposed by Bowden (1959) was used by scientists in the North America and elsewhere for almost two decades (1965–1985), till Gupta and Baum (1986) questioned this classification. Later, van Slageren (1994) also recognized *Triticum* and *Aegilops* as two independent valid genera, with the result that during the last more than two decades now, once again *Aegilops* has been recognized as valid genus and the use of new names of *Aegilops* species within the genus *Triticum* was discontinued. This old classification has been followed in listing alien species in Tables 1.3 and 1.4). One major disagreement is the treatment of *Amblyopyrum* (formerly *Ae. mutica*) as a separate genus (van Slageren 1994); in

Table 1.3 Some examples of alien species and the genes that were successfully utilized for wheat improvement

Alien species	Wheat cultivar	Gene/chromosome	Reference
Th. ponticum		*Sr43* (7el/7DL)	Kim et al. (1993)
Th. intermedium	L693 KSUD27, MWG684 SFR1	YrL693 (–/1B)	
	T4 (TA5504)	*Lr38*, 7a [7Ai#2L (7D)]	Friebe et al. (1993, 1996a, b)
Th. elongatum	Agatha	*Lr19*, 7D (T7DS_7DL-7Ae#1L)	
	Agent, Sears' translocations P83-171.1-12 (TA3475); P84-171.5-7 (TA3476); P73-231.1b-2, P75-231-1 (TA3506)	*Lr24*, 3D, PSR1203	McIntosh et al. (1977), Jiang et al. (1994a, b)
	Sears' translocation P75-271.3-2 (TA3494)	*Lr29*, 7D (T7DL-7Ae#1L_7Ae#1S)	Friebe et al. (1996a, b)
E. trachycaulus (TA12052)	WGRC45	Undesignated, 1B (T1B_1HtS)	Friebe et al. (2005)
Ae. umbellulata	Transfer	*Lr9*, 6BL (T6BS_6BL-6U#1L)	Sears (1956, 1961), Schachermayr et al. (1994), Autrique et al. (1995), Friebe et al. (1996a, b)
Ae. ventricosa	VPM1	*Lr37*, 2AS	Bariana and McIntosh (1993)
Ae. kotschyi	Line S14	*Lr54*/*Yr37*, 2D	Marais et al. (2005)
Ae. sharonensis	Line 0352-4	*Lr56*/*Yr38*, 6A	Marais et al. (2006)
Ae. geniculata PAU-T756 (TA10437)	TA5602 [DS WL711 $5M^g$(5D)]	*Lr57*, 5DS (T5DL_5DS-$5M^g$S)	Kuraparthy et al. (2007a)
Ae. triuncialis (TA10438)	TA5605 (WL711*4/TA10438)	*Lr58*, 2BL (T2BS_2BL-2tL)	Kuraparthy et al. (2007b)
Ae. peregrina	Line 0306	*Lr59*	G. F. Marais, unpublished
Ae. speltoides	CS 2A/2M#4/2, RL6079	*Lr28*, 4A (T4AS_4AL-7S#2S), OPJ-02378	McIntosh et al. (1982), Friebe et al. (1996a, b), Naik et al. (1998)
Ae. speltoides/T. monococcum amphiploid	RL5711	*Lr35*, 2B (T2B/2S#2)	Kerber and Dyck (1990), Friebe et al. (1996a, b)

(continued)

Table 1.3 (continued)

Alien species	Wheat cultivar	Gene/chromosome	Reference
Pop. No. 2'	Line 2-9-2	*Lr36*, 6BS (T6BL_6BS-6BS#2S)	Dvorak and Knott (1990)
	CI17882, CI17884, CI17885, KS90H450	*Lr47*, 7A (Ti7AS-7S#1S-7AS_7AL)	Friebe et al. (1991)
TA1836	WGRC47	Unknown	Brown-Guedira et al. (2003)
Ae. tauschii	Uruguay	*Lr1*, 5DL, PSR567 GLK621	McIntosh et al. (1965), Ling et al. (2004)
	Festiguay, Kenya W1483	*Lr2, Lr15*, 2DS	Luig and McIntosh (1968)
RL5289 (TA1599), TA1649, TA1670, TA1691, TA2378, TA2468, TA2470, TA2472, TA2527, TA2528, TA2529, TA2530	RL5406, WGRC7, WGRC16	*Lr21*, 1DS, KSUD14 (*Lr40* is identical to Lr21)	Kerber and Dyck (1969), Huang and Gill (2001)
RL5271	RL5404	*Lr22a*, 2DS	Dyck and Kerber (1970)
RL5497-1	RL5713	*Lr32*, 3DS, BCD1278 CDO395	Kerber (1987), Autrique et al. (1995)
Not found in *Ae. Tauschii*	RL6058, PI58548	*Lr34/Yr18*/Bdv1, 7DS, SWM10	Dyck (1977), Bossolini et al. (2006), Lagudah et al. (2006)
TA1675, TA2460, TA2470, TTCC295 (*Ae. cylindrica)*	WGRC2, WGRC10, WGRC16, WX930249-4-1	*Lr39*, 2DS, GDM35 (*Lr41* is identical to *Lr39*)	Raupp et al. (2001), Singh et al. (2003)
	WGRC11	*Lr42*, 1DS	Cox et al. (1994a, b)
	WGRC16	*Lr43, Lr21, Lr39*	G.L. Brown-Guedira, unpublished
RL5683, RL5686, RL5688, RL5689, RL5778, RL5688; RL5662, RL5764, RL5766, RL5767, RL5781-1,RL5782-1 RL5766, RL5767 RL5662	RL5865 (TC/RL5766), RL5866 (TC/RL5767), RL5867 (TC/RL5662), RL5869, RL5868	*LrA*, 2DS (not allelic to *Lr39*) *LrB*, 5D; LrC (not allelic to *Lr39*, suppressed at the hexaploid level); LrD (not allelic to *Lr39*, suppressed at the hexaploid level); ALrA ALrB ALrC	Innes and Kerber (1994)
TA2541, TA1661, TA1683, TA1585, TA1583, TA1672, TA1665, TA1667, TA1677, TA2482	WGRC12	ALrD ALrE ALrF ALrG ALrH ALrI ALrJ ALrK ALrL	Miller (1991)

(continued)

Table 1.3 (continued)

Alien species	Wheat cultivar	Gene/chromosome	Reference
Triticum timopheevii	FTF, Sabikei 12, Timvera W1308, PI170925 (Red Egyptian type)	*Lr18*, T5BS_5BL-5G#1L	McIntosh (1983), Yamamori (1994), Friebe et al. (1996a, b)
TA145, TA870, TA874, TA895, TA1520	WGRC36	*Lr50*, 2BL, GWM382	Brown-Guedira et al. (1999, 2003)
TA28, TA913, TA1538	WGRC35	Unknown	Brown-Guedira et al. (1999)
TA30		Unknown	Brown-Guedira et al. (1999)

Table 1.4 Examples of using FISH, GISH, and DNA-based molecular markers for mapping alien segments in derived lines of wheat

Alien species/genome	Wheat cultivar	Alien gene/chromosome	FISH/GISH/marker type used	Reference
Ae. tauschii (KB) *Ae. tauschii* (Lr21)	PBW343	1D, 2D, 4D, 6D, 1D	SSR SNP	Chhuneja et al. (2008), Neelam et al. (2013)
Th. ponticum	Jinan177		FISH pSc119.2, pAs1, GISH	Wang et al. (2005)
Th. intermedium	Z-148; MY11	1St-1D	PLUG; SCAR	Hu et al. (2012)
Ag. elongatum (*Lr19* + *Yp*); Ae. *longissima (Pm13)*	4*x* wheat	7Ag-7A 3S-3B	FISH, GISH	Ceoloni et al. (2000)
Ag. elongatum (*Lr29*)	RL6080	7Ag-7D	SCAR	Procunier et al. (1995)
Secale cereale (*Lr25*)	RL6084	2R-4A	SCAR	Procunier et al. (1995)
Ae. speltoides (*Sr39*)		2S-2B		Niu et al. (2011)
Th. intermedium				
Ae. geniculata (*Lr57, Yr40*)	TA5601 TA5602	5M^{g}S-5DL	SNP	Tiwari et al. (2014)
Ae. caudata (*LrAC* = *Lr57*)	T291-2 (PBW343)	5C-5D	SSR	Riar et al. (2012)

Table 1.4, it is retained within the genus *Aegilops* as *Ae. mutica*. (ii) Another important and unfortunate proposal made by Dewey (1984) and Love (1984) was the genomic system of classification, where no two taxa with the same genomic constitution could belong to the same genus. As a consequence, a large number of new genera were created, where *Aegilops–Triticum* group was split into as many as

16 genera (5 genera from *Triticum* and 11 genera from *Aegilops*). Consequently, diploid, tetraploid, and hexaploid wheats were assigned to different and independent genera based on their genomic constitutions. This classification was also criticized by us (Baum et al. 1987) and was never put to usage by geneticists and plant breeders.

1.7 Summary and Future Perspectives

Cytogenetic manipulation in polyploid wheats has been relatively easier due to the presence of homoeologous chromosomes carrying duplicate or triplicate genes, so that loss of one copy is tolerated due to compensating effect of the other inserted chromosomes, segments, or genes. It is also for this reason that the genetic variation from the primary, secondary, and tertiary gene pools has been extensively utilized for wheat improvement with remarkable success. The work involving this activity has been adequately covered in several earlier reviews, so that most of this earlier work has not been included in this chapter. Readers may like to consult these earlier reviews for details about the earlier work. However, a summary of this earlier work and some of the recent work on this subject has been covered in this chapter. In recent years, some of the limitations in the transfer of alien chromosome segments carrying desirable segments have now been overcome with the availability of gene targeting approaches and the DNA-based molecular markers; these have been briefly described.

With the availability of the newer approaches including those of gene targeting (not covered in this chapter) and the use of DNA-based molecular markers, a renewed interest in the transfer of alien genetic variation for wheat improvement and in the study of wheat cultivars or genetic stocks carrying alien genetic variation has been witnessed in recent years. More such studies will be conducted in future. For instance, the BBSRC-funded WISP (Wheat Improvement Strategic Program) consortium has been launched in UK with major emphasis on the use of alien genetic variation for wheat improvement. It has been recognized that the alien species carry wealth of genetic resource, which has only been marginally utilized so far. Major part of this genetic resource remains still unexploited and will be the subject of future research in the field of molecular cytogenetics and molecular-/genomics-based plant breeding.

References

Autrique E, Singh RP, Tanksley SD, Sorrells ME (1995) Molecular markers for four leaf rust resistance genes introgressed into wheat from wild species. Genome 38:75–83

Bariana HS, McIntosh RA (1993) Cytogenetic studies in wheat. XV. Location of rust resistance genes in VPM1 and its genetic linkage with other disease resistance genes in chromosome 2A. Genome 36:476–482

Baum BR, Estes JR, Gupta PK (1987) Assessment of the genomic system of classification in the *Triticeae*. Amer J Bot 74:1388–1395

Bossolini E, Krattinger SG, Keller B (2006) Development of simple sequence repeat markers specific for the *Lr34* resistance region of wheat using sequence information from rice and *Aegilops tauschii*. Theor Appl Genet 113:1049–1062

Bowden WM (1959) The taxonomy and nomenclature of the wheats, barleys, and ryes and their wild relatives. Can J Bot 37:657–684

Brown-Guedira GL, Cox TS, Gill BS, Sears RG (1999) Registration of KS96WGRC35 and KS96WGRC36 leaf rust-resistant hard red winter wheat germplasms. Crop Sci 39:595

Brown-Guedira GL, Gill BS, Fritz AK (2003) Performance and mapping of leaf rust resistance transferred to wheat from *Triticum timopheevii* subsp. armeniacum. Phytopathology 93:784–789

Cabrera A, Friebe B, Jiang J, Gill BS (1995) Characterization of *Hordeum chilense* chromosomes by C-banding and in situ hybridization using highly repeated DNA probes. Genome 38:435–442

Calderini DF, Ortiz-Monasterio I (2003) Grain position affects grain macronutrient and micronutrient concentrations in wheat. Crop Sci 43:141–151

Cao A, Xing L, Wang X, Yang X, Wang W, Sun Y, Qian C, Ni J, Chen Y, Liu D, Chen P (2011) Serine/threonine kinase gene *Stpk-V*, a key member of powdery mildew resistance gene *Pm21*, confers powdery mildew resistance in wheat. Proc Natl Acad Sci USA 108:7727–7732

Ceoloni C, Forte P, Ciaffi M, Nenno M, Bitti A, De Vita P, D'Egidio MG (2000) Chromosomally engineered durum wheat: the potential of alien gene introgressions affecting disease resistance and quality. In: Royo C et al. (eds) Durum wheat improvement in the Mediterranean region: new challenges. CIHEAM, Zaragoza, pp 363–371

Chen Q (2005) Detection of alien chromatin introgression from *Thinopyrum* into wheat using S genomic DNA as a probe—a landmark approach for *Thinopyrum* genome research. Cytogenet Genome Res 109:350–359

Chhuneja P, Kaur S, Goel RK, Aghaee-Sarbarzeh M, Prashar M, Dhaliwal HS (2008) Transfer of leaf rust and stripe rust resistance from *Aegilops umbellulata* Zhuk to bread wheat (*Triticum aestivum* L.). Genet Res Crop Evol 55:849–859

Cooper JK (2013) Synthetic hexaploid wheat as a source of improvement for winter wheat (*Triticum aestivum* L.) in texas. A&M University, Texas

Cooper JK, Ibrahim A, Rudd J, Malla S, Hays DB, Baker J (2012) Increasing hard winter wheat yield potential via synthetic wheat: I. Path-coefficient analysis of yield and its components. Crop Sci 52:2014–2022

Cox TS, Raupp WJ, Gill BS (1994a) Leaf rust-resistance genes *Lr41*, *Lr42*, and *Lr43* transferred from *Triticum tauschii* to common wheat. Crop Sci 34:339–343

Cox TS, Sears RG, Gill BS, Jellen RN (1994b) Registration of KS91WGRC11, KS92WGRC15, and KS92WGRC23 leaf rust resistant hard red winter wheat germplasms. Crop Sci 34:546

Dewey DR (1984) The genomic system of classification as a guide to intergeneric hybridization with perennial Triticeae. In: Gustafsson JP (ed) Gene manipulation in plant improvement. Plenum, New York, pp 209–279

Driscoll CJ, Jensen NF (1964) Characteristics of leaf rust transferred from rye to wheat. Crop Sci 4:372–374

Dvorak J, Knott DR (1990) Location of a *Triticum speltoides* chromosome segment conferring resistance to leaf rust in *Triticum aestivum*. Genome 33:892–897

Dyck PL (1977) Genetics of leaf rust reaction in three introductions of common wheat. Can J Genet Cytol 19:711–716

Dyck PL, Kerber ER (1970) Inheritance in hexaploid wheat of adult-plant leaf rust resistance derived from *Aegilops squarrosa*. Can J Genet Cytol 12:175–180

Evans LE (1964) Genome construction within the Triticinae. I. The synthesis of hexaploid ($2n = 42$) having chromosomes of *Agropyron* and *Aegilops* in addition to the A and B genomes of *Triticum durum*. Can J Genet Cytol 6(1):19–28

Evans LE, Jenkins BC (1960) Individual *Secale cereale* chromosome additions to *Triticum aestivum*. I. The addition of individual "Dakold" fall rye chromosomes to "Kharkov" winter wheat and their subsequent identification. Can J Genet Cytol 2:205–215

Farkas A, Molnar I, Dulai S, Rapi S, Oldal V, Cseh A, Kruppa K, Molnar-Lang M (2014) Increased micronutrient content (Zn, Mn) in the 3Mb(4B) wheat-*Aegilops biuncialis* substitution and 3Mb.4BS translocation identified by GISH and FISH. Genome 57:61–67

Friebe B, Mukai Y, Dhaliwal HS, Martin TJ, Gill BS (1991) Identification of alien chromatin specifying resistance to wheat streak mosaic virus and greenbug in wheat germplasm by C-banding and in situ hybridization. Theor Appl Genet 81:381–389

Friebe B, Mukai Y, Gill BS, Cauderon Y (1992a) C-banding and in situ hybridization analyses of *Agropyron intermedium,* a partial wheat × *Ag. intermedium* amphiploid, and six derived chromosome addition lines. Theor Appl Genet 84:899–905

Friebe B, Zeller FJ, Mukai Y, Forster BP, Bartos P, Mclntosh RA (1992b) Characterization of rest-resistant *wheat-Agropyron intermedium* derivatives by C-banding, in situ hybridization and isozyme analysis. Theor Appl Genet 83:775–782

Friebe B, Mukai Y, Gill BS, Cauderon Y (1992c) C-banding and in-situ hybridization analyses of *Agropyron intermedium,* a partial wheat × *Ag. intermedium* amphiploid, and six derived chromosome addition lines. Theor Appl Genet 84:899–905

Friebe B, Tuleen N, Jiang J, Gill BS (1993) Standard karyotype of *Triticum longissimum* and its cytogenetic relationship with *T. aestivum*. Genome 36:731–742

Friebe B, Jiang J, Tuleen N, Gill BS (1995a) Standard karyotype of *Triticum umbellulatum* and the characterization of derived chromosome addition and translocation lines in common wheat. Theor Appl Genet 90:150–156

Friebe B, Jiang J, Tuleen N, Gill BS (1995b) Standard karyotype of *Triticum umbellulatum* and the characterization of derived chromosome addition and translocation lines in common wheat. Theor Appl Genet 90:150–156

Friebe B, Zhang W, Porter DR, Gill BS (1995c) Nonhomoeologous wheat-rye translocations conferring resistance to greenbug. Euphytica 84:121–125

Friebe B, Jiang J, Raupp WJ, McIntosh RA, Gill BS (1996a) Characterization of wheat-alien translocations conferring resistance to diseases and pests: current status. Euphytica 91:59–87

Friebe B, Tuleen NA, Badaeva ED, Gill BS (1996b) Cytogenetic identification of *Aegilops peregrinum* chromosomes added to common wheat. Genome 39:272–276

Friebe B, Raupp WJ, Gill BS (1998) Alien sources for disease and pest resistance in wheat improvement. In: Lelley T (ed) Current topics in plant cytogenetics related to plant improvement '97, WUV-Universitatsverlag, Vienna, Austria, pp 63–71. Proceedings of an international symposium, Tulln, Austria, 21–22 Feb 1997

Friebe B, Qi LL, Nasuda S, Zhang P, Tuleen NA, Gill BS (2000) Development of a complete set of *Triticum aestivum-Aegilops spelotoides* chromosome addition lines. Theor Appl Genet 101:51–58

Friebe B, Zhang P, Linc G, Gill BS (2005) Robertsonian translocations in wheat arise by centric misdivision of univalents at anaphase I and rejoining of broken centromeres during interkinesis of meiosis II. Cytogenet Genome Res 109:293–297

Gill BS, Kimber G (1974) The Giemsa C-banded karyotype of rye. Proc Natl Acad Sci 71:1247–1249

Gill BS, Gill KS, Friebe B (1996) Expanding genetic maps: reevaluation of the relationship between chiasmata and crossovers. Chromosomes Today 12:283–298

Gupta PK (1995) *Cytogenetics*. Rastogi Publications, Meerut, India

Gupta PK, Baum BR (1986) Nomenclature and related taxonomic issues in wheats, triticales and some of their wild relatives. Taxon 35:144–149

Hu L, Li G, Zhan H, Liu C, Yang Z (2012) New St-chromosome-specific molecular markers for identifying wheat–*Thinopyrum intermedium* derivative lines. J Genet 91:e69–e74

Huang L, Gill BS (2001) An RGA-like marker detects all known *Lr21* leaf rust resistance gene family members in *Aegilops tauschii* and wheat. Theor Appl Genet 103:1007–1013

Huang L, Brooks SA, Li W, Fellers JP, Trick HN, Gill BS (2003) Map-based cloning of leaf rust resistance gene *Lr21* from the large and polyploid genome of bread wheat. Genetics 164:655–664

Huang HP, Newman M, Seager R, Kushnir Y, Participating CMIP2 + Modeling Groups (2004) Relationship between tropical Pacific SST and global atmospheric angular momentum in coupled models, LDEO Report, Lamont-Doherty Earth Observatory of Columbia University, Palisades, New York USA, p 43

Innes RL, Kerber ER (1994) Resistance to wheat leaf rust and stem rust in *Triticum tauschii* and inheritance in hexaploid wheat of resistance transferred from *T. tauschii*. Genome 37:813–822

Islam AKMR, Shepherd KW, Sparrow DHB (1981) Isolation and characterization of euplasmic wheat-barley chromosome addition lines. Heredity 46(2):161–174

Islam AKMR, Shepherd KW (1991) Alien genetic variation in wheat improvement. In: Gupta PK, Tsuchiya T (eds) Chromosome engineering in plants: genetics, breeding, evolution. Part A. Elsevier, Amsterdam, Oxford, New York, and Tokyo, pp 291–312

Jauhar PP, Peterson TS (2013) Synthesis and characterization of advanced durum wheat hybrids and addition lines with *Thinopyrum* chromosomes. J Hered 104:428–436

Jenkin M, Shepherd KQ, Brown AHD (1984) Isozyme variation associated with the 9A-1 *Agropyron* translocation in wheat and genetic mapping of its breakpoint. Waite Agricultural Research Institute Biennial Report 1984–85, Adelaide, p 72

Jiang J, Friebe B, Gill BS (1994a) Recent advances in alien gene transfer in wheat. Euphytica 73:199–212

Jiang J, Morris KLD, Gill BS (1994b) Introgression of *Elymus trachycaulus* chromatin into common wheat. Chrom Res 2:3–13

Joppa LR (1987) Aneuploid analysis in tetraploid wheat. In: Heyne EG (ed) Wheat and wheat improvement. American Society of Agronomy, Madison, WI, USA, pp 255–267

Joppa LR (1993) Chromosome engineering in tetraploid wheat. Crop Sci 33:908–913

Kazi M, Van Ginkel M (2004) Wild wheat relatives help boost genetic diversity. CIMMYT News

Kerber ER (1987) Resistance to leaf rust in wheat: *Lr32*, a third gene derived from *Triticum tauschii*. Crop Sci 27:204–206

Kerber ER, Dyck PL (1969) Inheritance in hexaploid wheat of leaf rust resistance and other characters derived from *Aegilops squarrosa*. Can J Genet Cytol 11:639–647

Kerber ER, Dyck PL (1990) Transfer to hexaploid wheat of linked genes for adult-plant leaf rust and seedling stem rust resistance from an ampliploid of *Aegilops speltoides*, "*Triticum monococcum*". Genome 33:530–537

Kim SG, Ashe J, Hendrich K, Ellermann JK, Merkle H, Ugrubil K, Georgopolus AP (1993) Functional magnetic resonance imaging of motor cortex: hemispheric asymmetry and handedness. Science 161:615–617

Knott DR (1971) The transfer of genes for disease resistance from alien species to wheat by induced translocations. In: Proceedings of a panel organized by FAO/IAEA on mutation breeding for disease resistance, Vienna, pp 67–77

Knott DR (1987) Transferring alien segments to wheat. In: Heyne EG (ed) Wheat and wheat improvement, 2nd edn, Monograph 13 Amer Soc Agron Md Wisconsin, pp 462–471

Koebner RMD, Shepherd KW (1986) Controlled introgression to wheat of genes from rye chromosome arm 1RS by induction of allosyndesis I. Isolation of recombinants. Theor Appl Genet 73:197–208

Kumar S, Kumar N, Balyan HS, Gupta PK (2003) 1BL.1RS translocation in some Indian bread wheat genotypes and strategies for its use in future wheat breeding. Caryologia 56(1):23–30

Kuraparthy V, Chhuneja P, Dhaliwal HS, Kaur S, Bowden RL, Gill BS (2007a) Characterization and mapping of cryptic introgression from *Ae. geniculata* with new leaf rust and stripe rust resistance genes *Lr57* and *Yr40* in wheat. Theor Appl Genet 114:1379–1389

Kuraparthy V, Chhuneja P, Dhaliwal HS, Kaur S, Bowden RL, Gill BS (2007b) Characterization and mapping of *Aegilops geniculata* introgressions with novel leaf rust and stripe rust resistance genes *Lr57* and *Yr40* in wheat. Theor Appl Genet 114:1379–1389

Kuraparthy V, Sood S, Chhuneja P, Dhaliwal HS, Kaur S, Bowden RL, Gill BS (2007c) A cryptic wheat–*Aegilops triuncialis* translocation with leaf rust resistance gene *Lr58*. Crop Sci 47:1995–v2003

Lage J, Skovmand B, Andersen S (2004) Field evaluation of emmer wheat-derived synthetic hexaploid wheat for resistance to Russian wheat aphid (homoptera: Aphididae). J Econ Ent 97:1065–1070

Lagudah ES, McFadden H, Singh RP, Huerta-Espino J, Bariana HS, Spielmeyer W (2006) Molecular genetic characterization of the *Lr34/Yr18* slow rusting resistance gene region in wheat. Theor Appl Genet 114:21–30

Lange W, Jochemsen G (1992) Use of the gene pools of *Triticum turgidum* ssp. dicoccoides and *Aegilops squarrosa* for the breeding of common wheat (*T. aestivum*), through chromosome-doubled hybrids. Euphytica 59:197–212

Ling HQ, Qiu JW, Singh RP, Keller B (2004) Identification and genetic characterization of an *Aegilops tauschii* ortholog of the wheat leaf rust disease resistance gene Lr1. Theor Appl Genet 109:1133–1138

Liu W, Rouse M, Friebe B, Jin Y, Gill BS, Pumphrey MO (2011) Discovery and molecular mapping of a new gene conferring resistance to stem rust, *Sr53*, derived from *Aegilops geniculata* and characterization of spontaneous translocation stocks with reduced alien chromatin. Chromosome Res 19:669–682

Love A (1984) Conspectus of the *Triticeae*. Feddes Repertorium 95:425–521

Luig NH, McIntosh RA (1968) Location and linkage of genes on wheat chromosome 2D. Can J Genet Cytol 10:99–105

Marais GF, McCallum B, Snyman JE, Pretorius ZA, Marais AS (2005) Leaf rust and stripe rust resistance genes *Lr54* and *Yr37* transferred ro wheat from *Aegilops kotschyi*. Plant Breed 124:538–541

Marais GF, McCallum B, Marais AS (2006) Leaf rust and stripe rust resistance genes derived from *Aegilops sharonensis*. Euphytica 149:373–380

McIntosh RA (1983) Genetic and cytogenetic studies involving *Lr18* resistance to *Puccinia recondita*. In: Sakamoto S (ed) Proceedings of the 6th international wheat genetics symposium, Plant Germ-Plasm Institute, Kyoto, Japan, pp 777–783

McIntosh RA, Baker EP, Driscoll CJ (1965) Cytogenetic studies in wheat. I. Monosomic analysis of leaf rust resistance in cultivars Uruguay and Transfer. Aust J of Biol Sci 18:971–977

McIntosh RA, Dyck PL, Green GJ (1977) Inheritance of leaf and item rust resistance in wheat cultivars Agent and Agatha. Aust J of Agri Res 28:37–45

McIntosh RA, Miller TE, ChapmanV (1982) Cytogenetical studies in wheat. XII. *Lr28* for resistance to *Puccinia recondita* and *Sr34* for resistance to *P. graminis tritici*. Zeitschrift für Panzenzü chtung 92:1–14

Miller D (1991) Genetic analysis of leaf rust resistance in *Triticum tauschii*, the D-genome progenitor of wheat. MS thesis, Kansas State University, Manhattan, USA

Molnár I, Molnár-Láng M (2010) GISH reveals different levels of meiotic pairing with wheat for individual *Aegilops biuncialis* chromosomes. Biol Plant 54:259–264

Molnar I, Nenavente E, Molnar-Lang M (2009) Detection of intergenomic chromosome rearrangements in irradiated *Triticum aestivum-Aegilops biuncialis* amphiploids by multicolour genomic in situ hybridization. Genome 52:156–165

Moorthy JN, Venkatakrishnan P, Savithaa G, Weiss RG (2006) Cis → trans and trans → cis isomerizations of styrylcoumarins in the solid state. Importance of the location of free volume in crystal lattices. Photochem Photobiol Sci 5:903–913

Mujeeb-Kazi A, Rajaram S (2002) Transferring alien genes from related species and genera for wheat improvement. In: Curtis BC, Rajaram S, Gómez Macpherson H (eds) Bread wheat: improvement and protection FAO plant production and protection series. FAO, Rome

Mukai Y, Friebe B, Gill BS (1992) Comparison of C-banding patterns hybridization sites using highly repetitive and total genomic rye DNA probes of 'Imperial' rye chromosomes added to 'Chinese Spring' wheat. Jpn J Genet 67:71–83

Mukai Y, Yamamoto M (1998) Application of multicolor fluorescence in situ hybridization to plant genome analysis. In: Gupta PK, Singh SP, Balyan HS, Sharma PC, Ramesh B (eds) Genetics and biotechnology in crop improvement. Rastogi Publications, Meerut, pp 14–23

Naik S, Gill KS, Prakasa VS, Gupta VS, Tamhanka SA, Pujar S, Gill BS, Ranjekar PK (1998) Identification of a STS marker linked to the *Aegilops speltoides*-derived leaf rust resistance gene *Lr28* in wheat. Theor Appl Genet 97:535–540

Naz AA, Kunert A, Lind V, Pillen K, Leon J (2008) AB-QTL analysis in winter wheat: II. Genetic analysis of seedling and field resistance against leaf rust in a wheat advanced backcross population. Theor Appl Genet 116:1095–1104

Neelam K, Brown-Guedira G, Huang L (2013) Development and validation of a breeder-friendly KASPar marker for wheat leaf rust resistance locus Lr21. Mol Breed 31:233–237

Nichols PGH (1983) Investigations of the amount of *Agropyron* chromatin in wheat cultivars 'Eagle' and 'Kite'. Honours thesis, University of Adelaide, p 94

Niu ZX, Klindworth DL, Friesen TL, Chao SM, Jin Y, Cai XW, Xu SS (2011) Targeted introgression of a wheat stem rust resistance gene by DNA marker assisted chromosome engineering. Genetics 187:1011–1021

Ochoa V, Madrid E, Said M, Rubiales D, Cabrera A (2015) Molecular and cytogenetic characterization of a common wheat-*Agropyron cristatum* chromosome translocation conferring resistance to leaf rust. Euphytica 201:89–95

Ogbonnaya FC, Ye G, Trethowan R, Dreccer F, Lush D, Shepperd J, Van Ginkel M (2007) Yield of synthetic backcross-derived lines in rainfed environments of Australia. Euphytica 157:321–336

Ogbonnaya FC, Abdalla O, Mujeeb-Kazi A, Kazi AG, Xu SS, Gosman N, Lagudah ES, Bonnett D, Sorrells ME (2013) Synthetic hexaploid in wheat improvement. In: Janick J (ed) Plant breeding reviews, vol 37. pp 35–122

Olson EL et al (2013) Introgression of stem rust resistance genes *SrTA10187* and *SrTA10171* from *Aegilops tauschii* to wheat. Theor Appl Genet 126:2477–2484

Procunier JD, Townley-Smith TF, Fox S, Prashar S, Gray M, Kim WK, Czarnecki E, Dyck PL (1995) PCR-based RAPD/DGGE markers linked to leaf rust resistance genes *Lr29* and *Lr25* in wheat (*Triticum aestivum* L.). J Genet Breed 49:87–92

Qi LL, Wang SL, Chen PD, Liu DJ, Friebe B, Gill BS (1997) Molecular cytogenetic analysis of *Leymus racemosus* chromosomes added to wheat. Theor Appl Genet 95:1084–1091

Rattey A, Shorter R, Chapman S, Dreccer F, Van HA (2009) Variation for and relationships among biomass and grain yield component traits conferring improved yield and grain weight in an elite wheat population grown in variable yield environments. Crop Pasture Sci 60:717–729

Rattey A, Shorter R, Chapman S (2011) Evaluation of CIMMYT conventional and synthetic spring wheat germplasm in rainfed sub-tropical environments. I. Grain yield components and physiological traits. Field Crop Res 124:195–204

Raupp WJ, Singh S, Brown-Guedira GL, Gill BS (2001) Cytogenetic and molecular mapping of the leaf rust resistance gene *Lr39* in wheat. Theor Appl Genet 102:347–352

Riar AK, Kaur S, Dhaliwal HS, Singh K, Chhuneja P (2012) Introgression of a leaf rust resistance gene from *Aegilops caudata* to bread wheat. J Genet 91:155–161

Riley R (1960) The meiotic behaviour, fertility and stability of wheat-rye chromosome addition lines. Heredity 14:89–100

Riley R (1966) Cytogenetics and wheat breeding. Contemp Agric 11–12:107–117

Riley R, Chapman V, Johnson R (1968) Introduction of yellow rust resistance of *Aegilops comosa* into wheat by genetically induced homoeologous recombination. Nature 217:383–384

Schachermayr R, Siedler H, Gale MD, Winzeler H, Winzeler M, Keller B (1994) Identification and localization of molecular markers linked to *Lr9* leaf rust resistance gene of wheat. Theor Appl Genet 88:110–115

Schachtman D, Munns R, Whitecross M (1991) Variation in sodium exclusion and salt tolerance in *Triticum tauschii*. Crop Sci 31:992–997

Schachtman D, Lagudah E, Munns R (1992) The expression of salt tolerance from *Triticum tauschii* in hexaploid wheat. Theor Appl Genet 84:714–719

Schneider A, Linc G, Molnár I, Molnár-Láng M (2005) Molecular cytogenetic characterization of *Aegilops biuncialis* and its use for the identification of 5 derived wheat—*Aegilops biuncialis* disomic addition lines. Genome 48(6):1070–1082

Schwarzacher T, Jonsson KA, Harrison GE (1992) Genomic in situ hybridization to identify alien chromosomes and chromosome segments in wheat. Theor Appl Genet 84:778–786

Sears ER (1956) The transfer of leaf rust resistance from *Aegilops umbellulata* to wheat. Brookhaven Symp Biol 9:1–22

Sears ER (1961) Identification of the wheat chromosome carrying leaf rust resistance from *Aegilops umbellulata*. Wheat Inf Serv 12:12–13

Sears ER (1972) Chromosome engineering in wheat. In: Stadler Symposia, vol 4. University of Missouri, Columbia, USA, pp 23–38

Sears ER (1973) *Agropyron*-wheat transfers induced by homoeologous pairing. In: Sears ER, Sears ML (eds) Proceedings of 4th international wheat genetics symposium, University of Missouri, Colombia, MO, pp 191–199

Sears ER (1981) Transfer of alien genetic material to wheat. In: Evans LT, Peacock WJ (eds) Wheat science-today and tomorrow. Cambridge University Press, Cambridge, pp 75–89

Sharma HC, Gill BS (1983) Current status of wide hybridisation in wheat. Euphytica 32:17–31

Sharma D, Knott DR (1966) The transfer of leaf rust resistance from *Agropyron* to *Triticum* by irradiation. Can J Genet Cytol 8:137–143

Sharma S, Xu S, Ehdaie B, Hoops A, Close JT, Lukaszewski JA, Waines JG (2011) Dissection of QTL effects for root traits using a chromosome arm-specific mapping population in bread wheat. Theor Appl Genet 122:759–769

Shearman V, Sylvester-Bradley R, Scott R, Foulkes M (2005) Physiological processes associated with wheat yield progress in the UK. Crop Sci 45:175–185

Singh S, Franks CD, Huang L, Brown-Guedira GL, Marshall DS, Gill BS, Fritz A (2003) *Lr41*, *Lr39*, and a leaf rust resistance gene from *Aegilops cylindrica* may be allelic and are located on wheat chromosome 2DS. Theor Appl Genet 108:586–591

Tahir R, Bux H, Kazi AG, Rasheed A, Napar AA, Ajmal SU, Mujeeb-Kazi A (2014) Evaluation of Pakistani elite wheat germplasm for T1BL.1RS chromosome translocation. J Agr Sci Tech 16:421–432

Talbot SJ (2011) Introgression of genetic material from primary synthetic hexaploids into an Australian bread wheat (*Triticum aestivum* L.). School of Agriculture, Food and Wine, University of Adelaide

Tiwari VK, Wang S, Sehgal S, Vrana J, Friebe B, Kubalakova M, Chhuneja P, Dolezel J, Akhunov E, Kalia B, Sabir J, Gill BS (2014) SNP Discovery for mapping alien introgressions in wheat. BMC Genomics 15:273

Trethowan R, Mujeeb-Kazi A (2008) Novel germplasm resources for improving environmental stress tolerance of hexaploid wheat. Crop Sci 48:1255–1265

Van Ginkel M, Ogbonnaya F (2007) Novel genetic diversity from synthetic wheats in breeding cultivars for changing production conditions. Field Crop Res 104:86–94

van Slageren MW (1994) Wild wheats: a monograph of *Aegilops* L. and *Amblyopyrum* (Jaub. & Spach) Eig (Poaceae), vol 7. Wageningen Agriculture University papers, p 513

Wang MC, Bohmann D, Jasper H (2005) JNK extends life span and limits growth by antagonizing cellular and organism-wide responses to insulin signaling. Cell 121(1):115–125

Yamamori M (1994) An N-band marker for gene *Lrl8* for resistance to leaf rust in wheat. Theor Appl Genet 89:643–646
Yang WY, Liu DL, Li J, Zhag LQ, Wei HT, HU XR, Zheng YL, Zou YC (2009) Synthetic hexaploid wheat and its utilization for wheat genetic improvement in China. J Genet Genomics 36:539–546

Chapter 2
Quantitative Trait Loci Mapping in Plants: Concepts and Approaches

Deepmala Sehgal, Richa Singh and Vijay Rani Rajpal

Abstract The narrow genetic base of modern crop cultivars is a serious obstacle to sustain and improve crop productivity due to rapidly occurring vulnerability of genetically uniform cultivars to potentially new biotic and abiotic stresses. Plant germplasm resources, originated from a number of historical genetic events as a response to environmental stresses and selection, are the important reservoirs of natural genetic variations that can be exploited to increase the genetic base of the cultivars. However, many agriculturally important traits such as productivity and quality, tolerance to environmental stresses, and some of forms of disease resistance are quantitative (also called polygenic, continuous, multifactorial, or complex traits) in nature. The genetic variation of a quantitative trait is controlled by the collective effects of numerous genes, known as quantitative trait loci (QTLs). Identification of QTLs of agronomic importance and its utilization in a crop improvement requires mapping of these QTLs in the genome of crop species using molecular markers. This review will focus on the basic concepts and a brief description of existing methodologies for QTL mapping and their merits and demerits including traditional biparental mapping and the advanced linkage disequilibrium (LD)-based association mapping. Examples of some of the recent studies on association mapping in various crop species are provided to demonstrate the merits of high-resolution association mapping approach over traditional mapping methods. This review thus will provide non-expert readers of crop breeding community an opportunity to develop a basic understanding of dissecting and exploiting natural variations for crop improvement.

D. Sehgal (✉)
International Maize and Wheat Improvement Center (CIMMYT), Km.45, Carretera, Mex-Veracruz El Batan, Texcoco CP 56130, Edo. de México, Mexico
e-mail: d.sehgal@cgiar.org

R. Singh
Department of Molecular Biology and Biotechnology, Chaudhary Charan Singh Haryana Agricultural University (CCSHAU), Hisar, Haryana 125004, India
e-mail: richabiotech77@gmail.com

V.R. Rajpal
Department of Botany, HansRaj College, University of Delhi, Delhi 11007, India
e-mail: vijayrani2@gmail.com

V.R. Rajpal et al. (eds.), *Molecular Breeding for Sustainable Crop Improvement*, Sustainable Development and Biodiversity 11,
DOI 10.1007/978-3-319-27090-6_2

Keywords Linkage map · Quantitative trait loci · Linkage disequilibrium · Association mapping

2.1 Introduction

As most of the modern genotypes cultivated today have descended from a relatively small number of landraces, the genes controlling important traits have reduced diversity compared to the gene pool of landraces and wild relatives. In view of such threats of constant genetic erosion, gene banks have been established and maintained in order to mainly preserve wild plant accessions as well as landraces. Gene bank collections are the important reservoirs of natural genetic variations originating from a number of historical genetic events as a response to environmental stresses (Hoisington et al. 1999). Unlocking biodiversity held in gene banks and mobilizing useful variation to breeding programs are required for the genetic improvement of crops and to meet the overarching goal of diversification of the adapted gene pools. However, many agriculturally important variations such as productivity and quality, tolerance to environmental stresses, and some of forms of disease resistance are controlled by polygenes which complicate the breeding process since phenotypic performances only partially reflects the genetic values of individuals. These complex traits are referred to as quantitative traits (also called as polygenic or multifactorial traits) and the regions within genomes that contain genes associated with a particular quantitative trait are known as quantitative trait loci (QTLs).

It is challenging to identify QTLs based on only traditional phenotypic evaluation. Identification of QTLs of agronomic importance and its utilization in a crop improvement requires mapping of these QTLs in a genome of crop species using molecular markers. Identification of QTLs with genetically linked DNA-markers is useful for incorporating genes into improved cultivars via marker-assisted selection (MAS), map-based cloning of the tagged genes, and for a better understanding of the genetics of complex traits (Asins 2002). Linkage analysis and association mapping are the two most commonly used methods for QTL mapping. This review provides an overview of the two QTL mapping methods with special focus on mapping population types, linkage map construction, and marker–trait association analysis using different statistical methods and software programs. The contents of the review will provide non-expert readers of crop breeding community an opportunity to develop a basic understanding of dissecting and exploiting natural variations for crop improvement.

2.2 Principle of QTL Analysis

Identifying a QTL or a gene within a plant genome is like finding the proverbial needle in a haystack. However, QTL analysis can be used to divide the haystack in manageable piles and systematically search them. In simple terms, QTL analysis is based on the principle of detecting an association between phenotype and the genotype of markers. Markers are used to partition the mapping population into different genotypic groups based on the presence or absence of a particular marker locus and to determine whether significant differences exist between groups with respect to the trait being measured (Tanksley 1993; Young 1996). A significant difference between phenotypic means of the groups (either 2 or 3), depending on the marker system and type of population, indicates that the marker locus being used to partition the mapping population is linked to a QTL controlling the trait.

2.3 Traditional QTL Mapping (Linkage Mapping)

The general steps involved in a traditional QTL mapping experiment are as follows: (1) select two parental strains that have differences between them in the alleles that affect variation in a trait. The parents need not be different in the mean phenotypic value of the trait as different allelic combinations can yield the same phenotypic mean; (2) develop an appropriate mapping population by crossing the selected parents; (3) phenotype the mapping population for the trait(s) of interest (morphological characters, agronomic traits, disease and pest scores, drought resistance, etc.) under greenhouse, screen-house, and/or field conditions; (4) generate the molecular data on the population with adequate number of uniformly spaced polymorphic markers; (5) construct a genetic map; and (6) identify molecular markers linked to the trait(s) of interest using statistical programs.

2.3.1 Mapping Populations Used in QTL Mapping Experiments

Various types of mapping population may be produced from the heterozygous F_1 hybrids:

1. Double haploid lines (DHLs): plants are regenerated from pollen (which is haploid) of the F_1 plants and treated to restore diploid condition in which every locus is homozygous. Since the pollen population has been generated by meiosis, the DHLs represent a direct sample of the segregating gametes.
2. Backcross (BC) population: the F_1 plants are backcrossed to one of the parents.
3. F_2 population: F_1 plants are selfed.

4. $F_{2:3}/F_{2:4}$ lines: $F_{3/4}$ plants tracing back to the same F_2 plant, also called F_2 families.
5. Recombinant inbred lines (RILs): inbred generation derived by selfing individual F_2 plants and further single seed descent. A population of RILs represents an 'immortal' or permanent mapping population.

Each of the above mapping populations possesses advantages and disadvantages. Hence, the choice of the type of mapping population depends on many factors such as the plant species, type of marker system used, and the trait to be mapped later on. F_2 populations, derived from F_1 hybrids, and BC populations, derived by crossing the F_1 hybrid to one of the parents, are the simplest types of mapping populations developed for self-pollinating species. Their main advantages are that they are easy to construct and require only a short time to produce. However, such populations are not fixable due to their inherent heterozygous genetic constitution. This restricts their wide utility in QTL analysis. The length of time needed for producing RIL population is the major disadvantage, because usually six to eight generations are required. Development of a DH population takes much less time than RIL; however, the production of DH populations is only possible in species that are amenable to tissue culture (e.g., cereal species such as rice, barley, and wheat). The major advantages of RIL and DH populations are that they produce homozygous or 'true-breeding' lines that can be multiplied and reproduced without genetic change occurring. This allows for the conduct of replicated trials across different locations and years. Furthermore, seed from individual RIL or DH lines may be transferred between different laboratories for further linkage analysis and the addition of markers to existing maps. Information provided by co-dominant markers is best exploited by an F_2 population, while information obtained by dominant marker systems can be maximized by using RILs or DHLs. Double haploids, F_2 or F_3 families, or RILs are advantageous if the trait to be mapped cannot be accurately measured on a single-plant basis but must be assessed in replicated field experiments.

2.3.2 *Construction of Genetic/Linkage Maps*

A linkage map may be thought of as a 'road map' of the chromosomes derived from two different parents. Linkage maps indicate the position and relative genetic distances between markers along chromosomes. Construction of a linkage map, using genotyping data generated on any of the above-mentioned mapping populations, is an important step before initiating any QTL analysis. In a segregating mapping population, there is a mixture of parental and recombinant genotypes. The frequency of recombinant genotypes is used to calculate recombination fractions, which is then used to infer the genetic distance between markers. By analyzing the segregation of markers, the relative order and distances between markers can be determined; the lower the frequency of recombination between two markers, the closer they are situated on a chromosome (conversely, the higher the frequency of recombination between two markers, the further away they are situated on a chromosome). Two commonly used mapping functions that convert recombination frequency into centimorgan (cM) distance are the Kosambi mapping function, which assumes that recombination events influence the occurrence of adjacent recombination events, and the Haldane mapping function, which assumes no interference between crossover events. Linkage between markers is usually calculated with an odds ratio (i.e., the ratio of linkage versus no linkage). This ratio is more conveniently expressed as the logarithm of the ratio and is called a logarithm of odds (LOD) value or LOD score (Risch 1992). LOD values of >3 are typically used to construct linkage maps. LOD values may be lowered in order to detect linkage over a greater distance or to place additional markers within maps constructed at higher LOD values (Collard et al. 2005). Linked markers are grouped together into linkage groups, which represent chromosomal segments or entire chromosomes.

It is important to note that distance on a linkage map is not directly related to the physical distance of DNA between genetic markers, but depends on the genome size of the plant species (Paterson 1996). Furthermore, the relationship between genetic and physical distance varies along a chromosome (Tanksley et al. 1992; Young 1994; Künzel et al. 2000). For example, there are recombination 'hot spots' and 'cold spots,' which are chromosomal regions in which recombination occurs more frequently or less frequently, respectively (Faris et al. 2000; Ma et al. 2001; Yao et al. 2002). Commonly used software programs for constructing linkage maps include Mapmaker/EXP (Lander et al. 1987; Lincoln et al. 1993), MapManager QTX (Manly et al. 2001), and THREaD Mapper Studio (Cheema et al. 2010) which are freely available from the Internet. JoinMap is another commonly used program for constructing linkage maps (Stam 1993).

2.3.3 Detection of QTLs

Four widely used methods for detecting QTLs are single-marker analysis, interval mapping by maximum likelihood, interval mapping by regression, and composite interval mapping.

2.3.3.1 Single-Marker Analysis (Point Analysis)

The traditional method to detect a QTL in the vicinity of a marker is studying single-genetic markers one at a time. The phenotypic means for progeny of each marker class are compared (e.g., means of the marker classes AA, Aa, aa). The difference between two means provides an estimate of the phenotypic effect of substituting an *A* allele by an *a* allele at the QTL. To test whether the inferred phenotypic effect is significantly different from zero, a simple statistical test, such as *t*-test or *F*-test, is used. A significant value indicates that a QTL is located in the vicinity of the marker. Single-point analysis does not require a complete molecular linkage map. The farther a QTL is from the marker, the less likely it is to be detected statistically due to crossover events between the marker and the gene.

2.3.3.2 Interval Mapping by Maximum Likelihood

QTL interval mapping is the most common method of QTL analysis. The principle behind interval mapping is to test a model for the presence of a QTL at many positions between two mapped marker loci. This model is a fit, and its goodness is tested using the method of maximum likelihood. For example, if it is assumed that a QTL is located between two markers, the 2-loci marker genotypes (i.e., AABB, AAbb, aaBB, aabb for DH progeny) each contain mixtures of QTL genotypes. Maximum likelihood involves searching for QTL parameters that give the best approximation for quantitative trait distributions that are observed for each marker class. Models are evaluated by computing the likelihood of the observed distributions with and without fitting a QTL effect. The map position of a QTL is determined as the maximum likelihood from the distribution of likelihood values (LOD scores: ratio of likelihood that the effect occurs by linkage to likelihood that the effect occurs by chance), calculated for each locus.

2.3.3.3 Interval Mapping by Regression

Interval mapping by regression (Haley and Knott 1992) was developed primarily as a simplification of the maximum likelihood method. It is essentially the same as the method of basic QTL analysis (regression on coded marker genotypes) except that phenotypes are regressed on QTL genotypes. Since the QTL genotypes are unknown, they are replaced by probabilities estimated from the nearest flanking markers.

2.3.3.4 Composite Interval Mapping

One of the factors that weakens interval mapping is fitting the model for a QTL at only one location. There are two problems with this approach: (a) the effects of additional QTL will contribute to sampling variance and (b) if two QTLs are linked, their combined effects will cause biased estimates. The method of composite interval mapping (CIM) was proposed as solution (Jansen and Stam 1994; Utz and Melchinger 1996; Zeng 1994). CIM will perform the analysis in the usual way, except that the variance from other QTLs is accounted for by including partial regression coefficients from markers (‘cofactors’) in other regions of the genome. CIM gives more power and precision than simple interval mapping (SIM) because the effects of other QTLs are not present as residual variance. CIM can remove the bias that can be caused by QTLs that are linked to the position being tested.

2.4 QTL Mapping Softwares

There are over 100 genetic analysis software packages (linkage analysis and QTL mapping). Here, we list some features of the most commonly used software packages.

2.4.1 MapMaker/QTL (ftp://genome.wi.mit.edu/pub/mapmaker3/)

A user-friendly, freely distributed software program runs on almost all platforms. It analyzes F_2 or backcross data using standard interval mapping.

2.4.2 MQTL

MQTL is a computer program for CIM in multiple environments. It can also perform SIM. Currently, MQTL is restricted to the analysis of data from homozygous progeny (double haploids, or RILs). Progeny types with more than two marker classes (e.g., F_2) are not handled.

2.4.3 PLABQTL (http://www.uni-hohenheim.de/~ipspwww/soft.html)

PLABQTL is a freely distributed computer program for CIM and SIM of QTL. Its main purpose is to localize and characterize QTL in mapping populations derived from a biparental cross by selfing or production of double haploids. Currently, this program is the easiest software for composite interval mapping.

2.4.4 QTL Cartographer (http://statgen.mcsu.edu/qtlcart/cartographer.html)

QTL Cartographer is a QTL software written for either UNIX, Macintosh, or Windows. It performs single-marker regression, interval mapping, and composite interval mapping. It permits analysis from F_2 or backcross populations. It displays map positions of QTLs using the GNUPLOT software.

2.4.5 MapQTL (http://www.cpro.dlo.nl/cbw/)

MapQTL is a licensed software program. It performs Kruskal–Wallis test (single-marker analysis), CIM, and multiple interval mapping on almost all kinds of mapping populations.

2.4.6 Qgene

Qgene is a QTL mapping and marker-aided breeding package written for Macintosh. It has a user-friendly graphical interface and produces graphical outputs. QTL mapping is conducted by either single-marker regression or interval regression.

2.4.7 SAS

SAS is a general statistical analysis software. It can detect QTL by identifying associations between marker genotype and quantitative trait phenotype by single-marker analysis approach such as ANOVA, *t*-test, GLM, or REG.

Table 2.1 summarizes QTL mapping studies in three most important staple crops, viz. wheat, maize, and rice, for various traits using different marker systems, analysis procedure, and software programs.

2.5 Why Association Mapping?

The precision of QTL mapping largely depends on the genetic variation (or genetic background) covered by a mapping population, the size of a mapping population, and number of marker loci used. Once QTLs affecting a trait of interest are accurately tagged using above-outlined approach, marker tags are the most effective tools in a crop improvement that allows the mobilization of the genes of interest from donor lines to the breeding material through MAS. Although traditional QTL mapping will continue being an important tool in gene tagging of crops, it is a ‘now classical approach’ and suffers from a number of limitations. First, allelic variation in each cross is usually restricted because typically only two parents are used to create a QTL mapping population. Second, since early generation crosses are used, the number of recombination events per chromosome is usually small. Third, a typical QTL identified from a cross consisting of a few hundred offspring can span anywhere between a few to tens of centimorgans encompassing several megabases. Such large genome regions contain, typically, hundreds if not thousands of genes,

Table 2.1 QTL mapping studies in three important crop species

Species	Trait	QTLs	Population	Linkage groups covered	Markers linked to QTL	Software used	References
Rice	Root penetration ability	4 QTLs—number of penetrated roots 19 QTLs—total root number 6 QTLs—root penetration index 10 QTLs—tiller number	RIL 'Indica (CO39) × Japonica (Moroberekan)'	Chr1, Chr9, Chr11, Chr12	RFLP	MAPMAKER/QTL	Ray et al. (1996)
	Submergence tolerance	Major gene, 4 QTLs	RIL 'Indica (IR74 × FR13A)'	Major gene on Chr9, other QTLs on Chr6, Chr7, Chr11, Chr12	RFLP, AFLP	MAPMAKER/QTL	Nandi et al. (1997)
	Phosphorus deficiency resistance	Major QTLs	RIL 'IR20 × IR55178-3B-9-3'	Chr12	AFLP, RFLP	MAPMAKER	Ni et al. (1998)
	Blast resistance	3 QTLs	F_4 'Nipponbare × Owarihatamochi'	Chr4, Chr9, Chr12	RFLP	MAPMAKER	Fukuoka and Okuno (2001)
	Milling quality, protein content, and color characteristics	2 major QTLs	RIL 'Zhenshan 97 × Minghui 63'	Chr3, Chr5	RFPL, SSRs	MAPMAKER/QTL	Tan et al. (2001)
	Drought tolerance	77 QTLs	DH 'Japonica (CT9993 × Indica (IR62266)'	Chr3, Chr4, Chr8	RFLP, AFLP, SSRs	MAPMAKER, MQTL	Lanceras et al. (2004)
	Basal root thickness	Major QTLs	RIL 'IRAT109 × Yuefu'	Chr4	SSRs	MAPMAKER/EXP, QTLMAPPER	Liu et al. (2008)
	Grain quality	16 QTLs	RIL 'Indica (Chuan7) × Japonica (Nanyangzhan)'	Chr1, Chr2, Chr3, Chr6, Chr7, Chr9	SSRs	MAPMAKER, QTLMapper	Lou et al. (2009)
	Drought resistance	22 QTLs	RIL 'Indica (IR20 × Nootripathu)'	Chr1, Chr2, Chr4, Chr5, Chr6, 11	SSRs	MAPMAKER/EXP MS-DOS, WINQTLCART	Gomez et al. (2010)
	Leaf width and grain number	2 major QTLs	F_2 'Indica (HP) × Nipponbare'	Chr1	SSRs, InDel	QTL IciMapping v 3.1	Tian et al. (2014)

(continued)

Table 2.1 (continued)

Species	Trait	QTLs	Population	Linkage groups covered	Markers linked to QTL	Software used	References
Wheat	Drought-induced abscisic acid production	1 QTL	DH 'Chinese spring × SQ1'	Chr 5AL	RFLP	MAPMAKER, MAPMAKER-QTL	Quarrie et al. (1994)
	Preharvest sprouting tolerance	One major QTL	RIL 'SPR8198 × HD2329'	Chr 3A	SSRs, AFLP, SAPML	MAPMAKER/EXP, QTL cartographer	Kulwal et al. (2005)
	Growth response to exogenously applied stress-induced hormones	9 QTLs	DHR 'Chinese Spring (CS) × Synthetic 6A (S6A)'	Chr 6AS	SSRs	QGENE software	Castro et al. (2008)
	Stripe rust resistance	2 QTLs	RIL 'Stephens × Michigan Amber'	Chr 6BS	SSRs	Mapmaker MacIntosh, Win QTL Cart software	Santra et al. (2008)
	Aluminum toxicity	2 major QTLs	'CS'/'CS (Synthetic 3B)' DH lines	Chr 4D, Chr 3B	SSRs	MAPMAKER, QGENE software	Navakode et al. (2009)
	Senescence-related traits under high temperature	16 QTLs	RIL 'Ventnor × Karl 92'	Chr2A, Chr 6A, Chr 6B, Chr3A, Chr 3B, Chr 7A	SSRs, AFPL	MAPMAKER/EXP, QTL cartographer	Vijayalakshmi et al. (2010)
	Terminal heat tolerance	3 major QTLs	RIL 'NW1014XHUW468'	Chr2B, Chr7B, Chr7D	SSRs	MAPMAKER, QTL cartographer	Paliwal et al. (2012)
	Grain yield under multi-environment	One major QTL	DH 'RAC875 × Kukri' RIL 'RAC875 × Kukri'	Chr 3B	SSRs, ISBP	MapManager Version QTXb20, MET QTL model	Bonneau et al. (2013)
	Powdery mildew resistance	4 QTLs	DH 'Pingyuan 50 × Mingxian169'	Chr2BS, Chr3BS, Chr5AL, Chr3BL	SSRs	MapManager QTXb20, QTL cartographer	Asad et al. (2014)
	Root penetration ability	13 QTLs	DH 'Cranbrook × Halberd'	Chr 2D, Chr 4A, Chr 6B, Chr 7B	AFLP, DArT	MultiQTL	Acuña et al. (2014)

(continued)

Table 2.1 (continued)

Species	Trait	QTLs	Population	Linkage groups covered	Markers linked to QTL	Software used	References
Maize	Drought tolerance	5 QTLs	SD34 × SD35 F3 families	Chr1, Chr3, Chr5, Chr6, Chr8	RFLP	Mapmaker, Mapmaker/QTL	Agrama and Moussa (1996)
	Downy mildew resistance	3 QTLs	RIL 'G62 × G58'	Chr1, Chr9	RFLP	MAPMAKER/EXP, single factor analysis for QTL identification	Agrama et al. (1999)
	Early plant vigor in chilly environments	17 QTLs	DH lines 'SL and TH'	Chr2, Chr3, Chr4, Chr5, Chr6, Chr7, Chr8, Chr10	SSRs	PLABQTL	Presterl et al. (2007)
	Head smut resistance	5 QTLs	$F_{2:3}$ families 'Mo17 × Huangzao4'	Chr1, Chr2, Chr3, Chr8	SSRs, AFLP	Mapmaker/EXP, Winqtl cartographer	Li et al. (2008)
	Reaction to phaeosphaeria leaf spot	6 QTLs	F_2 'L14-04B × L08-05F'	Chr1, Chr3, Chr4, Chr6, Chr8	SSRs	MAPMAKER/EXP, multiple interval mapping to map QTL	Moreira et al. (2009)
	Leaf temperature responses to drought	9 QTLs	RIL 'Zong3 × 87-1'	Chr1, Chr2, Chr9, Chr10	SSRs	QTL cartographer	Liu et al. (2011)
	Stay green traits	3 major QTLs; total 22 QTLs	F_2 'A150-3-2 × Mo17'	Chr1, Chr2, Chr3, Chr4, Chr5, Chr6, Chr9	SSRs	MAPMAKER/EXP, QTL mapping by QTLNetwork-2.0 using mixed model based composite interval mapping	Wang et al. (2012a, b)
	Gray leaf spot resistance	4 QTLs	Inbred line 'Y32 × Q11'	Chr1, Chr2, Chr5, Chr8	SSRs	Mapmaker, QTL cartographer	Zhang et al. (2012)
	Root architectural traits	15 QTLs	RIL 'Oh43 × W64a (OhW), Ny821 × H99 (NyH)'	Chr1, Chr2, Chr3, Chr4, Chr5, Chr7, Chr8, Chr9	SNPs	QTL package of R	Burton et al. (2014)
	Yield and agronomic traits	14 QTLs	F2 '02S6140 × KSS22'	Chr1, Chr2, Chr3, Chr7, Chr8, Chr10	SSRs	Mapmaker, QTL mapper	Park et al. (2014)

making the process of identifying the causal gene in a QTL region a tedious and quite time-consuming task through map-based cloning (Price 2006). In addition, for many organisms the generation of mapping populations is either not possible or at least very time-consuming. For instance, the long generation time of most forest trees have completely prevented any progress in elucidating the genetic basis of complex traits using QTL mapping experiments (Neale and Savolainen 2004).

These limitations, however, can be reduced with the use of 'association mapping' (Zhu et al. 2008). Turning the gene-tagging efforts from biparental crosses to natural populations (or germplasm collections) and from traditional QTL mapping to linkage disequilibrium (LD)-based association mapping can lead to the most effective utilization of ex situ conserved natural genetic diversity of worldwide crop germplasm resources. This approach has many major advantages over conventional QTL mapping. First, a much larger and more representative gene pool can be surveyed. Second, it bypasses the expense and time of mapping studies and enables the mapping of many traits in one set of genotypes. Third, a much finer mapping resolution can be achieved, resulting in small confidence intervals of the detected loci compared to classical mapping, where the identified loci need to be fine-mapped. Finally, it has the potential not only to identify and map QTLs but also to identify the causal polymorphism within a gene that is responsible for the difference in two alternative phenotypes.

2.5.1 General Steps in Association Mapping

The performance of association mapping includes the following general steps: (1) selection of a group of individuals from a natural population or germplasm collection with wide coverage of genetic diversity; (2) measuring the phenotypic characteristics (yield, quality, tolerance, resistance, etc.) in the population, preferably, in different environments and multiple replication/trial design; (3) genotyping the mapping population individuals with molecular markers; (4) quantification of the extent of LD of a chosen population genome using molecular marker data; (5) assessment of the population structure (the level of genetic differentiation among groups within a sampled population individuals) and kinship (coefficient of relatedness between pairs of each individuals within a sample); and (6) based on information gained through quantification of LD and population structure, correlation of phenotypic and genotypic data with the application of an appropriate statistical approach that reveals 'marker tags' positioned within close proximity of targeted trait of interest.

2.5.2 Selection of Association Mapping Panels

Various kinds of populations can be used for association mapping; gene bank collections, elite breeding materials and/or specialized populations (e.g., NAM or MAGIC). In the case of gene bank materials, core collections are expected to represent most of the genetic variability with a manageable number of accessions and thus are suitable for association studies. In addition, the process of selecting a minimum sample size with maximum variation has a normalizing effect that is expected to reduce population structure and decrease LD, thus creating a situation favorable for association mapping. Core collections are useful for mapping qualitative traits, such as disease resistance or quality characteristics. Their broad genetic variability makes them often unsuitable for analysis of quantitative traits because accessions are usually unadapted to growing conditions, resulting in poor precision of trait measurement.

In plant breeding programs, a large body of phenotypic data is accumulated for elite lines from replicated field experiments over locations and years, thereby saving time to develop a panel. The use of those data for association mapping requires statistical models accounting for covariances introduced both by experimental design (years, locations, replicates) and polygenic effects. Moreover, those data are often unbalanced because new lines are included in field trials each year, while other lines are discarded. Population structure and higher LD can be prominent in elite material because it is common for closely related lines to be admitted to advanced trials. However, if pedigrees are known, the relationships among the lines can be determined and used to control for polygenic effects. Although association mapping in elite lines may not offer much improved resolution compared with QTL analysis in biparental mapping populations, there are at least two important advantages: a substantially higher level of polymorphism and detection of favorable alleles directly in the target population. Elite lines might be more desirable materials for mapping low-heritability traits such as grain yield, as the material is genetically more stable and are well adapted to normal growing conditions.

To increase the power and mapping resolution of marker–trait associations, some specialized populations have been constructed utilizing a joint strength of QTL mapping and AM. For example, nested association mapping (NAM) populations and multiparent advanced generation intercross (MAGIC) populations have been developed in many crops (Kover et al. 2009; McMullen et al. 2009; Huang et al. 2012). NAM populations are developed by crossing a set of diverse lines (up to 25) to one reference line. F_1s of each cross are then selfed to develop RILs for each population. MAGIC populations are created by several generation of intercrossing among multiple founder lines, for example, four or eight lines. Multiple founders similar to a NAM population capture more allelic diversity than biparental mapping populations, whereas the multiple cycles of intercrossing give greater opportunity of recombination and hence greater precision of QTL mapping. However, it should be kept in mind that generating such specialized populations requires and a lot of effort, time, and investment.

2.5.3 *Concept and Calculation of Linkage Disequilibrium (LD)*

The terms linkage and LD are also often confused. Linkage refers to the correlated inheritance of loci through the physical connection on a chromosome, whereas LD refers to the correlation between alleles in a population (Flint-Garcia et al. 2003) but not necessarily on the same chromosome. As a starting point for association mapping, it is important to gain knowledge of the patterns of LD for genomic regions of the ‘target’ organisms and the specificity of the extent of LD among different populations or groups to design and conduct unbiased association mapping. The two most commonly used statistics to measure LD are r^2 (square of the correlation coefficient) and D' (disequilibrium coefficient). The statistics r^2 and D' reflect different aspects of LD and perform differently under various conditions. The r^2 is affected by both mutation and recombination, while D' is affected by more mutational histories.

There are freely available softwares such as GOLD (Abecasis and Cookson 2000), TASSEL (www.maizegenetics.net), and Powermarker (Liu and Muse 2005) to depict the structure and pattern of LD. One can estimate an average genome-wide decay of LD by plotting LD values (r^2 values) obtained from a data set covering an entire genome (i.e., with more or less evenly spaced markers across all chromosomes in a genome) against distance. When such a LD decay plot generated, the usual practice is to look for distance point where LD value (r^2) decreases below 0.1 or half strength of D' ($D' = 0.5$) based on the curve of nonlinear logarithmic trend line. This gives the rough estimates of the extent of LD for association study, but for more accurate estimates, highly significant threshold LD values ($r^2 \geq 0.2$) are also used as a cutoff point. The decrease of the LD within the genetic distance indicates that the portion of LD is conserved with linkage and proportional to recombination (Gupta et al. 2005).

2.5.4 *Types of Association Mapping*

Association mapping generally falls into two broad categories: (1) candidate-gene-based association mapping, which relates polymorphisms in selected candidate genes that have purported roles in controlling phenotypic variation for specific traits and (2) genome-wide association mapping (GWAM), which surveys genetic variation in the whole genome to find signals of association for various complex traits (Risch and Merikangas 1996; Zhu et al. 2008). The absolutely most important aspect when deciding between a candidate gene approach and a whole-genome study is the extent of LD (see section on LD below) in the organism of interest, because the extent of LD determines not only the mapping resolution that can be achieved, but also the numbers of markers that are needed for an adequate coverage of the genome in a genome-wide study (Whitt and Buckler 2003). In species where

LD extends over long physical distances, relatively few markers are needed to ensure adequate genome coverage. For example, in species like *Arabidopsis thaliana* or in inbred lines of barley, LD can extend for tens or even hundreds of kilo base pairs, thus allowing GWAM with a relatively low number of evenly spaced markers (Aranzana et al. 2005; Rostoks et al. 2006). However, in many predominantly or obligately outcrossing organisms, such as maize (Remington et al. 2001) and many forest trees (Neale and Ingvarsson 2008), LD only extends a few hundred base pairs. For these species, several million markers would have to be genotyped to ensure adequate genome-wide coverage. Hence, in such species before several million markers (mainly single nucleotide polymorphisms) are generated, candidate-gene-based association mapping can be equally useful.

A candidate-gene-based association study is more hypothesis-driven than a genome-wide study because mapping is restricted to genes thought to be good candidates for controlling the trait of interest (Neale and Savolainen 2004; Hall et al. 2010). Although the selection of candidate genes is not always straightforward, genes are usually selected based on information obtained from, for instance, genetic, biochemical, mutation, physiology, or expression studies in both model and non-model plant species. The construction of molecular linkage maps based on the genes [e.g., expressed sequence tags (ESTs), EST-SSRs)] is another way of identifying the candidate genes underlying QTL, instead of time-consuming fine mapping (Sehgal and Yadav 2009). Standard neutrality tests applied to DNA sequence variation data can also be used to select candidate genes or amino acid sites that are putatively under selection for association mapping. This is one of the effective alternative strategies in association mapping that allow reducing the total amount of marker genotyping in less number of individuals. This increases the power and precision of the trait–marker correlations. However, it is important to remember that a candidate gene approach is limited by the choice of candidate genes that are identified and hence always runs the risk of missing out on identifying causal mutations located in non-identified candidate genes.

2.5.5 Candidate-Gene-based Association Studies in Plants

The candidate gene strategy has shown promise for bridging the gap between quantitative genetic and molecular genetic approaches to study complex traits (Cattivelli et al. 2008; Ingvarsson and Street 2011). Along this line, important studies on association mapping with the candidate gene approach are summarized as follows. Vernalization requirement in wheat is controlled by four major genes, viz. *VRN1*, *VRN2*, *VRN3*, and *VRN4*, with *VRN1* gene copies *VRN-A1*, *VRN-B1*, and *VRN-D1* located on the long arms of chromosomes 5A, 5B, and 5D, respectively (Yoshida et al. 2010). An association mapping study conducted by Rousset et al. (2011) on 235 hexaploid wheat collections revealed the effects of the flowering time candidate genes in modulating flowering time in wheat. In that study, genetic variation in *VRN-A1*, *VRN-B1*, and *VRN-D1* genes explained a large part of phenotypic variation in growth habit. Huang and Brûlé-Babel (2012) studied genetic diversity, haplotype structure, and association of genes involved in starch biosynthesis in wheat. Genes encoding granule-bound starch synthase (GBSSI, also known as waxy or *Wx* genes) and soluble starch synthase (SSIIa) were selected for nucleotide diversity and SNP density study. None of the SNPs within the three *SSIIa* genes and *Wx-D1* gene was associated with yield-related traits. However, both SNPs and haplotypes within the *Wx-A1* gene were associated with seed number per spike, seed weight per spike and thousand kernel weight. Another study on grain size of wheat also demonstrated the association of haplotype of a grain size gene (*TaGW2*) with larger grain size, earlier heading date and maturity in hexaploid wheat (Su et al. 2011). Similarly, transcription factors such as the gibberellin-regulated Myb factor (GAMYB), the barley leucine zippers 1 and 2 (*BLZ1*, *BLZ2*), and the barley prolamin box binding factor (BPBF) were evaluated for their association with agronomic traits in barley (Haseneyer et al. 2010). SNPs within *BLZ1* were associated with days to flowering, and its haplotype was also associated with both days to flowering and plant height. The haplotype of *BLZ2* was associated with thousand kernel weight, while the haplotype of the *BPBF* gene was associated with both crude protein and starch in barley endosperm (Haseneyer et al. 2010). However, the candidate genes explained only a small portion of the total genetic variation. Similarly for sorghum and rice, candidate genes involved in starch biosynthesis were associated with the expected traits and the results were in agreement with QTL studies (Bao et al. 2006; Figueiredo et al. 2010).

More than 20 studies of candidate gene association analysis in maize have been published to date. These studies used candidate genes from well characterized and relatively simple metabolic pathways (Wilson et al. 2004; Harjes et al. 2008; Yan et al. 2010) or those with extensive prior evidence for the role of the candidate gene (s) in the control of the phenotype of interest such as map-based cloning studies (Salvi et al. 2007; Ducrocq et al. 2008; Zheng et al. 2008; Buckler et al. 2009). The candidate genes were also selected based on information in closely related species (Li et al. 2010a, b), and information from QTL mapping studies and/or expression results (Krill et al. 2010). The most comprehensive candidate gene association results have been recently reported in maize for SNPs identified from 540 genes putatively involved in accumulation of carbohydrate and ABA metabolites during drought stress (Setter et al. 2010). In this study, the SNP from a homologue of an *Arabidopsis* MADS-box gene was significantly associated with phaseic acid in ears of irrigated plants, while a SNP in pyruvate dehydrogenase kinase was significantly associated with silk sugar concentrations. Similarly, a SNP from an aldehyde oxidase gene was associated with ABA levels in silk under non-irrigated conditions.

The candidate gene association mapping approach has been widely applied in forest tree genetics studies as developing a biparental population is practically unfeasible for most conifers. González-Martínez et al. (2006) studied the pattern of polymorphisms of 18 drought-responsive candidate genes in 32 *Pinus taeda* L. individuals. LD within the sequenced gene regions varied from low to high depending on the candidate gene locus. A total of 196 SNPs and 82 LD blocks were obtained in 18 candidate gene loci. By constructing LD blocks, 94 haplotype SNPs were identified to improve the LD values and were successfully used in detecting significant r^2 values for LD blocks study. The same authors evaluated the association of four candidate genes belonging to different functional classes with carbon isotope discrimination (CID) at two locations. The genes were general protection factor (*dhn-1*), antioxidants (*sod-chl*), transcription factor (*wrky-like*), and putative cell wall protein (*lp5-like*). Antioxidant (*sod-ch1*) and *Cu/Zn superoxide dismutase* genes showed significant association with CID at both locations. However, none of the significant associations explained a substantial amount of phenotypic variance in CID.

2.5.6 Confounding Effects of Population Structure in Association Mapping

One of the main hurdles for using association mapping to dissect the genetic architecture of complex traits in plants is the risk of incurring false positives due to population structure (Pritchard et al. 2000). The problem of population structure arises because any phenotypic trait that is also correlated with the underlying population structure at neutral loci will show an inflated number of positive associations resulting in Type I errors. Among many methods developed to deal with this problem, 'genomic control' (GC) method (Devlin and Roeder 1999) estimates association using a large number of putative neutral markers or markers not thought to be involved in controlling the trait of interest. The distribution of the test statistic of interest is then calculated from these associations and a critical value corresponding to the desired Type I error rate is chosen from this distribution. Another method that is commonly used to control for population structure is structured associations (SA) (Pritchard et al. 2000). SA first searches a population for closely related clusters/subdivisions using Bayesian approach and then uses the clustering matrices (Q) in association mapping (by a logistic regression) to correct the false associations. Population structure and shared co-ancestry coefficients between individuals of subdivisions of a population can be effectively estimated with *STRUCTURE* program (Pritchard et al. 2000) using several models for linked and unlinked markers.

Principal component analysis (PCA) was recently suggested as a fast and effective way to diagnose population structure (Zhu and Yu 2009). The PCA method summarizes variation observed across all markers into a number of underlying component variables and these components, typically the first few, can then be used to replace Q to adjust for population structure. The PCA method makes it computationally feasible to handle a large number of markers (tens of thousands) and correct for subtle population stratification. There are many programs that can be used to calculate PCA such as DARwin (Perrier and Jacquemoud-Collet 2006) and EIGENSTRAT (Price et al. 2006).

However, incorporating only population structure information in the analysis is not good enough itself when highly structured population with some degree of related individuals are used in the association mapping. A mixed linear model (MLM) that combines both population structure information (*Q*-matrix or PCA) and level of pairwise relatedness coefficients (kinship-matrix) should be used in the analysis. Although computationally intensive, MLM approach is very effective in removing the confounding effects of the population in association mapping (Yu et al. 2005).

2.5.7 Association with Raw Data or BLUPs or Residuals

In general, raw data can be used directly in association analysis provided it is available for all entries and for all replicates in different locations/years. For cases where phenotypes are not evaluated for all individuals and replicates due to large sample size, best linear unbiased predictors (BLUPs) from a mixed model may be substituted as the dependent variable. In such cases, the association analysis using BLUPs can be performed with many fewer observations and require much less time.

More recently, researchers have also used residuals instead of raw data. The rationale is that after removing all the effects except the marker, including the polygenic genetic variance captured by the BLUPs, the signal due to marker association is still contained in the residuals. Signal from the marker will be removed only to the extent that it is correlated with the other effects. The residual approach performs as well as the approach using raw phenotype directly for low-heritability traits (Aulchenko et al. 2007). Because the association test using residuals is performed without including the polygenic random effect, tests of individual markers run quickly. The mixed model equations with thousands of individuals only need to be solved once for any particular phenotype. After that, the millions of association tests for individual markers can then be performed using simple *t*-tests or *F*-tests of the marker classes.

2.5.8 Association Analysis Programs

Public, freely available software suitable for association analysis using mixed models in plants includes TASSEL and EMMA/R. Both analyze moderately large data sets in a reasonable amount of time but only allow a single effect (samples or taxa) to be fit as a random effect. All other effects are treated as fixed. EMMA relies on the R for data management and visualization, whereas TASSEL handles those functions itself. Several commercial software packages available for association studies include ASREML, JMP Genomics, SAS, and GenStat. ASREML and JMP Genomics are specifically engineered for genetic analysis and can handle more complex models, whereas general purpose packages such as SAS Proc Mixed and GenStat can perform association analysis but require more expertise and programming on the part of the user.

From the user's perspective, clearly freely available software such as TASSEL plays an important role in scientific investigation. Another advantage of using TASSEL is that both graphical user interface (GUI) and command line interface (CLI) versions exist. In the GUI, the plug-ins are invoked by clicking buttons on the interface. With the CLI, the plug-ins are used in a predetermined pipeline that passes the output from one step to the input of another. Hence, scientists can use these versions depending on their expertise and consistent results are achieved independent of the interface. In the latest version of TASSEL (TASSEL 5.0), compressed MLM method is available for computing large data sets with up to 500,000 markers.

2.5.9 Significance Threshold

A threshold is set to declare significant associations. Any of the two statistical methods can be used to correct for multiple comparisons: false discovery rate (FDR) and Bonferroni correction. The correction is needed whenever one would like to test multiple hypotheses simultaneously. FDR controls the expected proportion of false positives among significant results by determining a threshold from the observed p value distribution in the data, whereas Bonferroni corrections control the chance of any false positives (Benjamini and Hochberg 1995). Given the aims of the study, one may consider a high FDR for some projects (e.g., investigating the genetic architecture of a trait) and a low FDR for others (e.g., identifying candidate loci for follow-up studies).

2.5.10 Validation of Association Results

Table 2.2 summarizes genome-wide association studies conducted in three major crops. It clearly shows the increase in association studies in wheat, maize, and rice for dissecting various complex traits. As the number of studies documenting alleles showing significant associations with quantitative trait variation, there is an increasing need to replicate findings before marker information is incorporated in selection decisions, or before large sums are invested into identification of causal factors and gene cloning. The most straightforward way is to compare the association mapping results with previous results published for the trait, for example, using biparental populations. If markers in close proximity (within 10 cM) to previously reported QTLs/genes are identified, the result will not only be validated

Table 2.2 Association mapping studies in three important crops

Crop	Marker system	Trait analyzed	Marker–trait association	Program used for genetic structure and association analysis	References
Rice	SSRs	Grain yield (GY), kernel length (KL), kernel width (KW), kernel length/width ratio (LWR), 1000 kernel weight (TKW)	GY-5 markers, KL-6 markers, KW-5 markers, LWR-5 markers, TKW-4 markers	STRUCTURE, PowerMarker 3.23, Arlequin 3.0, TASSEL	Agrama et al. (2007)
	SSRs, InDel	Heading date surveyed in 2006 (HD06), heading date in 2006 (HD07), plant height in 2006 (PH06), plant height in 2007 (PH07), panicle length in 2006 (PL06), panicle length in 2007 (PL07)	HD06-3 markers, HD07-3 markers, PH06-3 markers, PH07-2 markers, PL06-2 markers, PL07-1 marker	STRUCTURE, NYSYS-pc, SPAGeDi, TASSEL	Wen et al. (2009)
	SSRs, InDel	Single stigma exsertion (SStgE), dual exsertion (DStgE), total stigma exsertion (TStgE), spikelet length (SpkL), spikelet width (SpkW), spikelet length/width ratio (SpkL/W)	SStgE-4 markers, DStgE-6 markers, TStgE-5 markers, SpkL-4 markers, SpkW-3 markers, SpkL/W-7 markers	STRUCTURE, GenAlex, SPAGeDi, TASSEL	Yan et al. (2009)
	SSRs	Grain yield (YLD), Tiller number per plant (TILN), Panicle number per plant (PANN), Yield from ratiining (RYLD), Amylose content (AC), Head milled rice (MR)	**Low-land accessions**-PANN (2004)-1 marker, AC (2004, 2005)-1 marker, MR (2005)-2 markers, YLD (2005)-4 markers **Upland accession**-AC (2004, 2005)-1 marker	STRUCTURE, FSTAT 2.9.3.2, Genetix 4.03, SPAGeDi, TASSEL	Borba et al. (2010)
	SSRs	Sheath blight (ShB) resistance	ShB resistance-10 markers	STRUCTURE, PowerMarker, MEGA 4.0, NTSYSpc, TASSEL	Jia et al. (2012)
	SSRs, InDel	Harvest index traits in temperate (Arkansas) and subtropical climates (Texas)	Harvest index traits in Arkansas-36 markers, harvest index traits in Texas-28 markers	STRUCTURE, SPAGeDi, PDMIX800 SAS macro	Li et al. (2012)
	SSRs	Grain yield per plant, weight of 100 grains, panicle length	Grain yield per plant-3 markers, weight of 100 grains-1 marker, panicle length-5 markers	DARWin, STRUCTURE, TASSEL	Vanniarajan et al. (2012)
	SSRs	Cold tolerance	Colt tolerance-total 24 markers, in Yunnan (natural low temperature)-5 markers, in Jilin (cold water irrigation)-19 markers	STRUCTURE, Arlequin, SPAGeDi, NTSYSpc, TASSEL	Cui et al. (2013)
	SSRs	Seed vigor evaluated by root length (RL), shoot length (SL), shoot dry weight (SDW)	RL-7 markers, SL-12 markers, SDW-8 markers	STRUCTURE, Arlequin, SPAGeDi, TASSEL	Dang et al. (2014)
	SNPs	Genotype × Environment interactions for agronomic traits	10 agronomic traits-23 putative loci	STRUCTURE, TASSEL	Xu et al. (2014)

(continued)

Table 2.2 (continued)

Crop	Marker system	Trait analyzed	Marker–trait association	Program used for genetic structure and association analysis	References
Wheat	SSRs	Kernel size and milling quality	Kernel size and milling quality-62 loci	STRUCTURE, TASSEL	Breseghello and Sorrells (2006)
	DArT	Resistance to leaf rust (LR), powdery mildew (PM), stripe rust (SR), and yellow (stripe) rust (YR), plus grain yield (GY)	LR-87 markers, PM-61 markers, SR-63 markers, YR-122 markers, GY-213 markers	STRUCTURE, DARWIN, TASSEL	Crossa et al. (2007)
	SSRs	Russian wheat aphid resistance	Leaf chlorosis-28 markers, Leaf rolling-8 markers	STRUCTURE, TASSEL	Peng et al. (2009)
	SSRs	Drought-adaptive traits and grain yield in different environment different for water availability	Grain yield-56 markers in one environment, 20 markers in two environments, 3 markers in 5 environments, only one marker in six environments	STRUCTURE, TASSEL	Maccaferri et al. (2010)
	SSRs	Plant height under well water and drought stress condition	Plant height and drought tolerance coefficient-23 markers	PowerMarker, STRUCTURE, TASSEL	Zhang et al. (2011)
	SSRs	Preharvest sprouting tolerance (PHST)	PHST-30 markers	STRUCTURE, TASSEL	Jaiswal et al. (2012)
	STS, DArT	Stem rust resistance	Stem rust resistance-12 markers	SAS PROC PRINCOMP, TASSEL	Yu et al. (2012)
	DArT	Pest resistance—hessian fly (HF), Russian wheat aphid (RWA), sunn pest (SP), wheat stem saw fly (WSSF), and cereal leaf beetle (CLB)	HF-13 markers, RWA-13 markers, SP-17 markers, WSSF-5 markers, CLB-5 markers	STRUCTURE, TASSEL	Joukhadar et al. (2013)
	SNPs	Soil-borne wheat mosaic virus (SBWMV) resistance	SBWMV resistance-6 markers	STRUCTURE, TASSEL	Liu et al. (2014)
	SNPs	Stripe rust caused by (*Puccinia striiformis* f. sp. *tritici; Pst),* resistance	Resistance to *pst* at seedling stage-27 markers, resistance to *pst* at adult plant stage-38 markers	STRUCTURE, TASSEL	Zegeye et al. (2014)

(continued)

Table 2.2 (continued)

Crop	Marker system	Trait analyzed	Marker–trait association	Program used for genetic structure and association analysis	References
Maize	SNPs	Plant height	Plant height-204 markers	PLINK, STRUCTURE, TASSEL	Weng et al. (2011)
	SSRs	Downy mildew resistance	Downy mildew resistance-3 markers	STRUCTURE, TASSEL	Phumichai et al. (2012)
	SSRs, SNPs	Flowering time (FT), northern corn leaf blight resistance (NCLB), NCLB resistance corrected for FT	FT-7 markers, NCLB-4 markers, NCLB resistance corrected for FT-4 markers	STRUCTURE, ASREML release, EMMA	Van Inghelandt et al. (2012)
	SNPs, InDels	Head smut resistance	Head smut resistance-19 markers and 18 candidate genes	STRUCTURE, SPAGeDi, NTSYSpc, TASSEL	Wang et al. (2012a, b)
	SNPs	Chilling tolerance	Chilling tolerance-43 markers	STRUCTURE, SPAGeDi, NTSYSpc, TASSEL	Huang et al. (2013)
	SNPs	Oil biosynthesis in maize kernels	Kernel oil concentration and fatty acid composition-26 SNPs and candidate genes associated with oil concentration	TASSEL	Li et al. (2013)
	SNPs	Agronomic traits under well watered and water stresses conditions	126 trait × environment × treatment combinations-42 markers	STRUCTURE, TASSEL	Xue et al. (2013)
	SNPs	Waterlogging tolerance	Traits under normal and waterlogged condition (seeding height, root length, shoot fresh weight, root fresh weight, soot dry weight, root dry weight)-47 markers	STRUCTURE, SPAGeDi, NTSYSpc, TASSEL	Zhang et al. (2013)
	SNPs	Gray leaf spot (GLS) resistance	GLS resistance-51 markers	STRUCTURE, TASSEL	Shi et al. (2014)
	SNPs	Agronomic traits in an enlarged maize association panel	Ear leaf width, Ear length, kernel width, plant height and tassel main axis length-19 markers	TASSEL	Yang et al. (2014)

but also increase the confidence to pursue the new genomic target identified for the trait. Secondly, results can be validated in different populations. This is more reliable as the probability of observing false positives becomes small if significant associations are confirmed in two or more validation populations. Third, if association studies point to alleles with opposite effects on a trait of interest, one can generate multiple F_2 populations from parents that harbor contrasting alleles and determine whether differences in phenotype co-segregate with the locus in question. Once markers tightly linked to the target trait are validated, they provide several magnitudes of return on investment through increased speed and cost efficiency of breeding programs.

However, failure to replicate a previously documented association can occur because of a large number of issues, both in the initial and the replication study, including factors such as difficulties in replicating the environment, small sample size, poor study design, or lack of rigorous phenotype scoring (Manolio et al. 2009). Another concern is that allelic effects of previously documented associations usually decline in replication studies. This phenomenon is known as the 'Beavis effect' (Beavis 1994) in the QTL mapping literature and occurs because significant associations are reported only when test statistics exceed a predetermined critical threshold. The estimated effects of the detected associations are therefore sampled from a truncated distribution, and the weaker the initial effect the more serious this overestimation is (Rockman 2008). The Beavis effect has also been shown to occur in association mapping studies (Ingvarsson et al. 2008). Hence, careful consideration of the power of the prospective association study should be taken early on in the experiment, so that things like the Beavis effect can be minimized or eliminated.

2.6 Conclusions

Comparison of linkage analysis and association mapping for QTL detection revealed that linkage mapping is more useful for genome-wide scan for QTLs, while association mapping gives more precise location of an individual QTL. Therefore, linkage analysis may be preferred for preliminary location of QTLs and then use association mapping for more precise location. Association mapping is prone to the identification of false positives, especially if the experimental design is not rigorously controlled. For example, population structure has long been known to induce many false positives and accounting for population structure has become one of the main issues when implementing association mapping in plants. Also, with increasing numbers of genetic markers used, the problem becomes separating true from false positive and this highlights the need for independent validation of identified association. The examples of association mapping studies performed in three most important crops' germplasm largely demonstrate the flourish of crop genomics era with the utilization of powerful LD-based association mapping tool. Currently, a number of such studies are in progress for various other crops in many laboratories worldwide. The near-future completion of genome sequencing projects of crop species, powered with more cost-effective sequencing technologies, will certainly create a basis for application of whole-genome-association studies. This will provide with more powerful association mapping tool(s) for crop breeding and genomics programs in tagging true functional associations conditioning genetic diversities, and consequently, its effective utilization.

References

Abecasis GR, Cookson W (2000) GOLD—graphical overview of linkage disequilibrium. Bioinformatics 16:182–183

Acuña TB, Rebetzke G, He X et al (2014) Mapping quantitative trait loci associated with root penetration ability of wheat in contrasting environments. Mol Breed 1–12

Agrama HA, Moussa ME (1996) Mapping QTLs in breeding for drought tolerance in maize (*Zea mays* L.). Euphytica 91:89–97

Agrama H, Moussa M, Naser M et al (1999) Mapping of QTL for downy mildew resistance in maize. Theor Appl Genet 99:519–523

Agrama H, Eizenga G, Yan W (2007) Association mapping of yield and its components in rice cultivars. Mol Breed 19:341–356

Aranzana MJ, Kim S, Zhao K et al (2005) Genome-wide association mapping in Arabidopsis identifies previously known flowering time and pathogen resistance genes. PLoS Genet 1:e60

Asad MA, Bai B, Lan C et al (2014) Identification of QTL for adult-plant resistance to powdery mildew in Chinese wheat landrace Pingyuan 50. Crop J 2:308–314

Asins M (2002) Present and future of quantitative trait locus analysis in plant breeding. Plant Breed 121:281–291

Aulchenko YS, De Koning D-J, Haley C (2007) Genomewide rapid association using mixed model and regression: a fast and simple method for genomewide pedigree-based quantitative trait loci association analysis. Genetics 177:577–585

Bao J, Corke H, Sun M (2006) Microsatellites, single nucleotide polymorphisms and a sequence tagged site in starch-synthesizing genes in relation to starch physicochemical properties in nonwaxy rice (*Oryza sativa* L.). Theor Appl Genet 113:1185–1196

Beavis WD (1994) The power and deceit of QTL experiments: lessons from comparative QTL studies. In: Proceedings of the forty-ninth annual corn and sorghum industry research conference pp 250–266

Benjamini Y, Hochberg Y (1995) Controlling the false discovery rate: a practical and powerful approach to multiple testing. J Roy Stat Soc Ser B (Methodol) 289–300

Bonneau J, Taylor J, Parent B et al (2013) Multi-environment analysis and improved mapping of a yield-related QTL on chromosome 3B of wheat. Theor Appl Genet 126:747–761

Borba TCDO, Brondani RPV, Breseghello F et al (2010) Association mapping for yield and grain quality traits in rice (*Oryza sativa* L.). Genet Mol Biol 33:515–524

Breseghello F, Sorrells ME (2006) Association mapping of kernel size and milling quality in wheat (*Triticum aestivum* L.) cultivars. Genetics 172:1165–1177

Buckler ES, Holland JB, Bradbury PJ et al (2009) The genetic architecture of maize flowering time. Science 325:714–718

Burton AL, Johnson JM, Foerster JM et al (2014) QTL mapping and phenotypic variation for root architectural traits in maize (*Zea mays* L.). Theor Appl Genet 1–19

Castro A, Tacaliti M, Giménez D et al (2008) Mapping quantitative trait loci for growth responses to exogenously applied stress induced hormones in wheat. Euphytica 164:719–727

Cattivelli L, Rizza F, Badeck F-W et al (2008) Drought tolerance improvement in crop plants: an integrated view from breeding to genomics. Field Crops Res 105:1–14

Cheema J, Ellis TH, Dicks J (2010) THREaD mapper studio: a novel, visual web server for the estimation of genetic linkage maps. Nucl Acids Res 38:188–193
Collard B, Jahufer M, Brouwer J et al (2005) An introduction to markers, quantitative trait loci (QTL) mapping and marker-assisted selection for crop improvement: the basic concepts. Euphytica 142:169–196
Crossa J, Burgueno J, Dreisigacker S et al (2007) Association analysis of historical bread wheat germplasm using additive genetic covariance of relatives and population structure. Genetics 177:1889–1913
Cui D, Xu C-Y, Tang C-F et al (2013) Genetic structure and association mapping of cold tolerance in improved japonica rice germplasm at the booting stage. Euphytica 193:369–382
Dang X, Thi TGT, Dong G et al (2014) Genetic diversity and association mapping of seed vigor in rice (*Oryza sativa* L.). Planta 239:1309–1319
De Alencar Figueiredo L, Sine B, Chantereau J et al (2010) Variability of grain quality in sorghum: association with polymorphism in Sh2, Bt2, SssI, Ae1, Wx and O_2. Theor Appl Genet 121:1171–1185
Devlin B, Roeder K (1999) Genomic control for association studies. Biometrics 55:997–1004
Ducrocq S, Madur D, Veyrieras J-B et al (2008) Key impact of Vgt1 on flowering time adaptation in maize: evidence from association mapping and ecogeographical information. Genetics 178:2433–2437
Faris JD, Haen KM, Gill BS (2000) Saturation mapping of a gene-rich recombination hot spot region in wheat. Genetics 154:823–835
Flint-Garcia SA, Thornsberry JM, Iv B (2003) Structure of linkage disequilibrium in plants*. Annu Rev Plant Biol 54:357–374
Fukuoka S, Okuno K (2001) QTL analysis and mapping of pi21, a recessive gene for field resistance to rice blast in Japanese upland rice. Theor Appl Genet 103:185–190
Gomez SM, Boopathi NM, Kumar SS et al (2010) Molecular mapping and location of QTLs for drought-resistance traits in indica rice (*Oryza sativa* L.) lines adapted to target environments. Acta Physiol Plant 32:355–364
González-Martínez SC, Ersoz E, Brown GR et al (2006) DNA sequence variation and selection of tag single-nucleotide polymorphisms at candidate genes for drought-stress response in *Pinus taeda* L. Genetics 172:1915–1926
Gupta PK, Rustgi S, Kulwal PL (2005) Linkage disequilibrium and association studies in higher plants: present status and future prospects. Plant Mol Biol 57:461–485
Haley CS, Knott SA (1992) A simple regression method for mapping quantitative trait loci in line crosses using flanking markers. Heredity 69:315–324
Hall D, Tegström C, Ingvarsson PK (2010) Using association mapping to dissect the genetic basis of complex traits in plants. Briefings in functional genomics elp048
Harjes CE, Rocheford TR, Bai L et al (2008) Natural genetic variation in lycopene epsilon cyclase tapped for maize biofortification. Science 319:330–333
Haseneyer G, Stracke S, Piepho H-P et al (2010) DNA polymorphisms and haplotype patterns of transcription factors involved in barley endosperm development are associated with key agronomic traits. BMC Plant Biol 10:5
Hoisington D, Khairallah M, Reeves T et al (1999) Plant genetic resources: what can they contribute toward increased crop productivity? Proc Natl Acad Sci 96:5937–5943
Huang BE, George AW, Forrest KL, Kilian A, Hayden MJ, Morell MK, Cavanagh CR (2012) A multiparent advanced generation inter-cross population for genetic analysis in wheat. Plant Biotechnol J. 2012 0(7):826-39
Huang X-Q, Brûlé-Babel A (2012) Sequence diversity, haplotype analysis, association mapping and functional marker development in the waxy and starch synthase IIa genes for grain-yield-related traits in hexaploid wheat (*Triticum aestivum* L.). Mol Breed 30:627–645
Huang J, Zhang J, Li W et al (2013) Genome-wide association analysis of ten chilling tolerance indices at the germination and seedling stages in maize. J Integr Plant Biol 55:735–744
Ingvarsson PK, Street NR (2011) Association genetics of complex traits in plants. New Phytol 189:909–922

Ingvarsson PK, Garcia MV, Luquez V et al (2008) Nucleotide polymorphism and phenotypic associations within and around the phytochrome B2 locus in European aspen (*Populus tremula*, Salicaceae). Genetics 178:2217–2226

Jaiswal V, Mir R, Mohan A et al (2012) Association mapping for pre-harvest sprouting tolerance in common wheat (*Triticum aestivum* L.). Euphytica 188:89–102

Jansen RC, Stam P (1994) High resolution of quantitative traits into multiple loci via interval mapping. Genetics 136:1447–1455

Jia L, Yan W, Zhu C et al (2012) Allelic analysis of sheath blight resistance with association mapping in rice. PLoS ONE 7:e32703

Joukhadar R, El-Bouhssini M, Jighly A et al (2013) Genome-wide association mapping for five major pest resistances in wheat. Mol Breed 32:943–960

Kover PX, Valdar W, Trakalo J et al (2009) A multiparent advanced generation inter-cross to fine-map quantitative traits in Arabidopsis thaliana. PLoS Genet 5:e1000551

Krill AM, Kirst M, Kochian LV et al (2010) Association and linkage analysis of aluminum tolerance genes in maize. PLoS ONE 5:e9958

Kulwal P, Kumar N, Gaur A et al (2005) Mapping of a major QTL for pre-harvest sprouting tolerance on chromosome 3A in bread wheat. Theor Appl Genet 111:1052–1059

Künzel G, Korzun L, Meister A (2000) Cytologically integrated physical restriction fragment length polymorphism maps for the barley genome based on translocation breakpoints. Genetics 154:397–412

Lanceras JC, Pantuwan G, Jongdee B et al (2004) Quantitative trait loci associated with drought tolerance at reproductive stage in rice. Plant Physiol 135:384–399

Lander ES, Green P, Abrahamson J et al (1987) MAPMAKER: an interactive computer package for constructing primary genetic linkage maps of experimental and natural populations. Genomics 1:174–181

Li X, Wang Z, Gao S et al (2008) Analysis of QTL for resistance to head smut (*Sporisorium reiliana*) in maize. Field crops Res 106:148–155

Li Q, Li L, Yang X et al (2010a) Relationship, evolutionary fate and function of two maize co-orthologs of rice GW2 associated with kernel size and weight. BMC Plant Biol 10:143

Li Q, Yang X, Bai G et al (2010b) Cloning and characterization of a putative GS3 ortholog involved in maize kernel development. Theor Appl Genet 120:753–763

Li X, Yan W, Agrama H et al (2012) Unraveling the complex trait of harvest index with association mapping in rice (*Oryza sativa* L.). PLoS ONE 7:e29350

Li H, Peng Z, Yang X et al (2013) Genome-wide association study dissects the genetic architecture of oil biosynthesis in maize kernels. Nat Genet 45:43–50

Lincoln S, Daly M. Lander E (1993) Constructing genetic linkage maps with MAPMAKER/EXP. Version 3.0. Whitehead Institute for Biomedical Research Technical Report, 3rd edn

Liu K, Muse SV (2005) Powermarker: an integrated analysis environment for genetic marker analysis. Bioinformatics 21:2128–2129

Liu L, Mu P, Li X et al (2008) Localization of QTL for basal root thickness in japonica rice and effect of marker-assisted selection for a major QTL. Euphytica 164:729–737

Liu Y, Subhash C, Yan J et al (2011) Maize leaf temperature responses to drought: thermal imaging and quantitative trait loci (QTL) mapping. Environ Exp Bot 71:158–165

Liu S, Yang X, Zhang D et al (2014) Genome-wide association analysis identified SNPs closely linked to a gene resistant to Soil-borne wheat mosaic virus. Theor Appl Genet 127:1039–1047

Lou J, Chen L, Yue G et al (2009) QTL mapping of grain quality traits in rice. J Cereal Sci 50:145–151

Ma X-F, Ross K, Gustafson J (2001) Physical mapping of restriction fragment length polymorphism (RFLP) markers in homoeologous groups 1 and 3 chromosomes of wheat by in situ hybridization. Genome 44:401–412

Maccaferri M, Sanguineti MC, Demontis A et al (2010) Association mapping in durum wheat grown across a broad range of water regimes. J Exp Bot erq287

Manly KF, Cudmore RH Jr, Meer JM (2001) Map manager QTX, cross-platform software for genetic mapping. Mamm Genome 12:930–932

Manolio TA, Collins FS, Cox NJ et al (2009) Finding the missing heritability of complex diseases. Nature 461:747–753
Mcmullen MD, Kresovich S, Villeda HS et al (2009) Genetic properties of the maize nested association mapping population. Science 325:737–740
Moreira JUV, Bento DAV, De Souza AP et al (2009) QTL mapping for reaction to phaeosphaeria leaf spot in a tropical maize population. Theor Appl Genet 119:1361–1369
Nandi S, Subudhi P, Senadhira D et al (1997) Mapping QTLs for submergence tolerance in rice by AFLP analysis and selective genotyping. Mol Gen Genet MGG 255:1–8
Navakode S, Weidner A, Lohwasser U et al (2009) Molecular mapping of quantitative trait loci (QTLs) controlling aluminium tolerance in bread wheat. Euphytica 166:283–290
Neale DB, Ingvarsson PK (2008) Population, quantitative and comparative genomics of adaptation in forest trees. Curr Opin Plant Biol 11:149–155
Neale DB, Savolainen O (2004) Association genetics of complex traits in conifers. Trends Plant Sci 9:325–330
Ni J, Wu P, Senadhira D et al (1998) Mapping QTLs for phosphorus deficiency tolerance in rice (*Oryza sativa* L.). Theor Appl Genet 97:1361–1369
Paliwal R, Röder MS, Kumar U et al (2012) QTL mapping of terminal heat tolerance in hexaploid wheat (*T. aestivum* L.). Theor Appl Genet 125:561–575
Park KJ, Sa KJ, Kim BW et al (2014) Genetic mapping and QTL analysis for yield and agronomic traits with an F2: 3 population derived from a waxy corn× sweet corn cross. Genes Genom 36:179–189
Paterson AH (1996) Making genetic maps. Genome mapping in plants (biotechnology intelligence unit) 23–39
Peng J, Bai Y, Haley S et al (2009) Microsatellite-based molecular diversity of bread wheat germplasm and association mapping of wheat resistance to the Russian wheat aphid. Genetica 135:95–122
Perrier X, Jacquemoud-Collet J (2006) DARwin software
Phumichai C, Chunwongse J, Jampatong S et al (2012) Detection and integration of gene mapping of downy mildew resistance in maize inbred lines though linkage and association. Euphytica 187:369–379
Presterl T, Ouzunova M, Schmidt W et al (2007) Quantitative trait loci for early plant vigour of maize grown in chilly environments. Theor Appl Genet 114:1059–1070
Price AH (2006) Believe it or not, QTLs are accurate! Trends in Plant Science 11:213–216
Price AL, Patterson NJ, Plenge RM et al (2006) Principal components analysis corrects for stratification in genome-wide association studies. Nat Genet 38:904–909
Pritchard JK, Stephens M, Donnelly P (2000) Inference of population structure using multilocus genotype data. Genetics 155:945–959
Quarrie S, Gulli M, Calestani C et al (1994) Location of a gene regulating drought-induced abscisic acid production on the long arm of chromosome 5A of wheat. Theor Appl Genet 89:794–800
Ray J, Yu L, Mccouch S et al (1996) Mapping quantitative trait loci associated with root penetration ability in rice (*Oryza sativa* L.). Theor Appl Genet 92:627–636
Remington DL, Thornsberry JM, Matsuoka Y et al (2001) Structure of linkage disequilibrium and phenotypic associations in the maize genome. Proc Natl Acad Sci 98:11479–11484
Risch N (1992) Genetic linkage: interpreting LOD scores. Science 255:803–804
Risch N, Merikangas K (1996) The future of genetic studies of complex human diseases. Science 273:1516–1517
Rockman MV (2008) Reverse engineering the genotype–phenotype map with natural genetic variation. Nature 456:738–744
Rostoks N, Ramsay L, Mackenzie K et al (2006) Recent history of artificial outcrossing facilitates whole-genome association mapping in elite inbred crop varieties. Proc Natl Acad Sci 103:18656–18661

Rousset M, Bonnin I, Remoué C et al (2011) Deciphering the genetics of flowering time by an association study on candidate genes in bread wheat (*Triticum aestivum* L.). Theor Appl Genet 123:907–926

Salvi S, Sponza G, Morgante M et al (2007) Conserved noncoding genomic sequences associated with a flowering-time quantitative trait locus in maize. Proc Natl Acad Sci 104:11376–11381

Santra D, Chen X, Santra M et al (2008) Identification and mapping QTL for high-temperature adult-plant resistance to stripe rust in winter wheat (*Triticum aestivum* L.) cultivar 'Stephens'. Theor Appl Genet 117:793–802

Sehgal D, Yadav R (2009) Molecular markers based approaches for drought tolerance. In: Molecular techniques in crop improvement. Springer, Berlin, pp 207–230

Setter TL, Yan J, Warburton M et al (2010) Genetic association mapping identifies single nucleotide polymorphisms in genes that affect abscisic acid levels in maize floral tissues during drought. J Exp Bot erq308

Shi L, Lv X, Weng J et al (2014) Genetic characterization and linkage disequilibrium mapping of resistance to gray leaf spot in maize (*Zea may* L.). The Crop J 2:132–143

Stam P (1993) Construction of integrated genetic linkage maps by means of a new computer package: join map. Plant J 3:739–744

Su Z, Hao C, Wang L et al (2011) Identification and development of a functional marker of TaGW2 associated with grain weight in bread wheat (*Triticum aestivum* L.). Theor Appl Genet 122:211–223

Tan Y, Sun M, Xing Y et al (2001) Mapping quantitative trait loci for milling quality, protein content and color characteristics of rice using a recombinant inbred line population derived from an elite rice hybrid. Theor Appl Genet 103:1037–1045

Tanksley SD (1993) Mapping polygenes. Annu Rev Genet 27:205–233

Tanksley S, Ganal M, Prince J et al (1992) High density molecular linkage maps of the tomato and potato genomes. Genetics 132:1141–1160

Tian Y, Zhang H, Xu P et al (2014) Genetic mapping of a QTL controlling leaf width and grain number in rice. Euphytica 1–11

Utz H, Melchinger A (1996) PLABQTL: a program for composite interval mapping of QTL. J Quant Trait Loci 2:1–5

Van Inghelandt D, Melchinger AE, Martinant J-P et al (2012) Genome-wide association mapping of flowering time and northern corn leaf blight (*Setosphaeria turcica*) resistance in a vast commercial maize germplasm set. BMC Plant Biol 12:56

Vanniarajan C, Vinod K, Pereira A (2012) Molecular evaluation of genetic diversity and association studies in rice (*Oryza sativa* L.). J Genet 91:9–19

Vijayalakshmi K, Fritz AK, Paulsen GM et al (2010) Modeling and mapping QTL for senescence-related traits in winter wheat under high temperature. Mol Breed 26:163–175

Wang A-Y, Li Y, Zhang C-Q (2012a) QTL mapping for stay-green in maize (*Zea mays*). Can J Plant Sci 92:249–256

Wang M, Yan J, Zhao J et al (2012b) Genome-wide association study (GWAS) of resistance to head smut in maize. Plant Sci 196:125–131

Wen W, Mei H, Feng F et al (2009) Population structure and association mapping on chromosome 7 using a diverse panel of Chinese germplasm of rice (*Oryza sativa* L.). Theor Appl Genet 119:459–470

Weng J, Xie C, Hao Z et al (2011) Genome-wide association study identifies candidate genes that affect plant height in Chinese elite maize (*Zea mays* L.) inbred lines. PLoS ONE 6:e29229

Whitt SR, Buckler Iv ES (2003) Using natural allelic diversity to evaluate gene function. In: Plant functional genomics. Springer, Berlin, pp 123–139

Wilson LM, Whitt SR, Ibáñez AM et al (2004) Dissection of maize kernel composition and starch production by candidate gene association. Plant Cell Online 16:2719–2733

Xu F-F, Tang F-F, Shao Y-F et al (2014) Genotype× environment interactions for agronomic traits of rice revealed by association mapping. Rice Sci 21:133–141

Xue Y, Warburton ML, Sawkins M et al (2013) Genome-wide association analysis for nine agronomic traits in maize under well-watered and water-stressed conditions. Theor Appl Genet 126:2587–2596

Yan WG, Li Y, Agrama HA et al (2009) Association mapping of stigma and spikelet characteristics in rice (*Oryza sativa* L.). Mol Breed 24:277–292

Yan J, Kandianis CB, Harjes CE et al (2010) Rare genetic variation at *Zea mays* crtRB1 increases [beta]-carotene in maize grain. Nat Genet 42:322–327

Yang N, Lu Y, Yang X et al (2014) Genome wide association studies using a new nonparametric model reveal the genetic architecture of 17 agronomic traits in an enlarged maize association panel. PLoS Genet 10:e1004573

Yao H, Zhou Q, Li J et al (2002) Molecular characterization of meiotic recombination across the 140-kb multigenic a1-sh2 interval of maize. Proc Natl Acad Sci 99:6157–6162

Yoshida T, Nishida H, Zhu J et al (2010) Vrn-D4 is a vernalization gene located on the centromeric region of chromosome 5D in hexaploid wheat. Theor Appl Genet 120:543–552

Young ND (1994) Constructing a plant genetic linkage map with DNA markers. In: DNA-based markers in plants. Springer, Berlin, pp 39–57

Young N (1996) QTL mapping and quantitative disease resistance in plants. Annu Rev Phytopathol 34:479–501

Yu J, Pressoir G, Briggs WH et al (2005) A unified mixed-model method for association mapping that accounts for multiple levels of relatedness. Nat Genet 38:203–208

Yu L-X, Morgounov A, Wanyera R et al (2012) Identification of Ug99 stem rust resistance loci in winter wheat germplasm using genome-wide association analysis. Theor Appl Genet 125:749–758

Zegeye H, Rasheed A, Makdis F et al (2014) Genome-wide association mapping for seedling and adult plant resistance to stripe rust in synthetic hexaploid wheat. PLoS ONE 9:e105593

Zeng Z-B (1994) Precision mapping of quantitative trait loci. Genetics 136:1457–1468

Zhang J, Hao C, Ren Q et al (2011) Association mapping of dynamic developmental plant height in common wheat. Planta 234:891–902

Zhang Y, Xu L, Fan X et al (2012) QTL mapping of resistance to gray leaf spot in maize. Theor Appl Genet 125:1797–1808

Zhang X, Tang B, Yu F et al (2013) Identification of major QTL for waterlogging tolerance using genome-wide association and linkage mapping of maize seedlings. Plant Mol Biol Rep 31:594–606

Zheng P, Allen WB, Roesler K et al (2008) A phenylalanine in DGAT is a key determinant of oil content and composition in maize. Nat Genet 40:367–372

Zhu C, Yu J (2009) Nonmetric multidimensional scaling corrects for population structure in association mapping with different sample types. Genetics 182:875–888

Zhu C, Gore M, Buckler ES et al (2008) Status and prospects of association mapping in plants. Plant Genome 1:5–20

Chapter 3
Developing and Deploying Abiotic Stress-Tolerant Maize Varieties in the Tropics: Challenges and Opportunities

B.M. Prasanna

Abstract Maize (*Zea mays* L.) is a preferred staple for 900 million consumers in the developing world, including 120–140 million poor farm families. The demand for maize is growing sharply due to its multifaceted uses (food, feed, fodder, specialty corn, and industrial uses). However, maize yields in the tropical rainfed environments, especially in sub-Saharan Africa (SSA) and South Asia, are affected by an array of abiotic and biotic stresses, thereby limiting national maize yields to 1–3 tons per hectare (t/ha), while the global average is around 5 t/ha. Therefore, developing and deploying high-yielding, climate-resilient maize (with tolerance to drought, heat, waterlogging, and biotic stresses), coupled with climate-smart agricultural practices, are critical for improving maize yields and reducing the high risk and vulnerability of the maize-growing smallholder farmers in the tropics to the climate variability. International Maize and Wheat Improvement Center (CIMMYT) has been intensively engaged since 1970s in breeding elite tropical maize germplasm with tolerance to important abiotic stresses, especially drought, using managed-stress screening and selection for key secondary traits. This formed the base for successful development, testing, and deployment of CIMMYT-derived abiotic stress-tolerant maize varieties in SSA, Latin America, and Asia, in partnership with an array of public and private sector institutions. Notable among the projects with strong focus on development and delivery of abiotic stress-tolerant tropical maize germplasm are the following: Drought Tolerant Maize for Africa (DTMA), Water Efficient Maize for Africa (WEMA), Improved Maize for African Soils (IMAS), and Heat Tolerant Maize for Asia (HTMA). Increasing genetic gains and breeding efficiency, especially in developing high-value abiotic stress-tolerant maize germplasm, requires: (a) carefully undertaken field-based phenotyping at several relevant sites as well as under technically demanding managed-stress screens; (b) better understanding of the genetic architecture of traits; and (c) utilization of modern breeding tools/strategies, including genome-wide association

B.M. Prasanna (✉)
Global Maize Program, CIMMYT (International Maize and Wheat Improvement Center), Nairobi, Kenya
e-mail: b.m.prasanna@cgiar.org

V.R. Rajpal et al. (eds.), *Molecular Breeding for Sustainable Crop Improvement*, Sustainable Development and Biodiversity 11,
DOI 10.1007/978-3-319-27090-6_3

studies (GWAS), genomic selection, and doubled haploid (DH) technology for rapid development of improved products. Multi-institutional efforts, especially public–private alliances, are also key to ensure that the improved varieties effectively reach the farming communities vulnerable to climate change.

Keywords Maize · Climate resilience · Phenotyping · Genetic gains · Seed delivery

3.1 Introduction

Maize is the key crop for food and feed security and income generation for millions of smallholder farmers in sub-Saharan Africa (SSA), Asia, and Latin America. It is a major source of calories in the diets of nearly 230 million inhabitants of developing countries—81 million in SSA, 141 million in South Asia, and 8 million in Latin America. Annual per capita maize consumption averages 36, 10, and 23 kg, respectively, in these regions, but this masks significant variation and per capita food consumption of maize. In Mesoamerica, annual maize consumption exceeds 80 kg per capita in Guatemala, Honduras, and El Salvador, rising to 125 kg in Mexico (Shiferaw et al. 2011).

Maize accounts for almost half of calories and protein consumed in eastern and southern Africa (ESA) and one-fifth of calories and protein consumed in West Africa. Maize consumption levels exceed 130 kg per capita per year in Lesotho, Malawi, and Zambia. The highest amounts of maize consumed are found in southern Africa, at 85 kg/capita/year, as compared to 27 kg/capita/year in East Africa and 25 kg/capita/year in West and Central Africa (Shiferaw et al. 2011). In South and South-east Asia, where direct maize consumption on an annual average is estimated to be only 6 and 16 kg per capita, respectively, there are several areas (especially in the highlands and tribal regions) where maize is consumed directly at much higher rates (Babu et al. 2013).

In Asia, countries such as China, India, and Indonesia have recorded impressive growth rates in maize production (in the range of 5–6 % per year). Maize is now the crop with the largest cultivated area in China, with nearly 33.5 million hectares (Hu and Zimmer 2013); globally, China's maize acreage and production are next only to the USA. Maize yields have registered impressive increases in China, from 4.5 to 5.75 t/ha (+0.9 % per annum). The Corn Belt of China, stretching from the northeast to south-west of the country, cuts across 11 provinces (Heilongjiang, Jilin, Hebei, Henan, Shandong, Inner Mongolia, Liaoning, Shanxi, Yunnan, Sichuan, and Shaanxi) accounts for 81 % of area under maize and nearly 83 % of total maize produced (Hu and Zimmer 2013).

Although developed countries, particularly the USA, contribute predominantly to maize production, demand for maize in developing countries is expected to surpass the demands for both wheat and rice by the year 2020 (Pingali and Pandey 2001). The growth in demand for human consumption of maize in the developing world is predicted to be 1.3 % per annum until 2020. Moreover, rising incomes are expected to result in a doubling of consumption of meat across the developing world (Naylor et al. 2005), leading to a predicted growth in demand for feed maize of 2.9 % per annum. Hence, there is need for at least a 2 % per annum increase in maize production to meet this growth in global human population and shift in dietary preferences (Ortiz et al. 2010). Maize demand is projected to see 87 % rise by 2020 as compared to its demand in 1995 (IFPRI 2003). An array of factors are contributing to this sharp increase in maize demand, including increase in per capita income, changing diets, and a rapidly growing poultry sector (Shiferaw et al. 2011). For instance, India's maize demand has been forecast to grow by 36 per cent in the next four years touching 30 million tons in 2017 and double within the next nine years to touch about 44 million tons by 2022.

3.2 The Challenge of Improving Maize Productivity in the Tropics

Both production and productivity have to be significantly improved if the developing world has to successfully meet the rapidly growing demand for maize. The average maize yields in several of the African countries are still below 1 t/ha, while many countries have only 1–2 t/ha, due mainly to poor soils and farmers' limited access to fertilizer or improved maize seed. Similarly, maize yields in many of Asian countries remain low, with India, Nepal, and the Philippines achieving ≈2 t/ha, Indonesia and Vietnam ≈3.5 t/ha, Thailand almost 4 t/ha, and China 5 t/ha, compared to the world average of 4.7 t/ha in 2005 and current USA average of 9.4 t/ha (Prasanna et al. 2010).

Several factors, including overdependence on rainfall, frequent droughts, yield losses due to pre- and post-harvest pathogens and insect–pests, weeds, poor agronomic management, and lack of access to quality seed, continue to affect maize production and productivity in the developing world, particularly in SSA, Asia, and Latin America. It is notable that eight major maize-producing countries—China, India, Indonesia, Nepal, Pakistan, Philippines, Thailand, and Vietnam—taken together, produce 98 % of Asia's maize and 28 % of global maize. In most of these countries, maize is predominantly grown under rainfed conditions by the smallholder, resource-poor farmers. Increasing maize yield by even 1 t/ha in the low-yielding countries in Africa and Asia could deliver a much higher relative impact than does the same increase in the high-yielding environments.

3.2.1 Drought Stress

With most maize production dependent on rainfall, especially in the developing world, maize is particularly vulnerable to drought and its yields fluctuate more widely from year to year than is the case for rice and wheat, which are more commonly irrigated. Thus, drought is recognized as the most important constraint across the rainfed lowland and upland environments in the developing world. For instance, over 80 % of maize grown in South and South-east Asia is rainfed, with an average yield that is less than half of the irrigated maize. There is further increase in rainfed maize area at 1.8 % per year, which is six times more than the irrigated area (Edmeades 2007). The decline in the irrigated area is mostly due to the diminishing groundwater table that puts the irrigated area under threat. This situation is likely to exacerbate in the coming decades due to climate change, often leading to inadequate and/or uneven incidence of rainfall in the crop season alongside temperature changes (IPCC 2007; ADB 2009). Alleviating the effects of drought alone could increase average maize yields by 35 % across Asia-7 (excluding China) and by 28 % in south-west China (Gerpacio and Pingali 2007).

3.2.2 Waterlogging Stress

Waterlogging is a major problem for maize production in several maize agro-ecologies where rainfall is erratic and intense, and the soil drainage capacity is poor. Over 18 % of the total maize production area in South and South-east Asia is frequently affected by floods and waterlogging problems, causing production losses of 25–30 % annually (Cairns et al. 2012). The problem of waterlogging during the crop cycle is exacerbated due to climate change in some maize-growing regions in the developing world; for example, the distribution patterns of rainfall rather than total annual rainfall are predicted to change in South Asia and in many areas in SSA (IPCC 2007). Flood and waterlogging frequently affect more than 18 % of the total maize production area in South and South-east Asia causing production losses of 25–30 % annually (Zaidi et al. 2010; Cairns et al. 2012).

Climate change effects are expected to further complicate the already difficult situation of uneven/poor rainfall distribution pattern. Countries in the Greater Himalayan region—Bangladesh, Bhutan, northern India, and Nepal—are facing increased frequency and magnitude of extreme weather events, resulting in flooding, landslides, and devastation of agricultural crops, besides negative impacts on ecological health. The coastal areas of Bangladesh, India, the Maldives, and Sri Lanka are at high risk from projected sea level rise that may cause displacement of human settlements, saltwater intrusion, loss of agricultural land and wetlands, and a negative impact on tourism and fisheries (Ahmad and Suphachalasai 2014). Flash floods occur not only during the seedling stage but also at the flowering and grain filling stages, often forcing the farmers (especially in Bangladesh and Eastern India) to harvest maize ears before physiological maturity.

3.2.3 *Heat Stress*

Maize is particularly vulnerable to the reproductive stage heat stress. Climate projections also suggest that elevated temperatures, especially in the drought-prone areas of SSA and rainfed areas in South Asia, are likely to result in significant crop yield losses (Cairns et al. 2013a). From analysis of over 20,000 historical maize trial yields in Africa, Lobell et al. (2011) reported a yield reduction of 1 and 1.7 % for every degree-day above 30 °C under optimal rainfed and drought conditions, respectively. Temperatures are expected to increase in SSA by an average of 2.1 °C by 2050 (Cairns et al. 2012). The most important effects of elevated temperatures on maize yield reduction include shortened life cycle, reduced light interception, and increased sterility (Cairns et al. 2012).

3.2.4 *Poor Soil Fertility*

Declining soil fertility and expanding soil acidity, low phosphorus availability, and aluminium toxicity affect maize yields on about 4 million hectares of land worldwide (Shiferaw et al. 2011). The problem of poor soil fertility is particularly severe in SSA where all the maize mega environments are affected (Pingali and Pandey 2001). Use of fertilizer and restorative crop management practices remains relatively low and inefficient in many developing countries, particularly in SSA (Smale et al. 2011).

3.3 Developing Climate-Resilient Maize Varieties: Some Major Initiatives

The future of maize production and, consequently, the livelihoods of several million smallholder farmers worldwide are based to a great extent on breeding for high-yielding and stress-resilient varieties. The technological opportunities for maize improvement have increased tremendously in recent years. Significant strides have been made particularly with regard to understanding the phenotypic and molecular diversity in the maize germplasm, identification of genes/QTLs influencing diverse traits, especially tolerance to important biotic and abiotic stresses, developing precision phenotyping protocols, and utilizing marker-assisted or genomics-assisted breeding strategies for improving stress resilience in maize. Some of the major initiatives on developing climate-resilient maize for the tropics are highlighted below.

3.3.1 Drought Tolerance

Understanding the environmental conditions that contribute to drought and effectively unravelling genetic variability for drought tolerance in appropriate environments are two critical factors for the success of breeding for drought tolerance. CIMMYT's work since 1970s on characterization of drought-prone environments in the tropics, identification of suitable secondary traits and trait donors in breeding for drought tolerance, optimizing procedures for undertaking managed drought stress phenotyping trials, developing drought-tolerant (DT) maize germplasm through extensive multi-location and multi-year experiments, and disseminating the stress-tolerant cultivars in partnerships with various public and private organizations, holds considerable significance for improving the livelihoods of the resource-poor farmers in the developing world.

CIMMYT is presently implementing an array of projects in SSA, Asia, and Latin America for developing and deploying climate-resilient varieties. Under the Drought Tolerant Maize for Africa (DTMA) project, jointly implemented by CIMMYT and IITA, in close collaboration with NARS and private sector institutions in 13 countries in Africa, nearly 180 drought-tolerant maize varieties have been released during 2007–2014, with close to 60 % of these being hybrids. These varieties perform as well as or better than the commercial varieties currently available on the market under optimum (no water deficit stress) conditions and outperform the best commercial checks by at least 25–30 % under drought stress and low-input conditions. DTMA has also facilitated production and delivery of about 52,000 tons of DT maize seed in 2014 in partnerships with about 90 seed companies, benefiting an estimated 5 million African households.

The DT varieties developed by CIMMYT typically have a combination of traits that confer them tolerance to drought conditions; these include reduced barrenness under drought stress, short anthesis-silking interval, reduced leaf senescence (as compared to susceptible germplasm), and longer leaf area duration during grain filling (Edmeades 2008; Bruce et al. 2002). Some of the DT varieties developed by CIMMYT and released in SSA have wide adaptation. For example, one of the most popular DT varieties (ZM521), developed at CIMMYT-Zimbabwe, is currently grown in several countries in eastern and southern Africa, including Angola, Burundi, Ethiopia, Kenya, Malawi, Mozambique, South Africa, Tanzania, Zambia, and Zimbabwe.

The Water Efficient Maize for Africa (WEMA) Project is another important public–private partnership, that is intensively engaged in developing and deploying drought-tolerant and insect-resistant white maize varieties in five target countries in SSA (Kenya, Tanzania, Uganda, Mozambique, and South Africa), through a combination of conventional breeding, marker-assisted breeding, and transgenes.

3.3.2 Heat Stress Tolerance

Compared to other abiotic stresses associated with climate change, especially drought stress, work on developing and deploying heat stress-tolerant tropical/subtropical maize is still in its infancy. Studies undertaken by the CIMMYT team to identify heat stress-tolerant tropical maize lines among the elite, DT maize germplasm developed in Mexico, Asia, and Africa revealed high vulnerability of most of the tropical maize germplasm, including commercial cultivars in South Asia, SSA, and Latin America, to reproductive stage heat stress. Several of the DT parents developed by CIMMYT and widely used in hybrid maize breeding in eastern and southern Africa were found to be highly susceptible to drought stress under elevated temperatures; a notable example is CML442 × CML444 that is used as the female parent in several commercial hybrids. Therefore, intensive efforts are required to ensure that the most widely used DT inbred lines and hybrids also possess tolerance to combined drought and heat stresses, especially for deployment in drought-prone areas where temperatures are predicted to increase.

Despite the above-mentioned limitation, a few CIMMYT inbred lines with high levels of tolerance to drought as well as combined drought and heat stress, most notably La Posta Sequia C7-F64-2-6-2-2 and DTPYC9-F46-1-2-1-2, have been identified. Such lines are presently being utilized in developing elite germplasm with tolerance to combined drought and heat stress (Cairns et al. 2012, 2013b). CIMMYT is presently implementing two major research projects, supported by USAID under the Feed the Future initiative, for developing and deploying heat-resilient maize for SSA and Asia. The Heat-Tolerant Maize for Asia (HTMA) Project, initiated in 2012, brings together public and private institutions based in South Asia (Bangladesh, India, Nepal, and Pakistan), besides Purdue University, USA, for accelerated development and deployment of heat stress-resilient maize germplasm.

3.3.3 Waterlogging Tolerance

Considerable variation was observed among maize inbreds in tolerance to waterlogging/flooding of older seedlings (Mano et al. 2006; Zaidi et al. 2010). At EMBRAPA-Brazil, recurrent selection over 12 cycles resulted in the development and subsequent release of the waterlogging-tolerant BRS4154 maize line, with 20 % yield advantage under waterlogging compared to the original source (Ferreira et al. 2007). Both additive and non-additive gene actions contribute to the expression of waterlogging tolerance (Zaidi et al. 2010). QTL mapping undertaken by CIMMYT-Asia team, using single-nucleotide polymorphisms (SNPs), revealed five QTLs on chromosomes 1, 3, 5, 7, and 10 conferring waterlogging tolerance; these QTLs together explained approximately 30 % of phenotypic variance for grain yield under waterlogging stress, with effects ranging from 520 to 640 kg/ha for individual genomic regions (Zaidi et al. 2015).

3.4 Improved Germplasm with Package(s) of Adaptive Traits

For developing climate-resilient maize germplasm, breeding programmes also need to incorporate, more efficiently, packages of traits such as abiotic and biotic stress tolerance. Climate change projections suggest more frequent weather extremes, and occasionally, more than one within one crop season (e.g. drought and waterlogging) could also happen, thereby increasing on the short-run the likelihood of crop failures and on the long-run major production declines (Zaidi and Cairns 2011).

In southern Africa and South Asia, maize farmers are likely to require varieties with tolerance to drought stress at elevated temperatures. Similarly, tolerance to both drought and waterlogging is becoming increasingly important for some areas in South and South-east Asia and may be required by farmers in small areas in SSA (Cairns et al. 2013a). Under the BMZ-funded Abiotic stress-Tolerant Maize for Asia (ATMA) project led by CIMMYT, and with partners in India, Bangladesh, the Philippines, Indonesia, and Vietnam, significant progress was made towards development of improved maize germplasm adapted to South and South-east Asia, and with enhanced levels of tolerance to drought, waterlogging, or combined stress tolerance. New hybrid combinations were developed by crossing promising stress-tolerant lines (DT and/or waterlogging tolerance) and evaluated across moisture regimes, including managed drought and waterlogging stresses, and optimal conditions. A set of approximately 50 promising hybrids are at advanced stage and ready for large-scale adaptive trials.

For effectively developing climate-resilient maize varieties with packages of adaptive traits, breeding programmes need to be reoriented for simultaneous selection under combinations of stresses. This requires establishment of a strong network of managed-stress phenotyping/screening sites and use of standardized protocols for specific combinations of stresses predicted in the target environments. Future maize genotypes should also be equipped with a more efficient rooting system to improve water-use efficiency. From the breeding perspective, this will require developing high-throughput root screening systems (rhizotronics) under both field and controlled conditions and identifying key root traits associated with improved water and nutrient capture in the field. Such efforts could lead to identification of suitable trait donors for improving water and nutrient use efficiency of maize varieties.

3.5 Approaches for Enhancing Genetic Gains in Stress-Prone Environments

The ability to develop, in a cost- and time-efficient manner, elite maize hybrids with high yield potential and necessary adaptive traits (abiotic and biotic stress resilience) will be critical for the improved productivity and diversification of cropping

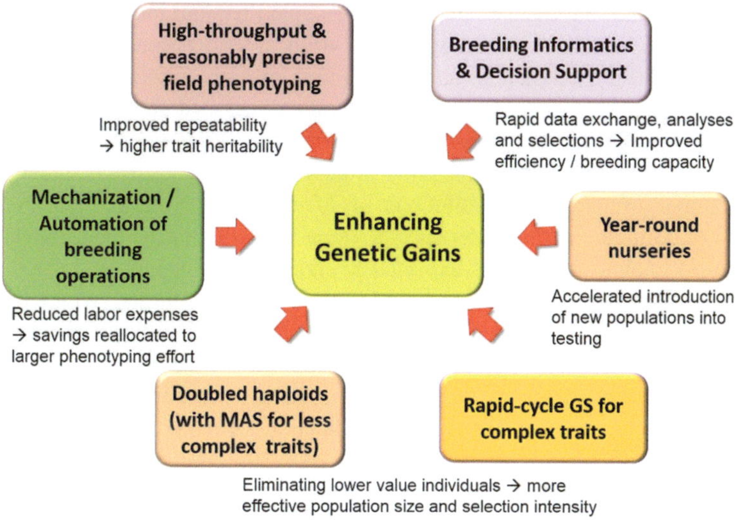

Fig. 3.1 Components for enhancing genetic gains in maize

systems. While conventional breeding has been successful in developing an array of elite maize hybrids, rapid advances in breeding tools and techniques, especially doubled haploidy (DH), high-throughput phenotyping for traits of interest, mechanization of breeding operations (to the extent possible), molecular marker-assisted breeding, and decision support tools/systems (Fig. 3.1) offer excellent opportunities for improving genetic gains and enhancing breeding efficiency. A few of these important components will be briefly discussed here.

3.5.1 Doubled Haploid (DH) Technology in Maize Breeding

Greater access to low-cost hybrid seed and more rapid development of improved hybrids are vital to increase maize productivity and enhance income opportunities to maize growers, while meeting the demands of food, feed, and nutrition security. Development of stable and productive inbred lines to produce hybrid seed is the cornerstone of successful and affordable hybrid maize technology. The DH technology is now a powerful tool to accelerate development, identification, and use of elite breeding lines. DH not only significantly reduces the time and resources required for generating homozygous lines, but also enhances "forward breeding" (Geiger and Gordillo 2009; Prasanna et al. 2012).

The in vivo haploid induction using temperate haploid inducers (genetic stocks with high haploid induction capacity) has been adapted by an array of commercial maize breeding programmes in Europe, North America, and more recently in Asia (especially China), but the lack of tropically adapted haploid inducer lines impeded the application of DH technology in the tropical maize breeding programmes (Prigge et al. 2012). Tropically, adapted haploid inducers with a haploid induction rate (HIR) of up to 10 % have been recently developed through collaboration between CIMMYT and the University of Hohenheim (Prigge et al. 2011; Prasanna et al. 2012). These tropicalized haploid inducers have already been shared with a large number of interested institutions in Africa, Latin America, and Asia, for research or commercial use under specific terms and conditions. The availability of tropicalized haploid inducers is expected to significantly enhance the efficiency of DH line production, increase seed set and rates of induction, and reduce the costs of inducer line maintenance and seed production. CIMMYT is also in the process of further optimizing the DH protocol and developing second-generation haploid inducer lines adapted to specific tropical environments, especially SSA, Asia, and Latin America, to further widen the application of DH technology in maize breeding programmes.

While DH technology is the primary mode of deriving new inbred lines by several large private sector breeding programmes, National Agricultural Research Systems (NARS) and small and medium enterprise (SME) seed companies in several Asian countries have so far not derived benefits out of maize DH technology for various reasons. CIMMYT, in partnership with Kenya Agricultural and Livestock Research Organization (KALRO), has also established a centralized maize DH facility at Kiboko (Kenya) for strengthening maize breeding programmes, including those of NARS and SME seed companies in SSA. A similar facility is being planned for Asia, in partnership with Asian institutions.

3.5.2 High-Throughput and Reasonably Precise Phenotyping

Field-based phenotyping still remains a major bottleneck for future breeding progress. Phenotyping capacity of several institutions in Asia is lagging far behind the capacity to generate genomic information. Phenotyping capacity is constrained in many countries, limiting our ability to breed better cultivars with higher grain yield and stress resilience (Prasanna et al. 2013b; Araus and Cairns 2014). Field-based phenotyping of appropriately selected traits, using low-cost, easy-to-handle tools, is now possible and should become an integral and key component in the maize breeding programmes. There is also a distinct need for the public and private institutions to come together and establish "phenotyping networks" for comprehensive and efficient characterization of genetic resources and breeding materials for important target traits.

Molecular breeding strategies, such as genome-wide association studies (GWAS), marker-assisted recurrent selection (MARS), and genome-wide selection (GS), are being implemented by several institutions worldwide. However, genotypic predictions for both MARS- and GS-based strategies depend heavily on a single phenotyping cycle, thus increasing the need for reliable phenotyping methodologies (Cobb et al. 2013).

Combining advances in aeronautics and high-performance computing is paving the way for the development of field-based phenotyping platforms (Araus and Cairns 2014); such platforms could range from ground-based to the aerial. Recently, under the MAIZE CGIAR Research Program, the Crop Breeding Institute (Zimbabwe), University of Barcelona (Spain), AirElectronics (Spain), Consejo Superior de Investigaciones Científicas (Spain), and CIMMYT developed an UAV (unmanned aerial vehicle), called "Skywalker". The "Skywalker" is able to carry a payload of up to 1 kg and can carry several sensors including thermal, multispectral, and digital cameras. This UAV is currently being used to identify genotypic variability in plant water status under drought stress and biomass production and senescence under drought, heat, and low N stresses (Cairns et al. 2012b) at CIMMYT-Harare, with promising results.

Beyond such technological advances, there is also an immense need for measuring and reducing the effects of field variability, thereby increasing the genetic signal-to-noise ratio to detect real differences between genotypes. CIMMYT is making intensive efforts for characterizing field variability at the key phenotyping sites worldwide and for improving field-based phenotyping through various approaches, such as monitoring soil moisture using neutron probes/time-domain reflectometer (TDR), non-destructive estimation of biomass (using NDVI or Normalized Differential Vegetation Index), and analysing canopy behaviour using Infrared thermography and functional aspects of roots using Rhizotronics (Prasanna et al. 2013b).

Breeding programmes of majority of the NARS and SME seed companies in the developing countries have limited capacity for undertaking precision phenotyping, particularly under repeatable and representative levels of abiotic stresses in the field. Intensive efforts are therefore required to build the capacity of the institutions on methods to characterize and control field site variation (for improving repeatability), adopting appropriate experimental designs, selection of "right" traits for phenotyping, proper integration, analysis, and application of heterogeneous data sets, in addition to generating better awareness of technological advances with regard to phenotyping.

3.5.3 Genomics-Assisted Breeding

Molecular marker-assisted or genomics-assisted breeding is the way forward in effectively meeting the greater challenge of developing cultivars with combinations of relevant adaptive traits, including biotic and abiotic stress tolerance, besides nutritional quality. With the rapid reduction in genotyping costs, new genomic selection technologies (Bernardo and Yu 2007, Heffner et al. 2009) have become available that allow the maize breeding cycle to be greatly reduced, facilitating inclusion of information on genetic effects for multiple stresses in selection decisions.

High-density genotyping using platforms such as genotyping by sequencing (GBS) is now an integral part of CIMMYT's maize molecular breeding strategies. GWAS is being implemented by CIMMYT-GMP for identification of genomic regions associated with an array of important traits, especially abiotic stress tolerance and disease resistance, coupled with validation of the significant genomic leads in an array of tropical/subtropical biparental populations (Prasanna et al. 2014). MARS and GS are being implemented by CIMMYT and partners through several projects in the tropics, especially for the improvement of complex traits. A recent comparative study of pedigree selection, MARS, and GS undertaken across 8–10 biparental populations demonstrated the superior performance of hybrids derived from Cycle 3 of both MARS and GS schemes over pedigree selection in most populations compared with the Cycle 0. The overall gain per year for MARS and GS under managed drought and well-watered environments was two-to threefold higher than the gain achieved via pedigree selection (Beyene et al. 2014).

CIMMYT is also employing joint GWAS and linkage mapping approach for identifying breeder-ready markers for resistance to major diseases affecting tropical maize such as Turcicum leaf blight (TLB), grey leaf spot (GLS), maize lethal necrosis (MLN), common rust, ear rots, and corn stunt complex. A recent example is with regard to the maize streak virus (MSV), a major disease that affects maize productivity in several countries in SSA. CIMMYT Maize Program has fine-mapped and identified SNP markers for a major QTL for MSV resistance (*msv1*) and validated these markers on a set of DH lines that have been phenotyped for responses to MSV in different locations in SSA (Sudha Nair et al., manuscript in preparation). Forward breeding and MABC are ongoing using a three-marker haplotype for *msv1* selection. Simultaneous with the marker discovery and validation, strategies for incorporating validated markers in breeding pipelines, through both conventional and DH-based breeding schemes, are also being developed and implemented (Prasanna et al. 2014).

3.5.4 Breeding Informatics and Decision Support Tools

A careful balance of many diverse elements is required to design and implement an appropriate decision support system that provides an optimal combination of time, cost, and genetic gain (Xu et al. 2012). Such a system would need to include the following: (a) managing and analysing large amounts of genotype, pedigree, phenotype, and environment data; (b) selecting desirable recombinants through an optimum combination (in time and space) of phenotypic and genotypic information; and (c) developing breeding systems that minimize population sizes, number of generations, and overall costs while maximizing genetic gain for traditional and novel target traits (Prasanna et al. 2014).

Effective management of product and trait pipelines in breeding programmes requires effective management of pedigree, phenotypic, and genotypic databases, accurate forecasting of genotyping and phenotyping services, as well as optimized decision-making tools/system. Standardized software tools for forecasting, project management, and conventional and molecular breeding data review can streamline the process from initial discovery to final deployment of products through coordinated workflows.

3.5.5 Tapping the Vast Genetic Diversity in Maize

Although maize hybrids represent the most economically important portion of the species, breeding populations, open pollinated varieties (OPVs), and landraces contain the majority of the allelic diversity, much of which has never been incorporated into improved maize cultivars. The CIMMYT Gene Bank holds ~27,000 maize entries, of which ~24,000 are landraces/OPVs collected from diverse regions in Latin America, Africa, and Asia, and held in trust since several decades (Ortiz et al. 2010; Prasanna 2012). Many favourable alleles for an array of useful traits, including tolerance to biotic and abiotic stresses and nutritional quality, are available in these invaluable genetic resources, following natural and farmer's selections over the decades/centuries.

Maize has enormous genetic diversity that offers incredible opportunities for genetic enhancement. There is no lack of favourable alleles in the global maize germplasm that contribute to higher yield, abiotic stress tolerance, disease resistance, or nutritional quality improvement. However, these desirable alleles are often scattered over a wide array of landraces or populations. Our ability to broaden the genetic base of maize and to breed climate-resilient and high-yielding cultivars adaptable to diverse agro-ecologies where maize is grown will undoubtedly depend on the efficient and rapid discovery and introgression of novel/favourable alleles and haplotypes (Prasanna 2012).

A well-characterized and well-evaluated germplasm collection would have greater chances of contributing to the development of novel and improved varieties and, consequently, greater realization of benefits for the resource-poor farmers. Simultaneously with the wider adoption of high-throughput molecular tools, there is a distinct need to establish global phenotyping network for comprehensive and efficient characterization of genetic resources and breeding materials for an array of target traits, particularly for biotic and abiotic stress tolerance and nutritional quality. This would significantly accelerate genomics-assisted breeding, diversification of the genetic base of elite breeding materials, creation of novel varieties, and countering the effects of global climate changes.

Seeds of Discovery (*SeeD*) is a novel project, funded by the Mexican Government, which aims to discover the extent of allelic variation in the genetic resources of maize and wheat, through high-density genotyping/resequencing, multi-location phenotyping for prioritized traits, and novel bioinformatics tools for discovery and use of favourable alleles and haplotypes associated with important traits (Kevin Pixley, personal communication).

3.6 Delivering Climate-Resilient Maize Varieties to the Smallholder Farmers

Developing and releasing climate-resilient varieties is by itself not adequate to lift communities from climate vulnerability. Affordability and access of smallholder farmers to quality seed of stress-resilient maize varieties in the vulnerable agro-ecologies is highly important. This certainly warrants innovative approaches and partnerships to reach the unreached and to make a difference to their livelihoods.

In recognition that a common constraint for SME seed companies and of rapid scale-up of new varieties is parental line maintenance and foundation (basic) seed production, CIMMYT has been providing appropriate technical support for these activities, at least in the initial phases of variety commercialization, on a case-by-case basis in SSA, Latin America, and Asia. The basis for determining this support is the "seed road maps" that are developed with partner institutions for effective scale-up, promotion, and delivery of improved varieties to the smallholders in the target geographies.

Experiences of CIMMYT strongly indicate that besides strengthening the seed sector (especially the SME seed companies), appropriate government policies and adoption of progressive seed laws and regulations are vital for improving smallholder farmers' access to improved seed and for overcoming key bottlenecks affecting maize seed value chain. This is particularly important in the areas of policy, credit availability, seed production, germplasm, and marketing. A proactive approach that combines promising technological, institutional, and policy solutions to manage the risks within vulnerable communities implemented by institutions operating at different levels (community, subnational, and national) is considered to be the way forward for managing climate variability and extremes (Shiferaw et al. 2014). Geographic information system (GIS) could play an integral role in targeting breeding programmes by predicting regions of vulnerability, targeting germplasm movement, and identifying future climates for agricultural production environments (Cairns et al. 2013a).

CIMMYT's recent initiative of establishing the International Maize Improvement Consortium (IMIC) in Asia and in Latin America, in partnership with nearly 80 SME seed companies, is a huge step forward. The underlying principles of this partnership include research prioritization that is client-determined, a more focused, demand-driven approach for product development, while drawing synergies through a collaborative testing network for targeted impacts.

3.7 Conclusions

Intensive multidisciplinary and multi-institutional efforts are required to develop and deliver climate-resilient tropical/subtropical maize germplasm for the benefit of smallholder farmers. Genetics and breeding alone cannot solve the complex challenge of enhancing maize productivity at the smallholder farm level, especially in the face of depleting/degrading natural resources and changing climates. There is a distinct need for effective complementation of improved maize cultivars by suitable conservation agriculture practices as well as institutional and policy innovations that support maize growth and development. This includes understanding the smallholder farmers' affordability and access to quality seed, measures to overcome constraints in adoption of high-yielding, stress-resilient, and nutritionally enriched maize varieties, and partnerships and policies to significantly enhance maize production and utilization.

The use of higher spatial resolution modelling is essential for the identification of high-priority geographic areas for development and deployment of improved germplasm suited to the future climates. Temperature thresholds for current cereal varieties and the interaction of heat stress with other components of climate change (especially drought and biotic stresses) must also be considered. The application of biophysical and economic models in maize improvement, decision support, and foresight requires implementing harmonized procedures for data acquisition, incorporating diverse and actual data sets (cultivar-specific data, climatic data, soil data, important macro- and micro- nutrients, pests/pathogens data, crop management practices, and socio-economic data) in a meaningful way for reliable predictions and practical utility (Prasanna et al. 2013a).

Recent research has led to the development of a suite of soil and crop management practices for increasing resource-use efficiency while maintaining soil health, and mitigating greenhouse gas emissions (Govaerts et al. 2009). Increased use of natural resource management is required to reduce agricultural impacts and to increase efficient water use. However, one must recognize the fact that, in general, most of the modern high-yielding varieties were developed using conventional tillage and crop establishment practices in high-input environments. There could be significant variety x management interaction, with variability in the response of current maize germplasm to resource conservation technologies. A new generation of maize cultivars, suitable for conservation agriculture-based practices, need to be bred. In addition, achieving increased adaptation action will necessitate integration of climate change-related issues with other risk factors, such as climate variability and market risk, and with other policy domains, such as sustainable development (Howden et al. 2007).

References

ADB (2009) Food security under threat from climate change. http://reliefweb.int/report/

Ahmed M, Suphachalasai S (2014) Assessing the costs of climate change and adaptation in South Asia. Asian Development Bank, Mandaluyong City, Philippines

Araus JL, Cairns J (2014) Field high-throughput phenotyping—the new crop breeding frontier. Trends Plant Sci 19:52–61

Babu R, Palacios NP, Gao S et al (2013) Validation of the effects of molecular marker polymorphisms in LcyE and CrtRB1 on provitamin A concentrations for 26 tropical maize populations. Theor Appl Genet 126:389–399

Bernardo R, Yu J (2007) Prospects for genome-wide selection for quantitative traits in maize. Crop Sci 47:1082–1090

Beyene Y, Semagn K, Crossa J, et al (2014) Genetic gains in grain yield through genomic selection in eight biparental maize populations evaluated under managed drought stress environments in sub-Saharan Africa. Crop Science (in Press)

Bruce WB, Edmeades GO, Barker TC (2002) Molecular and physiological approaches to maize improvement for drought tolerance. J Exp Bot 53:13–25

Cairns JE, Sonder K, Zaidi PH et al (2012) Maize production in a changing climate: impacts, adaptation and mitigation strategies. Adv Agron 114:1–58

Cairns JE, Hellin J, Sonder K et al (2013a) Adapting maize to climate change in sub-Saharan Africa. Food Sec 5:345–360

Cairns JE, Crossa J, Zaidi PH et al (2013b) Identification of drought, heat and combined drought and heat tolerant donors in maize. Crop Sci 53:1335–1346

Cobb JN, DeClerck G, Greenberg A, Clark R, McCouch S (2013) Next-generation phenotyping: requirements and strategies for enhancing our understanding of genotype–phenotype relationships and its relevance to crop improvement. Theor Appl Genet 126:867–887

Edmeades GO(2007) Maize in South and South East Asia—A Prospective Study of CIMMYT's Future Role in the Region, CIMMYT, Mexico D.F

Edmeades GO(2008) Drought tolerance in maize: an emerging reality. A feature In James C, 2008. Global Status of Commercialized Biotech/GM Crops: 2008. ISAAA BriefNo. 39. ISAAA: Ithaca, NY

Ferreira JL, Coelho CHM, Magalhães PC, Gomes e Gama EE, Borém A (2007) Genetic variability and morphological modifications inflooding tolerance in maize, variety BRS-4154. Crop Breeding and Appl Biotech 7:314–320

Geiger HH, Gordillo GA (2009) Doubled haploids in hybrid maize breeding. Maydica 54:485–499

Gerpacio RV, Pingali PL (2007) Tropical and subtropical maize in Asia: production systems, Constraints, and Research Priorities, CIMMYT, Mexico DF

Govaerts B, Verhulst N, Castellanos-Navarrete A, Sayre KD, Dixon J, Dendooven L (2009) Conservation agriculture and soil carbon sequestration: between myth and farmer reality. Crit Rev Plant Sci 28:97–122

Heffner EL, Sorrells ME, Jannink J-L (2009) Genomic selection for crop improvement. Crop Sci 49:1–12

Howden SM, Soussana J-F, Tubiello FN, Chettri N, Dunlop M, Meinke H (2007) Adapting agriculture to climate change. Proc Natl Acad Sci USA 104:19691–19696

Hu X, Zimmer Y (2013) China's corn production—where to establish agri benchmark farms in corn. Working paper 2013/3. http://www.agribenchmark.org/fileadmin/Dateiablage/B-Cash-Crop/Working-Paper/cc-1303-China-XH-YZ.pdf

IFPRI (2003) 2025 Projections. International model for policy analysis of agriculture commodities and trade (IMPACT) special project: Global Trend in Food Security & Demand, IFPRI, Washington, D.C

IPCC (2007) Climate change 2007: synthesis report. an assessment of the Inter-Governmental Panel on Climate Change

Lobell DB, Bänziger M, Magorokosho C, Vivek B (2011) Nonlinear heat effects on African maize as evidenced by historical yield trials. Nat Clim Change 1:42–45
Naylor R, Steinfeld H, Falcon W et al (2005) Losing the links between livestock and land. Science 310:1621–1622
Ortiz R, Taba S, Tovar VHC et al (2010) Conserving and enhancing maize genetic resources as global public goods—a perspective from CIMMYT. Crop Sci 50:13–28
Pingali PL, Pandey S (2001) Meeting world maize needs: technological opportunities and priorities for the public sector. In: Pingali PL (ed) CIMMYT 1999/2000 world maize facts and trends. CIMMYT, Mexico DF, pp 1–24
Prasanna BM (2012) Diversity in global maize germplasm and novel initiatives for characterization and utilization. J Biosci 37:843–855
Prasanna BM, Pixley KV, Warburton M, Xie C (2010) Molecular marker-assisted breeding for maize improvement in Asia. Mol. Breed 26:339–356
Prasanna BM, Chaikam V, Mahuku G (2012) Doubled haploid technology in maize breeding: theory and practice. CIMMYT, Mexico D.F., 50 p
Prasanna BM, Cairns J, Xu Y (2013a) Genomic tools and strategies for breeding climate resilient cereals. In: Kole C (ed) Genomics and breeding for climate resilient crops. Vol. I. Concepts and strategies. Springer, New York, pp 213–239
Prasanna BM, Araus JL, Crossa J et al (2013b) High-throughput and precision phenotyping in cereal breeding programs. In: Gupta PK, Varshney RK (eds) Cereal genomics-II. Springer, Dordrecht, pp 341–374
Prasanna BM, Babu R, Nair S et al (2014) Molecular marker-assisted breeding for tropical maize improvement. In: Bohn M, Lai J, Wusirika R (eds) Genetics, Genomics and Breeding of Maize. Science Publishers/CRC Press, Boca Raton, pp 89–118
Prigge V, Sanchez C, Dhillon BS et al (2011) Doubled haploids in tropical maize: I. Effects of inducers and source germplasm on in vivo haploid induction rates. Crop Sci 51:1498–1506
Prigge V, Xu X, Li L et al (2012) New insights into the genetics of in vivo induction of maternal haploids, the backbone of doubled haploid technology in maize. Genetics 190:781–793
Shiferaw B, Prasanna B, Hellin J, Banziger M (2011) Crops that feed the world 6. Past successes and future challenges to the role played by maize in global food security. Food Secur 3:307–327
Shiferaw B, Tesfaye K, Kassie M, Abate T, Prasanna BM, Menkir A (2014) Managing vulnerability to drought and enhancing livelihood resilience in sub-Saharan Africa: technological, institutional and policy options. Weather and Climate Extremes 3:67–79
Smale M, Byerlee D, Jane T (2011) Maize revolution in Sub-Saharan Africa. Policy research working paper 5659, World Bank
Xu Y, Xie CX, Wan J et al (2012) Marker-assisted selection: strategies and examples from cereals. In: Gupta PK, Varshney RK (eds) Cereal genomics-II. Springer, Dordrecht, pp 375–411
Zaidi PH, Cairns JE (2011) Enhancing climate-resilience in tropical maize. In: Zaidi PHet al. (eds) Addressing climate change effects and meeting maize demand for Asia. Book of extended summaries of the 11th Asian maize conference, Nanning, China, 7–11 Nov 2011. CIMMYT, Mexico DF, pp 13–16
Zaidi PH, Maniselvan P, Srivastava A, Yadav P, Singh RP (2010) Genetic analysis of water-logging tolerance in tropical maize (*Zea mays* L.). Maydica 55:17–26
Zaidi PH, Seetharam K, Vinayan MT, Nagesh P, Babu R, Vivek BS (2015) Abiotic stress-resilient maize for adaptation to climate change in the Asian tropics. In: Prasanna BM et al (eds). Book of Extended Summaries, 12th Asian Maize Conference and Expert Consultation on Maize for Food, Feed, Nutrition and Environmental Security. Bangkok, Thailand, October 30–November 1, 2014. CIMMYT, Mexico DF and APAARI, Bangkok, pp 104–115

Chapter 4
Harnessing Apomixis for Heterosis Breeding in Crop Improvement

Sazda Abdi, Shashi, Anuj Dwivedi and Vishnu Bhat

Abstract Apomixis is an asexual mode of reproduction through seeds where embryo develops without undergoing meiosis and fertilization of gametes. Majority of natural apomicts are polyploids and thought to have evolved through hybridization and polyploidization. Apomixis is highly desirable for agriculture as it fixes hybridity or heterosis. Apomicts form huge polyploid complexes in nature which are the results of their facultative nature. They harbor enormous amount of variability resulting in cytotypes. Majority of the crop plants do not reproduce through apomixis although few wild relatives of crop plants such as *Pennisetum glaucum* and *Zea mays* reproduce asexually. Harnessing apomixis for heterosis breeding of crop plants through introgression of this trait from tertiary to primary gene pool was not possible due to imprinting barriers. Deviation in endosperm balance number from the male and female parents during introgression caused poor seed set in *Pennisetum* and *Zea mays* hybrids. Apomicts exhibit three major developmental variations from normal sexual reproduction, viz. apomeiosis, parthenogenesis, and autonomous endosperm development. Initial studies indicated that all the three components are governed by a single or a few genes which was later refuted owing to recombinants showing independent events. Thus, genetics of apomixis is very complex and is often riddled with large-scale segregation distortions. In many apomictic grasses, transmission of apomixis is through a physically large, hemizygous, non-recombining genomic region. One of the genes from an apospory-specific genomic region (ASGR) of *Pennisetum squamulatum*, namely *BABY BOOM LIKE*, elicited parthenogenetic development of embryo in the sexual

S. Abdi · Shashi · A. Dwivedi · V. Bhat (✉)
Department of Botany, University of Delhi, Delhi 110007, India
e-mail: bhatv64@rediffmail.com

S. Abdi
e-mail: saz15abdi@gmail.com

Shashi
e-mail: shasbot.du@gmail.com

A. Dwivedi
e-mail: anujdwivedi2009@gmail.com

V.R. Rajpal et al. (eds.), *Molecular Breeding for Sustainable Crop Improvement*, Sustainable Development and Biodiversity 11,
DOI 10.1007/978-3-319-27090-6_4

pearl millet. Unraveling of genetic and molecular mechanisms controlling apomixis could revolutionize the way agriculture is practiced.

Keywords Apospory · Diplospory · Agamic complex · Apomeiosis · Parthenogenesis

4.1 Introduction

Deployment of novel technologies in agriculture leads to increased food production which is needed to cater to the requirement of ever increasing population. Recent crop improvement programs thus include generation of hybrids in many crops to exploit heterosis or hybrid vigor. In spite of their popularity, poor and marginal farmers can not afford buying hybrid seeds due to their higher cost. The process of development of hybrids is very laborious, especially in the absence of availability of male sterile lines. Apomixis, an asexual mode of reproduction through seeds, can help fix the hybridity as it can fix the genotype of an individual plant.

Apomixis is a reproductive process where embryos develop without the occurrence of meiosis and fertilization (Bhat et al. 2005). The endosperm development could be either autonomous or pseudogamous when the unreduced central cell undergoes division without fertilization or when the unreduced central cell fuses with a sperm cell to form primary endosperm nucleus that undergoes normal mitotic cell division to form endosperm, respectively. The term 'apomixis' was coined by Winkler in 1908 where 'apo' means 'away from' while 'mixis' means the act of mixing. Nogler (1984) later termed it as agamospermy. The first report of occurrence of apomixis in higher plants was in 1841 by Smith who observed that a female plant of *Alchornia ilicifolia* belonging to family Euphorbiaceae formed seeds in the absence of a male parent.

Apomixis is classified into two types, viz. gametophytic and sporophytic. In gametophytic apomixis, either a nucellar cell (apospory) or the megaspore mother cell (MMC) (diplospory) forms the unreduced embryo sac bypassing meiosis or the unreduced egg cell develops into embryo parthenogenetically. In sporophytic apomixis, the nucellar or integumentary cells form adventive embryos directly without meiosis. The later is also termed as adventive embryony, and when multiple embryos develop from both nucellar cells and MMC-derived reduced egg cell from a sexual embryo sac, it is termed as polyembryony.

Apomixis involves a kind of somatic manipulation (Bicknell and Catanach 2015) which is a potentially invaluable tool for plant breeders to exploit heterosis. It also helps in maintaining uniformity among F_1 hybrid seeds and when autonomous endosperm development happens, even difficulties associated with cross-compatibility and pollen availability could be overcome. Since apomixis helps in fixing an individual's genotype, potentially even one elite individual could be released as a variety. This has immense use in agriculture as seeds are the major propagules used for multiplication as well as for human consumption.

Mainly, there are three components of apomixis which include 'apomeiosis,' i.e., bypassing or undertaking incomplete or modified meiosis, 'parthenogenesis' where an unreduced (2n) egg cell develops into an embryo without fertilization and 'autonomous or pseudogamous endosperm development,' where the central cell develops into endosperm without fertilization or after fertilization with a sperm cell, respectively. Although these three events occur in a coordinated manner, it has been found out that they can also be uncoupled where either of these events along with a sexual reproductive event can occur. Such instances are less frequent in a population.

Several attempts were made toward introgression of apomixis from secondary and tertiary gene pools to primary gene pool (Dujardin and Hanna 1989; Hanna et al. 1998; Petrov et al. 1984; Kindiger et al. 1996). None of the crop plants exhibit this phenomenon, while some of their wild relatives reproduce asexually. Although apomixis is transferable from one gene pool to another, problems related to its penetrance and expressivity caused seed sterility. Due to ploidy barriers, endosperm development was either absent or incomplete leading to abnormal seeds or lower seed set. But, several studies on the genetics of inheritance of apomixis were conducted in many taxa which initially indicated monogenic, dominant inheritance pattern which was later refuted as components of apomixis could be uncoupled. During segregation analysis, three developmental events segregated among the progeny implicating independent segregation. Hence, genetics of apomixis is very complex and in some taxa such as *Pennisetum* and *Cenchrus* which reproduce through apospory, an apospory-specific genomic region (ASGR) which is a physically large, hemizygous, non-recombining chromosomal region is known to control apomixis (Ozias-Akins et al. 1998).

Many apomictic taxa form agamic complexes in nature due to hybridization between diploid parental species and polyploid derivatives (Hojsgaard and Hörandl 2015a). Since majority of the apomicts are polyploids, they are the reservoirs of variation due to many cytological aberrations associated with hybridization. Majority of the apomictic taxa are facultative in nature inhabiting residual sexuality. Thus, formation of agamic complexes is mainly due to polyploidy, apomixis, and hybridization which could eventually form microspecies (Hojsgaard and Hörandl 2015a).

4.2 Distribution of Apomicts Among Angiosperms

Apomixis is widespread among angiosperms as nearly 32 orders, 78 families, and 293 genera were reported to contain apomictic species. Among various types of apomixis, adventitious embryony occurs in majority of the genera (148) followed by apospory (110) and diplospory (68). Adventive embryony is mostly observed in fabids, malvids, and lamiids. More than 75 % of apomicts-containing genera are reported from three large families, Poaceae, Asteraceae, and Rosaceae (Hojsgaard et al. 2014a). But the orders containing these three families also contain multiple families in which apomixis is not reported. Diversity among apomicts is similar to general biodiversity pattern as the majority of apomicts-containing genera occur in the tropical region. There has been a tendency for apomicts to colonize larger areas as many apomictic species are found in multiple ecological zones covering wider geographical areas. Most of the apomicts also showed invasiveness (Peters 2001; Chapman et al. 2003; Brock 2004; Hao et al. 2011).

Whether apomixis causes genetic diversity among taxa is still not clear. Generally, apomixis is considered as an evolutionary dead end as there would be loss of genotype heterogeneity due to fixation of alleles (Darlington 1939; Stebbins 1950; Grant 1981) in the population. Due to the loss of genotypic plasticity, apomicts may have reduced potential to adapt to environmental variations. According to this view, majority of the agamic complexes should have become extinct, which is not the case as per recent evidence (Hojsgaard and Hörandl 2015a, b; Hörandl and Pan 2007). Instead, it is widely believed that apomixis is a facilitator of diversification due to polyploidy, facultative apomixis, and reversions to sexuality. The details about various circumstances leading to diversification within an agamic complex will be discussed later in this chapter.

The phylogenetic distributions of apomictic taxa in angiosperms revealed a broadly scattered taxonomic distribution (Hörandl and Hojsgaard 2012). Apomixis occurs in ancestors of some of the major clades of angiosperms indicating its possible ancient descent, while its absence in *Amborella* contradicts this conclusion.

4.3 Evolution of Apomixis

Understanding evolution of asexual reproduction in plants may help in elucidating its mechanism. Three major developmental events during apomixis, namely apomeiosis, parthenogenesis, and autonomous endosperm development may have evolved independently or together. Similarly, the presence of various types of apomixis which are widely scattered along many angiosperm families also implied multiple independent evolution of apomixis (Carman 1997; Van Dijk and Vijverberg 2005). While in most of the apomictic taxa, apomeiosis and parthenogenesis were reported to be controlled separately by two unlinked loci, inheritance studies in apomictic grasses have indicated single dominant locus for both the events (Ozias-Akins and van Dijk 2007). One locus controlling both the events could be an evolutionarily advanced phenomenon for obvious advantage of non-segregation of the locus from generation to generation.

A characteristic feature of majority of apomicts is that they are polyploids which could have happened after hybridization between genetically divergent individuals. Added to this, many apomicts are also allopolyploids containing more than one genome. Carman (1997) used this observation to propose a theory that asynchronous expression of duplicate genes coming from different genomes during hybridization caused apomixis. According to this theory, there is no specific gene for apomixis, instead the duplicate genes controlling various developmental events such as megasporogenesis, megagametogenesis, double fertilization, embryo, and endosperm formation express asynchronously to cause unreduced embryo sac formation and parthenogenesis. Asynchrony in gene expression was reported in *Bothriochloa holboellii* where heterochrony, gene duplication, and parent of origin effects were observed in transcriptome (Bicknell and Catanach 2015).

The major components of apomixis might have originated by mutation (Nogler 1984). It is generally believed that deregulation of developmental events in the sexual reproductive pathway results in apomixis as both are closely related pathways (Tuckers et al. 2003). There is also a strong correlation between occurrence of apomixis and polyploidy. Exceptions such as *Boechera* (Kantama et al. 2007) exist, although diploid apomicts are generally weaker individuals. But, polyploidy increases the tolerance level of deleterious mutations which are accumulated in apomicts. It can buffer negative effects of the mutational load (Muller's ratchet) (Otto and Whitton 2000).

Apomixis also leads to allelic sequence divergence, called Meselson effect (Mark Welch and Meselson 2000) which is due to the gain of neutral allelic differences. Recent study of transcriptomes of *Ranunculus auricomus* complex, consisting of diploid sexuals and hexaploid apomicts, indicated Meselson-like sequence divergence. But there was no mutation accumulation (Pellino et al. 2013). This could be ascribed to relatively younger population of apomicts in *Ranunculus* (approx. 70,000 years) or it may have facultative sexuality because of which deleterious mutations are purged-out (Hojsgaard and Hörandl 2015a).

In many apomictic taxa, a specific genomic region called ASGR is known to control apomixis (Ozias-Akins and van Dijk 2007). A single chromosome carrying this region is responsible for transmitting apomixis (Akiyama et al. 2011). At least between *Pennisetum squamulatum* and *Cenchrus ciliaris*, ASGR is known to be highly conserved based on syntenic relationships between chromosomal sequences identified by BAC probes, shared cytological features such as hemizygosity and their heterochromatic nature (Ozias-Akins et al. 1998). The origin and evolution of ASGR was studied using fluorescence in situ hybridization (FISH) and *ndhF* gene and *trnL-F* sequence analysis in *Pennisetum* and *Cenchrus*. Low rates of sequence variation at the ASGR suggested a recent origin of ASGR (Akiyama et al. 2011).

Another recent study on *Ranunculus* species complex based on crosses between sexual diploid and tetraploid *Ranunculus auricomus* species also strengthened the earlier view that hybridization, rather than polyploidy, is responsible for the ovule developmental alterations observed in their hybrids (Hojsgaard et al. 2014b). Emergence of apospory bypassing meiosis was evident after sexual hybridization and polyploidization. In this study, polyploidy was a key factor for functional apomixis as it stabilized deviations in paternal to maternal genome contributions in the endosperm for successful seed formation.

It is interesting to know that asexuality is rare in higher organisms when compared to sexuality. Asexual females could transmit their genes twice as compared to sexual females as they produce half of their progeny as males. This gives asexual females an advantage because of which they should be able to invade the population replacing sexual individuals which is generally observed in case of natural apomicts (de Meeûs et al. 2007).

4.4 Genetics of Apomixis

Apomixis is a heritable trait and inheritance of apomixis in grasses indicated it to be controlled by a single dominant locus (Savidan 2000). In grasses such as *Panicum maximum* (Savidan 1982), *Brachiaria* hybrids (Valle and Glenke 1993), *Cenchrus ciliaris* L. (Sherwood et al. 1994; Jessup et al. 2002), and *Poa pratensis* (Barcaccia et al. 1998), segregation pattern indicated one or a few Mendelian factors (Aguilera et al. 2015). Whereas in *Taraxacum* (Van Dijk and Bakx-Schotman 2004), *Hypericum* (Schallau et al. 2010), *Erigeron* (Noyes and Rieseberg 2000), and *Hieracium* (Catanach et al. 2006), apomeiosis and parthenogenesis are controlled by two different loci as evident from independent segregation of each component. On the contrary, Asker (1980) reported apospory under polygenic control. In Citrus, a single locus of approx. 300 kb is reported to be responsible for polyembryony (Nakano et al. 2012). Studies of inheritance of apomixis in many taxa indicated simplex genotypes in both monocots and dicots (Table 4.1).

In *Pennisetum squamulatum* and *Cenchrus ciliaris*, an ASGR has been reported as exclusively associated with apomixis (Akiyama et al. 2004). This region is highly heterochromatic and hemizygous residing in a single chromosome covering 50-Mb genomic region. It is recombinationally suppressed and contains high copy reterotransposons. Recent reports of recombination within ASGR of *C. ciliaris* (Conner et al. 2013; Yadav et al. 2012) could help in dissecting out this region to identify genes associated with apomeiosis and parthenogenesis. One of the genes from ASGR, *BABY BOOM LIKE* from apomictic *P. squamulatum*, has been characterized which elicits parthenogenetic embryo formation from unfertilized eggs in a sexual pearl millet plant (Conner et al. 2015). This has categorically demonstrated the role of ASGR in apomictic development. Similarly, in *Paspalum* species, linkage mapping studies revealed segregation distortion and lack of recombination, synteny between apomixis-associated markers and rice map, and narrowing of the chromosomal apomixis controlling region (ACR) through comparative mapping (Ortiz et al. 2013). Several markers such as RAPD, AFLP, and RFLPs linked to apospory locus were detected in *P. simplex* (Labombarda et al. 2002) and *P. notatum* (Martinez et al. 2003; Pupilli et al. 2004). Based on mapping results, one of the apomixis-linked BAC clones from *P. simplex* was sequenced which revealed 10 % noncoding sequences, 13 sequences pertained to transposons and retrotransposons and four putative genes (Ortiz et al. 2013). Out of these genes, two genes co-segregated with apomixis in several *Paspalum* species which have shown similarity to protein kinase and protein of the ERD1/XPR1/SYG1 family.

In *Hieracium*, a series of deletion mutants were generated which showed two distinct loci, first, *LOSS OF APOMEIOSIS* (*LOA*) and, the second, *LOSS OF*

Table 4.1 The genotypic constitution of various apomictic taxa along with the number of loci controlling apomixis (modified from Ozias-Akins and van Dijk 2007)

Species	Family	Apomixis type	Loci	Genotype	References
Monocots					
Brachiaria brizantha	Poaceae	Apospory, pseudogamous endosperm	1	Aaaa	Miles and Escandon (1997), Pessino et al. (1997, 1998)
Cenchrus ciliaris	Poaceae	Apospory, pseudogamous endosperm	1	Aaaa	Roche et al. (1999), Jessup et al. (2002)
Panicum maximum	Poaceae	Apospory	1	Aaaa	Ebina et al. (2005), Savidan (1980)
Paspalum notatum	Poaceae	Apospory, pseudogamous endosperm	1	Aaaa	Martinez et al. (2001, 2003), Stein et al. (2004)
Paspalum simplex	Poaceae	Apospory, pseudogamous endosperm	1	Aaaa	Caceres et al. (2001), Calderini et al. (2006), Labombarda et al. (2002), Pupilli et al. (2001, 2004)
Pennisetum squamulatum	Poaceae	Apospory, pseudogamous endosperm	1	Aaaa	Roche et al. (2001), Ozias-Akins et al. (1993)
Poa pratensis	Poaceae	Apospory	2	Aaaa Pppp	Albertini et al. (2001a, 2001b), Barcaccia et al. (1998), Porceddu et al. (2002)
Tripascum dactyloides	Poaceae	Diplospory, mitotic, pseudogamous endosperm	1?	Dddd	Grimanelli et al. (1998), Leblanc et al. (1995)
Eudicots					
Ranunculus auricomus	Ranunculaceae	Apospory, pseudogamous endosperm	1	Aaaa	Nogler (1984)

(continued)

Table 4.1 (continued)

Species	Family	Apomixis type	Loci	Genotype	References
Taraxacum officinale	Asteraceae	Diplospory, meiotic, autonomous endosperm	3	Ddd Ppp	van Dijk (2003), Vijverberg et al. (2004)
Erigeron annuus	Asteraceae	Diplospory, mitotic, autonomous endosperm	2	D/dd*) Fff	Noyes et al. (2007), Noyes and Rieseberg (2000)
Hieracium caespitosum	Asteraceae	Apospory, autonomous endosperm	2	Aaaa Pppp	Bicknell et al. (2000), Catanach et al. (2006)

* The only genomic region for which a disomic inheritance model gives a better fit than trisomic models

PARTHENOGENESIS (*LOP*) (Hand and Koltunow 2014). Using these mutants, Catanach et al. (2006) identified mutations in apomixis loci through mapping using AFLP markers. *LOA* and *LOP* could be distinctly mapped. Similarly, the third component of autonomous endosperm development could be separated from parthenogenesis in *Hieracium* (Ogawa et al. 2013).

While the forward genetic approaches are slowly elucidating the complexity of apomixis, several reverse genetic approaches using functional genomic tools have also generated interesting results. In *Arabidopsis*, mutants reflecting apomeiosis during diplospory could be obtained by inducing mutations in *DYAD* allele of *SWITCH1*. The *dyad* mutants produced unreduced gametes which fertilized to produce triploid seeds (Ravi et al. 2008). Later, high-frequency unreduced gametes were produced by mutating three meiotic genes, viz. *osd1*, *rec8*, and *spo11* in *Arabidopsis* and these mutants were named as *MiMe* (mitosis instead of meiosis) (d'Erfurth et al. 2009). In another independent study, deletion of centromeric histone protein (CenH3) in either male or female gamete eliminated the haploid set of chromosomes during fertilization resulting in haploids (Ravi and Chan 2010). This clearly demonstrated uniparental transmission through egg. When *dyad* mutant was crossed with *cenh3* mutant, majority of the progeny were of maternal origin (Marimuthu et al. 2011). This study indicated that apomixis could be synthesized using mutants.

The third component of apomixis, namely autonomous endosperm development, could be induced in *Arabidopsis* by manipulating a gene belonging to Polycomb group of genes. The fertilization-independent seed development (FIS) complex consists of four genes, namely *MEDEA* (*MEA*) (Grossniklaus et al. 1998), fertilization-independent seed2 (*FIS2*) (Chaudhury et al. 1997), fertilization-independent endosperm (*FIE*) (Ohad et al. 1999), and multicopy suppressor of *IRA1* (*MSI1*). Mutation in *MEA* or *FIS2* or *FIE* induces autonomous endospermy, while mutation in *MSI1* induces parthenogenetic embryos from egg cell (Guitton and Berger 2005). This complex has repressive function and represses downstream genes responsible for fertilization-independent seed development when they are transcriptionally active. The above studies have independently established the genetic mechanisms controlling three components of apomixis, viz apomeiosis, parthenogenesis, and autonomous endosperm development.

4.5 Agamic Complexes

The fact that majority of apomicts occur naturally in facultative form, they tend to form larger agamic complexes (Babcock and Stebbins 1938) which originate through continuous extension of an apomictic species via hybridization. Even in clonally propagated species, variation is created by the process of gene conversion, mitotic recombination, and epigenetic drift (Hojsgaard and Hörandl 2015b). The presence of diploids and polyploids within agamic complex results in various polyploid derivatives through backcrossing (Koch et al. 2003; Guo et al. 2004; Hörandl 2009; Sochor et al. 2015). In agamic polyploid complexes, newly formed polyploids reproduce asexually (Babcock and Stebbins 1938; Grant 1981). Evolution of agamic complexes is hastened due to polyploidy, hybridization, and apomixis which result in 'microspecies.' These microspecies exhibit distinct morphological, cytological, and genetic variations (Hojsgaard and Hörandl 2015b).

Following Carman's hypothesis (Carman 1997), variation within the agamic complexes could be caused due to duplicate genes governing developmental events during the evolution of asexual neo-polyploids. There is also a possibility of reversal to sexuality, and eventually, cycles of agamospermous polyploids and sexually reproducing species facilitate diversification (Hojsgaard and Hörandl 2015b). Moreover, polyploidy helps apomicts to adapt to diverse ecological niches by preserving genetic variability even when meiotic recombination and genetic segregation due to fertilization are lacking.

In addition to ploidy variation, asexuals are also helped by epigenetic variation (Hardesty et al. 2012; Roiloa et al. 2014) which increases their ability to acclimatize newer environments. In general, agamic complexes exhibit two types of reproductive series in Panicoid grasses, viz. first, the diploid–tetraploid–(di) haploid cycle and, second, the diploid–tetraploid–diploid cycle. The first type is observed in *Bothriochloa–Dichanthium* agamic complex (de Wet 1968; de Wet and Harlan 1970a, b) and in *Panicum* (Savidan and Pernès 1982). The second type is reported in *Paspalum* (Hanna and Burton1986; Quarin 1992). They are mostly consisting of diploids, facultative apomictic tetraploids, triploids, and aneuploids (Naumova et al. 1999).

The best-known examples of agamic complexes are *Amelanchier* (Campbell et al. 1999), *Boechera* (Schranz et al. 2005), *Antennaria* (Bayer 1987), *Crataegus* (Talent and Dickinson, 2005), *Crepis* (Babcock and Stebbins 1938, Whitton et al. 2008), *Taraxacum* (Richards 1973; van Oostrum et al. 1985, den Nijs and Menken 1996), *Capillipedium–Dichanthium–Bothriochloa* (de Wet and Harlan 1970a, b), *Panicum maximum* (Pernès 1975), *Tripsacum* (Moreno-Perez et al. 2009), *Ranunculus auricomus* (Hörandl et al. 1997; Hörandl and Greilhuber 2002;

Cosendai and Hörandl 2010), and *Paspalum* (Norrmann et al. 1989; Urbani et al. 2002; Daurelio et al. 2004). Few agamic complexes are described in this chapter.

4.5.1 *Taraxacum* Agamic Complex (Asteraceae)

The genus *Taraxacum* (Dandelions) belongs to the family Asteraceae and the tribe Cichorieae, consisting of some 2000 species, assembled in 30 sections of which 90 percentage are agamospermous (Ford and Richards 1985; Mogie and Ford 1988). The taxa have been distinguished as macrospecies and microspecies by Dudman and Richards (1997). Due to the presence of various breeding systems and ploidy levels, the taxonomy of *Taraxacum* is complicated (Richards 1986). It reproduces mainly by diplosporous type of gametophytic apomixis with autonomous endosperm development (Asker and Jerling 1992). In this complex, both facultative apomicts and sexually reproducing plants coexist, sexual diploids cover approximately 13 % of the species and are generally self-incompatible (Richards 1973). On the contrary, polyploids reproduce through diplospory and their chromosome number ranges from 3x to 10x (Battaglia 1948; Richards1973). Strains of sexual diploids are generally widespread in central Europe where they coexist with diverse ploidy range of apomictic accessions, mostly triploids (3x = 24) (Ozias-Akins and van Dijk 2007). Section *Erythrosperma* covers two *Taraxacum* species which shares parallel habitat and closely related taxonomic features, i.e., *Taraxacum lacistophyllum* (Dahlst.) and *T. brachyglossum* (Dahlst) Dahlst. (Ford and Richards 1985). Different sections contain diverse *Taraxacum* accessions such as *Taraxacum officinale* (Section *Ruderalia*, agamospermous triploid species), *Taraxacum laevigatum* (Section *Erythrosperma*, agamospermous triploid species), *Taraxacum subnaevosum* and *Taraxacum nordstedtii* (Section *Spectabilia*, agamospermous species), and *Taraxacum platycarpum* and *Taraxacum japonicum* (Section *Mongolica*, sexual diploid species), (Ford and Richards 1985; Mitsuyuki et al. 2014).

4.5.2 *Capillipedium–Dichanthium–Bothriochloa* Agamic Complex (Poaceae)

Among all the agamic complexes, *Capillipedium–Dichanthium–Bothriochloa* agamic complex is always fascinating. Harlan and de Wet (1963) studied in detail about the genera of *Capillipedium*, *Dichanthium*, and *Bothriochloa*. The genera is broadly scattered over Europe, the Mediterranean region, Asia, Australia, and the New World (Harlan and de Wet (1963); de Wet and Harlan 1966). De Wet (1968) described a possible evolution in the genus *Dichanthium* based on ploidy cycles involving diploids, tetraploids, and haploids. Mehra (1961) reported the basic chromosome number of *Dichanthium* as 10 and mentioned about the existence of diverse ploidy levels ranging from diploid to hexaploid. *Dichanthium annulatum, Dichanthium caricosum*, and *Dichanthium aristatum* occur in southern Asia (Celarier et al. 1958). In the genus *Capillipedium*, *C. assimile* generally are tetraploids broadly scattered over southern Asia, India and adjoining islands. Another hexaploid species, *C. parviflorum*, is distributed in Southeast Asia, India, Africa, and Australia. However, diploid strains were only found in Hong Kong and in tropical India (de Wet and Harlan 1970a). The Eurasian *B. ischaemum* is classified as a tetraploid, pentaploid, and hexaploid strains mainly occur as apomictic tetraploid in Europe, Near East, and central Asia (Celarier and Harlan 1958) and as auto-hexaploids in Turkey (de Wet and Harlan 1970b). Generally, in this agamic complex, gene flow takes place in numerous directions with a minor occurrence of incompatibility barriers. Hybridization between *Capillipedium* and *Dichanthium* takes place in the presence of *Bothriochloa* which acts as a bridge species (Fig. 4.1) (de Wet and Harlan 1970b). *Bothriochloa intermedia* (R.Br.) A. Camus [equivalent to *B. bladhii* (Retz.) S.T. Blake (Quattrocchi

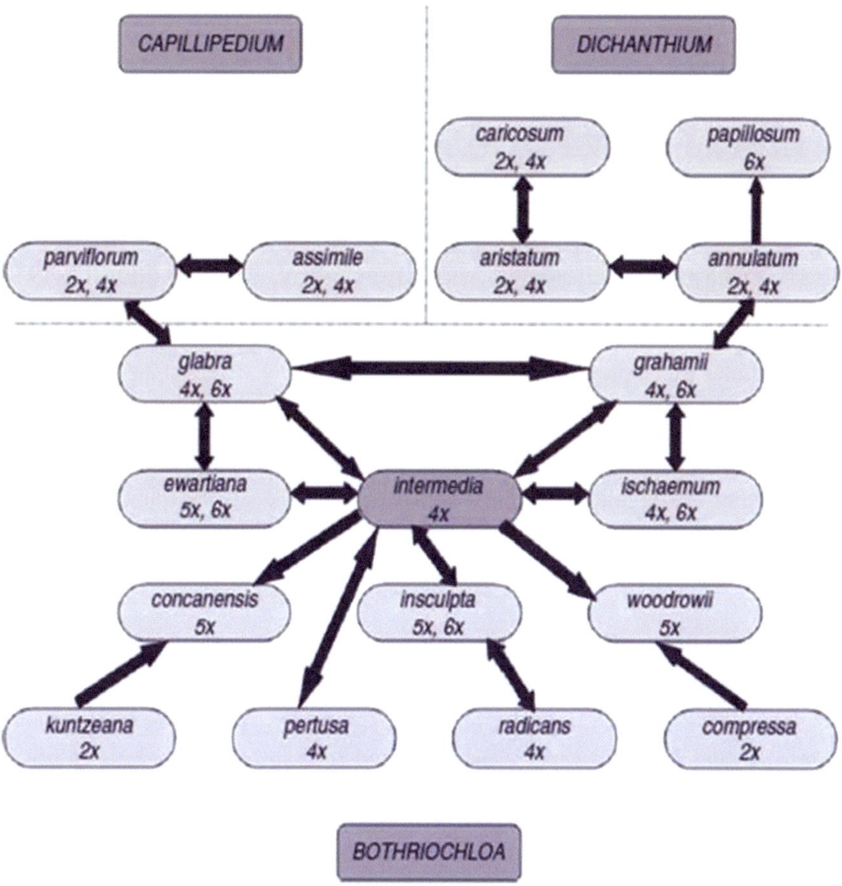

Fig. 4.1 Diagrammatic representation showing relationship among the members of *Capillipedium–Dichanthium–Bothriochloa* agamic complex (adopted from Berthoud 2001)

2008)] mainly acts as a bridge species between *Dichanthium* and *Capillipedium* species (de Wet and Harlan 1970b; Berthaud 2001).

In this agamic complex, *B. intermedia* (R.Br.) A. Camus hybridizes with *B. ewartiana* (Domin) C.E. Hubbard, *B. ischaemum* (Linn.) Keng, *Capillipedium parvifolium* (R.Br.) Stapf, and *D. annulatum* (Forssk.) Stapf. Based on this, it was concluded that *B. intermedia* is a compilospecies, one which carries genes from several other species of *Bothriochloa* O. Kuntze, *Capillipedium* Stapf., and *Dichanthium* Willemet (Harlan and de Wet 1963; de Wet and Harlan 1966, 1970a, b). Because of this, de Wet and Harlan (1968) suggested that all the three taxa could be grouped under *Dichanthium* genera.

4.5.3 *Tripsacum* Agamic Complex (Poaceae)

The genus *Tripsacum* is a member of Poaceae family, subtribe Maydeae and broadly distributed in the American continents, extremely variable including 16 species consisting of an agamic complex with different chromosome numbers ranging from diploids ($2n = 2x = 36$) to hexaploids ($2n = 6x = 108$) (Mangelsdorf and Reeves 1931; Petrov et al. 1979; Randolph 1970; de Wet et al. 1981). It is a wild relative of maize belonging to secondary gene pool and widely spread in Mexico (Berthaud and Savidan 1989). Farquharson (1955) reported for the first time the phenomena of apomixis in polyploid strains of *T. dactyloides* (L.) and observed both facultative apomixis and polyembryony. Brown and Emery (1958) reported the occurrence of diplosporous megasporogenesis in *T. dactyloides* clone, while de Wet et al. (1973) described aposporous type in maize × *T. dactyloides* hybrids. Later on, diplospory in *Tripsacum* was clearly demonstrated by Burson et al. (1990) based on embryological investigations using sectioned and stained ovaries from two triploid and one tetraploid *T. dactyloides* accessions collected from the USA. Leblanc et al. (1995) also analyzed *Tripsacum* germplasm and revealed that all diploid accessions were sexual, while all polyploid races were diplosporous of the Antennaria type.

4.5.4 *Hieracium* and *Pilosella* Agamic Complex (Asteraceae)

Hieracium and *Pilosella* genera, earlier treated as *Hieracium* subgenus *Pilosella*, contain closely related species which have evolved through hybridization and polyploidization (Hand et al. 2015). They contain mostly polyploid species and few diploid species (Fehrer et al. 2007a, b). While *Hieracium* reproduces through diplospory, *Pilosella* reproduces through apospory. In both these genera, embryo and endosperm develop independently. Apomixis is reported in 13 species of *Hieracium* (Bergman 1941; Skawinska 1963). Due to the problems associated with male sterility in the hybrids resulting in poor seed set, *Hieracium* species have not been used in apomixis research. On the other hand, *Pilosella* has been used as a model system extensively due to the occurrence of apospory combined with autonomous endosperm development. Various mutants and accessions were used for studying the mechanisms of apomixis (Koltunow et al. 2013). The mode of apomixis in *Pilosella* is facultative which results in three non-apomictic types of progeny, viz. 2n + 0, n + 0, and 2n + n types which represent female and male gamete ploidy (Hand et al. 2015). In an agamic complex when more than two species or cytotypes occur within a population, there will be enormous number of hybrid forms with different ploidy levels (Krahulcová et al. 2009).

According to taxonomic classification of *Hieracium* and *Pilosella*, there are basic and intermediate species where basic species reproduce through either sexual or apomictic modes, while the intermediate species are derived from hybridizations between basic species and they are polyploids. *Pilosella* contains two loci, namely *LOSS OF APOMEIOSIS* (*LOA*) and *LOSS OF PARTHENOGENESIS* (*LOP*), where *LOA* is required for apospory including differentiation of AI cells, while *LOP* is required for autonomous embryo and endosperm development (Hand et al. 2015). Apomixis coexists with sexuality in *Pilosella,* while in 16 species of *Hieracium* surveyed, 1–7 % of ovules showed residual sexuality. The *LOA* and *LOP* markers were not conserved between *Pilosella* and *Hieracium* supporting independent evolution of both the loci (Fig. 4.2). The origins of intermediate species and the most recent hybrids derived from basic species of *Pilosella* plants are explained in Fig. 4.2.

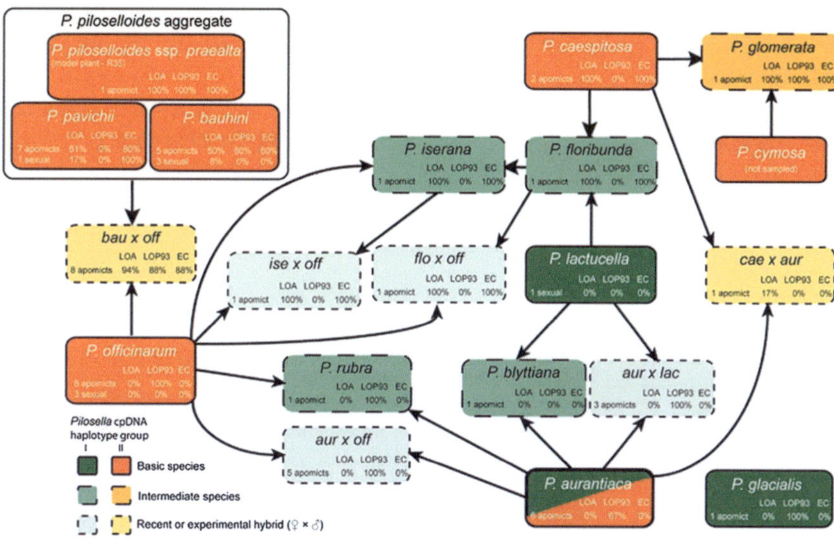

Fig. 4.2 *Pilosella* species forming agamic complex after putative hybridization between basic species resulting in intermediate species, marked by *arrows*. *Green* (*Pilosella I*) and *orange* (*Pilosella II*) colors represent major chloroplast haplotype group. The number of individuals sampled, reproductive modes, the occurrence of hemizygous elongated chromosome (EC) and the conservation of *LOA*- and *LOP*-associated markers are mentioned for each taxon (adopted from Hand et al. 2015)

4.6 Conclusions

Asexual reproduction in plants is a potential tool to fix heterosis which will reduce the cost of hybrid seeds. Although natural occurrence of apomixis is reported in many families of angiosperm taxa, its transfer to cultivated crops has been unsuccessful. Based on the observation that majority of gametophytic apomicts are polypoids, apomixis is thought to have evolved through hybridization and polyploidization. The genetics of apomixis is very complex as evident from segregation distortions and the major developmental events such as apomeiosis, parthenogenesis, and autonomous endosperm development seem to segregate independently within a population. Three different mechanisms have been identified so far regulating those developmental events. While mutations in a meiotic gene such as *DYAD* could produce unreduced gametes within an embryo sac, an egg cell-specific gene of apomictic origin such as *BABY BOOM LIKE* could cause parthenogenetic embryo formation when it is expressed in a sexual plant. Autonomous endosperm development could be induced by silencing any member of Polycomb group genes such as *MEDEA*. Although these three groups of genes control three different developmental events of apomictic pathway, they seem to be independently governing the components, thus the master regulator of apomixis as a whole is still unknown. Synthesizing an apomictic plant was still possible by hybridization between *dyad* and *cenh* mutants which could be adopted in other crops as well. Many apomicts form polyploid complexes in nature called agamic complexes which harbor enormous amount of variability due to the presence of residual sexuality. When apomixis technology becomes a tool for plant breeders, it may eventually result in greater diversification and speciation in plants. This could also result in more hazardous plant types such as weeds which become invasive. With the development of more robust techniques of genomics and transcriptomics, apomixis may become a reality for crop improvement in the near future.

References

Aguilera PM, Galdeano F, Quarin CL, Pablo J, Ortiz A, Espinoza F (2015) Inheritance of aposporous apomixis in interspecific hybrids derived from sexual *Paspalum plicatulum* and apomictic *Paspalum guenoarum*. Crop Sci 55:1947–1956

Akiyama Y, Conner JA, Goel S, Morishige DT, Mullet JE, Hanna WW, Ozias-Akins P (2004) High-resolution physical mapping in *Pennisetum squamulatum* reveals extensive chromosomal heteromorphism of the genomic region associated with apomixis. Plant Physiol 134:1733–1741

Akiyama Y, Goel S, Conner JA, Hanna WW, Yamada-Akiyama H, Ozias-Akins P (2011) Evolution of the apomixis transmitting chromosome in *Pennisetum*. BMC Evol Biol 11:289

Albertini E, Barcaccia G, Porceddu A, Sorbolini S, Falcinelli M (2001a) Mode of reproduction is detected by Parth1 and Sex1 SCAR markers in a wide range of facultative apomictic Kentucky bluegrass varieties. Mol Breed 7:293–300

Albertini E, Porceddu A, Ferranti F, Reale L, Barcaccia G et al (2001b) Apospory and parthenogenesis may be uncoupled in *Poa pratensis*: a cytological investigation. Sex Plant Reprod 14:213–217

Asker S (1980) Gametophytic apomixis: elements and genetic regulation. Hereditas 93:277–293

Asker S, Jerling L (1992) Apomixis in plants. CRC Press, London

Babcock EB, Stebbins GL (1938) The American species of Crepis: Their relationships and distribution as affected by polyploidy and apomixis. Carnegie Inst Wash Publ 504:1–200

Barcaccia G, Mazzucato A, Albertini E, Zethof J, Gerats A et al (1998) Inheritance of parthenogenesis in *Poa pratensis* L.: auxin test and AFLP linkage analyses support monogenic control. Theor Appl Genet 97:74–82

Battaglia E (1948) Ricerche sulla parameiosi restitzionale nel genere *Taraxacum*. Caryologia 1:1–47

Bayer RJ (1987) Evolution and phylogenetic relationships of the *Antennaria* (Asteraceae: Inuleae) polyploid agamic complex. Biologie Zentralblatt 106:683–698

Bergman B (1941) Studies on the embryo sac mother cell and its development in *Hieracium* (subg.) *Archieracium*. Svensk Bot Tidskr 35:1–41

Berthaud J (2001) Apomixis and the management of genetic diversity. In: Savidan Y, Carman JG, Dresselhaus T (eds) The flowering of apomixis: from mechanisms to genetic engineering. CIMMYT, IRD, Mexico DF, pp 8–23

Berthaud J, Savidan Y (1989) Genetic resources of *Tripsacum* and gene transfer to maize. In: Mujeeb-Kazi A, Sitch LA (eds) Review of advances in plant biotechnology, 1985–1988. 2nd international symposium on genetic manipulation in crops. CIMMYT and IRRI, Mexico DF, Mexico and Manila Philippines, pp 121–130

Bhat V, Dwivedi KK, Khurana JP, Sopory SK (2005) Apomixis: an enigma with potential applications. Curr Sci 89:1879–1893

Bicknell R, Catanach A (2015) Apomixis: the asexual formation of seed. In: Li X-Q, et al (eds) Somatic genome manipulation. LLC 2015. Springer Science & Business Media, Berlin. doi:10.1007/978-1-4939-2389-2_7

Bicknell RA, Borst NK, Koltunow AM (2000) Monogenic inheritance of apomixis in two *Hieracium* species with distinct developmental mechanisms. Heredity 84:228–237

Brock MT (2004) The potential for genetic assimilation of a native dandelion species, *Taraxacum ceratophorum* (Asteraceae), by the exotic congener *T. officinale*. Am J Bot 91:656–663

Brown WF, Emery WH (1958) Apomixis in the Gramineae: panicoideae. Am J Bot 45:253–263

Burson BL, Voigt PW, Sherman RA, Dewald CL (1990) Apomixis and sexuality in Eastern Gamagrass. Crop Sci 30:86–89

Caceres ME, Matzk F, Busti A, Pupilli F, Arcioni S (2001) Apomixis and sexuality in *Paspalum simplex*: characterization of the mode of reproduction in segregating progenies by different methods. Sex Plant Reprod 14:201–206

Calderini O, Chang S, de Jong H, Busti A, Paolocci F et al (2006) Molecular cytogenetics and DNA sequence analysis of an apomixis-linked BAC in *Paspalum simplex* reveal a non pericentromere location and partial microcolinearity with rice. Theor Appl Genet 112:1179–1191

Campbell CS, Alice LA, Wright WA (1999) Comparisons of within population genetic variation in sexual and agamospermus *Amelanchier* (Rosaceae) using RAPD markers. Plant Syst Evol 215:157–167

Carman JG (1997) Asynchronous expression of duplicate genes in angiosperms may cause apomixis, bispory, tetraspory, and polyembryony. Biol J Linn Soc 61:51–94

Catanach AS, Erasmuson SK, Podivinsky E, Jordan BR, Bicknell R (2006) Deletion mapping of genetic regions associated with apomixis in *Hieracium*. Proc Natl Acad Sci USA 103:18650–18655

Celarier RP, Harlan JR (1958) The cytogeography of the *Bothriochloa ischaemum* complex. I. Taxonomy and geographic distribution. J Linn Soc London Bot 55:755–760

Celarier RP, Mehra KL, Wulf ML (1958) Cytogeography of the *Dichanthium annulatum* complex. Brittonia 10:59–72

Chapman H, Houliston GJ, Robson B, Ilne I (2003) A case of reversal: the evolution and maintenance of sexuals from parthenogenetic clones in *Hieracium pilosella*. Int J Plant Sci 164:719–728

Chaudhury AM, Ming L, Miller C, Craig S, Dennis ES, Peacock WJ (1997) Fertilization-independent seed development in *Arabidopsis thaliana*. Proc Natl Acad Sci USA 94:4223–4228

Conner JA, Gunawan G, Ozias-Akins P (2013) Recombination within the apospory specific genomic region leads to the uncoupling of apomixis components in *Cenchrus ciliaris*. Planta 238:51–63

Conner JA, Mookkan M, Huo H, Chae K, Ozias-Akins P (2015) A parthenogenesis gene of apomict origin elicits embryo formation from unfertilized eggs in a sexual plant. Proc Natl Acad Sci USA 112(36):11205–11210

Cosendai AC, Horandl E (2010) Cytotype stability, facultative apomixis and geographical parthenogenesis in *Ranunculus kuepferi* (Ranunculaceae). Ann Bot 105:457–470

d'Erfurth I, Jolivet S, Froger N, Catrice O, Novatchkova M, Mercier R (2009) Turning meiosis into mitosis. PLOS Biol 7:e1000124. 16:1–128

Darlington CD (1939) The evolution of genetic systems. Cambridge University Press, Cambridge, p 254

Daurelio LD, Espinoza F, Quarin CL, Pessino SC (2004) Genetic diversity in sexual diploid and apomictic tetraploid populations of *Paspalum notatum* situated in sympatry or allopatry. Plant Syst Evol 244:189–199

de Meeûs T, Prugnolle F, Agnew P (2007) Asexual reproduction: genetics and evolutionary aspects. Cell Mol Life Sci 64:1355–1372

den Nijs JCM, Menken SBJ (1996) Relations between breeding system, ploidy level, and taxonomy in some advanced sections of *Taraxacum*. In: Hind DJN, Beentje HJ (eds) Compositae: systematics. Proceedings of the international compositae conference. Royal Botanic Gardens, Kew, pp 665–677

de Wet JMJ, Harlan JR (1970a) Apomixis, polyploidy and speciation in *Dichanthium*. Evolution 24:270–277

de Wet JMJ (1968) Diploid-tetraploid-haploid cycles and the origin of variability in *Dichanthium*. Evolution 22:394–397

de Wet JMJ, Engle LM, Grant CA (1973) Breeding behaviour of maize-*Tripsacum* hybrids. Crop Sci 13:254–256

de Wet JMJ, Harlan JR (1966) Morphology of the compilospecies *Bothriochloa intermedia*. Amer J Bot 53:94–98

de Wet JMJ, Harlan JR (1970b) *Bothriochloa intermedia*—a taxonomic dilemma. Taxon 19:339–340

de Wet JMJ, Timothy DH, Hilu KW, Fletcher GB (1981) Systematics of South American *Tripsacum* (Gramineae). Am J Bot 68:269–276

Dudman AA, Richards AJ (1997) Dandelions of Great Britain and Ireland. Botanical Society of the British Isles, London

Dujardin M, Hanna WW (1989) Developing apomictic pearl millet: characterization of a BC3 plant. J Genet Breed 43:145–151

Ebina M, Nakagawa H, Yamamoto T, Araya H, Tsuruta S et al (2005) Co-segregation of AFLP and RAPD markers to apospory in Guineagrass (*Panicum maximum* Jacq.). Grassland Sci 51:71–78

Farquharson LI (1955) Apomixis and polyembryony in *Tripsacum dactyloides*. Am J Bot 42: 737–743

Fehrer J, Gemeinholzer B, Chrtek J Jr, Bra¨utigam S (2007a) Incongruent plastid and nuclear DNA phylogenies reveal ancient intergeneric hybridization in *Pilosella* hawkweeds (*Hieracium, Cichorieae*, Asteraceae). Mol Phylogenet Evol 42:347–361

Fehrer J, Krahulcová A, Krahulec F, Chrtek Jr J, Rosenbaumova´ R, Bräutigam S (2007b) Evolutionary aspects in *Hieracium* subgenus *Pilosella*. In: Hörandl E, Grosniklaus U, van

Dijk P, Sharbel T (eds) Apomixis: evolution, mechanisms and perspectives. Koeltzn, Königstein, pp 359–390

Ford H, Richards AJ (1985) Isozyme variation within and between *Taraxacum* agamospecies in a single locality. Heridity 55:289–291

Grant V (1981) Plant speciation. Columbia University Press, New York, p 435

Grimanelli D, Leblanc O, Espinosa E, Perotti E, Gonzales de Leon D, Savidan Y (1998) Mapping diplosporous apomixis in tetraploid *Tripsacum*: one gene or several genes? Heredity 80:33–39

Grossniklaus U, Koltunow A, van Lookeren Campagne M (1998) A bright future for apomixis. Trends Plant Sci 3:415–441

Guitton E-A, Berger F (2005) Loss of function of MULTICOPY SUPPRESSOR OF IRA 1 produces nonviable parthenogenic embryos in *Arabidopsis*. Curr Biol 15:750–754

Guo M, Rupe MA, Zinselmeier C, Habben J, Bowen BA, Smith OS (2004) Allelic variation of gene expression in maize hybrids. Plant Cell 16:1707–1716

Hörandl E, Pan O (2007) Patterns and sources of genetic diversity in apomictic plants: implications for evolutionary potentials In: Hörandl E, Grossniklaus U, Van Dijk PJ, Sharbel T (eds) Apomixis: evolution, mechanisms and perspectives. International Association for Plant Taxonomy, Ruggel, pp 169–194

Hand ML, Koltunow AM (2014) The genetic control of apomixis: asexual seed formation. Genetics 197:441–450

Hand ML, Vı´t P, Krahulcová A, Johnson SD, Oelkers K, Siddons H, Chrtek J, Fehrer Jr. J, Koltunow AMG (2015) Evolution of apomixis loci in *Pilosella* and *Hieracium* (Asteraceae) inferred from the conservation of apomixis-linked markers in natural and experimental populations. Hered 114:17–26

Hanna WW, Burton GW (1986) Cytogenetics and breeding behavior of an apomictic triploid in bahiagrass. J Hered 77:457–459

Hanna WW, Roche D, Ozias-Akins P (1998) Use of apomixis in crop improvement. In: Virmani SS, Siddiq ED, Muralidharan K (eds) Advances in hybrid rice technology. Proceedings of the 3rd international symposium on hybrid rice. International Rice Research Institute, Manila, pp 283–296

Hao JH, Qiang S, Chrobock T, van Kleunen M, Liu QQ (2011) A test of Baker's law: breeding systems of invasive species of Asteraceae in China. Biol Invasions 13:571–580

Hardesty BD, Le Roux JJ, Rocha OJ, Meyer JY, Westcott D, Wieczorek AM (2012) Getting here from there: testing the genetic paradigm underpinning introduction histories and invasion success. Divers Distrib 18:147–157

Harlan JR, de Wet JMJ (1963) The compilospecies concept. Evolution 17:497–501

Hörandl E, Dobes C, Lambrou M (1997) Chromosomen-und Pollenuntersuchungen an osterreichischen Arten des apomiktischen *Ranunculus auricomus*-Komplexes. Bot Helv 107:195–209

Hörandl E, Greilhuber J (2002) Diploid and autotetraploid sexuals and their relationship to apomicts in the *Ranunculus cassubicus* group: insights from DNA content and isozyme variation. Plant Syst Evol 234:85–100

Hörandl E, Hojsgaard D (2012) The evolution of apomixis in angiosperms: a reappraisal. Plant Biosyst 146:681–693

Hojsgaard D, Hörandl E (2015a) A little bit of sex matters for genome evolution in asexual plants. doi:10.3389/fpls.2015.00082

Hojsgaard D, Hörandl E (2015b) Apomixis as a facilitator of range expansion and diversification in plants. In: From Genotype to phenotype. Springer International Publishing, Switzerland. doi:10.1007/978-3-319-19932-0_16

Hojsgaard D, Klatt S, Baier R, Carman JG, Hörandl E (2014a) Taxonomy and biogeography of apomixis in angiosperms and associated biodiversity characteristics. Crit Rev Plant Sci 33:414–427

Hojsgaard DH, Greilhuber J, Pellino M, Paun O, Sharbel TF, Hörandl E (2014b) Emergence of apospory and by pass of meiosis via apomixis after sexual hybridization and polyploidization. New Phytol 204:1000–1012

Hörandl E (2009) A combinational theory for maintenance of sex. Heredity 103:445–457

Jessup RW, Burson BL, Burow GB, Wang YW, Chang C, Li Z, Paterson AH, Hussey MA (2002) Disomic inheritance, suppressed recombination, and allelic interactions govern apospory in buffel grass as revealed by genome mapping. Crop Sci 42:1688–1694

Kantama L, Sharbel TF, Eric Schranz M, Mitchell-Olds T, De Vries S, De Jong H (2007) Diploid apomicts of the *Boechera holboelli* complex display large scale chromosome substitutions and aberrant chromosomes. Proc Natl Acad Sci USA 104:14026–14031

Kindiger B, Sokolov V, Dewald C (1996) A comparison of apomictic reproduction in eastern gamagrass (*Tripsacum dactyloides* L.) and maize-*Tripsacum* hybrids. Genetica 97:103–110

Koch MA, Dobeš C, Mitchell-Olds T (2003) Multiple hybrid formation in natural populations: concerted evolution of the internal transcribed spacer of nuclear ribosomal DNA (ITS) in North American *Arabis divaricarpa* (Brassicaceae). Mol Biol Evol 20:338–350

Koltunow AM, Ozias-Akins P, Siddiqi I (2013) Apomixis. In: Becraft PW (ed) Seed genomics. Wiley, London, pp 83–110

Krahulcová A, Rotreklová O, Krahulec F, Rosenbaumová R, Plačková I (2009) Enriching ploidy level diversity: the role of apomictic and sexual biotypes of *Hieracium* subgen. *Pilosella* (Asteraceae) that coexist in polyploid populations. Folia Geobot 44:281–306

Labombarda P, Busti A, Caceres ME, Pupilli F, Arcioni S (2002) An AFLP marker tightly linked to apomixis reveals hemizygosity in a portion of the apomixis-controlling locus in *Paspalum simplex*. Genome 45:513–519

Leblanc O, Grimanelli D, González-de-León D, Savidan Y (1995) Detection of the apomictic mode of reproduction in maize-*Tripsacum* hybrids using maize RFLP markers. Theor Appl Genet 90:1198–1203

Mangelsdorf PC, Reeves RG (1931) Hybridization of maize, *Tripsacum* and *Euchlaena*. J Hered 22:329–343

Marimuthu MPA, Jolivet S, Ravi M, Pereira L, Davda JN et al (2011) Synthetic clonal reproduction through seeds. Science 331:876

Markwelch D, Meselson M (2000) Evidence for the evolution of *Bdelloid rotifers* without sexual reproduction or genetic exchange. Science 288:(5468)1211–1215

Martinez EJ, Hopp HE, Stein J, Ortiz JPA, Quarin CL (2003) Genetic characterization of apospory in tetraploid *Paspalum notatum* based on the identification of linked molecular markers. Mol Breed 12:319–327

Martinez EJ, Urbani MH, Quarin CL, Ortiz JPA (2001) Inheritance of apospory in bahiagrass, *Paspalum notatum*. Hereditas 135:19–25

Mehra KL (1961) Chromosome number, geographical distribution and taxonomy of the *Dichanthium annulatum* complex. Cytologia 17:176

Miles JW, Escandon ML (1997) Further evidence on the inheritance of reproductive mode in *Brachiaria*. Can J Plant Sci 77:105–107

Mitsuyuki MC, Hoya A, Shibaike H, Watanabe M, Yahara T (2014) Formation of a hybrid triploid agamosperm on a sexual diploid plant: evidence from progeny tests in *Taraxacum platycarpum*. Plant Syst Evol 300:863–870

Mogie M, Ford H (1988) Sexual and asexual *Taraxacum* species. Biol J Linnean Soc 35:155–168

Moreno-Perez E, Garcia-Velazquez A, Avendano-Arrazate CH (2009) Estudio citolo´gico en poblaciones diploides y poliploides del ge´nero *Tripsacum*. Interciencia 34:791–795

Nakano M, Shimada T, Endo T, Fujii H, Nesumi H, Kita M, Ebina M, Shimizu T, Omura M (2012) Characterization of genomic sequence showing strong association with polyembryony among diverse *Citrus* species and cultivars, and its synteny with *Vitis* and *Populus*. Plant Sci 183:131–142

Naumova TN, Hayward MD, Wagenvoort M (1999) Apomixis and sexuality in diploid and tetraploid accessions of *Brachiaria decumbens*. Sex Plant Reprod 12:43–52

Nogler GA (1984) Gametophytic apomixis. In: Johri BM (ed) Embryology of angisoperms. Springer, Berlin, pp 475–518

Norrmann GA, Quarín CL, Burson BL (1989) Cytogenetics and reproductive behavior of different chromosome races in six *Paspalum* species. J Heredity 80:24–28

Noyes RD, Baker R, Mai B (2007) Mendelian segregation for two-factor apomixis in *Erigeron annuus* (Asteraceae). Heredity 98:92–98

Noyes RD, Rieseberg LH (2000) Two independent loci control agamospermy (apomixis) in the triploid flowering plant *Erigeron annuus*. Genetics 155:379–390

Ogawa D, Johnson SD, Henderson ST, Koltunow AM (2013) Genetic separation of autonomous endosperm formation (AutE) from two other components of apomixis in *Hieracium*. Plant Reprod 26:113–123

Ohad N, Yadegari R, Margossian L, Hannon M, Michaeli D, Harada JJ, Goldberg RB, Fischer RL (1999) Mutations in *FIE*, a WD polycomb group gene, allow endosperm development without fertilization. Plant Cell 11:407–416

Ortiz JPA, Quarin CL, Pessino SC, Acuna CA, Martinez EJ, Espinoza F, Hojsgaard DH, Sartor ME, Caceres ME, Pupilli F (2013) Harnessing apomictic reproduction in grasses: what we have learned from *Paspalum*. Ann Bot (Lond) 112:767–787

Otto SP, Whitton J (2000) Polyploidy incidence and evolution. Annu Rev Genet 34:401–437.

Ozias-Akins P, Lubbers EL, Hanna WW, Mc Nay JW (1993) Transmission of the apomictic mode of reproduction in *Pennisetum*: co-inheritance of the trait and molecular markers. Theor Appl Genet 85:632–638

Ozias-Akins P, van Dijk PJ (2007) Mendelian genetics of apomixis in plants. Annu Rev Genet 41:509–537

Ozias-Akins PE, Roche D, Hanna WW (1998) Tight clustering and hemizygosity of apomixis linked molecular markers in *Pennisetum squamulatum* implies genetic control of apospory by a divergent locus that may have no allelic form in sexual genotypes. Proc Natl Acad Sci USA 95:5127–5132

Pellino M, Hojsgaard D, Schmutzer T, Scholz U, Hörandl E, Vogel H et al (2013) Asexual genome evolution in the apomictic *Ranunculus auricomus* complex: examining the effects of hybridization and mutation accumulation. Mol Ecol 22:5908–5921.

Pernès J (1975) Organization evolution d'un groupe agamique: la section des Maximae du genre. *Panicum* (Gramineae). ORSTOM, Paris, pp 1–106

Pessino SC, Evans C, Ortiz JPA, Armstead I, do Valle CB, Hayward MD (1998) A genetic map of the apospory-region in *Brachiaria* hybrids: identification of two markers closely associated with the trait. Hereditas 128:153–158

Pessino SC, Ortiz J, Leblanc O, do Valle CB, Hayward MD (1997) Identification of a maize linkage group related to apomixis in *Brachiaria*. Theor Appl Genet 94:439–444

Peters HA (2001) *Clidemia hirta* invasion at the Pasoh Forest Reserve: an unexpected plant invasion in an undisturbed tropical forest. Biotropica 33:60–68

Petrov DF, Belousov NI, Fokina ES, Laikova LI, Yatsenko RM, Sorokina TP (1984) Transfer of some elements of apomixis from *Tripsacum* to maize. In: Petrov DF (ed) Apomixis and its role in evolution and breeding. Oxonian Press, New Delhi, pp 9–7

Petrov DF, Belousova NI, Fokina ES (1979) Inheritance of apomixis and its elements in maize-*Tripsacum* hybrids. Genetika 15:1827–1836

Porceddu A, Albertini E, Barcaccia G, Falistocco E, Falcinelli M (2002) Linkage mapping in apomictic and sexual Kentucky bluegrass (*Poa pratensis* L.) genotypes using a two way pseudotestcross strategy based on AFLP and SAMPL markers. Theor Appl Genet 104:273–280

Pupilli F, Labombarda P, Caceres ME, Quarin CL, Arcioni S (2001) The chromosome segment related to apomixis in *Paspalum simplex* is homoeologous to the telomeric region of the long arm of rice chromosome 12. Mol Breed 8:53–61

Pupilli F, Martinez EJ, Busti A, Calderini O, Quarin CL, Arcioni S (2004) Comparative mapping reveals partial conservation of synteny at the apomixis locus in *Paspalum* spp. Mol Genet Genom 270:539–548

Quarin CL (1992) The nature of apomixis and its origin in Panicoid grasses. Apomixis News 5:8–15

Quattrocchi U (2008) CRC World dictionary of grasses: common names, scientific names, eponyms, synonyms and etymology. Taylor and Francis Group, Boca Raton, pp 632–637

Randolph LF (1970) Variation among *Tripsacum* populations of Mexico and Guatemala. Brittonia 22:305–337

Ravi M, Chan SWL (2010) Haploid plants produced by centromere-mediated genome elimination. Nature 464:615–618
Ravi M, Marimuthu MPA, Siddiqi I (2008) Gamete formation without meiosis in *Arabidopsis*. Nature 451:1121–1124
Richards AJ (1973) The origin of *Taraxacum* agamospecies. Bot J Linn Soc 66:189–211
Richards AJ (1986) Plant breeding systems. Allen & Unwin, London
Roche D, Chen Z, Hanna W, Ozias-Akins P (2001) Non-Mendelian transmission of an apospory-specific genomic region in a reciprocal cross between sexual pearl millet (*Pennisetum glaucum*) and an apomictic F1 (*P. glaucum* x *P. squamulatum*). Sex Plant Reprod 13:217–223
Roche D, Cong P, Chen ZB, Hanna WW, Gustine DL et al (1999) An apospory-specific genomic region is conserved between buffelgrass (*Cenchrus ciliaris* L.) and *Pennisetum squamulatum* Fresen. Plant J 19:203–208
Roiloa SR, Antelo B, Retuerto R (2014) Physiological integration modifies d15 N in the clonal plant *Fragaria vesca*, suggesting preferential transport of nitrogen to water-stressed offspring. Ann Bot 114:399–411
Savidan Y (1980) Chromosomal and embryological analyses in sexual X apomictic hybrids of *Panicum maximum* Jacq. Theor Appl Genet 57:153–156
Savidan Y (1982) Nature et hérédité de l′apomixie chez *Panicum maximum* Jacq. ORSTOM Travaux et Documentos, 153. ORSTOM, Paris
Savidan Y (2000) Apomixis: genetics and breeding. Plant Breed Rev 18:13–86
Savidan Y, Pernès J (1982) Diploid-tetraploid-dihaploid cycles and the evolution of *Panicum maximum* Jacq. Evolution 36:596–600
Schallau A, Arzenton F, Johnston AJ, Hähnel U, Koszegi D et al (2010) Identification and genetic analysis of the APOSPORY locus in *Hypericum perforatum* L. Plant J 62:773–784
Schranz ME, Dobes C, Koch MA, Mitchell-Olds T (2005) Sexual reproduction, hybridization, apomixis, and polyploidization in the genus *Boechera* (Brassicaceae). Am J Bot 92:1797–1810
Sherwood RT, Berg CC, Young BA (1994) Inheritance of apospory in buffelgrass. Crop Sci 34:149–1494
Skawin′ska R (1963) Apomixis in *Hieracium alpinum* L. Acta Biol Cracov 5:7–14
Smith J (1841) Notice of a plant which produces perfect seeds without any apparent action of pollen. Trans Linnean Soc London 18:509–512
Sochor M, Vašut RJ, Sharbel TF, Trávníček B (2015) How just a few makes a lot: Speciation via reticulation and apomixis on example of European brambles (*Rubus* subgen. *Rubus*, Rosaceae) Mol Phyl Evol 89:13–27
Stebbins GL (1950) Variation and evolution in plants. Columbia University Press, New York, p 643
Stein J, Quarin CL, Martinez EJ, Pessino SC, Ortiz JPA (2004) Tetraploid races of *Paspalum notatum* show polysomic inheritance and preferential chromosome pairing around the apospory-controlling locus. Theor Appl Genet 109:186–191
Talent N, Dickinson TA (2005) Polyploidy in *Crataegus* and *Mespilus* (Rosaceae, Maloideae): evolutionary inferences from flow cytometry of nuclear DNA amounts. Can J Bot 83:1268–1304
Tucker MR, Araujo ACG, Paech NA, Hecht V, Schmidt EDL et al (2003) Sexual and apomictic reproduction in *Hieracium* subgenus *Pilosella* are closely interrelated developmental pathways. Plant Cell 15:1524–1537
Urbani MH, Quarin CL, Espinoza F, Penteado MIO, Rodrigues IF (2002) Cytogeography and reproduction of the *Paspalum* simplex polyploid complex. Plant Syst Evol 236:99–105
Valle CB, Glenke C (1993) Towards defining the inheritance of apomixis in *Brachiaria*. Apomixis Newslett 6:24–25
Van Dijk PJ (2003) Ecological and evolutionary opportunities of apomixis: insights from *Taraxacum* and *Chondrilla*. Philos Trans R Soc London Ser B 358:1113–1121
Van Dijk PJ, Bakx-Schotman JMT (2004) Formation of unreduced megaspores (diplospory) in apomictic dandelions (*Taraxacum*) is controlled by a sex-specific dominant gene. Genetics 166:483–492

Van Dijk PJ, Vijverberg K (2005) The significance of apomixis in the evolution of the angiosperms: a reappraisal. In: Bakker F, Chatrou L, Gravendeel B, Pelser PB (eds) Plant species-level systematics: new perspectives on pattern and process (Gantner, Ruggell, Liechtenstein), pp 101–116

Van Oostrum H, Sterk AA, Wijsman HJW (1985) Genetic variation in agamospermous microspecies of *Taraxacum* sect. *Erythrosperma* and sect. *Obliqua*. Heredity 55:223–228

Vijverberg K, Van der Hulst RGM, Lindhout P, Van Dijk PJ (2004) A genetic linkage map of the diplosporous chromosomal region in *Taraxacum officinale* (common dandelion; Asteraceae). Theor Appl Genet 108:725–732

Whitton J, Sears CJ, Bacck EJ, Otto SP (2008) The dynamic nature of apomixis in the angiosperms. Int J Plant Sci 169:169–182

Winkler H (1908) Über Parthenogenesis und Apogamie im Pflanzenreiche. Progr Rei Bot 2: 293–454

Yadav CB, Anuj, Kumar S, Gupta MG, Bhat V (2012) Genetic linkage maps of the chromosomal regions associated with apomictic and sexual modes of reproduction in *Cenchrus ciliaris*. Mol Breed 30(1):239–250

Chapter 5
Status and Opportunities of Molecular Breeding Approaches for Genetic Improvement of Tea

Rahul Kumar, Kuldip, Paramvir Singh Ahuja and Ram Kumar Sharma

Abstract Tea is the most popular perennial plantation crop in the Southeast Asian countries because of its attractive aroma, taste, and health benefits. Tea plantations provides an important agro-based, eco-friendly employment generating and export oriented industries in all the tea-growing countries. However, the future of tea industry depends on the availability of high-yielding and high-quality tea clones with greater tolerance to pest, diseases, and environmental stresses. Genetic improvement of tea involves identification, characterization, evaluation, domestication, maintenance, and utilization of germplasm for the development of superior plant material. Conventional breeding program in tea is, however, limited by long gestation period, outbreeding nature, and self-incompatibility. This chapter summarizes the status of emerging molecular genomic information that can expedite the genetic improvement in tea and hence the productivity too. This will also provide a background for possibilities of modern tea breeding together with some current efforts for the development of sequence-based markers such as microsatellites, single-nucleotide polymorphisms (SNPs) and links genetic diversity of existing gene pools for the identification of diverse parental groups and efficient phenotyping to support operational breeding. Preliminary attempts on quantitative trait locus (QTL) mapping in tea were also reviewed, and prospectives are also provided on power of association genetics to dissect quantitative traits. Challenges and opportunities to integrate advancement and advent of next-generation sequencing (NGS) technologies to generate genome-wide makers and to integrate genomic information into directional selective breeding are also discussed.

Keywords *Camellia sinensis* · Genetic diversity · Linkage mapping · Molecular markers · Simple sequence repeat

---

R. Kumar · Kuldip · P.S. Ahuja · R.K. Sharma (✉)
Biotechnology Division, CSIR-Institute of Himalayan Bioresource Technology, (Council of Scientific & Industrial Research), Post Box 6, Palampur 176061, Himachal Pradesh, India
e-mail: ramsharma@ihbt.res.in

R. Kumar
DAV University, Jalandhar 144012, India

V.R. Rajpal et al. (eds.), *Molecular Breeding for Sustainable Crop Improvement*, Sustainable Development and Biodiversity 11,
DOI 10.1007/978-3-319-27090-6_5

5.1 Introduction

Tea, *Camellia sinensis* (L.) O. Kuntze, is one of the most popular non-alcoholic and low-cost beverages prepared by adding processed leaves of the tea plant in hot water and is consumed by a two-third of the world population. It is believed that tea has been discovered accidentally in 2700 BC. During the same time, the Han emperor defined Ti as buds taken from the plant t'u which is now known as tea or ch'a. It can be, therefore, assumed that tea was being used as a "drink" for nearly 5000 years. Presently, tea is well known for its heritable brew, flavor, taste, and refreshing values worldwide. It provides different types of beverages that include the most popular non-alcoholic black, green, and oolong tea and contribute about 78, 20, and 2 %, respectively (Wu and Wei 2002; Basu 2002–2003). Tea contains catechins, vitamins, amino acids, caffeine, and antioxidant properties which complement it with medicinal values and makes it beneficial for health. Growing wild, a tea plant can reach heights of 15 m (50 feet); however, under cultivation, shrubs are maintained at 60–100 cm for harvesting. Plantations start to flower after three to eight years, and commercial production is mainly through the selection of superior plant materials followed by clone cultivations. Once planted, these plantations can be harvested for more than a 100 years (Fang et al. 2006; Kwong-Robbins 2005; Chen et al. 2008).

5.2 Classification

The tea plant (*Camellia sinensis*) belongs to the family Theaceae in the section Thea (Prince and Parks 2001). Sealy (1958) gave the complete designation as *Camellia sinensis* (L.) O. Kuntze. He further classified the genus into 12 subgeneric sections and reported 82 species in the genus *Camellia.* Chang and Bartholomew (1984), however, recognized over 200 species after revising earlier classifications. Some other taxonomic classification systems have also been proposed in the recent past by Chang (1981, 1998) who recognized 284 species and Ming (2000) who identified 119 species. All the proposed classifications so far are summarized in Table 5.1.

Table 5.1 Existing classifications of *Camellia* complex

Classification systems	Species proposed
Sealy (1958)*	12 sub-subgenera; 82 species
Wight (1962)*	82 species
Chang and Bartholomow (1984)	200 species
Chang (1981, 1998)	200 species
Ming (2000)	119 species

*Sealy (1958) and White (1962) have not designated several intergrades as separate species, which is the most acceptable

5.3 Tea Cytogenetics

Tea, generally a diploid plant species has clumpy chromosomes with $2n = 2x = 30$ (Longley and Tourje 1959; Bezbaruah 1971) and the estimated haploid genome size is 4.0 Gb (Tanaka and Taniguchi 2007). However, few polyploids, such as triploids [TV-29, HS-10 A, UPASI-3, UPASI-20 ($2n = 3x = 45$)], tetraploids ($2n = 4x = 60$), pentaploids ($2n = 5x = 75$), and aneuploids ($2n \pm 1$ to 29), have also been reported (Devarumath et al. 2002; Singh 1980; Zhan et al. 1987). *C. sasanqua* has been shown to form stable polyploidy series of tetraploid ($2n = 4x = 60$) and hexaploid ($2n = 6x = 90$) plants, whereas triploids ($2n = 3x = 45$) are also identified in *C. rosiflora* (Bezbaruah 1971). Additionally, some natural polyploids have also been reported in tea (Simura 1935; Bezbaruah 1971; Wachira and Kiplang'at 1991). Although these polyploids have desirable characters like high vigor and resistance to various environmental stresses, particularly winter hardiness, they do not always contribute to high yield (Bezbaruah 1968). Karyotypic studies revealed that the tea accessions/clones belonging to the three main races (Assam, China, and Cambod) of tea are diploid and did not show any major differences in their gross morphology and karyotype, although minor differences existed (Bezbaruah 1971). The meiotic divisions in Assam and hybrid teas were regular with 15 bivalents and regular segregation. However, hybrids showed several abnormalities in the pachytene chromosome, suggesting differences in the gene content. Banding techniques and differential staining of chromosomes to study the karyotypic differences within and between the species are of immense importance in the future cytological studies in tea.

5.4 Tea Gene Bank in India

For the maintenance of a heterogeneous gene pool at the national level, regional germplasm collection centers have been established in predominant tea-growing areas. Tea germplasm is now maintained at the national level at three different geographical locations, which include Tea Research Association, Tocklai Experimental Station (TRA, TES), Jorhat, Assam, northeastern India; United Planters' Association of Southern India (UPASI), Valparai, Coimbatore, South India, and CSIR Institute of Himalayan Bioresource Technology (IHBT), Palampur, which is known for collections from northwestern India. It is estimated that presently more than 2532 accessions are held at TRA, TES, Jorhat, Assam (Singh 1999). Large numbers of commercially important accessions are also available at germplasm repositories at UPASI and CSIR-IHBT. The number apparently is quite impressive, but when one critically considers the number of collections for a species like tea, which is self-incompatible, outcrossing, and the rate at which the germplasm is being lost, one can realize that the representation is far from adequate.

Considering the facts, a selection scheme was started during 1971–1972 to collect and conserve the valuable and diverse tea genotypes from old seed–grown sections in commercial tea plantation, which may be considered as "gold mine" for future genetic improvement of tea.

Considering the cross-compatibility across the *Camellia* species, cultivated species of tea plants bred with wild *Camellia* species, namely *C. flava*, *C. petelotii*, and *C. lutescens* (Wight 1962; Sharma and Venkataramani 1971). Additionally, *C. irrawadiensis* and *C. taliensis*, having morphological proximity with cultivated tea, have also contributed to the existing gene pool of tea (Banerjee 1992). Therefore, tea cultivated at the national level is highly heterogeneous, having several intergrades, introgressants, and putative hybrids. Tea quality and yield largely remain the major global breeding objectives; however, depending on the local needs, type of manufactured tea, and climatic zone, the breeding objectives vary significantly in different tea-producing countries (Mondal 2009). While the breeding objectives of black tea-producing countries, namely India, Kenya, Sri Lanka, Bangladesh, and Indonesia, are biased toward the development of high-yielding and high-quality clones, countries away from the equator, such as Japan and China, are focused on the development of cold- and frost-resistant varieties. Horizontal increase of production by extension planting is a prime focus of tea breeding worldwide. Additionally, significant efforts have also been made to combat abiotic and biotic stresses in tea. Among the abiotic stresses, drought has significantly affected the tea productivity in all tea-growing regions of the world. In India, few drought-tolerant clones/accessions (TV-1, TV-17, TV-19, TV-20, UPASI-2, UAPSI-9, UPASI-20, ATK-1, BSS2) have been recommended for drought-prone area. Another important trait is water logging, which limits and reduces productivity during rainy season mainly in northeastern India. Winter dormancy is another abiotic stress causing no leaf production during winter months. Winter dormancy also causes significant annual production loss in Japan and China. In India, tea accessions/clones, namely C-1, CR-6017, UPASI-15, UPASI-16, and UPASI-19 and some new collections from the abandoned tea gardens of North-East India are recommended for frost-prone areas.

Among the biotic stresses, blister blight caused by the pathogen *Exobasidium vexans* Wasee is one of the most serious diseases of tea (Arulpragasam 1992), which inflicted severe crop loss in countries like Sri Lanka, Indonesia, and India. Yield loss and quality are significantly affected by disease incidence (Gulati et al. 1993). Even though the nature and basis of resistance of the disease are not known, certain South Indian tea clones are known to manifest resistance to blister blight. Debnath and Paul (1994) observed no correlation between anatomical and morphological characters of 17 susceptible tea clones with various levels of disease severity. However, pre- and postinflectional biochemical and physical changes in the host plants play a vital role in triggering mechanism that imparts resistance to the disease. Jayaramraja et al. (2006) reported that a popular South Indian tea clone SA-6 is found to be highly resistant to blister blight infection, while the TES-34 was reported to be highly susceptible. Higher amount of epicuticular wax, thickened epidermis, upper and lower cuticles, stomatal frequency on both the surfaces, and compactness of the palisade cells might have cumulative effects in providing resistance against blister blight infection in SA-6. Further, chitinase assay as well as Western blotting studies confirmed that the constitutive level of chitinase expression was higher in the resistant clones when compared with the susceptible tea clone. Increased content of epicatechin and lower concentration of epigallocatechin were also observed in tea cultivars resistant to blister blight leaf disease in comparison with susceptible cultivars. The resistance of TRI 2043, the purple green-leafed cultivar, is attributed to the additional catechin source provided by the presence of high levels of anthocyanins.

5.5 Commercial Tea Production

Botanically, only *Camellia sinensis* var *sinensis*—"China type," *C. sinensis* var *assamica* as *C. assamica* (Masters) Kitamura—"Assam type," and *Camellia assamica* ssp. *lasiocalyx* (Masters) Wight—"Cambod type," have contributed significantly to the entire genetic pool of cultivated tea worldwide. However, apart from these, several other *Camellia* species are being used as beverages in parts of China and Indo-China, including *C. taliensis*, *C. irrawadiensis*, *C. gradibractiata*, *C. kwangsiensis*, *C. gymnogyna*, *C. crassicolumna*, *C. tachangensis*, and *C. ptilophylia* (Chang and Bartholomew 1984; Banerjee 1992). Some others like *C. fraternal* are exploited for seed oil, which are used for cooking and also in pharmaceutical and cosmetic industries. Additionally, many *Camellia* species like *C. japonica, C. rosiflora, and C. irrawadiensis* are of great ornamental value with beautiful leaves and flower structures, besides some desirable traits combined with commercially important *camellia* species through hybrid breeding.

Morphologically, *C. sinensis* var. *sinensis*, the China-type tea is dwarf and slow-growing shrub, tolerant to a cold climate. The relatively small, thick, and leathery leaves have stomata that appear to be sunken in the lamina. Petioles are short and stout, 3–7 in numbers, and provide the leaf an erect stature. The flowers can be born singly or in pairs in the axils having 6–10 mm long pedicels with 2–3 subopposite scars. The flowers can be further characterized by 7–8 cup-shaped, 1.5–2.0 cm long broad oval to suborbicular petals with about 3–5 styles that are generally free for greater part of their length but occasionally free up to the base of the ovary. The capsules have 1, 2, or 3 locules, each containing 1–3 nearly spherical seeds of about 10–15 mm diameter. Assam type (*C. assamica*) is a large, tall, and quick-growing tree which prefers a semitropical climate, indigenous to the Assam region in India (Kaundun and Matsumoto 2003). These varieties have a higher caffeine and catechin content than the China varieties. The Assam tea varieties are having large, thin, and glossy leaf with more or less acuminate apex, very distinct marginal veins, and broadly elliptic leaf blades that are usually 8–20 cm long and 3.5–7.5 cm wide. The Cambod or Indo-China (*Camellia assamica* ssp *lasiocalyx*) is thought to be an intermediate between the China- and Assam-type tea. This is an upright tree (6–10 m tall), with several, almost equally developed branches and more or less erect, glossy, light green to coppery-yellow or pinkish red leaves, the size of which is intermediate between *C. sinensis* and *C. assamica*. Although the flowers are more or less similar to the Assam type yet, they have 4 or more bracteoles, 3–4 ovules with 5 locules, and 3–5 styles that are free up to half the length.

5.6 Conventional Breeding Efforts

Tea is a cross-pollinated plant; hence, the progenies obtained are highly heterogeneous in nature. Selfing either does not set seeds or results into a very low germination ability. Development of pure lines in perennial crop like tea is time-consuming and attributed to practical difficulties. Self- and cross-pollination experiments carried out in 40 years at Tocklai give a clear picture of the extent of pollination behavior in all three commercial tea plants and their progenies. Background knowledge of improved seed stocks, selection of parental groups, and their genetic constituents is among the major requirement of successful breeding in tea. In contrast, clonal selection and multiplication are comparatively easy which does not require any knowledge of plant genetic makeup. Better understanding of the factors involved in flowering and fruit set in tea will be helpful in tea-breeding program.

Conventional tea breeding through seeds, cuttings, and grafting is although well established and has contributed significantly to tea improvement over the past several decades, but proved to be time-consuming and labor-intensive. Further, perennial nature, long gestation periods, high inbreeding depression, self-incompatibility, unavailability of distinct mutant of different biotic and abiotic stresses, low success rate of hand pollination, short flowering time (2–3 months), long duration for seed maturation (12–18 months), and clonal difference of flowering time and fruit-bearing capability are among the major bottlenecks for the implementation of conventional breeding in tea.

5.7 DNA Markers and Their Utilization for Tea Genetic Diversity Characterization

Three types of genetic markers, namely morphological markers, protein-based markers, and DNA-based markers, have been used in genomic diversity analysis and plant breeding. Variations among genotypes within a species are the raw materials for genomic analysis. To get a genetic marker, the marker locus has to show experimentally detectable variation (polymorphism) among individuals in the test population. The variations can be considered at different biological levels, from the simple heritable phenotype to the detection of variation at the single nucleotide. Once the variation is identified and the genotypes of all individuals are known, the frequency of recombination events between loci is used to estimate linkage distances between markers (Liu 1998). According to Liu (1998), genomic analysis using genetic markers should be based on well-established genetic models. If the underlying genetics of a marker is not clear, then the analysis may be misleading. More importantly, the marker assay should be repeatable at different times in the same or different laboratories.

The application of DNA markers in crop breeding includes the DNA polymorphic assays for genetic mapping, marker-assisted plant breeding, genomic analysis, parasite diagnosis, and genotyping (Mignouna et al. 1996). These molecular technologies include restriction fragment length polymorphism (RFLP), randomly amplified polymorphic DNA (RAPD), amplified fragment length polymorphism (AFLP), simple sequence repeat polymorphism (SSRP), sequence characterized amplified regions (SCARs), sequence tagged sites (STSs), cleaved amplified polymorphic regions (CAPs), single-strand conformational polymorphism (SSCP), double-strand conformational polymorphism (DCSP), and single-nucleotide polymorphism (SNP). Since different DNA marker technologies detect different types of variations, the choice of marker(s) is crucial. The polymerase chain reaction (PCR)-based DNA amplification techniques are in general more advantageous than classical RFLP markers, but have limitations. The two most important characteristics of good markers include the multiplex ratio (the number of markers that can be generated in a single reaction) and the polymorphism information content (the effective number of alleles that can be detected per marker in a set of individuals) (Gysel et al. 1996).

5.7.1 Randomly Amplified Polymorphic DNA (RAPD)

The RAPD technique is based on the amplification of random DNA sequences by the PCR with arbitrary primers. The assay is technically simple and fast and requires only small quantities of DNA preparations (Williams et al. 1990). The RAPD markers have been widely used in tea for the construction of linkage maps (Hackett et al. 2000) and estimation of genetic diversity and population differentiation (Chalmers et al. 1992; Sharma et al. 1995; Wachira et al. 1995, 2001; Wachira 2002). The advantage of RAPD over RFLP is of comparatively high multiplex ratio and can yield up to 20 informative markers per primer (Powell et al. 1996). However, considering sensitivity of in repeated experiments, RAPD markers require optimization amplification conditions to avoid inconsistencies (Williams et al. 1993). Moreover, Mendelian dominant genetic inheritance and homoplasy (indistinguishable fragments size may not represent homologous loci, Weising et al. 2005) limits the utility of RAPD markers in comparative mapping projects.

5.7.2 Amplified Fragment Length Polymorphism (AFLP)

The AFLP technique combines the robustness of RFLP analysis with the reliability of stringent PCRs and allows the simultaneous amplification of a very large number of fragments of which 10–50 may be polymorphic, depending on the genome size and target species. The total number of fragments amplified can be adjusted by altering the total number of arbitrary nucleotides used in the primers. Additionally, AFLP allows dominant as well as codominant markers to be analyzed. However, scoring different alleles of a particular locus is not obvious, implying that information content is rather low, but this problem is overcome by other molecular marker systems, such as the simple sequence repeat (SSR). AFLP has been used to establish the extent of genetic variation and population differentiation in different plant species including tea (Vos et al. 1995; Paul et al. 1997; Loh et al. 2000; Hackett et al. 2000; Wachira et al. 2001; Sharma et al. 2010; Raina et al. 2012). AFLP is particularly useful owing to their high multiplex ratio (Wachira et al. 2001). AFLP fingerprinting of commercially important Indian tea accessions (123 accessions/clones) revealed that most of the diversity is restricted in among tea populations [85 % genetic variations (Fig. 5.1)]. Cluster analysis, principal component analysis, and analysis of molecular variance (AMOVA) detected only a limited genetic variation (15 %) among the populations, suggesting their origin from a similar genetic pool (Sharma et al. 2010).

Further, as a largest diversity characterization efforts, for the first time, AFLP analysis of 1644 accessions and clones of Indian tea with known 15 morphotypes revealed that tea cultivated at the Indian subcontinent is highly heterogeneous and at least six gene pools can be predicted based on PCoA and NJ analyses (Fig. 5.2), with one group in each, constituted mostly by China hybrid, Assam–China hybrid, and Assam hybrid morphotypes, of distinct genetic identity (Raina et al. 2012).

5.7.3 Simple Sequence Repeat (SSR) Marker Resource in Tea

The ubiquity of microsatellite or SSRs in eukaryotic genomes (Hamada et al. 1982; Tautz and Renz 1984) and their usefulness as genetic markers have been well established over the last decade. Microsatellites are motifs of 2–6 bp in length (Jacob et al. 1991) and mainly characterized by high frequency, codominance,

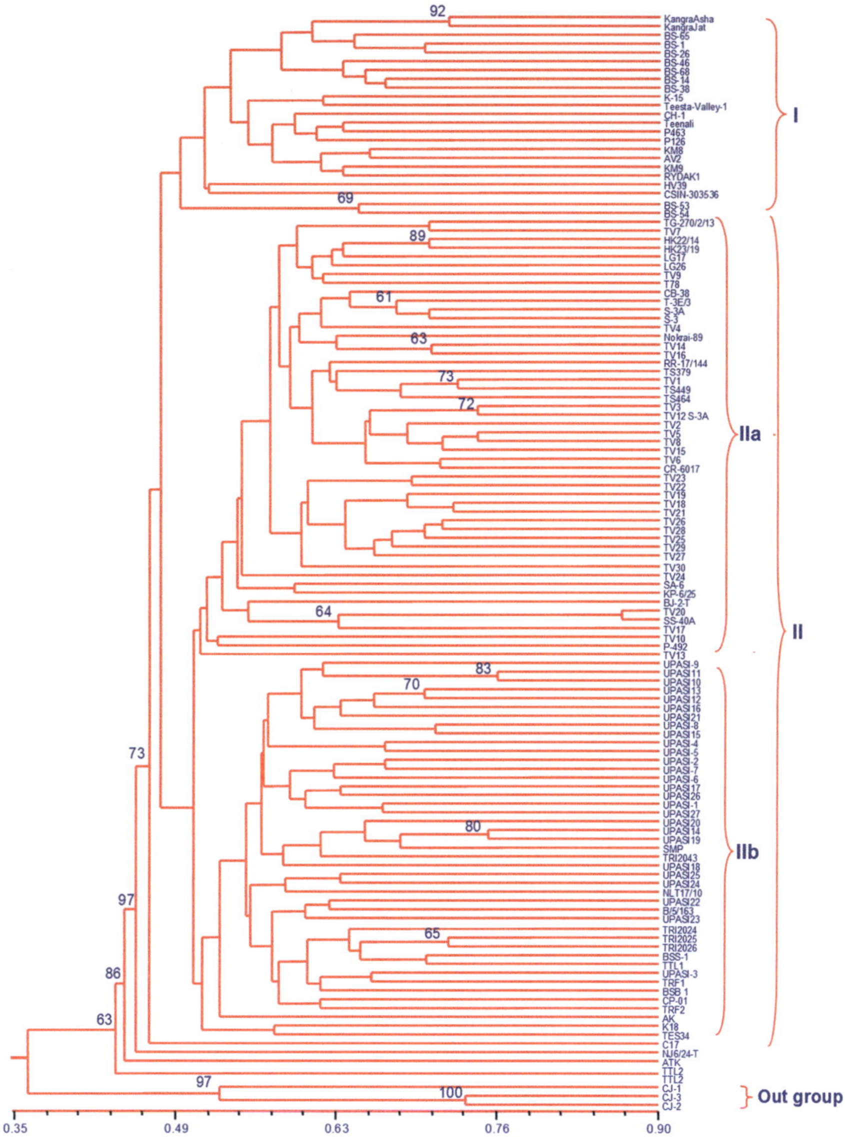

Fig. 5.1 AFLP-based genetic relationships of 123 tea accessions. Bootstrap values greater than 60 % are indicated (Sharma et al. 2010)

multiallelic nature, reproducibility, extensive genome coverage, and ease of detection by PCR with unique primer pairs that flank the repeat motif (Gupta and Varshney 2000). As a result of these characteristics, microsatellites have become the most favored genetic markers for plant breeding and genetic applications such as assessment of genetic diversity, constructing framework genetic maps,

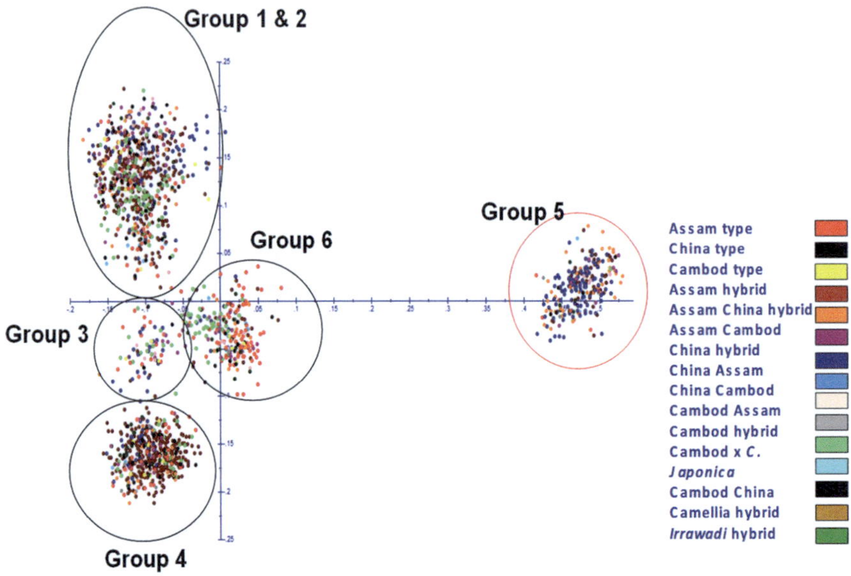

Fig. 5.2 Principal coordinate analysis of 1644 tea accessions based on AFLP analysis (Raina et al. 2012)

marker-aided selection, and comparative mapping studies (Wu and Tanksley 1993; Powell et al. 1996). Owing to their codominant nature, SSRs are particularly useful in determining parentage.

In general, SSRs are identified from either genomic DNA or cDNA sequences. The standard method for developing microsatellite markers involves the creation of small-insert genomic library, subsequent hybridization with tandem repeated oligonucleotides, sequencing of candidate clones, and designing of primers flanking the SSRs (Kijas et al. 1994; Edward et al. 1996). Different enrichment strategies have been used by different workers to develop SSR markers. Freeman et al. (2004) identified 4 marker, but later on, a set of 11 SSR markers were developed through PCR-based isolation of microsatellite array (PIMA)-based method (Hung et al. 2008). However, comparatively larger repertoire of 150 novel genomic microsatellite markers was identified from (GA)n-enriched genomic libraries of Assam tea (Bhardwaj et al. 2013). Availability and continuous enrichment of expressed sequence tags (ESTs) database in tea can serve as the alternative strategy for the identification and development of microsatellite markers. There are about 212 SSR markers, including genomic and EST-based SSRs, reported in tea (Zhao et al. 2008; Ma et al. 2010; Sharma et al. 2009, 2011a, b; Yang et al. 2009). SSR markers developed in the recent past reported high level of heterozygosity in tea and these marker loci recorded very high transferability rate in related *Camellia* spp; therefore, they have wider utility for genotyping studies in tea complex (Sharma et al. 2009, 2011a, b). In the recent past, with the help of different next-generation sequencing

Table 5.2 Status of microsatellite marker development in tea

Approach	No. of SSR markers	Reference
Enriched genomic libraries	15	Freeman et al. (2004)
	158	Bhardwaj et al. (2013)
Public EST database	112	Sharma et al. (2011a, b)
	74	Ma et al. (2010)
	61	Sharma et al. (2009)
	24	Zhao et al. (2008)
	10	Ueno and Tsumura (2009).
PCR amplification of microsatellite array	11	Hung et al. (2008)
Transcriptome data	3767 EST-SSRs (36 validated SSR markers)	Wu et al. (2013)

technologies, efforts have been made for the development of sequence-based maker resources in tea (Wu et al. 2013). Multiple transcriptome sequencing by our laboratory has generated large array of different sequence-based marker resources in tea (unpublished data). SSR markers identified in tea by different groups are summarized in Table 5.2.

5.8 Molecular Breeding Approaches

Marker-assisted selection (MAS) is the selection and development of superior plants using DNA markers. This approach relies on the concept that a particular allele of a gene or multiple alleles are responsible for explaining superior phenotypical diversity. Therefore, the phenotypic characteristic can be predicted very early in the breeding program by the presence of the desired major alleles. An additional advantage of MAS is that the evaluation of the phenotype is based on the genotype and environmental interactions. The regions on the genome where these alleles are located are referred to as quantitative trait loci (QTLs). Agronomical important traits, for example yield and quality, can usually be linked to more than such one gene/QTL and can be identified through genetic analysis. During genetic analysis, a particular size band of a marker can sometimes be associated with morphological traits by coincidence. Because of the long juvenile period and generation interval, it is arguable that MAS has more to offer for the genetic improvement of perennial tree species including tea than for short-lived annuals (Bradshaw 1998). For example, if MAS proves useful not just in identifying clones for propagation in the current generation, but also in choosing superior parents with complementary multilocus genotypes for the next generation, 10 years or more might be saved in each breeding cycle (Bradshaw 1998).

Bulked segregant analysis (Michelmore et al. 1991; Giovanonni et al. 1992) is one of the best mapping approaches for rapidly identifying markers linked to any specific gene or genomic region. Two bulked DNA samples can be generated from a segregating population derived from a single cross. Each pool, or bulk, contains individuals that are identical for a particular trait or genomic region but arbitrary at all unlinked regions. Two bulks are therefore genetically dissimilar in the selected region but seemingly heterozygous at all other regions. Thus, the bulks are homogeneous for a particular trait (e.g., resistant or susceptible to a specific pathogen, respectively), but heterogeneous at all unlinked regions. These bulks are screened for differences using DNA marker approach to identify DNA markers in linked to trait of interest. Any differences between these bulks (e.g., presence vs. absence of a band on a gel) represent a candidate for a marker linked closely to the trait of interest. Since the bulks are supposed to contrast for alleles contributing positive and negative effects, any marker polymorphism between the two bulks indicates the linkage between the marker and trait of interest, which then has ultimately to be verified in a segregating population. BSA is dependent on accurate phenotyping and has been proved popular method in marker-assisted breeding in different plant species, which include lettuce, *Rhycosporium secalis* (Michelmore et al. 1991; Barua et al. 1993; Ardiel et al. 2002). However, this method has been applied mostly to crop plant species and is yet to be widely applied for those crops wherein classical mapping procedures are complicated by large genomes, outbreeding, and long gestation period, such as forest trees. In the first for a tree species, a study by Yang et al. (1997) found a PCR marker, RAPD-OPK/1300, linked to scab resistance gene *Vf* in apples. The marker was further cloned and sequenced leading to the development of a SCAR, which has been used to identify individuals resistant to the disease. Not only in higher eukaryotes like angiosperms but also in other lower eukaryotes like *Saccharomyces cerevisiae*, bulk segregant analysis by high-throughput sequencing has revealed a novel xylose utilization gene (Wenger et al. 2010).

In tea, Kamunya et al. (2010) constructed two bulked DNA samples for yield obtained by using equal amounts of DNA from ten top- and low-performing progeny. Screening of 252 random 10-mer primers, 96 AFLP primer combinations, and 15 SSR primer pairs on the two parents and respective bulks rapidly identify site-specific QTL markers. However, following exploratory genotyping of the individual progeny with informative primers, it was discovered that although some marker loci could discriminate the bulks and their respective parents during the screening process, tea being a outbreeding crop, in most cases the pattern was not reproducible upon genotyping. The patterns were mostly confounded by the appearance of recombinants in either of the two classes. Similar kinds of limitations and problems were also confounded in our study using two bulks for dissecting blister blight disease resistance and susceptible progenies utilizing SSR, RAPD, and AFLP markers.

5.8.1 Genome Mapping and QTL Analysis

In full-sib families of outcrossing species, the construction of genetic maps is far more complicated than that of progenies derived from the intercross of pure lines. The main difference is the number of marker alleles and the segregation pattern of marker genotypes which may vary from locus to locus in outcrossing species (Jansen 2005; Lu et al. 2004; Maliepaard et al. 1997; Van Ooijen 2011). To overcome these problems of linkage analysis in outcrossing species, "two-way pseudo-testcross" strategy as proposed by Grattapaglia and Sederoff (1994) in two parents of an interspecific full-sib cross of *E. grandis* and *E. urophylla* using dominant RAPD markers has been well adopted and is being utilized for linkage mapping in outbred crops like tea. A population of type CP (cross-pollinated) in two-way pseudo-testcross is the product of two independent and distinct meioses, i.e., the meiosis of the two parents. The meioses are distinct in the fact that some loci are segregated in the first parent, some in the second, and the others in both. For the construction of genome map in a pseudo-testcross population, marker loci segregated in 1:1 ratio should be differentiated from those of 3:1. For the population which has been constructed by a cross between two heterozygous parents and subsequently used for genome mapping, it is desirable to screen markers with backcross configurations when we are dealing with dominant markers. By doing this, the well-known problem associated with dominant markers in the repulsion linkage phase in F_2 configuration can be avoided. The relative linkage information content for different markers' system was reviewed as early as in 1956 by Allard. The use of dominant (single-dose) markers that are segregated in "testcross" configuration in heterozygous individuals (Gebhardt et al. 1989; Ritter et al. 1990; Carlson et al. 1991) gave rise to the use of "pseudo-testcross" mapping strategy in several outbred plant species (Sobral and Honeycutt 1993; Hemmat et al. 1994).

The use of dominant RAPD markers in this full-sib family resulted in three types of segregating markers: (a) markers heterozygous in male and homozygous null in female and subsequently segregating in testcross pattern (1:1 segregation ratio) and hence inherited from the male parent, (b) markers heterozygous in female and homozygous null in male and further segregating again in testcross pattern (1:1 segregation ratio), but this time inherited from the female parent, and (c) markers segregating in intercross pattern (3:1 segregation ratio) inherited from both parents. It is obvious from the above classification that based on the parental source of the testcross markers, the first two testcross marker sets are used to construct two different single-tree genetic maps of the two parental trees. The name "two-way pseudo-testcross" was given to the approach because the testcross configuration of individual markers cannot be inferred a priori as in true testcrosses and because the posterior inference has to be extended to both parents (Grattapaglia and Sederoff 1994). In the last two decades, the "two-way pseudo-testcross" mapping approach has been applied in a wide range of forest tree species, initially in conjunction with RAPD or AFLP marker analysis (Verhaegen and Plomion 1996; Marques et al. 1998; Arcade et al. 2000; Lerceteau et al. 2000; Wu et al. 2000; Cervera et al. 2001;

Chagne et al. 2002; Chen et al. 2010) and gradually with codominant SSRs (Brondani et al. 2002, 2006; Shepherd et al. 2003; Freeman et al. 2006; Bundock et al. 2008; Plomion et al. 2011; Alves et al. 2012) as well as by DArT markers (Kullan et al. 2012). In some of such pseudo-testcross studies, it has been possible to use the intercross markers to establish homology or large-scale synteny of the two testcross parental maps (Verhaegen and Plomion 1996; Barreneche et al. 1998; Marques et al. 1998; Wu et al. 2000). However, the size of the informative progenies gets reduced in such cases, and a maximum of only 25 % of mapping progeny could be found informative when dominantly scored intercross markers were used and mapped and resulted in very low power to map such markers in both parental maps (Liu 1998). This problem becomes even more trivial by the relatively low proportion of intercross markers commonly observed in full-sib progenies of forest trees. This problem can now be addressed by the use of codominant markers such as SSRs and gene-based markers (Barreneche et al. 1998; Brondani et al. 1998; Chagne et al. 2002; Yin et al. 2004).

5.8.2 *Molecular Marker Analysis*

Genome map construction is based on genome variation at locations which can be identified by molecular assays or traditional trait observations. Screening polymorphic markers is the first step of an efficient genome map construction. If a marker does not show the polymorphism for the set of progeny, then the marker is found to be non-informative monomorphic marker and hence cannot be included in downstream data analysis. Additionally, some non-parental bands reported in few studies (Hunt and Page 1992; Reineke and Karlovsky 2000; Riedy et al. 1992; Scott et al. 1992) probably represent artifactual heteroduplex molecules (Ayliffe 1994; Novy and Vorsa 1996). Such heteroduplex formation can occur when two allelic DNA segments differing by one or more base substitutions, insertions, and/or deletions are amplified during the PCR. Alternatively, some parental bands may also be observed to be completely absent from the progeny (Halldén et al. 1996; Heun and Helentjaris 1993). These observations are probably a consequence of competition for target sequences.

5.8.3 *Segregation Distortion*

High rate of segregation distortion in paternal markers in "two-way pseudo-testcross mapping" is attributed to the multiple-pollen hypothesis. Segregation distortion is a common phenomenon wherein the genotypic frequency of a locus does not follow a typical Mendelian ratio (Xu and Hu 2009). Segregation distortion (between 25 and 35) has been found comparable in many studies of tea (26.5–32.9 % at $P < 0.01$) (Hackett et al. 2000; Huang et al. 2005, 2006). The percentage of loci showing significant segregation distortion varies greatly, depending on the plant species, population type, and specific cross (Cloutier et al. 1997). Higher segregation distortion is usually found in distantly related species or subspecies (Jenczewski et al. 1997; Virk et al. 1998). Possible explanations of segregation distortion include chromosomal rearrangements, gametic or zygotic selection (Perfectti and Pascual 1996), and inbreeding depression (Husband and Schemske 1996). However, causes of the higher distorted segregation for tea are still unknown. Segregation distortion is known to have a bias estimation of recombination fractions, either overestimation (Lashermes et al. 2001) or underestimation (Cloutier et al. 1997), and it may affect distance estimation and the order of markers in linkage groups (Van Os 2005). As a result, markers showing obvious segregation distortion can be excluded from the map.

5.8.4 Linkage Mapping and Marker Trait Association in Tea

Linkage map may be thought of as a "road map" of the chromosomes derived from two different parents. Linkage maps indicate the position and relative genetic distances between markers along chromosomes, which are analogous to signs or landmarks along a highway. The most important use of linkage maps is to identify chromosomal locations containing genes and QTLs associated with traits of interest (Collard et al. 2005).

As in other plant species, the future of tea industry depends on the availability of high-yielding and quality tea genotypes tolerant to biotic and abiotic stresses. Highly heterogeneous gene pool due to outbreeding nature coupled with long gestation period (Sharma et al. 2009, 2010; Raina et al. 2012; Karthigeyan et al. 2008; Bhardwaj et al. 2014), genetic improvement of tea in India is largely attributed to development and selection of high-yielding better quality tea clones. However, such selections are largely offset due to gradual replacement of most of the pioneer plantations, therefore leading to narrow genetic base. Further, considering harvestable yield of tea which is confined to the terminal two top leaves and a bud, leaf characteristics are therefore considered as the basis for the selection of the two most important agronomic genotypes with higher yield and quality. However, these traits require several years to develop stable traits, irrespective of the environment and climate changes. Thus, the selection based on one or more molecular markers linked to QTLs could be potentially important to expedite the breeding efforts.

Most of the agronomic traits of tea are quantitative in nature and therefore not amenable to easy manipulations in breeding programs without elaborate and long-term field testing in at least more than one environment in order to determine their inheritance, adaptability, and stability. Only a few efforts have been made for QTL mapping in tea. In early 1990s, first linkage map and putative RAPD makers associated with theanine content, date of bud sprouting, resistance to anthracnose, and tolerance to cold were detected (Tanaka 1996). Molecular markers are better utilized in QTL analysis if they are already placed in linkage maps, which enable precise positioning of QTLs. Hackett et al. (2000) using RAPD and AFLP mapped 126 markers (1349.7 cM) spanning an average distance of 11.7 cM between loci on the map. A few linkage maps have been constructed in the recent past in tea by Taniguchi et al. (2012) and Hu et al. (2013). In a RAPD study by Mishra and Sen-Mandi (2004), a band associated with drought tolerance in tea was identified (at 1400 bp with OPAH02 primer). QTL mapping for mulberry scale, MSR-1 (a single dominate gene), has been developed into a MAS marker (Tanaka and Taniguchi 2007). Mulberry scale is a polyphagous scale insect that causes major damage to tea trees in Japan. The development of MAS markers to detect resistance against mulberry scale insects can play a crucial role in tea breeding (Kaneko et al. 2006; Takeda 2004; Tanaka and Taniguchi 2007). Kamunya et al. (2010) developed a map containing 30 (19 maternal and 11 paternal) linkage groups that spanned 1411.5 cM with a mean interval of 14.1 cM between loci. Based on the map, QTL analysis was performed on five-year yield data across the two sites. Twenty-three putative QTLs were detected, 16 in five different linkage groups for Timbilil, two in two groups for Kangaita, and the rest associated with unassigned markers. At least two unassigned markers associated with yield at Kangaita over the whole study period, suggesting potential candidate markers for site-specific MASs (Fig. 5.3).

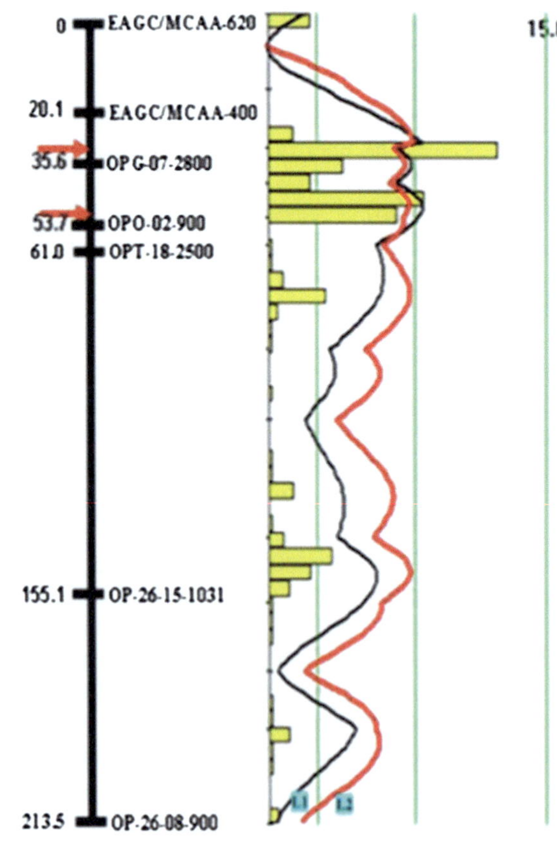

Fig. 5.3 Linkage group 1 depicting the position of multiple-yield QTLs at 2 cM each from marker OPG-70-2800 and OPO-70-90. The LRS line (L1) obtained in composite interval sent the mapping. *Bar* represents the estimated confidence interval by bootstrap resampling in Map Manager QTX (Kamunya et al. 2010)

Recently, QTL analysis using a moderately saturated genetic map of 406 SSR markers in a pseudo-testcross population of 183 individuals identified 25 QTLs associated with catechin content. Nine of them remained stable. QTLs over two measurement years were mapped on LG03, LG11, LG12, and LG15 (Ma et al. 2014). In tea, similar to other out-crossing plant species, a population derived from two known non-inbred parents is scored for any molecular markers like RAPD, AFLP and SSR markers, in order to develop a linkage map. However, a very high proportion of the markers exhibiting unexpected segregation ratios in light of their configurations in the parents have been found, which can most easily be explained by the hypothesis of three male parents contributing pollen to this cross as revealed and discussed by Hackett et al. (2000) wherein linkage genome map with 15 maternal linkage groups with at least three or more markers was constructed. However, paternal map could not be constructed because mix of half-sib populations confirmed multiple-pollen theory.

Parent-wise segregation pattern in the study by Kamunya et al. (2010) was found to be a bit less deviated. The markers from the paternal parent have the highest proportion of distortions, and those from the maternal parent have the least. Hackett et al. (2000) and Kamunya et al. (2010) used 90 and 42 offspring, respectively, for their maternal linkage map. While Hackett et al. (2000) constructed the female map with the help of 112 markers, Kamunya et al. (2010) constructed the map with 100 markers. Other linkage maps in tea have also been created by some workers like Taniguchi et al. (2012) where they used 54 F_1 clones following a three-line reference map approach with the help of AntMap (Iwata and Ninomiya 2006). All the genome and QTL mapping efforts are summarized in Table 5.3.

Table 5.3 Genome and QTL mapping status in tea

QTL/genetic mapping	Marker type	Description	Reference
Theanine content, date of bud sprouting, and tolerance to cold	RAPD		Tanaka (1996)
Genetic map	RAPD and AFLP	126 markers covering 1349.7 cM	Hackett et al. (2000)
Genetic map/yield trait	RAPD, AFLP, SSR, ISSR	Genetic map with spanned 1411.5 cM with a mean interval of 14.7 cM	Kamunya et al. (2010)
		Twenty-three putative yield QTLs for site species mapping	
Genetic map	RAPD, SSR, CAPS, STS	1124 markers with a core map length of 1218 cM	Taniguchi et al. (2012)
Genetic map	RAPD, AFLP, ISSR, STS, SSR, CAPS	367 linked markers covering 4482.9 cM (map density 12.2 cM/marker)	Hu et al. (2013)
Genetic map/catechin	SSR	25 QTLs associated with catechins. 9 are stable throughout the year	Ma et al. (2014)

5.9 Current Opportunities and Future Prospects

Modern high throughput genotyping technologies based on cost effective next generation sequencing, development of genome wide markers and saturated linkage maps will enable us to understand the genetic structures, establishing marker-trait association of agronomic traits and improve crops rapidly and precisely. However, considering genome complexity ($\sim$4 Gb), availability of genetic resources for *C. sinensis* is relatively scarce. Till January 2014, only 159034 nucleotide sequences and 49759 ESTs of *C. sinensis* were available in GenBank. In the past decade, hundreds of SSR markers for *C. sinensis* have been reported and some of these have been genetically mapped (Sharma et al. 2009, 2011a, b; Taniguchi et al. 2012). Still, therefore, there is an urgent need of developing various sequence-based genome-wide coverage makers preferably mapped to tea genome for the effective application of genetic mapping and molecular breeding programs for *C. sinensis* and related species.

NGS platforms (Illumina and 454) are revolutionary techniques that can produce millions of sequences at a relatively low cost compared with traditional methods. Recently, few efforts have been made for tissue- and trait-specific transcriptome sequencing in tea (Shi et al. 2011). These NGS-derived public databases can also be explored for the development and utilization of sequence-based molecular markers like SSRs and SNPs and can fasten the genome mapping and molecular breeding in tea.

5.9.1 Association Genetics and Linkage Disequilibrium Mapping

Quantitative trait marker association based on population-wide linkage disequilibrium has significant potential as a method to identify genetic linkage at higher resolution than traditional mapping allows. Linkage (or gametic phase) disequilibrium (LD) is the non-random association between alleles, usually at linked loci. Association genetics identifies DNA marker alleles that are differentially abundant in the individuals carrying alternative QTL alleles. The principle of association genetics is similar to linkage mapping in a segregating family and to the concept of using markers linked to QTLs to select superior individuals. However, complex genotypic patterns due to mixture of more allelic genotypes than in a typical controlled cross greatly affact resolution and applicability. The resolution of a QTL on a linkage map depends on the size of the segregating population and marker density.

The biology of woody perennials can provide us with indications about nucleotide diversity and LD structure relative to other plant groups. Most tree species are outbreed in nature therefore require large number of loci for the evaluation of LD structure. Tea being a tree crop which is typically outcrossing where pollen and seeds are normally dispersed over long distances influenced by anthropogenic activities is expected to retain high genetic variation and disperse new alleles, reducing the possibility of allele loss by genetic drift. With the availability of NGS technologies, large-scale linkage disequilibrium studies could be initiated to find genome-wide and large-scale marker and trait association in tea.

5.9.2 Genomic Selection

Genomic selection is a form of MAS in which genetic markers covering the whole genome are used so that all QTLs in linkage disequilibrium with at least one marker can be predicted. Simulation results and limited experimental results suggest that breeding values can be predicted with high accuracy using genetic markers alone. Implementation of genomic selection is likely to have major implications for genetic evaluation systems and for genetic improvement programs in tea.

5.10 Conclusions

In conclusion, current genetic improvement programs in tea, needs to fastened, includes saturation of genetic linkage maps using sequence-based SSR and SNP markers and dissection of important traits using well-defined, accurate, and refined populations. Marker and trait association can be achieved through either traditional QTL mapping approach or modern-day association mapping approach which is more suitable for an outbred crop like tea. Marker-assisted selection could be fastened through modern-day tools of genomic selection in tea.

Acknowledgment We acknowledged funding from CSIR-PLOMICS and DBT Tea projects. The Director of CSIR-IHBT is acknowledged for providing necessary facilities. This is IHBT communication No. 3863.

References

Alves A, Rosado C, Faria D, Guimarães L, Lau D, Brommonschenkel S, Grattapaglia D, Alfenas A (2012) Genetic mapping provides evidence for the role of additive and non-additive QTLs in the response of inter-specific hybrids of Eucalyptus to *Puccinia psidii* rust infection. Euphytica 183:27–38

Arcade A, Anselin F, Faivre Rampant P, Lesage MC, Paques LE, Prat D (2000) Application of AFLP, RAPD and ISSR markers to genetic mapping of European and Japanese larch. Theor Appl Genet 100:299–307

Ardiel GS, Grewal TS, Deberdt P, Rossnagel BG, Scoles GJ (2002) Inheritance of resistance to covered smut in barley and development of a tightly linked SCAR marker. Theor Appl Genet 104:457–464

Arulpragasam PV (1992) Disease contrl in Asia. In: Wilson KC, Cliffor MN (eds) Tea cultivation to consumption. Chapman and Hall, London, pp 353–374
Ayliffe MA, Lawrence GJ, Ellis JG, Pryor AJ (1994) Heteroduplex molecules formed between allelic sequences cause nonparental RAPD bands. Nucleic Acids Res 22:1632–1636
Banerjee B (1992) Botanical classification of tea. In: Wilson KC, Clifford MN (eds) Tea: cultivation to consumption. Chapman and Hall, London, pp 25–52
Barreneche T, Bodénès C, Lexer C, Trontin JF, Fluch S, Streiff R, Plomion C, Roussel G, Steinkellner H, Burg K, Favre JM, Glössl J, Kremer A (1998) A genetic linkage map of *Quercus robur* L. (pedunculate oak) based on RAPD, SCAR, microsatellite, minisatellite, isozyme, and 5S rDNA markers. Theor Appl Genet 97:1090–1103
Barua UM, Chalmers KJ, Hackett CA, Thomas WTB, Powell W, Waugh R (1993) Identification of RAPD markers linked to *Rhyncosporium secalis* resistance locus in barley using isogenic lines and bulked segregant analysis. Heredity 71:177–184
Basu B (2002–2003) Drink tea and keep healthy. Int J Tea Sci 2(3):5–7
Bezbaruah HP (1968) Genetic improvement of tea in Northeast India. Its problems as well as possibilities. Indian J Genet 28:126–134
Bezbaruah HP (1971) Cytological investigation in the family Theaceae-I: chromosome numbers in some Camellia species and allied genera. Caryologia 24:421–426
Bhardwaj P, Kumar R, Sharma H, Ahuja PS, Sharma RK (2013) Development and utilization of genomic and genic microsatellite markers in Assam tea (*Camellia assamica* spp. *assamica*) and related *Camellia* species. Plant Breed 132:748–763
Bhardwaj P, Sharma RK, Kumar R, Sharma H, Tewari R, Ahuja PS (2014) SSR marker based DNA fingerprinting and diversity assessment in superior tea germplasm
Bradshaw HD Jr (1998) Case history in genetics of long-lived plants. Molecular approaches to domestication of a fast-growing forest tree. *Populus*. In: Paterson AH (ed) Molecular dissection of complex. CRC Press, New York, pp 219–228
Brondani RPV, Brondani C, Tarchini R, Grattapaglia D (1998) Development, characterization and mapping of microsatellite markers in *Eucalyptus grandis* and *E. urophylla*. Theor Appl Genet 97:816–827
Brondani RPV, Brondani C, Grattapaglia D (2002) Towards a genuswide reference linkage map for Eucalyptus based exclusively on highly informative microsatellite markers. Mol Genet Genomics 267:338–347
Brondani RPV, Williams ER, Brondani C, Grattapaglia D (2006) A microsatellite-based consensus linkage map for species of *Eucalyptus* and a novel set of 230 microsatellite markers for the genus. BMC Plant Biol 6:20
Bundock PC, Potts BM, Vaillancourt RE (2008) Detection and stability of quantitative trait loci (QTL) in *Eucalyptus globulus*. Tree Genet Genomes 4:85–95
Carlson JE, Tulsieram LK, Glaubitz JC, Luk VWK, Kauffeld C, Rutledge R (1991) Segregation of random amplified DNA markers in F1 progeny of conifers. Theor Appl Genet 83:194–200
Cervera MT, Stormel V, Ivens B, Gusmão J, Liu BH, Hostyn V, Van Slyckenc J, Van Montagu M, Boerjan W (2001) Dense genetic linkage maps of three *Populus* species (*Populus deltoides*, *P. nigra* and *P. trichocarpa*) based on AFLP and microsatellite markers. Genetics 158:787–809
Chagne D, Lalanne C, Madur D, Kumar S, Frigerio J-M, Krier C, Decroocq S, Savoure Al, Bou-Dagher-Karrat M, Bertocchi E, Brach J, Plomion C (2002) A high density linkage map for maritime pine based on AFLPs. Ann For Sci 59:627–636
Chalmers KJ, Waugh R, Sprent JI, Powell W (1992) Detection of genetic variation between and within populations of *Gliricidia sepium* and *G. maculata* using RAPD markers. Heredity 69:465–472
Chang HT (1981) Taxonomy of the genus Camellia. Acta Sci Nat Univ Sunyatseni Monogr Ser 1:1–180
Chang HT (1998) Flora of Reipublicae Popularis Sinicae, Delectis Florae Republicae Popularis Sinicae, Agendae Academiae Sinicae Edita. Tomus 49:101–113
Chang HT, Bartholomew B (1984) Camellias. Timber Press, Portland

Chen L, Yu FL, Yao MZ, Lu B, Yang K, Du YY (2008) Preparation of the UPOV guidelines for the conduct of tests for distinctness, uniformity and stability-Tea plant [*Camellia sinensis* (L.) O. Kuntze]. Agric Sci China 7(2):224–231

Chen M-M, Feng F, Suil X, Han S (2010) Genetic linkage maps of *Pinus koraiensis* Sieb. et Zucc. based on AFLP markers. Afr J Biotechnol 9(35):5659–5664

Cloutier S, Cappadocia M, Landry BS (1997) Analysis of RFLP mapping inaccuracy in *Brassica napus* L. Theor Appl Genet 95:83–91

Collard BCY, Jahufer MZZ, Bronwer JB, Pang ECK (2005) An introduction to markers, quantitative trait loci (QTL) mapping and marker-assisted selection for crop improvement. The basic concepts. Euphytica 142:169–196

Debnath S, Paul AK (1994) Susceptibility of tea cultivars to blister blight disease and some of their anatomical and morphological characters. Two Bud 41:48–49

Devarumath RM, Nandy S, Rani V, Marimuthu S, Muraleedharan N, Raina SN (2002) RAPD, ISSR and RFLP fingerprints as useful markers to evaluate genetic integrity of micropropagated plants of three diploid and triploid elite tea clones representing *Camellia sinensis* (China type) and *C. assamica* ssp. Assamica (Assam-India type). Plant Cell Rep 21:166–173

Edwards KJ, Barker JHA, Daly A, Jones C, Karp A (1996) Microsatellite libraries enriched for several microsatellite sequences in plants. Biotechniques 20:758

Fang WP, Jlang CJ, Yu M, Ye AH, Wang ZX (2006) Differentially expression of Tua1, a Tubulin-encoding gene, during flowering of tea plant *Camellia sinensis* (L.) O. Kuntze using cDNA Amplified Fragment Length Polymorphism technique. Acta Biochim Biophys Sin 38 (9):653–662

Freeman S, West J, James C, Lea V, Mayes S (2004) Isolation and characterization of highly polymorphic microsatellites in tea (*Camellia sinensis*). Mol Ecol Notes 4:324–326

Freeman J, Potts BM, Shepherd M, Vaillancourt RE (2006) Parental and consensus linkage maps of Eucalyptus globulus using AFLP and microsatellite markers. Silvae Genetica 55:202–217

Gebhardt CET, Debeneu R, Schachtschabel U, Walkemeier B et al (1989) RFLP analysis and linkage mapping in *Solanum tuberosum*. Theor Appl Genet 78:65–75

Giovanonni J, Wing R, Tanksley SD (1992) Isolation of molecular markers from specific chromosomal intervals using DNA pools from existing mapping populations. Nucleic Acids Res 19:6553–6558

Grattapaglia D, Sederoff R (1994) Genetic linkage maps of *Eucalyptus grandis* and *E. urophylla* using a pseudo-testcross mapping strategy and RAPD markers. *Genetics* 137:1121–1137

Gulati A, Gulati A, Ravindranath SD, Chakrabarty DN (1993) Economic yield losses caused by Exobasidium vexansin tea planta-tions. Indian Phytopathol 46:155–159

Gupta PK, Varshney RK (2000) The development and use of microsatellite markers for genetic analysis and plant breeding with emphasis on bread wheat. Euphytica 113:163–185

Gysel AV, Montagu MA, Breyne P (1996) Applications of AFLP in marker-assisted breeding plant genetics. In: Crouch JH, Tenkouano A (eds) DNA markerassisted improvement of the staple crops of Sub-Saharan Africa. Proceedings of the workshop on DNA markers at IITA held by the crop improvement division, IITA, Ibadan, Nigeria, 21–22 Aug 1996, pp 16–21

Hackett CA, Wachira FN, Paul S, Powell W, Waugh R (2000) Construction of a genetic linkage map for *Camellia sinensis* (tea). Heredity 85:346–355

Halldén C, Hansen M, Nilsson N-O, Hjerdin A, Säll T (1996) Competition as a source of errors in RAPD analysis. Theor Appl Genet 93:1185–1192

Hamada H, Petrino MG, Kakunaga T (1982) A novel repeated element with Z-DNA forming potential is widely found in evolutionarily diverse eukaryotic genomes. Proc Nat Acad Sci USA 79:6465–6469

Hemmat M, Weeden NF, Manganaris AG, Lawson DM (1994) Molecular marker linkage map for apple. J Hered 85:4–11

Heun M, Helentjaris T (1993) Inheritance of RAPDs in F1 hybrids of corn. Theor Appl Genet 85:961–968

Hu C-Y, Lee T-C, Tsai H-T, Tsai Y-Z, Lin S-F (2013) Construction of an integrated genetic map based on maternal and paternal lineages of tea (*Camellia sinensis*). *Euphytica*. doi: 10.1007/s10681-013-0908-0

Huang JA, Li JX, Huang YH, Luo JW, Gong ZH, Liu ZH (2005) Construction of AFLP molecular markers linkage map in tea plant. J Tea Sci 25:7–15

Huang FP, Liang YR, Lu JL, Chen RB (2006) Genetic mapping of first generation of backcross in tea by RAPD and ISSR markers. J Tea Sci 26:171–176

Hung CY, Wang KH, Huang CC, Gong X, Ge XJ, Chiang TY (2008) Isolation and characterization of 11 microsatellite loci from *Camellia sinensis* in Taiwan using PCR-based isolation of microsatellite arrays (PIMA). Conserv Genet 9:945–947

Hunt GJ, Page RE (1992) Patterns of inheritance with RAPD molecular markers reveal novel types of polymorphism in the honey bee. Theor Appl Genet 85:15–20

Husband BC, Schemske DW (1996) Evolution of magnitude and timing of inbreeding depression in plants. Evolution 50:554–570

Iwata H, Ninomiya S (2006) AntMap: Constructing genetic linkage maps using an ant colony optimization algorithm. Breed Sci 56:371–377

Jacob HJ, Lindpainter K, Lincoln SE, Kusumi K, Bunnker RK, Mao YP, Ganter D, Dzau VJ, Lander ES (1991) Genetic mapping of a gene causing hypertension in the stroke-prone spontaneously hypertensive rat. Cell 67:213–224

Jansen J (2005) Construction of linkage maps in full-sib families of diploid outbreeding species by minimizing the number of recombinations in hidden inheritance vectors. Genetics 170:2013–2025

Jayaramraja PR, Pius PK, Manian S, Nithya MS (2006) Certain factors associated with blister blight resistance in *Camellia sinensis* (L.) O. Kuntze. Physiol Molec Plant Patho 67:291–295

Jenczewski E, Gherardi M, Bonnin I, Prosperi JM, Olivieri I, Huguet T (1997) Insight on segregation distortions in two intraspecific crosses between annual species of Medicago (Leguminosae). Theor Appl Genet 94:682–691

Kamunya SM, Wachira FN, Pathak RS, Korir R, Sharma V, Kumar R, Bhardwaj P, Chalo R, Ahuja PS, Sharma RK (2010) Genomic mapping and testing for quantitative trait loci in tea (*Camellia sinensis* (L.) O. Kuntze). Tree Genet Genomes 6:915–929

Kaneko S, Ozawa A, Saito T, Tatara A, Katayama H, Doi M (2006) Relationship between the seasonal prevalence of the predacious coccinellid Pseudoscymnus hareja (Coleoptera: Coccinellidae) and the mulberry scale *Pseudaulacaspis pentagona* (Hemiptera: Diaspididae) in tea fields: monitoring using sticky traps. Appl Entomol Zool 41(4):621–626

Karthigeyan S, Rajkumar S, Sharma RK, Gulati A, Sud RK, Ahuja PS (2008) High level of genetic diversity among the selected accessions of tea (*Camellia sinensis*) from abandoned tea gardens in western Himalaya. Biochem Genet 46:810–819

Kaundun SS, Matsumoto S (2003) Development of CAPS markers based on three key genes of the phenylpropanoid pathway in tea, *Camellia sinensis* (L.) O. Kuntze and differentiation between assamica and sinensis varieties. Tree Genet Genomes 106:375–383

Kijas JMH, Fowler JCS, Garbett CA, Thomas MR (1994) Enrichment of microsatellites from the citrus genome using biotinylated oligonucleotide sequences bound to streptavidin coated magnetic particles. Biotechniques 16:656–662

Kullan ARK, van Dyk MM, Jones N, Kanzler A, Bayley A, Myburg AA (2012) High-density genetic linkage maps with over 2400 sequence-anchored DArT markers for genetic dissection in an F2 pseudo-backcross of *Eucalyptus grandis* × *E. urophylla*. Tree Genet Genomes 8:163–175

Kwong-Robbins C (2005) Tea time: have you had your tea yet? US Pharmacol 10:47–50

Lashermes P, Combes MC, Prakash NS, Trouslot P, Lorieux M, Charrier A (2001) Genetic linkage map of *Coffea canephora*: effect of segregation distortion and analysis of recombination rate in male and female meioses. Genome 44:589–596

Lerceteau E, Plomion C, Andersson B (2000) AFLP mapping and detection of quantitative trait loci (QTLs) for economically important traits in *Pinus sylvestris*: a preliminary study. Mol Breed 6:451–458

Loh PJ, Kiew R, Set O, Gan LH, Gan YY (2000) A study of genetic variation and relationships within the Bamboo sub-tribe *Bambusinae* using amplified fragment length polymorphism. Ann Bot 85:607–612

Longley AE, Tourje EC (1959) Chromosome numbers of certain *Camellia* species and allied genera. *Am Camellia Yb*, 33–39

Lu Q, Cui Y, Wu R (2004) A multilocus likelihood approach to joint modeling of linkage, parental diplotype and gene order in a full-sib family. BMC Genet 5:20

Lui B (1998) Statistical genomics: linkage, mapping and QTL analysis. CRC Press, Boca Raton

Ma JQ, Zhou YH, Ma CL, Yao MZ, Jin JQ, Wang XC, Liang C (2010) Identification and characterization of 74 novel polymorphic EST-SSR markers in the tea plant, *Camellia sinensis* (Theaceae) Am J Bot. doi:10.3732/ajb.1000376

Ma JQ, Yao MZ, Ma CL, Wang XC, Jin JQ, Wang XM, Chen WL (2014) Construction of a SSR-based genetic map and identification of QTLs for catechins content in tea plant (*Camellia sinensis*). PLoS ONE 9(3):93131

Maliepaard C, Jansen J, Van Ooijen JW (1997) Linkage analysis in a full-sib family of an outbreeding species: overview and consequences for applications. Genet Res 70:237–250

Marques CM, Araújo JA, Ferreira JG, Whetten R, O'Malley DM, Liu B-H, Sederoff R (1998) AFLP genetic maps of *Eucalyptus globulus* and *E. tereticornis*. Theor Appl Genet 96:727–737

Michelmore R, Paran I, Kesseli RV (1991) Identification of marker linked to disease resistance gene by bulk segregant analysis: a rapid mathod to detect markers in specific genomic regions using segregating populations. Proc Nat Acad Sci USA 88:9828–9832

Mignouna HD, Fatokun CA, Thottappilly G (1996) Choice of DNA marker system. In: Crouch JH, Tenkouano A (eds) DNA marker-assisted improvement of the staple crops of Sub-Saharan Africa. Proceedings of the workshop on DNA markers at IITA held by the crop improvement division, IITA, Ibadan, Nigeria, 21–22 Aug 1996

Ming T (2000) Monograph of the genus *Camellia*. Chinese Academy of Sciences, Yunnan Science and Technology Press, Kunming Institute of Botany, Kunming

Mishra RK, Sen-Mandi S (2004) Molecular profiling and development of DNA marker associated with drought tolerance in tea clones growing in Darjeeling. Curr Sci 87(1):60–66

Mondal TK (2009) Tea. Compendium of Transgenic. Crop Plant 99–116

Novy RG, Vorsa N (1996) Evidence for RAPD heteroduplex formation in cranberry: implications for pedigree and genetic-relatedness studies and a source of co-dominant RAPD markers. Theor Appl Genet 92:840–849

Paul S, Wachira FN, Powell W, Waugh R (1997) Diversity and genetic differentiation among populations of Indian and Kenyan tea (*Camellia sinensis* (L.) O. Kuntze) revealed by AFLP markers. Theor Appl Genet 94(2):255–263

Perfectti F, Pascual L (1996) Segregation distortion of isozyme loci in cherimoya (*Annona cherimola* Mill). Theor Appl Genet 93:440–446

Plomion C, Bousquet J, Cole CD (2011) Mapping in conifers. In: Genetics, genomics and breeding of conifer trees, Edenbridge Science Publishers and RC Press, New York, pp 196–238

Powell W, Morgante M, Andre C, Hanafey M, Vogel J, Tingey S, Rafalski A (1996) The comparison of RFLP, RAPD, AFLP and SSR (microsatellite) markers for germplasm analysis. Mol Breed 2:225–238

Prince LM, Parks CR (2001) Phylogenetic relationships of Theaceae Inferred from chloroplast DNA sequence. Am J Bot 88(12):2309–2320

Raina SN, Ahuja PS, Sharma RK, Das SC, Bhardwaj P, Negi R, Sharma V, Singh SS, Sud RK, Kalia RK, Pandey V, Banik J, Razdan V, Sehgal D, Dar TH, Kumar A, Bali S, Bhat V, Sharma S, Prasanna BM, Goel S, Negi MS, Vijayan P, Tripathi SB, Bera B, Hazarika M, Mandal AKA, Kumar RR, Vijayan D, Ramkumar S, Chowdhury BR, Mandi SS (2012) Genetic structure and diversity of India hybrid tea. Genet Resour Crop Evol 59:1527–1541

Reineke A, Karlovsky P (2000) Simplified AFLP protocol: replacement of primer labeling by the incorporation of α-labeled nucleotides during PCR. Biotechniques 28:622–623

Riedy MF, Hamilton WJ III, Aquadro CF (1992) Excess of non-parental bands in offspring from known primate pedigrees assayed using RAPD PCR. Nucleic Acids Res 20:918

Ritter E, Gebhardt C, Salamin F (1990) Estimation of recombination frequencies and construction of linkage maps from crosses between heterozygous parents. Genetics 125:645–654

Scott MP, Haymes KM, Williams SM (1992) Parentage analysis using RAPD PCR. Nucleic Acids Res 20:5493

Sealy J (1958) A revision of the genus *Camellia*. Royal Horticultural Society, London

Sharma VS, Venkataramani KS (1971) The tea complex. 1. Taxonomy of tea clones. Proced Indian Acad Sci 53:178–187

Sharma VS, Dawson IK, Waugh R (1995) Relationships among cultivated and wild lentils revealed by RAPD analysis. Theor Appl Genet 91:647–654

Sharma RK, Bhardwaj P, Negi R, Mohapatra T, Ahuja PS (2009) Identification, characterization and utilization of unigene derived microsatellite markers in tea (*Camellia sinensis* L.). BMC Plant Biol 9:53

Sharma RK, Negi MS, Sharma S, Bhardwaj P, Kumar R, Bhattachrya E, Tripathi SB, Vijayan D, Baruah AR, Das SC, Bera B, Rajkumar R, Thomas J, Sud RK, Muraleedharan N, Hazarika M, Lakshmikumaran M, Raina SN, Ahuja PS (2010) AFLP-based genetic diversity assessment of commercially important tea germplasm in India. Biochem Genet 48:549–564

Sharma V, Bhardwaj P, Kumar R, Sharma RK, Sood A, Ahuja PS (2011a) Identification and cross-species amplification of EST derived SSR markers in different bamboo species. Conserv Genet 10:721–724

Sharma H, Kumar R, Sharma V, Kumar V, Bhardwaj P, Ahuja PS, Sharma RK (2011b) Identification and cross transferability of 112 novel unigene derived microsatellite markers in tea (*Camellia sinensis* L). Am J Bot 98:e133–e138

Shepherd M, Cross M, Dieters MJ, Henry R (2003) Genetic maps for *Pinus elliottii* var. *elliottii* and *P. caribaea* var. *hondurensis* using AFLP and microsatellite markers. Tree Genet Genomes 106:1409–1419

Shi CY, Yang H, Wei CL, Yu O, Zhang ZZ, Jiang CJ, Sun J, Li yY, Chen Q, Xia T, Wan XC (2011) Deep sequencing of the *Camellia sinensis* transcriptome revealed candidate genes for major metabolic pathways of tea-specific compounds. BMC Genome 12:131

Simura T (1935) Cytological investigations in tea plant. A preliminary report. Proc Crop Sci Soc Jpn 7:121–133

Singh ID (1980) Tea germplasm in India. Plant Genet Res News Lett 43:12–16

Singh ID (1999) Plant improvement. In: Jain NK (ed) Global advances in tea science, pp 427–448

Sobral BWS, Honeycut RJ (1993) High output genetic mapping of polyploids using PCR-generated markers. Theor Appl Genet 86:105–112

Takeda M (2004) Effects of temperature on oviposition in overwintering females and hatch in first-generation larvae of *Pseudaulacaspis pentagona* (Hemiptera: Diaspididae). Appl Entomol Zool 39(1):15–26

Tanaka J (1996) RAPD linkage map of tea plant and the possibility of application in tea genetics and breeding. Tea Res J 84:44–45

Tanaka J, Taniguchi F (2007) Genome mapping and molecular breeding in plant, vol 6. Chapter 6: Tea. ISBN: 13 978-3-540-34537-4, pp 119–126

Taniguchi F, Furukawa K, Ota-Metoku S, Yamaguchi N, Ujihara T, Kono I, Fukuoka H, Tanaka J (2012) Construction of a high-density reference linkage map of tea (*Camellia sinensis*). Breed Sci 62:263–273

Tautz D, Renz M (1984) Simple sequences are ubiquitous repetitive components of eukaryotic genomes. Nucleic Acids Res 12:4127–4138

Ueno S, Tsumura Y (2009) Development of microsatellite and amplicon length polymorphism markers for Camellia japonica L. from tea plant (Camellia sinensis) expressed sequence tags. Mol Ecol Resour 9:814–816

Van Ooijen JW (2011) Multipoint maximum likelihood mapping in a full-sib family of an outbreeding species. Genet Res 93:343–349

Van Os H (2005) The construction of an ultra-dense genetic linkage map of potato. Wageningen University, The Netherlands

Verhaegen D, Plomion C (1996) Genetic mapping in *Eucalyptus urophylla* and *Eucalyptus grandis* using RAPD markers. Genome 39:1051–1061

Virk PS, Ford-Lloyd BV, Newbury HJ (1998) Mapping AFLP markers associated with subspecific differentiation of *Oryza sativa* (rice) and an investigation of segregation distortion. Heredity 81:613–620

Vos P, Hogers R, Bleeker M, Reijans M, Van de Lee T, Hornes M, Frijters A, Pot J, Peleman J, Kuiper M, Zabeau M (1995) AFLP: a new technique for DNA fingerprinting. Nucleic Acids Res 23:4407–4414

Wachira FN (1991) Newly identified Kenyan polyploid tea strains. Tea. 12:10–13

Wachira FN (2002) Genetic mapping of tea. A review of achievements and opportunities. Tea 23(2):91–102

Wachira FN, Waugh R, Hackett CA, Powell W (1995) Detection of genetic diversity in tea (*Camellia sinensis*) using RAPD markers. Genome 38:201–210

Wachira FN, Tanaka J, Takeda Y (2001) Genetic variation and differentiation in tea (*Camellia sinensis*) germplasm revealed by RAPD and AFLP variation. J Hortic Sci Biotechnol 76:557–563

Weising K, Nybom H, Wolff K, Kahl G (2005) DNA fingerprinting in plants. Principles, methods, and applications, 2nd edn. CRC Press, Taylor and Group, 444 p

Wenger JW, Schwartz K, Sherlock G (2010) Bulk segregant analysis by high-throughput sequencing reveals a novel xylose utilization gene from *Saccharomyces cerevisiae*. PLoS Genet 6:e1000942

Wight W (1962) Tea classification revised. Curr Sci 31:298–299

Williams JG, Kubelik AR, Livak KJ, Rafalski JA, Tingey SV (1990) DNA polymorphisms amplified by arbitrary primers are useful as genetic markers. Nucleic Acids Res 18:6513–6653

Williams JGK, Rafalski JA, Tingey SV (1993) Genetic analysis using RAPD markers. In: Wu R (ed) Methods in enzymology, vol 218. Academic Press, London, pp 704–740

Wu K, Tanksley SD (1993) Abundance, polymorphism and genetic mapping of microsatellites in rice. Mol Gen Genet 241:225–235

Wu CD, Wei GX (2002) Tea as a functional food for oral health. Nutrition 18:443–444

Wu L, Joshi CP, Chiang VL (2000) A xylem specific cellulose synthase gene from aspen (*Populus tremuloides*) is responsive to mechanical stress. Plant J 22:495–502

Wu H, Chen D, Li J, Yu B, Qiao X et al (2013) De novo characterization of leaf transcriptome using 454 sequencing and development of EST-SSR markers in tea (*Camellia sinensis*). Plant Mol Biol Rep 31:524–538

Xu S, Hu Z (2009) Mapping quantitative trait loci using distorted markers. Int J Plant Genomics 2009:1–11

Yang HY, Korban SS, Kruger J, Schmidt H (1997) The use of a modified bulk segregant analysis to identify a molecular marker linked to a scab resistance gene in apple. Euphytica 94:175–182

Yang JB, Yang J, Li HT, Zhao Y, Yang SX (2009) Isolation and characterization of 15 microsatellite markers from wild tea plant (*Camellia taliensis)* using FIASCO method. Conserv Genet 10:1621–1623

Yin TM, DiFazio SP, Gunter LE, Riemenshneider D, Tuskan GA (2004) Large scale heterospecific segregation distortion in *Populus* revealed by a dense genetic linkage map. Theor Appl Genet 109:451–463

Zhan Z, Ke N, Chen B (1987) The cytology of tea clonal cultivars fujian shuixian and their infertile mechanism. In: Proceedings of international tea quality. Human health symposium, China, p 46

Zhao LP, Liu Z, Chen EL, Yao EMZ, Wang EXC (2008) Generation and characterization of 24 novel EST derived microsatellites from tea plant (*Camellia sinensis*) and cross-species amplification in its closely related species and varieties. Conserv Genet 9:1327–1331

Chapter 6
Molecular Cytogenetic Approaches in Exploration of Important Chromosomal Landmarks in Plants

Santosh Kumar Sharma, Maki Yamamoto and Yasuhiko Mukai

Abstract Multicolored fluorescence-based chromosome biology or 'molecular cytogenetics' in common continue to flourish and make essential contributions to elucidate the plant gene regulation, genome architecture, and organization by revealing essential chromosomal landmarks. Fluorescence in situ hybridization (FISH) and its modifications, such as extended DNA fiber-FISH, bacterial artificial chromosome (BAC)-FISH, multicolor-FISH (McFISH), and super-stretched pachytene-FISH, allow the study of minute details of chromosome structure and subsequently permit sophisticated analyses of chromosomal behavior. Similarly, genomic in situ hybridization (GISH) facilitates genome-specific chromosome painting in hybrids and polyploids, analysis of recombination of partially homologous chromosomes in interspecific/generic natural hybrids, and detection of transgene and/or alien chromatin in synthetic hybrids. The global patterns of chromatin modification (e.g., DNA methylation and histone tail modifications) along with nuclear size and shape, relative content and distribution of hetero/euchromatin, and organization as well as structure of chromosomes (e.g., position and orientation) provide new insights into epigenomic evolution of the particular plant species. Molecular cytogenetics also provide information on gene pool diversity and relatedness of the plant to its wild relative that ultimately may serve as a baseline data for plant breeding programs. As more genomes become sequenced, such cytogenetic tools will play a greater role in investigating the function of those genomes. Attempts have been made to summarize the utility

S.K. Sharma · Y. Mukai (✉)
Laboratory of Plant Molecular Genetics, Division of Natural Sciences,
Osaka Kyoiku University, Kashiwara, Osaka, Japan
e-mail: ymukai@cc.osaka-kyoiku.ac.jp

M. Yamamoto
Department of Rehabilitation Sciences, Kansai University of Welfare Sciences,
Kashiwara, Osaka, Japan

V.R. Rajpal et al. (eds.), *Molecular Breeding for Sustainable Crop Improvement*, Sustainable Development and Biodiversity 11,
DOI 10.1007/978-3-319-27090-6_6

of molecular cytogenetic tools in exploration of important chromosomal landmarks in plants. The evolution of plant cytogenetic research from classical to molecular level and modern to next-generation era has been discussed.

Keywords FISH · GISH · Chromosome painting · Chromatin dynamics

6.1 Introduction

6.1.1 Classical Cytogenetics

The term cytogenetics is referred to the study of genetic consequences in terms of chromosome number, structure, and behavior vis-à-vis speciation and evolution. Cytogenetics has been proved to be an integral part of genome mapping projects owing to magnificent chromosomal dynamics during mitosis and meiosis. The field of plant cytogenetics was heavily induced by Barbara McClintock's pioneering work on maize (*Zea mays*) (McClintock 1929, 1932, 1938, 1941a, b, 1984). McClintock used carmine for staining and uniquely identified all of the individual chromosomes from a single meiotic nucleus with a combination of two metrics, i.e., the relative lengths and arm ratios of the chromosomes. Her studies on unequivocal identification of individual chromosomes established a milestone in the scientific community, which allowed neo-discoveries regarding the dynamic structure and behavior of the maize genome (McClintock 1929, 1932, 1938, 1941a, b, 1984). Further, development of chromosome-banding techniques greatly improved the usefulness of chromosome biology to understand the basic genome architecture. In this context, Caspersson et al. (1968) proposed Q-banding pattern using the fluorescent dye quinacrine on plant chromosomes. Vosa and Marchi (1972) compared Giemsa C-banding to Q-banding on the chromosomes of bean (*Vicia faba*), keeled garlic (*Allium carinatum*), and maize. Further, Giemsa staining technique also showed its utility to identify individual rice prometaphase chromosomes (Kurata and Omura 1978), karyotype development for diploid rye (*Secale cereale*) (Gill and Kimber 1974), and barley (*Hordeum vulgare*) (Linde-Laursen 1975). With the advent of information on DNA and its characteristics, modifications of DNA staining dyes and banding techniques were adapted and optimized for cytogenetic characterization of different plant species. These classical approaches have proven invaluable for chromosome characterization, but the development of in situ hybridization, which allows for direct visualization of specific DNA sequences on chromosomes, forms a quantum leap forward for cytogenetics by combining cytology with molecular biology (Gill and Friebe 1998; Harper and Cande 2000).

6.1.2 Molecular Cytogenetics

A combination of 'classical cytogenetics' and 'recombinant DNA technology' gave birth to a versatile multicolored fluorescence engineering-based chromosome biology called 'molecular cytogenetics.' During initial development, such technique had been performed by using radioactive nucleic acid probes for the detection of specific DNA or RNA sequences in metaphase chromosomes or interphase. Subsequently, in the late nineties, methods for labeling nucleic acids with non-radioactive haptens such as biotin became available and adopted widely (Jiang and Gill 1994). The advantages of non-radioactive probes over radioactive probes include increased stability, safe handling, rapid, precise spatial localization, less back ground, and most importantly the ability to use multiple colors on a single chromosome preparation.

The development of in situ hybridization (ISH) techniques opened up opportunities for cytogenetic analysis of any species, regardless of its inherent chromosome morphology (Gall and Pardue 1969; Pardue and Gall 1975; John et al. 1969). In plants, the use of radioactive tagged or modified nucleotides (labeled with biotin, digoxigenin, or fluorescent haptens) and FISH probes also permits microscopic visualization and localization of complementary sequences in cells/nuclei and on individual chromosomes (Mukai et al. 1991; Fransz et al. 1996a; Mukai and Yamamoto 1998). Basic FISH makes use of green and red fluorochromes for probe detection and DAPI (4,6-diamidino-2-phenylindole) for counterstaining the chromosomal DNA. Although FISH is commonly used to map unique or low-copy-number sequences, however it also showed its potential to localize repetitive sequence in order to produce chromosome-specific landmarks or explore genome relations in polyploidy/closely related plant species (Lysak et al. 2001, 2003; Kato et al. 2004; Lamb and Birchler 2006). FISH has been found most successful in mapping the repetitive and single-copy DNA sequences on prometaphase chromosomes, interphase nuclei, pachytene complements, chromatin fibers, and naked DNA molecules. Accurate localization of repetitive and tandem arrays plays a major role in chromosome identification and karyotype analysis in plants (Mukai and Yamamoto 1998). The broad applications of FISH in structural, comparative, and functional genomics place plant cytogenetics in an important place to complement, accelerate, or guide plant genome research (Lamb et al. 2007; Danilova and Birchler 2008; Nagaki et al. 2012b). On the other hand, genomic in situ hybridization (GISH) (Le et al. 1989; Mukai and Gill 1991), a special type of FISH that uses genomic DNA of a donor species as a probe in combination with an excess amount of unlabeled blocking DNA, provides a powerful technique to monitor chromatin introgression during interspecific hybridization. In addition, the GISH technique allows the study of genome affinity between polyploid species and their progenitors (Mukai et al. 1993b; Raina et al. 1998; Raina and Mukai 1999). GISH is thus a valuable supplemental technique to traditional genome analysis such as conventional meiotic pairing analysis.

Molecular cytogenetics has now become an indispensible tool and a conceptual foundation for modern genome projects by providing significant information on individual chromosome portfolio of the organism under investigation.

6.2 Advances in Molecular Cytogenetic Techniques

Rapid developments in genetics, molecular genetics, molecular biology, and genomics, together with molecular cytogenetics, have driven major conceptual advances in mitotic, meiotic analysis, chromosome structure, and chromosome manipulation. Along such development although the principal steps of the FISH technique have remained same, various technical developments have been adapted in plant molecular cytogenetics. The basic development was the use of several colors for labeling the probes which provide holistic view of genome structure at a single glance, i.e., McFISH and McGISH (Mukai et al. 1993b; Mukai 1996). Some of the recent developments in the field of plant molecular cytogenetics in order to understand genome architecture and organization at ultra-resolution are described below.

6.2.1 Tyr-FISH

Tyr-FISH was developed to improve the detection sensitivity of FISH experiments. This method allows signal amplification by using a peroxidase-conjugated antibody as the first layer of signal detection. Fluorochrome-labeled tyramides as peroxidase substrate are used to generate and deposit many fluorochromes close to the in situ bound peroxidase (Raap et al. 1995). The sensitivity of the basic FISH technique can be increased by 10–100 times using such modification. DNA probes smaller than 1 kb were successfully visualized on plant chromosomes using Tyr-FISH (Khrustaleva and Kik 2001; Stephens et al. 2004).

6.2.2 DNA Fiber-FISH

The DNA fiber-FISH technology is applied to characterize complex genomic arrangements in plant nuclei by using decondensed chromatin and highly extended intact DNA fibers on microscopic slides (Fransz et al. 1996a). The method involves the release of DNA molecules from lysed nuclei followed by spreading them on the surface of a microscope slide and the hybridization of probes using a standard FISH method. Applying FISH probes to the stretched DNA molecules provides the higher spatial resolution with increased detection sensitivity. DNA prepared from BAC clones or plant tissues extends approximately 2.5–3.5 kb/μm on slides and provides fine-mapping resolution of up to a few kilobases. The drawback of the technique is that chromosome identification requires control DNA sequences, since there is no chromosome structure. In plants, Fransz et al. (1996b) demonstrated the utility of the extended DNA fiber-FISH (EDF-FISH) technology to characterize *Arabidopsis thaliana* and tomato genome. Later, this method was applied on other plants (e.g., rye, rice, and maize) in order to characterize complex genomic arrangements (Nagaki et al. 2004; Jin et al. 2004; Nakano et al. 2005; Yamamoto and Mukai 2007). Fiber-FISH is particularly informative when the exact position and ordering of DNA clones are needed. It can also evaluate the distances and overlaps between neighboring sequences (Ersfeld 2004; Suzuki et al. 2004; Yamamoto and Mukai 2007). The minimum target DNA size that can be distinguished unambiguously in plants is 10 kb (representing a $\sim$3 μm fluorescent signal, de Jong et al. 1999); however, good flanking markers are crucial in order to differentiate and identify shorter DNA stretches.

6.2.3 Three-Dimensional (3D) FISH

The 3D-FISH technique had developed by Bass et al. (1997). Meiotic cells of maize were fixed in a buffer to preserve chromosome structure. Pollen mother cells were also gently extruded out of the fixed anthers and embedded in optically clear polyacrylamide for staining and imaging. Stacks of FISH images were taken and composed into a single 3D image. Individual chromosomes bearing the FISH signals were traced out and computationally straightened (Harper and Cande 2000). Since the chromosome structure can be preserved using this technique, it is advantageous for the identification of precise location of DNA probes on the chromosomes as well as within the nucleus.

6.2.4 FISH on Super-Stretched Chromosomes

Interphase nuclei, super-stretched mitotic metaphase chromosomes, and meiotic pachytene chromosome provide intermediate resolving power for FISH mapping. The relative positions of clone separated by <100 kb can be resolved on these cytological targets (Jiang et al. 1996; Wang et al. 2006). Pachytene chromosomes are particularly versatile targets for FISH mapping. Late pachytene chromosomes can be used to orient the telomere–centromere positions of the adjacent clones, whereas early pachytene chromosomes can be used to resolve even partially overlapped BAC clones. Nevertheless, pachytene chromosomes are not amenable for cytological analysis in many plant species.

On the other hand, flow-sorted plant chromosome at meiotic metaphase can be stretched to more than 100 times of their original size (Valarik et al. 2004). FISH on stretched chromosomes showed brighter signals than on the untreated control presumably as a result of better probe accessibility to the stretched chromatin. FISH on super-stretched metaphase chromosomes provides a mapping resolution of up to 70 kb (Valarik et al. 2004), similar to the resolution on meiotic pachytene chromosomes (Cheng et al. 2002). Thus, this modification of FISH provides an alternative mapping target for those plant species where meiotic pachytene chromosomes are not suitable for cytological analysis.

6.2.5 BAC-FISH

For the genome-wide sequencing project, a genomic library-holding large DNA fragments is an important tool for physical mapping or positional cloning of important chromosome landmarks. BAC-FISH, a unique tool of molecular cytogenetics, uses genomic DNA cloned in large-insert vectors such as bacterial artificial chromosomes (BACs) (Shizuya et al. 1992) in combination with FISH. This technique has shown its tremendous potential for physically mapping of specific DNA sequences and identifying individual chromosomes in plants (Suzuki and Mukai 2004). The BAC clones provide efficient resources for chromosome-specific FISH markers especially for plant species having small genomes such as rice, cotton, and sorghum. BAC-FISH favors the large clone as a probe for better resolution. The conventional FISH analysis on plant chromosomes employing probes containing over 10-kb insert DNA provides stable and distinct signals (Mukai and Yamamoto 1998; Suzuki et al. 2010). It is quite difficult to detect a single locus by using a plasmid clone of several kbs as the FISH probe. In this regard, the BAC clones containing around 50–100 kb fragments are suitable for probe of the FISH analysis.

6.3 Molecular Cytogenetics in Plant Genome Research

6.3.1 Physical Mapping and In Situ PCR

Plant genome are known for abundance of repeat sequences and cytogenetic or physical mapping of such repetitive DNA sequences decipher their genomic distribution and precisely identify the typical chromosome or set of chromosomes. These repeated rDNA gene clusters are being widely used and a common starting point for FISH-based mapping (Mukai et al. 1991; Yamamoto and Mukai 1991; Fransz et al. 1996a; Mukai and Yamamoto 1998; Sharma et al. 2012). The two types of ribosomal RNA genes (rDNA), 18S-5.8S-26S rDNA and 5S rDNA, have been extensively used as probes for physical mapping in higher plants due to their arrangement in tandem arrays (Mukai 1999). FISH mapping of rDNA clusters has provided a number of chromosomal markers that proved their efficacy in exploration of chromosome evolution and species interrelationships.

In hexaploid wheat, the six loci of 5S rRNA genes were identified on the short arm of the chromosomes of homoeologous group 1 and 5 (1A, 1B, 1D, 5A, 5B, and 5D) (Mukai et al. 1990), whereas 18S-5.8S-26S rDNA loci were mapped on the short arm of 1A, 1B, 6B, and 5D chromosomes and the long arm of 7D chromosome (Mukai et al. 1991) (Fig. 6.1). The rRNA genes are associated with the nucleolar organizing region (NOR), and the visualization of such repeat clusters at interphase represents the number of active rDNA loci. Multicolor FISH (McFISH) approach targeting repetitive DNA and rDNA probes also serves as chromosome identification markers in many plant species, for example, common wheat (Fig. 6.2). Similarly, Xu and Earle (1996) mapped the 45S rRNA DNA loci on to the tomato pachytene chromosomes, and Pedrosa et al. (2002) demonstrated the rDNA FISH for creating a karyotype of the model legume lotus. In addition, rDNA FISH in combination with other tandem repeats aids the generation of core cytogenetic maps, as demonstrated for maize, wheat (Jaing and Gill 1994), cotton (Hanson et al. 1996), tomato (Xu and Earle 1996), Pinus (Hizume et al. 2002), and Arabidopsis (Koornneef et al. 2003). The rDNA sequences are conserved across most plant species, but other tandem repeats exhibit variable degree of conservation.

Further, the chromosomal localization of rDNA has been widely used for comparative characterization of polyploid plant species. A comparison of FISH patterns of polyploid species with those of diploid progenitors of *Aegilops* revealed natural amphiplasty, in which the active rDNA sites either transformed to inactive or silent (deleted) during polyplodization event (Yamamoto 1994). Similarly, the U genome mostly suppresses the NOR activity of other genomes in tetraploids. On the other hand, the NOR activity of the D-genome chromosomes is completely suppressed by other genomes. In hexaploid species, all rDNA sites on the third genome remain active, reflecting time lapse after polyploid formation.

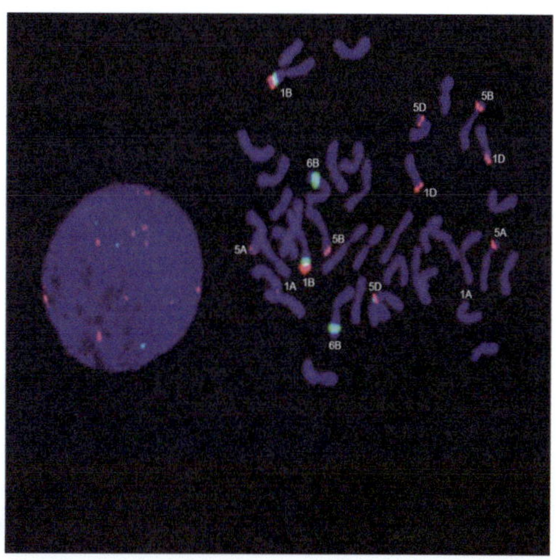

Fig. 6.1 Multicolor FISH mapping of 5S rRNA and 18S-5.8S-26S rRNA genes on the chromosomes of bread wheat (*Triticum aestivum*, $2n = 6x = 42$, AABBDD genome), and fluorescence signals can be seen for 5S (*red*) and 18S-5.8S-26S (*green*) rRNA genes, respectively (Mukai 2004)

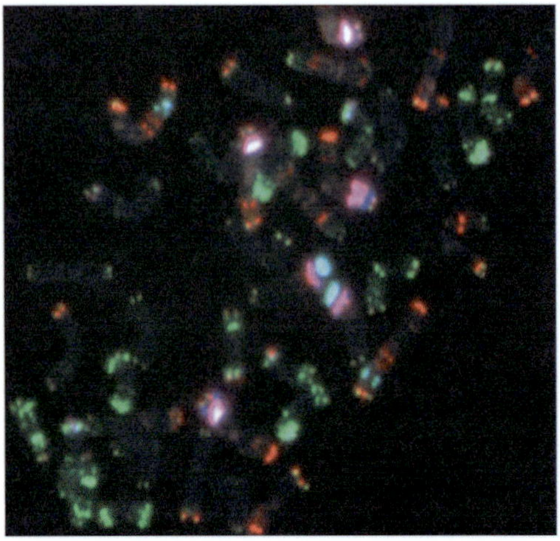

Fig. 6.2 Seven-color FISH on a metaphase cell of common wheat. Seven DNA sequences-pSc119.2, pSc74, pAs1, telomere, 18S-26S rDNA, 5S rDNA, and gliadin were detected by *red*, *bluish green*, *green*, *orange*, *pink*, *blue,* and *white* fluorescence, respectively, and the photographs were taken by triple exposures (Mukai 1996)

Simultaneously, Mukai and Apples (1996) invented the in situ polymerase chain reaction (in situ PCR)-FISH for mapping plant genes. This method uses the extreme temperature gradient sensitivity of PCR along with the cytological location of DNA sequences by means of in situ hybridization. The in situ locations of the rye-specific spacer region were determined on metaphase chromosomes. In such experiment, two pairs of primers for rye, i.e., Nor-R1 and rye 5S-Rrna-R1, were amplified in situ, which resulted in 386- and 107-bp amplified products, respectively. Rye NOR primers (45S) were localized on chromosome 1R and 4R, while 5S primers showed signals on the chromosome 1R and 3R. Interestingly, a previously described locus chromosome 5R did not show any signal in this experiment. It was concluded that the absence of a 5S site could be due to the sequence differences between the different 5S rDNA lineages. Several chromosome-specific sequences were also identified through primers specific to the chromosome. Thus, in situ PCR proved its utility in amplification of DNA sequences of specific plant chromosomes and for mapping low-copy genes of interest (Mukai and Yamamoto 1998).

Centromeric and telomeric sequences are also widely used in FISH mapping studies. Telomere repeats are highly conserved in plant species and occur in at least two major variants, i.e., (TTAGGG)n and (TTTAGGG)n (Lapitan et al. 1989; Adams et al. 1998; Fajkus et al. 2005). Similarly, the centromere associated 156-bp tandem repeat of maize, Cent C, was first discovered by Ananiev et al. (1998) and has become an invaluable cytogenetic milestone for maize and many related grass species. Cent O, a 155-bp centromere-specific satellite repeat sequence, the 180-bp satellite repeat, and CEN38, a 140-bp repeat sequence, have proven useful for labeling the primary constriction in rice, Arabidopsis, and sorghum, respectively

(Heslop-Harrison et al. 1999; Cheng et al. 2002; Nagaki et al. 2003; Kim et al. 2005).

Further, employing BAC clones as a probes in FISH experiments become revolutionizing inventory in the field of molecular cytogenetic and extensively used in many plant species including cotton (Hanson et al. 1995), rice (Jiang et al. 1995), tomato (Zhong et al. 1996), Arabidopsis (Fransz et al. 1996b), onion (Suzuki et al. 2001), and sorghum (Kim et al. 2005). This approach can also be used to acquire insight for ongoing genome-sequencing projects worldwide.

King et al. (2002) demonstrated a GISH-based approach for physical mapping to distinguish recombination events between chromosomes of *Festuca pratensis* and *Lolium perenne*. A similar approach has also been used for the integration of genetic and physical maps of two Allium chromosomes (Khrustaleva et al. 2005). This GISH-based mapping strategy is similar to physical mapping using deletion and translocation stocks. This approach overcomes the major drawback of the tedious and time-consuming process of developing a large number of deletion and translocation stocks.

On the other hand, DNA clones were also used as probes for comparative FISH mapping in relative species. Several cytogenetics researchers reported FISH mapping of *A. thaliana* BACs on chromosomes of *Brassica* species. Comparative FISH mapping between *Arabidopsis* and *Brassica* provided a direct visualization of the genome duplication within *Brassica* species (Howell et al. 2005; Lysak et al. 2005). In addition, comparative chromosome painting with pooled BAC probes was used to investigate ancestral relationships among species that diverged within the Brassicaceae (Lysak et al. 2001, 2003, 2005). Recently, Koo and Jiang (2009) developed a technique by stretching maize pachytene chromosomes mechanically more than 20 times longer than their original size. Such super-stretched pachytene chromosomes can be directly used in conventional as well as molecular cytogenetic experiments. Super-stretching of the chromosomes coupled with immunofluorescence in situ detection of DNA methylation can lead to a new dimension and higher resolving power to modern molecular cytogenetics research. Collectively, these studies revealed that such FISH-based plant cytogenetical tools are uniquely informative and beneficial for genome analysis.

6.3.2 Chromosome Identification Subject to Parentage, Hybridity, and Ploidy

Fluorescence signal allows the identification of chromosomes, specific sequences, segments, or whole set of chromosome to gain a genome-wide view at a single glance in order to understand the plant genome organization and behavior. Fluorescence signals of either a single repetitive DNA probe or a mixture of several probes can be utilized for hybridization independently to identify individual chromosomes within a species. Chromosome identification through FISH method has advantage over the traditional chromosome-banding techniques due to availability of several probes for a particular species. Many repetitive DNA elements can also generate specific FISH signal pattern on individual chromosome within a single species (Mukai et al. 1991; Mukai and Yamamoto 1998; Koo et al. 2005). In this context, Pedersen and Langridge (1997) demonstrated the identification of all 21 chromosomes of hexaploid wheat through fluorescence signals derived from two different repetitive DNA probes. Later, similar approach has been adopted in several plant species for chromosome identification (Franz et al. 1998; Hizume et al. 2002; Kato et al. 2004; Koo et al. 2004).

On the other hand, GISH provides a direct visual method for distinguishing parental genomes and analyzing genome organization in intra-/interspecific hybrids, allopolyploid species, and introgression lines. This technique has an incredible prospective to identify application in identifying alien chromatin introgression and to study chromosomal pairing and recombination between divergent genomes. GISH has validated its utility in recognizing synthetic *Hordeum chilense* × *Secale africanum* (Schwarzacher et al. 1989) and *Triticum aestivum* (wheat) × *S. cereale* (rye) (Le et al. 1989). Mukai and Gill (1991) showed that GISH optimally detects barley chromosomes in a wheat background and further identified A-, B-, and D-genomes of the common wheat (Mukai et al. 1993b) using the same approach (Fig. 6.3). Similarly, Raina et al. (1998) and Raina and Mukai (1999) conclusively revealed that *Coffea congensis* and *C. eugenioides*, and *Arachis villosa* and *A. ipaensis* are the diploid wild progenitors of allotetraploid *C. arabica* ($2n = 4x = 44$) and *A. hypogaea* ($2n = 4x = 40$), respectively, using GISH as a tool. GISH has also been widely used to characterize the genome constitution of natural hybrids and to identify the parental origin of specific loci. By following the same approach, Takahashi et al. (1999) categorized the ancestral genome donors in maize and

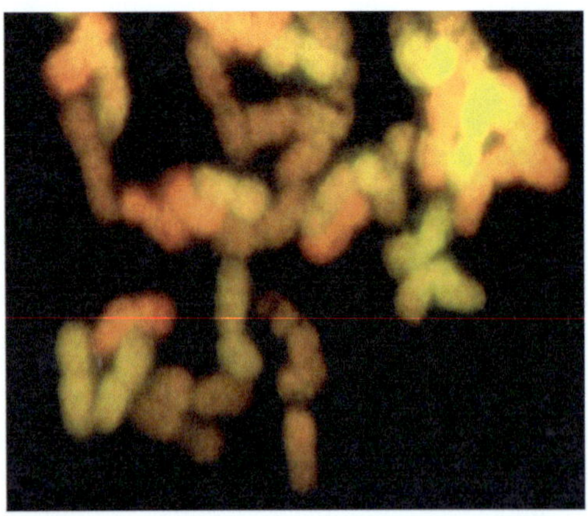

Fig. 6.3 Chromosome identification of *Triticum aestivum* ($2n = 6x = 42$). The AABBDD genome was simultaneously discriminated using GISH technique in which the diploid A genome progenitor *Triticum urartu* (*yellow*), diploid B genome progenitor *Aegilops speltoides* (*brown*), and diploid D genome progenitor *Aegilops squarrosa* (*orange*) have been identified precisely

examined inter-genomic translocations and homeologous chromosome pairing (Zwierzykowski et al. 2008), as well as chromosomal areas with large species-specific sequences (alien chromatin introgression) or translocation break points (Qi et al. 2008). Such versatile approach of molecular cytogenetics also provides insight into somaclonal variation, the origin of B chromosomes, control of chromosome pairing, and other aspects of chromosome evolution (Kato et al. 2005).

6.3.3 Karyotype and Phylogenetic Analysis

FISH-based chromosome identification systems could lead to precise karyotyping and to understand the evolution of particular plant taxa by means of speciation from wild to cultivated ones. For example, several repetitive DNA probes generate specific hybridization pattern on chromosomes of wheat and related species (Mukai et al. 1993a; Pederson and Langridge 1997). The FISH karyotypes from some repetitive DNA probes are similar to karyotypes based on C- or N-banding analysis (Cuadrado et al. 1995; Pederson and Langridge 1997). FISH-based karyotyping also specifies the phylogenetic view of related plant species (Lim et al. 2000). A number of repetitive DNA probes had utilized to develop FISH karyotypes of several diploid and polyploid *Triticum* and *Aegilops* species by Badaeva et al. (1996a, b). Similarly, comparative FISH mapping using several repetitive DNA probes in *Nicotiana* species found *N. tomentosiformis* to be the T-genome donor (Lim et al. 2000). Comparison of such karyotypes evidently revealed chromosomal landmarks to understand the evolutionary relationship between these species. Karyotyping using repetitive DNA probes can also visualize inter-genomic chromosome translocations in polyploid species. Since molecular cytogenetic techniques are often used to compare the ability of different genomes to hybridize (homology of genomes), together with the use of interspecific hybrids and allopolyploids, there by can serve as a powerful tool to understand phylogenetic relationships between species that is independent of nucleotide sequence-based approaches.

6.3.4 Chromosome Painting

The basic principle of FISH was further exploited to ‘paint’ individual plant chromosomes. The ‘chromosome painting’ is one of the most powerful molecular cytogenetic techniques to analyze nuclear organization and genome structure through visualization of specific cytogenetic target regions or entire chromosomes using this technique (Pinkel et al. 1986). Such technique involves the hybridization of fluorescence-tagged chromosome-specific composite probe pools (generally BAC clones) to various cytological preparations. Lysak et al. (2001) painted the chromosome of dicotyledonous model organism *A. thaliana* for the first time by employing selected BACs as differential labeled probes. However, in plants, mainly due to the presence of large amounts of repetitive DNA sequences such technique is remained limited (Jiang and Gill 2006). Such technique was found to be useful to identify individual chromosome in the interphase nuclei and could reveal the spatial arrangement and functional properties of individual chromatin domains. Further, Han et al. (2003, 2004) modified the McGISH to identify closely related wheat-Thinopyrum intermediates. Such chromosome painting provides insight into genome duplication/multiplication and karyotype evolution in closely related taxa. Arabidopsis chromosome and/or segment-specific probes were hybridized to 'paint' the chromosomes from species related to *A. thaliana* (Lysak et al. 2005). In later studies, the chromosome painting technique was applied successfully in related *Brassica* species (Lysak et al. 2010). These experiments proved that the technique is feasible for the detailed investigation of the pairing behavior of homologous chromosomes during early prophase I. Painting by this method is found to be feasible on small B chromosomes as well as alien chromosomes that possess chromosome-specific repeats (Houben et al. 1996). Comparative chromosome painting is an efficient and powerful approach to study the partial genome duplications and karyotype evolution. This advantage of the technique has been used to investigate the mechanisms of chromosome number reduction in *A. thaliana* and related Brassicaceae species.

Successful interspecific chromosome painting experiments were carried out between sorghum and maize (Koumbaris and Bass 2003). Ma et al. (2010) used *Brachypodium distachyon* BAC-clone to map the barley genome. Recently, the evolution and taxonomic split of the model grass *B. distachyon* were analyzed, and substantial phenotypic, cytogenetic, and molecular differences were detected between three cytotypes with the help of chromosome painting (Catalán et al. 2012). The development of comparative chromosome painting paves the way toward comparative chromosome mapping in several crop taxa including Triticeae hexaploid wheat, thereby facilitating the formulation of meaningful breeding program in light of the gene pool diversity.

6.3.5 Alien Chromatin and Transgene Detection

Schwarzacher et al. (1992) ascertained the alien chromatin incorporated from Leymus, Thinopyrum, Hordeum, or Secale in five bread wheat lines by GISH analysis. Friebe et al. (1991) also used GISH to locate the translocation chromosomes in different leaf rust-resistant wheat using GISH technology. Mukai et al. (1993a) also noticed the rye chromatin transfer in wheat. This technique has been effectively applied to detect genome donors in *Brassica* allopolyploids (Snowdon et al. 1997).

FISH has also analyzed the structure of the transgene loci on interphase nuclei, metaphase chromosomes, and on extended DNA fibers (Forsbach et al. 2003;

Chen et al. 2003). Particle bombardment often generates very large, high-copy-number transgenic arrays that can extend for megabases. Interestingly, earlier studies showed that dispersed metaphase FISH signals come together at interphase. By contrast, Agrobacterium transformation gives rise to lower transgene copy numbers and is usually characterized by single discrete FISH signals. Employing molecular cytogenetic approaches, transgenes have been identified in Arabidopsis, barley, and rice, respectively (Forsbach et al. 2003; Chen et al. 2003).

6.4 Modern Molecular Cytogenetics

A biological question may not be solved by a simple localization of DNA sequences in interphase nuclei or on chromosomes. However, physical localization of a DNA sequence together with its associated protein may dramatically enhance the power of FISH. The global patterns of chromatin modification (e.g., DNA methylation and histone tail modifications) along with nuclear size and shape, relative content and distribution of heterochromatin/euchromatin, and organization and structure of chromosomes (e.g., position and orientation) provides new insight into evolution of the particular plant species at chromosomal level. Therefore, it has acquired am important share in this newly developing research field of studying chromatin dynamics through localization of epigenetic signatures of histone/DNA modifications and methylation. It has also been emphasized that amino-terminal tails of histone proteins are targets for a series of posttranslational modifications (PTMs), including acetylation, phosphorylation, and methylation. These modifications regulate chromatin structure and gene expression (Jenuwein and Allis 2001).

6.4.1 Immuno-FISH

Several plant laboratories have developed techniques that combine FISH with immunoassay methods (Jasencakova et al. 2001; Zhong et al. 2002; Nagaki et al. 2005, 2012a, b; Lavania et al. 2012). Such modernization of cytogenetic technique involves an immunoassay of specific antibodies and cytological preparations followed by standard FISH procedure. Immuno-FISH has been used to reveal DNA methylation and histone modifications with specific genomic region. A number of antibodies are available for studying 5mC and histone modifications vis-à-vis chromatin status. Recently, several studies have been conducted on plants using immunohistochemical staining to elucidate chromosomal distribution pattern of the epigenetic marks including Arabidopsis (Zhang et al. 2008), *Allium* (Suzuki et al. 2010; Nagaki et al. 2012b), maize (Jin et al. 2008; Koo and Jiang 2009; Koo et al. 2011), rice (Yan et al. 2010), brassica (Wang et al. 2011), Barley (Sanei et al. 2011), tobacco (Nagaki et al. 2009), sugarcane (Nagaki et al. 2005), and other taxa

(Lavania et al. 2012). Most of the studies suggest that H3K4me1,2,3 mostly mark euchromatin, while H3K9me1 and H3K27me1 mostly target heterochromatin (Fuchs et al. 2006). While H3K9me2 and H3K27me2,3 showed diverse distribution pattern among angiosperms (Fuchs et al. 2006). On the other hand, centromere-specific histone H3 (CENH3) is one of the most fundamental centromeric proteins known to be involved in recruiting other centromeric proteins. CENH3 was first identified as CENP-A in humans (Earnshaw and Rothfield 1985) and subsequently found in a large number of plant species including Brassicaceae, Solanaceae, Leguminosae, Poaceae, and Juncaceae species (Zhong et al. 2002; Telbert et al. 2002; Nagaki et al. 2004, 2005, 2009, 2012a; Sanei et al. 2011; Tek et al. 2011; Wang et al. 2011; Neumann et al. 2012). Since CENH3 comprises part of the core histone that binds directly to DNA at centromeres, centromeric DNA has been isolated from several plant species using antibodies against CENH3 (Nagaki et al. 2003, 2004, 2009, 2011, 2012b; Nagaki and Murata 2005; Tek et al. 2011; Zhong et al. 2002; Neumann et al. 2012; Houben et al. 2007). Immunostaining of chromosomes of *Allium* species using anti-AfiCENH3 antibody has been shown in Fig. 6.4. Such studies suggest that these histone variants have immense potential to generate extensive information about chromosomal distribution pattern of the epigenetic marks in a wide range of plant species (Sharma et al. 2015).

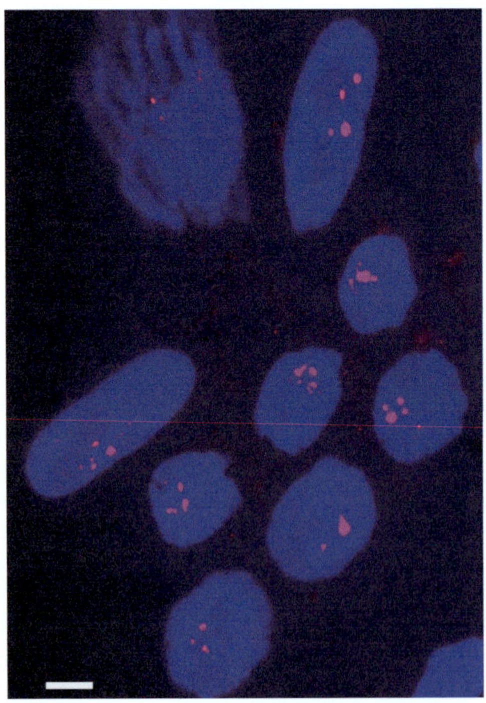

Fig. 6.4 Immunostaining of chromosomes of *Allium* species using anti-AfiCENH3 antibody: DAPIstained chromosomes (*blue*) and visualization of immunosignals of anti-AfiCENH3 antibody (*red*)

6.5 Future Prospects

Exciting advances in plant molecular cytogenetic tools and array-based techniques are changing the nature of chromosome biology, in both basic research and at molecular diagnostic levels. Cytogenetic analysis now extends beyond the simple description of the chromosomal status of a genome and allows the study of fundamental biological questions of chromosomal evolution underlying speciation and adaptation. One of the major challenges in plant cytogenetics includes the increment of the resolution power of in situ hybridization and immunostaining techniques to detect shorter nucleotide stretches or single antigen molecules reliably on metaphase chromosomes, extended chromatin fibers and/or in interphase nuclei. Further, improvement of efficient and effective fluorescent chromatin tags for in vivo studies is also needed. FISH may play a powerful role to delineate the structure and DNA composition of long track of highly repeated regions, for example, centromere as well as telomeric ends that are difficult to clone.

As discussed earlier in this article, DNA methylation, nucleosome remodeling (including histone modification and histone variants), and noncoding RNAs can organize chromatin into accessible (euchromatic) and inaccessible (heterochromatic) sub-domains. This extends the information potential of the genetic code, and one genome can generate many 'epigenomes' in time and space, during the life span of an organism. The implications of epigenetic research seek attention and efforts that should be targeted to epigenome in a variety of plant systems especially at chromosome inheritance level. In a recent study, it was shown that these epigenetic modifications are not as conserved as was once thought. Further, very little is known about histone/DNA methylation/modification in large genome plants (Houben et al. 2003), which make up the bulk of the angiosperms (Arumuganthan and Earle 1991). Immuno-FISH should be practiced worldwide that has potential to significantly increase the resolving power to reveal fine interaction between DNA and proteins.

6.6 Next-Generation FISH

Next-generation sequencing (NGS) technologies continue to develop at a fast pace, and whole genome sequence of several plants have either been released or to be released soon. NGS technologies of third-generation platforms could produce reads reaching up to a few kilobases, whereas read lengths presently range from 30 to 400 bp depending on the platform. NGS may also facilitate probe development for studies of chromosome using FISH. These genomic regions can be mapped on the chromosome for precise location information with reference to chromosome rearrangements and translocation events and to identify chromosome with/without physical gaps, if any. Further, transcriptome sequencing has also been engaged in construction of large datasets of nuclear genes. NGS is also making the rapid sequencing of complete nuclear genomes routine, thus transforming the genomics research field and opening up new avenues of systematic endeavor in comparative genomics. Further, research should be aimed at understanding the distribution, location, and copy number of the epigenetically inherited gene/genic regions identified through NGS data in several crop/plant species/families in order to shed light on the role of chromatin dynamics in speciation and evolution.

Acknowledgement We thank 'Japan Society for the Promotion of Science (JSPS)' and 'Osaka Kyoiku University,' Osaka, Japan, for providing research funds (SKS: P13399, 2013) and facilities for plant molecular cytogenetic research, respectively. Sincere thanks to Dr. Go Suzuki and all members of Plant Molecular Genetics Laboratory, Osaka Kyoiku University, Osaka, Japan, for their constant encouragement and help.

References

Adams Ananiev EV, Phillips RL, Rines HW (1998) Chromosome-specific molecular organization of maize (*Zea mays* L.) centromeric regions. Proc Natl Acad Sci USA 95:13073–13078

Arumuganthan K, Earle ED (1991) Nuclear DNA content of some important plant species. Plant Mol Biol Rep 9:208–218

Badaeva ED, Friebe B, Gill BS (1996a) Genome differentiation in *Aegilops*. 1. Distribution of highly repetitive DNA sequences on chromosomes of diploid species. Genome 39:293–306

Badaeva ED, Friebe B, Gill BS (1996b) Genome differentiation in *Aegilops*. 1. Physical mapping of 5S and 18S-26S ribosomal RNA gene families in diploid species. Genome 39:1150–1158

Bass HW, Marshall WF, Sedat JW et al (1997) Telomeres cluster de novo before the initiation of synapsis: a three-dimensional spatial analysis of telomere positions before and during meiotic prophase. J Cell Biol 137:5–18

Caspersson T, Farber S, Foley GE et al (1968) Chemical differentiation along metaphase chromosomes. Exp Cell Res 49:219–222

Catalán P, Müller J, Hasterok R et al (2012) Evolution and taxonomic split of the model grass *Brachypodium distachyon*. Ann Bot 109:385–405

Chen S, Jin W, Wang M et al (2003) Distribution and characterization of over 1000 T-DNA tags in rice genome. Plant J 36:105–113

Cheng Z, Buell CR, Wing RA, Jiang J (2002) Resolution of fluorescence in-situ hybridization mapping on rice mitotic prometaphase chromosomes, meiotic pachytene chromosomes and extended DNA fibers. Chromosome Res 10:379–387

Cuadrado A, Ceoloni C, Jouve N (1995) Variation in highly repetitive DNA composition of heterochromatin in rye studied by fluorescence in situ hybridization. Genome 38:1061–1069

Danilova TV, Birchler JA (2008) Integrated cytogenetic map of mitotic metaphase chromosome 9 of maize: resolution, sensitivity, and banding paint development. Chromosoma 117:345–356

de Jong J, Fransz P, Zabel P (1999) High resolution FISH in plants—techniques and applications. Trends Plant Sci 4:258–263

Earnshaw WC, Rothfield N (1985) Identification of a family of human centromere proteins using autoimmune sera from patients with scleroderma. Chromosoma 91:313–321

Ersfeld K (2004) Fiber-FISH: fluorescence in situ hybridization on stretched DNA. Methods Mol Biol 270:395–402

Fajkus J, Sykorova E, Leitch AR (2005) Telomeres in evolution and evolution in telomeres. Chromosome Res 13:469–479

Forsbach A, Schubert D, Lechtenberg B et al (2003) A comprehensive characterization of single-copy T-DNA insertions in the *Arabidopsis thaliana* genome. Plant Mol Biol 52:161–176

Fransz PF, Alonso-Blanco C, Liharska TB et al (1996a) High resolution physical mapping in *Arabidopsis thaliana* and tomato by fluorescence in situ hybridization to extended DNA fibers. Plant J 9:421–430

Fransz PF, Stam M, Montijn B et al (1996b) Detection of single-copy genes and chromosome rearrangements in *Petunia hybrida* by fluorescence in situ hybridization. Plant J 9:767–774

Fransz PF, Armstrong S, Alonso-Blanco C et al (1998) Cytogenetics for the model system *Arabidopsis thaliana*. Plant J 13:867–876

Friebe B, Mukai Y, Dhaliwal HS et al (1991) Identification of alien chromatin specifying resistance to wheat streak mosaic virus and greenbug in wheat germplasm by C-banding and in situ hybridization. Theor Appl Genet 81:381–389

Fuchs J, Demidov D, Houben A, Schubert I (2006) Chromosomal histone modification patterns—from conservation to diversity. Trends Plant Sci 11:199–208

Gall JG, Pardue ML (1969) Formation and detection of RNA-DNA hybrid molecules in cytological preparations. Proc Natl Acad Sci USA 63:378–383

Gill BS, Friebe B (1998) Plant cytogenetics at the dawn of the 21st century. Curr Opin Plant Biol 1:109–115

Gill BS, Kimber G (1974) Giemsa C-banding and the evolution of wheat. Proc Natl Acad Sci USA 71:4086–4090

Han FP, Fedak G, Benabdelmouna A et al (2003) Characterization of six wheat × *Thinopyrum intermedium* derivatives by GISH, RFLP and multicolor GISH. Genome 46:490–495

Han F, Liu B, Fedak G, Liu Z (2004) Genomic constitution and variation in five partial amphiploids of wheat-*Thinopyrum intermedium* as revealed by GISH, multicolor GISH and seed storage protein analysis. Theor Appl Genet 109:1070–1076

Hanson RE, Zwick MS, Choi S et al (1995) Fluorescent in situ hybridization of a bacterial artificial chromosome. Genome 38:646–651

Hanson RE, Islam-Faridi MN, Percival EA (1996) Distribution of 5S and 18S–28S rDNA loci in a tetraploid cotton (*Gossypium hirsutum* L.) and its putative diploid ancestors. Chromosoma 105:55–61

Harper LC, Cande WZ (2000) Mapping a new frontier; development of integrated cytogenetic maps in plants. Funct Integr Genomics 1:89–98

Heslop-Harrison JS, Murata M, Ogura Y (1999) Polymorphisms and genomic organization of repetitive DNA from centromeric regions of *Arabidopsis thaliana* chromosomes. Plant Cell 11:31–42

Hizume M, Shibata F, Matsusaki Y, Garajova Z (2002) Chromosome identification and comparative karyotypic analyses of four *Pinus* species. Theor Appl Genet 105:491–497

Houben A, Kynast RG, Heim U et al (1996) Molecular cytogenetic characterization of the terminal heterochromatic segment of the B-chromosome of rye (*Secale cereale*). Chromosoma 105:97–103

Houben A, Demidov D, Gernand D et al (2003) Methylation of histone H3 in euchromatin of plant chromosomes depends on basic nuclear DNA content. Plant J 33:967–973

Houben A, Demidov D, Caperta AD et al (2007) Phosphorylation of histone H3 in plants—a dynamic affair. Biochim Biophys Acta 1769:308–315

Howell EC, Armstrong SJ, Barker GC et al (2005) Physical organization of the major duplication on *Brassica oleracea* chromosome O6 revealed through fluorescence in situ hybridization with *Arabidopsis* and *Brassica* BAC probes. Genome 48:1093–1103

Jasencakova Z, Meister A, Schubert I (2001) Chromatin organization and its relation to replication and histone acetylation during the cell cycle in barley. Chromosoma 110:83–92

Jenuwein T, Allis C (2001) Translating the histone code. Science 293:1074–1080

Jiang J, Gill BS (1994) Non-isotopic in situ hybridization and plant genome mapping: the first 10 years. Genome 37:717–725

Jiang J, Gill BS (2006) Current status and the future of fluorescence in situ hybridization (FISH) in plant genome research. Genome 49:1057–1068
Jiang J, Gill BS, Wang GL et al (1995) Metaphase and interphase fluorescence in situ hybridization mapping of the rice genome with bacterial artificial chromosomes. Proc Natl Acad Sci USA 92:4487–4491
Jiang J, Hulbert SH, Gill BS, Ward DC (1996) Interphase fluorescence in situ hybridization mapping: a physical mapping strategy for plant species with large complex genomes. Mol Gen Genet 252:497–502
Jin WW, Melo JR, Nagaki K et al (2004) Maize centromeres: organization and functional adaptation in the genetic background of Oat. Plant Cell 16:571–581
Jin W, Lamb JC, Zhang W (2008) Histone modifications associated with both A and B chromosomes of maize. Chromosome Res 16:1203–1214
John HA, Birnstiel ML, Jones KW (1969) RNA-DNA hybrids at the cytological level. Nature 223:582–587
Kato A, Lamb JC, Birchler JA (2004) Chromosome painting using repetitive DNA sequences as probes for somatic chromosome identification in maize. Proc Natl Acad Sci USA 101:13554–13559
Kato A, Vega JM, Han F et al (2005) Advances in plant chromosome identification and cytogenetic techniques. Curr Opin Plant Biol 8:148–154
Khrustaleva LI, Kik C (2001) Localization of single-copy T-DNA insertion in transgenic shallots (*Allium cepa*) by using ultra-sensitive FISH with tyramide signal amplification. Plant J 25:699–707
Khrustaleva LI, de Melo PE, van Heusden AW, Kik C (2005) The integration of recombination and physical maps in a large-genome monocots using haploid genome analysis in a trihybrid *Allium* population. Genetics 169:1673–1685
Kim JS, Klein PE, Klein RR (2005) Molecular cytogenetic maps of sorghum linkage groups 2 and 8. Genetics 169:955–965
King J, Armstead IP, Donnison IS et al (2002) Physical and genetic mapping in the grasses *Lolium perenne* and *Festuca pratensis*. Genetics 161:315–324
Koo DH, Jiang JM (2009) Super-stretched pachytene chromosomes for fluorescence in situ hybridization mapping and immunodetection of DNA methylation. Plant J 59:509–516
Koo DH, Plaha P, Lim YP et al (2004) A high-resolution karyotype of *Brassica rapa* ssp. *pekinensis* revealed by pachytene analysis and multicolor fluorescence in situ hybridization. Theor Appl Genet 109:1346–1352
Koo DH, Choi HW, Cho J et al (2005) A high-resolution karyotype of cucumber (*Cucumis sativus* L 'Winter Long') revealed by C-banding, pachytene analysis, and RAPD-aided fluorescence in situ hybridization. Genome 48:534–540
Koo DH, Han F, Birchler JA, Jiang J (2011) Distinct DNA methylation patterns associated with active and inactive centromeres of the maize B chromosome. Genome Res 21:908–914
Koornneef M, Fransz P, De Jong H (2003) Cytogenetic tools for *Arabidopsis thaliana*. Chromosome Res 11:183–194
Koumbaris GL, Bass HW (2003) A new single-locus cytogenetic mapping system for maize (*Zea mays* L.): overcoming FISH detection limits with marker-selected sorghum (*S. propinquum* L.) BAC clones. Plant J 35:647–659
Kurata N, Omura T (1978) Karyotype analysis in rice, I. A new method for identifying all chromosome pairs. Jpn J Genet 53:251–255
Lamb JC, Birchler JA (2006) Retroelement genome painting: cytological visualization of retroelement expansions in the genera *Zea* and *Tripsacum*. Genetics 173:1007–1021
Lamb JC, Meyer JM, Corcoran B et al (2007) Distinct chromosomal distributions of highly repetitive sequences in maize. Chromosome Res 15:33–49
Lapitan NL, Ganal MW, Tanksley SD (1989) Somatic chromosome karyotype of tomato based on in situ hybridization of the TGR I satellite repeat. Genome 32:992–998

Lavania UC, Srivastava S, Lavania S (2012) Autopolyploidy differentially influences body size in plants, but facilitates enhanced accumulation of secondary metabolites, causing increased cytosine methylation. Plant J 71:539–549

Le HT, Armstrong KC, Miki B (1989) Detection of rye DNA in wheat-rye hybrids and wheat translocation stocks using total genomic DNA as a probe. Mol Biol Rep 7:150–158

Lim KB, Chung JD, Van Kronenburg BCE et al (2000) Introgression of *Lilium rubellum* Baker chromosomes into *L. Longiflorum* Thunb.: a genome painting study of the F1 hybrid, BC1 and BC2 progenies. Chromosome Res 8:119–125

Linde-Laursen I (1975) Giemsa C-banding of the chromosomes of 'Emir' barley. Hereditas 81:285–289

Lysak MA, Fransz PF, Ali HBM, Schubert I (2001) Chromosome painting in *Arabidopsis thaliana*. Plant J 28:689–697

Lysak MA, Pecinka A, Schubert I (2003) Recent progress in chromosome painting of Arabidopsis and related species. Chromosome Res 11:195–204

Lysak MA, Koch MA, Pecinka A, Schubert I (2005) Chromosome triplication found across the tribe Brassicaceae. Genome Res 15:516–525

Lysak MA, Mandakova T, Lacombe E (2010) Reciprocal and multi-species chromosome BAC painting in crucifers (Brassicaceae). Cytogenet Genome Res 129:184–189

Ma L, Vu GT, Schubert V et al (2010) Synteny between *Brachypodium distachyon* and *Hordeum vulgare* as revealed by FISH. Chromosome Res 18:841–850

McClintock B (1929) A cytological and genetical study of triploid maize. Genetics 14:180–222

McClintock B (1932) A correlation of ring-shaped chromosomes with variation in Zea mays. Proc Natl Acad Sci 18:677–681

McClintock B (1938) The fusion of broken ends of sister half-chromatids following breakage at meiotic anaphase. Mo Agric Exp Stat Res Bull 290:1–48

McClintock B (1941a) The association of mutants with homozygous deficiencies in *Zea mays*. Genetics 26:542–571

McClintock B (1941b) The stability of broken ends of chromosomes in *Zea mays*. Genetics 26:234–282

McClintock B (1984) The significance of responses of the genome to challenge. Science 226:792–801

Mukai Y (1996) Multicolor fluorescence in situ hybridization: a new tool for genome analysis. In: Jauhar PP (ed) Methods of genome analysis in plants. CRC Press, Boca Raton, pp 181–192

Mukai Y (1999) Molecular cytogenetic studies on chromosomes of useful plants. Breed Res 1:165–172 (Japanese)

Mukai Y (2004) Fluorescence in situ hybridization. In: Goodman RM (ed) Encyclopedia of plant and crop science. Marcel Dekker, New York, pp 468–471

Mukai Y, Appels R (1996) Direct chromosome mapping of plant genes by in situ polymerase chain reaction (in situ PCR). Chromosome Res 4:401–404

Mukai Y, Gill BS (1991) Detection of barley chromatin added to wheat by genomic in situ hybridization. Genome 34:448–452

Mukai Y, Yamamoto M (1998) Application of multicolor fluorescence in situ hybridization to plant genome analysis. In: Gupta PK (ed) Genetics and biotechnology in crop improvement. Rastogi Publication, Meerut, India, pp 14–23

Mukai Y, Endo TR, Gill BS (1990) Physical mapping of the 5S rRNA multigene family in common wheat. J Heredity 81:290–295

Mukai Y, Endo TR, Gill BS (1991) Physical mapping of the 18S-26S rRNA multigene family in common wheat: identification of a new locus. Chromosoma 100:71–78

Mukai Y, Friebe B, Hatchett JH, Yamamoto M, Gill BS (1993a) Molecular cytogenetic analysis of radiation-induced wheat-rye terminal and intercalary chromosomal translocations and the detection of rye chromatin specifying resistance to Hessian fly. Chromosoma 102:88–95

Mukai Y, Nakahara Y, Yamamoto M (1993b) Simultaneous discrimination of the three genomes in hexaploid wheat by multicolor fluorescence in situ hybridization using total genomic and highly repeated DNA probes. Genome 36:489–494

Nagaki K, Murata M (2005) Characterization of CENH3 and centromere—associated DNA sequences in sugarcane. Chromosome Res 13:195–203
Nagaki K, Talbert PB, Zhong CX (2003) Chromatin immunoprecipitation reveals that the 180-bp satellite repeat is the key functional DNA element of *Arabidopsis thaliana* centromeres. Genetics 163:1221–1225
Nagaki K, Cheng Z, Ouyang S et al (2004) Sequencing of a rice centromere uncovers active genes. Nat Genet 36:138–145
Nagaki K, Kashihara K, Murata M (2005) Visualization of diffuse centromeres with centromere-specific histone H3 in the holocentric plant *Luzula nivea*. Plant Cell 17:1886–1893
Nagaki K, Kashihara K, Murata M (2009) A centromeric DNA sequence colocalized with a centromere-specific histone H3 in tobacco. Chromosoma 118:249–257
Nagaki K, Shibata F, Suzuki G et al (2011) Coexistence of NtCENH3 and two retrotransposons in tobacco centromeres. Chromosome Res 19:591–605
Nagaki K, Shibata F, Kanatani A et al (2012a) Isolation of centromeric-tandem repetitive DNA sequences by chromatin affinity purification using a HaloTag7-fused centromere-specific histone H3 in tobacco. Plant Cell Rep 31:771–779
Nagaki K, Yamamoto M, Yamaji N et al (2012b) Chromosome dynamics visualized with an anti-centromeric histone H3 antibody in *Allium*. PLoS ONE 7(2):e51315
Nakano A, Suzuki G, Yamamoto M et al (2005) Rearrangements of large-insert T-DNA in transgenic rice. Mol Genet Genomics 273:123–129
Neumann P, Navratilova A, Schroeder-Reiter E et al (2012) Stretching the rules: monocentric chromosomes with multiple centromere domains. PLoS Genet 8(6):e1002777
Pardue ML, Gall JG (1975) Nucleic acid hybridization to the DNA of cytological preparations. Methods Cell Biol 10:1–16
Pederson C, Langridge P (1997) Identification of the entire chromosome complement of bread wheat by two-color FISH. Genome 40:589–593
Pedrosa A, Sandal N, Stougaard J et al (2002) Chromosomal map of the model legume *Lotus japonicus*. Genetics 161:1661–1672
Pinkel DT, Straume T, Gray JW (1986) Cytogenetic analysis using quantitative, high sensitivity, fluorescence hybridization. Proc Natl Acad Sci USA 85:2934–2938
Qi LL, Pumphrey MO, Friebe B et al (2008) Molecular cytogenetic characterization of alien introgressions with gene Fhb3 for resistance to *Fusarium* head blight disease of wheat. Theor Appl Genet 117:1155–1166
Raap A, van de Corput M, Vervenne R et al (1995) Ultra-sensitive FISH using peroxidase-mediated deposition of biotin- or fluorochrome tyramides. Hum Mol Genet 4:529–534
Raina SN, Mukai Y (1999) Genomic in situ hybridization identifies the diploid wild progenitors of cultivated (*Arachis hypogaea*) and related wild (*A. monticola*) peanut species. Pl Syst Evol 214:251–262
Raina SN, Mukai Y, Yamamoto M (1998) In situ hybridization identifies the diploid progenitor species of *Coffea arabica* (Rubiaceae). Theor Appl Genet 97:1204–1209
Sanei M, Pickering R, Kumke K et al (2011) Loss of centromeric histone H3 (CENH3) from centromeres precedes uniparental chromosome elimination in interspecific barley hybrids. Proc Natl Acad Sci USA 108:E498–E505
Schwarzacher T, Leitch AR, Bennett MD, Heslop-Harrison JS (1989) In situ localization of parental genomes in a wide hybrid. Ann Bot 63:315–324
Schwarzacher T, Anamthawat-Jónsson K, Harrison GE et al (1992) Genomic in situ hybridization to identify alien chromosomes and chromosome segments in wheat. Theor Appl Genet 84:778–786
Sharma SK, Mehra P, Kumaria S et al (2012) Physical localization and probable transcriptional activity of 18S–5.8 S–26S rRNA gene loci in some Asiatic Cymbidiums (Orchidaceae) from north-east India. Gene 499:362–366

Sharma SK, Yamamoto M, Mukai Y (2015) Immuno-cytogenetic manifestation of epigenetic chromatin modification marks in plants. Planta 241:291–301

Shizuya H, Birren B, Kim UJ (1992) Cloning and stable maintenance of 300-kilobase-pair fragments of human DNA in *Escherichia coli* using an F-factor-based vector. Proc Natl Acad Sci 89:8794–8797

Snowdon RJ, Kohler W, Kohler A (1997) Chromosomal localization and characterization of rDNA loci in the *Brassica* A and C genomes. Genome 40:582–587

Stephens JL, Brown SE, Lapitan NL, Knudson DL (2004) Physical mapping of barley genes using an ultrasensitive fluorescence in situ hybridization technique. Genome 47:179–189

Suzuki G, Mukai Y (2004) Plant BAC libraries as tools for molecular cytogenetics. In: Williams CR (ed) Focus on genome research. Nova Science Publishers, New York, pp 205–220

Suzuki G, Ura A, Saito N, Do GS, Seo BB, Yamamoto M, Mukai Y (2001) BAC FISH analysis in *Allium cepa*. Genes Genet Syst 76:251–255

Suzuki G, Tanaka S, Yamamoto M (2004) Visualization of the *S*-locus region in *Ipomoea trifida*: toward positional cloning of self-incompatibility genes. Chromosome Res 12:475–481

Suzuki G, Shiomi M, Morihana S (2010) DNA methylation and histone modification in onion chromosomes. Genes Genet Syst 85:377–382

Takahashi C, Marshall JA, Bennett MD, Leitch IJ (1999) Genomic relationships between maize and its wild relatives. Genome 42:1201–1207

Talbert PB, Masuelli R, Tyagi AP, Comai L, Henikoff S (2002) Centromeric localization and adaptive evolution of an *Arabidopsis histone* H3 variant. Plant Cell 14:1053–1066

Tek AL, Kashihara K, Murata M, Nagaki K (2011) Functional centromeres in *Astragalus sinicus* include a compact centromere-specific histone H3 and a 20-bp tandem repeat. Chromosome Res 19:969–978

Valarik M, Bartos J, Kovarova P et al (2004) High-resolution FISH on super-stretched flow-sorted plant chromosomes. Plant J 37:940–950

Vosa CG, Marchi P (1972) Quinacrine fluorescence and giemsa staining in plants. Nature New Biol 237:191–192

Wang CJ, Harper L, Cande WZ (2006) High-resolution single-copy gene fluorescence in situ hybridization and its use in the construction of a cytogenetic map of maize chromosome 9. Plant Cell 18:529–544

Wang G, He Q, Cheng Z (2011) Characterization of CENH3 proteins and centromere-associated DNA sequences in diploid and allotetraploid *Brassica* species. Chromosoma 120:353–365

Xu J, Earle ED (1996) High resolution physical mapping of 45S (5.8S, 18S and 25S) rDNA gene loci in the tomato genome using a combination of karyotyping and FISH of pachytene chromosomes. Chromosoma 104:545–550

Yamamoto M (1994) Molecular-cytogenetic analysis of the rDNA region in *Triticum* and *Aegilops*. Ph.D. thesis, Kyoto University, Japan, p 248

Yamamoto M, Mukai Y (1991) Fluorescence in situ hybridization: a new approach to study chromosome structure. In: Proceedings of 2nd international symposium on chromosome engineering in plants, Columbia, MO, Aug 13–15, pp 296–301

Yamamoto M, Mukai Y (2007) Extended DNA fiber FISH in plants: visible message from cell nuclei. The Nucleus 50:439–452

Yan H, Kikuchi S, Neumann P et al (2010) Genome-wide mapping of cytosine methylation revealed dynamic DNA methylation patterns associated with genes and centromeres in rice. Plant J 63:353–365

Zhang WL, Lee HR, Koo DH, Jiang JM (2008) Epigenetic modification of centromeric chromatin: hypomethylation of DNA sequences in the CENH3-associated chromatin in *Arabidopsis thaliana* and maize. Plant Cell 20:25–34

Zhong XB, Fransz PF, Wennekes-van E (1996) High-resolution mapping on pachytene chromosomes and extended DNA fibres by fluorescence in-situ hybridization. Plant Mol Biol Rep 14:232–242

Zhong CX, Marshall JB, Topp C et al (2002) Centromeric retroelements and satellites interact with maize kinetochore protein CENH3. Plant Cell 14:2825–2836

Zwierzykowski Z, Zwierzykowska E, Taciak M (2008) Chromosome pairing in allotetraploid hybrids of *Festuca pratensis* × *Lolium perenne* revealed by genomic in situ hybridization (GISH). Chromosome Res 16:575–585

Chapter 7
Technological Advances in Studying Gene Pool Diversity and Its Exploitation

Sapinder Bali, Shailendra Goel and Soom Nath Raina

Abstract Molecular biology-based plant breeding methodology has contributed significantly to crop improvement by creation of new improved varieties with superior genotypes. The advancements in genomic sciences led to the development of several new faster methods that have improved our understanding and accessibility of available gene pools, thus providing an efficient source of information. These developments have enabled researchers to develop richer gene pools that would be capable of meeting the challenges of increasing agriculture demands over the globe. Informative descriptions of gene pools may become available as the genomes are being sequenced, functionally characterized, and made available at the public domain. Modern technologies for large-scale marker surveys are capable of exploring various dimensions of gene pools and provide the raw information for understanding the extent of genetic variation in gene pools and its exploitation for crop improvement. High-throughput marker development methods provide several advantages as sufficient databases can be generated for various crop genomes assisting in the assessment of crop diversity which can be an efficient source of information for developing breeding schemes for crop improvement programmes in future.

Keywords Gene pool · Molecular markers · Crop diversity · Crop improvement

S. Bali · S. Goel (✉)
Department of Botany, University of Delhi, Chhatra Marg,
New Delhi 110007, India
e-mail: shailendragoel@gmail.com

S.N. Raina
Amity Institute of Biotechnology, Amity University, Sector 125,
Noida 201303, Uttar Pradesh, India

V.R. Rajpal et al. (eds.), *Molecular Breeding for Sustainable Crop Improvement*, Sustainable Development and Biodiversity 11,
DOI 10.1007/978-3-319-27090-6_7

7.1 Introduction

Crop gene pools have helped to sustain the agricultural demands of the world populations for thousands of years. But our knowledge about content, structure, distribution, and diversity of the gene pools had been very scarce (Lee 1998). Molecular marker-based techniques such as restriction fragment length polymorphism (RFLP), random amplified polymorphic DNA (RAPD), simple sequence repeats (SSR), and amplified fragment length polymorphism (AFLP) have been routinely used to perform taxonomical, phylogenetic, evolutionary and genetic studies at smaller scale in crop plants for decades. But due to low-throughput marker techniques, most of the germplasm has remained uncharacterized, unexplored, and underutilized for several years. With the advent of faster and cheaper technologies, a new class of advanced high-throughput techniques has emerged in the recent years, which are primarily derived from modification of earlier basic technology combining the advantageous features of several techniques. It is now possible to genotype hundreds of individuals across thousands of loci, which facilitates the genetic analysis of the available germplasm on larger scale (Agarwal et al. 2008; Glaszmann et al. 2010), thus generating broad gene pool collections of useful genetic resources for various domesticated crop plants. These methods have led to the increase in sensitivity and resolution for detecting genetic distinctiveness among closely related individuals (Royo and Galan 2009).

The gene pool diversity can be analysed in three phases: sampling the individuals, genotyping the individuals for large independent loci, and estimating the diversity patterns (Luikart et al. 2003). These well-characterized gene pools can be utilized for the development of new improved cultivars by initiating various crop improvement programmes in domesticated crops (Hajjar and Hodgkin 2007; Hamblin et al. 2011). The increasing ease of generating abundant genome-based markers spanning across the whole genome and ease of sampling numerous individuals in minimal time have made the genome-wide scanning an attractive approach for population genetics programmes as well (Hamblin et al. 2011).

High-throughput marker techniques can be used for the whole-genome and gene-targeted surveys, which provide insights into the population genomics and assist crop improvement programmes in many crop plants, viz. chickpea, grapevine, cacao, and banana (Glaszmann et al. 2010). In order to fully utilize the advantages of high-throughput genotyping, one needs to be vigilant in choosing the appropriate technology depending on the goals and the stage of the experiment to be performed, being cognizant of the number of samples and resources (Edenberg and Liu 2009).

Based on the nature of their origin in the genome, high-throughput markers can be classified into two major classes, viz. genomic markers and genic markers (Fig. 7.1). This classification although artificial provides a convenient way to organize various technologies available in the fathomable framework. Some of the markers, although explained under one category in the present chapter, can be classified under both the categories as indicated in Fig. 7.1.

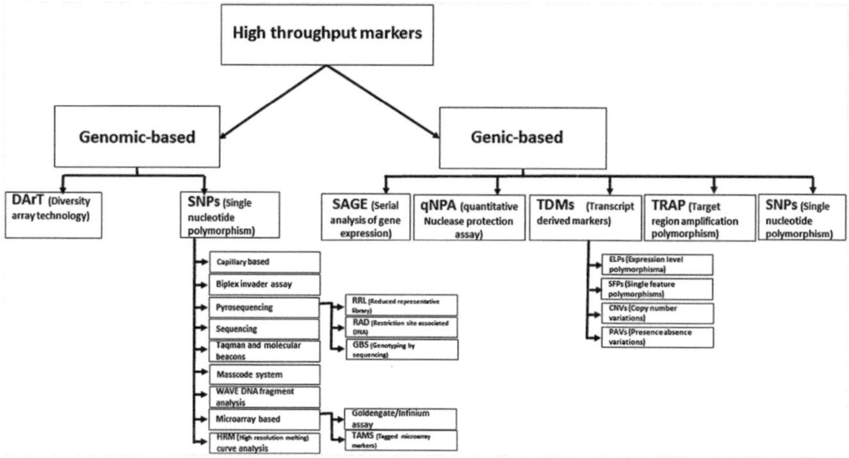

Fig. 7.1 Schematic representation of the types of genomic and genic high-throughput molecular markers

Genomic markers can be developed from the whole-genome sequences and are of two major types, viz. diversity array technology (DArT) and single-nucleotide polymorphisms (SNPs). SNP markers can be further classified into various types based on the technology used for genotyping. Likewise, genic markers are developed from the transcribed regions of the genome and can be of five main types, viz. serial analysis of gene expression (SAGE), quantitative nuclease protection assay (qNPA), transcript-derived markers (TDMs), target region amplification polymorphism (TRAP), and SNPs.

In this chapter, we will discuss various high-throughput techniques available for genomic and genic marker development.

7.2 High-Throughput Genomic Markers

Various genomic high-throughput markers have been discussed below.

7.2.1 Diversity Array Technology (DArT Markers)

DArT is a high-throughput marker technique that can detect DNA polymorphism spanning hundreds of genomic loci based on endonuclease restriction site variation between genotypes (Jaccoud et al. 2001; Wenzl et al. 2004). DArT does not require any prior sequence information and utilizes a microarray platform. Individual clones to be genotyped are prepared from a genomic representation library by amplifying the restriction fragments which are labelled and then hybridized to the arrays. The polymorphism is based on the presence versus absence of hybridization to individual array elements. The platform allows high-throughput screening of hundreds of molecular markers in parallel and is especially suited for the generation of genome-wide markers for large-scale germplasm characterization and genetic linkage mapping. Linkage maps can be further exploited for QTL analysis and map-based cloning of agronomically important genes. DArT enables efficient fingerprinting of germplasm, whole-genome screening, simultaneous detection of quantitative trait loci, marker-based selection of loci of interest, and introgression of preferable genomic regions. These markers have been utilized for diversity analysis and generating linkage maps in major crop plants such as rice (Xie et al. 2006), wheat (Akbari et al. 2006; Semagn et al. 2006; Crossa et al. 2007; White et al. 2008), barley (Wenzl et al. 2004, 2006; Alsop et al. 2011; Steffenson et al. 2007; Hearnden et al. 2007), sorghum (Jordan et al. 2010; Mace et al. 2008, 2009), common beans (Brinez et al. 2012), pigeonpea (Yang et al. 2011), oat (Tinker et al. 2009), ryegrass (King et al. 2013), willow (Przyborowski et al. 2013), and eucalyptus (Petroli et al. 2012).

7.2.2 Single-Nucleotide Polymorphisms (SNP) Markers

SNP markers are the individual nucleotide polymorphisms found in the genome. SNPs can be discovered from the sequence information using various platforms (Steemers and Gunderson 2007). The binary nature of SNPs and no known effects on gene expression or function have made them a marker of choice (Edenberg and Liu 2009). These markers can be utilized to scan one or more individuals for SNPs ranging from hundreds to thousands or allelotyping hundreds of individuals for one SNP locus. SNPs can be efficiently used for marker-assisted selection (MAS) programmes because of the high fidelity of their inheritance, strict biallelic nature, and extraordinary abundance in the genome (Gupta et al. 2001).

Several methods are available for the discovery of SNP markers. The most common method is to directly compare sequence obtained from public databases (Kwok and Chen 2003; Matukumalli et al. 2006) and amplification of target genome loci from various individuals for sequence comparison (Twyman 2004; Suh and Vijg 2005). Various methods can be followed for SNP pre-screening, viz. single-strand conformational polymorphism (SSCP) (Orita et al. 1989), overlapping regions in BACs and PACs (Miller et al. 1998), denaturation kinetics (Gundry et al. 2003), chemical cleavage (Ellis et al. 1998; Tabone et al. 2006), enzyme cleavage (Oleykowski et al. 1998; Goldrick 2001; Sokurenko et al. 2001; Till et al. 2004), array hybridization (Borevitz et al. 2003), mismatch repair detection (Fakhrai-Rad et al. 2004), and bacteriophage Mu DNA transposition (Yanagihara and Mizuuchi 2002; Orsini et al. 2007). The choice of technique used for SNP depends on the regions of interest in the genome and resources available.

The pre-screened polymorphic SNP markers are further utilized for individual genotyping. A number of platforms are currently available for high-throughput SNP genotyping (Kim and Misra 2007). No single technique is ideal for all applications, and the choice of the technique used is often influenced by the need of study. The criterion of choice includes speed of assay development, cost of the assays, and multiplexing potential of the assay (Kwok 2001 ; Kwok and Chen 2003). Various genotyping platforms available have been described below.

7.2.2.1 Genotyping System Based on Capillary Electrophoresis

There are three main multiplex systems available for SNP genotyping using capillary electrophoresis, viz. SNaPshot Multiplex System (Applied Biosystems), SNPlex genotyping system (Applied Biosystems), and LightTyper system (Roche Applied Science). SNaPshot Multiplex System is based upon primer extension, followed by detection using capillary electrophoresis (multiplexing ≤ 10 SNPs with 3 ng DNA). SNPlex genotyping system is based on oligonucleotide ligation assay (PCR-based) followed by capillary electrophoresis (multiplexing up to 48 SNPs) (Vega et al. 2005). LightTyper system (Roche Applied Science) is based upon melting curve analysis to discriminate individual SNPs. These methods use fluorescently labelled oligonucleotides for detection of the SNPs and can process approximately 1.5 million genotypes in just 5 days (Bennett et al. 2005).

7.2.2.2 TaqMan Assay

TaqMan Assay genotypes individual SNPs using a 5′ nuclease assay of *Taq* polymerase to generate a fluorescent signal during PCR (Vega et al. 2005). SNPs are run as sets of 96 or 384 samples, and the assay relies on sequence differences between alleles using different TaqMan probes designed for each allele. The detection is based on FRET assay in which two dyes (5′ reporter dye and a 3′ quencher dye) are covalently linked to the variant allele probes. The proximity between the two dyes suppresses the fluorescence in the intact probes. During annealing step in PCR, the TaqMan probe hybridizes to the target SNP site and the extension leads to the release of reporter and quencher dyes. This release is caused by the 5′ nuclease activity of the *Taq* polymerase which results in fluorescent signal of the reporter dye. Polymerase-based exonuclease activity takes place only in the perfectly hybridized probes (with perfect complementarity), whereas the probe with a mismatched base is not recognized by the polymerase, thus resulting in SNP allele calling. The fluorescent signal for the reporter dye and the quencher dye is measured and the ratio of the signals indicates SNP for the sample.

7.2.2.3 Pyro-sequencing for SNP Genotyping

Pyro-sequencing utilizes a synthesis reaction which results in flashes of light which are produced due to the incorporation of a nucleotide in DNA synthesis, thus detecting a SNP marker (Pourmand et al. 2002). Template generated by PCR is hybridized with a sequencing primer prior to pyro-sequencing. During the sequence elongation step, a single-nucleotide (dNTP) addition results in release of a pyrophosphate molecule, which converts adenosine phosphosulfate (APS) to adenosine triphosphate (ATP) by sulfurylase. This ATP is finally utilized by luciferase enzyme to generate signal in the form of light flashes. The reaction is completed in a few milliseconds, the light produced can be captured with a CCD camera, and the incorporation of the nucleotides is quantified (Royo and Galan 2009). This technique is flexible and accurate and possesses the advantage of parallel processing and automation. The technique also possesses the advantage of not using labelled primers or labelled nucleotides, and gel electrophoresis (Ronaghi 2003). This technology was developed by pyro-sequencing AB (SSE:PYRO) (Uppasala, Sweden). Nowadays, pyro-sequencing services are provided by Roche, Switzerland (454 sequencing).

7.2.2.4 Next-Generation Sequencing (NGS) Methods

Recently introduced next-generation sequencing (NGS) technologies have opened up efficient approaches to generate abundant sequence information as compared to former sequencing methods, viz. Sanger sequencing methods (Pareek et al. 2011; Berkman et al. 2012). This technology can sequence millions of DNA fragments per run and has been used for various sequencing applications, viz. de novo sequencing, resequencing to detect SNPs, transcriptome sequencing, immuno-precipitation-based protein–DNA or protein–RNA interaction mapping, and DNA methylation using bisulfite-mediated cytosine conversion. Initially, three major platforms were commercially used for sequencing, viz. Illumina Genome Analyzer (Solexa), Roche GS FLX Sequencer (454 technology, 454 Life Sciences, Roche), and Applied Biosystems SOLiD sequencer (SOLiD 3 System). The 454 sequencer uses pyro-sequencing methodology (explained earlier). It produces longer reads of ~ 1 kb, as compared to 250-bp reads on Illumina platform. Recently introduced more efficient and affordable sequencing platforms such as Ion Torrent PGM (Personal Genome Machine) (Life Technologies, USA), PacBio RS (Pacific Biosciences, USA), and Hiseq (Illumina, USA) platforms have further advanced the sequencing technology. Ion Torrent is based on semiconductor technology that detects the protons released during nucleotide incorporation (Rothberg et al. 2011), PacBio utilizes a process enabling single-molecule real-time (SMRT) sequencing (Eid et al. 2009), and Illumina Hiseq is based on sequencing by strand synthesis using fluorescently labelled reversible terminator nucleotides (Minoche et al. 2011). These technologies have revolutionized the molecular biology-based methods, thus enabling the genetic analysis of the genomes at high-throughput scales which were not possible earlier with conventional techniques. SNP detection by resequencing is not advisable for high-throughput SNP genotyping as high cost is involved to carry out sequencing on larger sets of genotypes (Lam et al. 2010). In order to reduce the cost, several alternate methods have been developed for SNP genotyping. These methods involve sequencing a fraction of the individual genome. There are three methods, namely reduced representation libraries (RRLs) or complexity reduction of polymorphic sequences (CRoPS), restriction site associated DNA (RAD) sequencing, and genotyping by sequencing (GBS) (Davey et al. 2011). These methods have been discussed below.

Reduced Representation Libraries (RRLs)

Reduced representation library is prepared for CRoPS. RRLs are prepared by restriction of the whole genome and size fractionating target regions which are deep sequenced using available platforms (Luca et al. 2011). It is an economical single-step method which can be utilized for SNP discovery, validation, and characterization. It is also well suited for de novo discovery of SNPs that may be applied to any species with a partially sequenced genome available (Tassell et al. 2008; Hyten et al. 2010). The only disadvantage of this approach is that the amount of DNA required is too high (10–50 μg).

Restriction Site Associated DNA (RAD)

RAD is based on next-generation sequencing for simultaneous discovery and genotyping of thousands of SNP markers in hundreds of individuals with minimal resource utilization (Etter et al. 2011). RAD sequencing is based on a combination of two concepts: restriction digestion of the genome and using molecular identifiers (MID) to associate sequence reads to different individuals. RADseq utilizes unique adapters that bind and amplify restriction sites only. It has recently been used for variety of applications including genetic mapping and QTL analysis (Rowe et al. 2011).

Genotyping by Sequencing (GBS)

Most of the modern sequencing platforms for SNP genotyping per sample is expensive for small-scale users. GBS approach is highly cost-effective for large-scale SNP discovery and genotyping. It involves a simple library preparation which is amenable for a large set of individuals, and it requires very small amounts of DNA ($\sim$100 ng). Restriction library preparation and selection step are introduced to reduce the genome complexity. The number of SNPs genotyped can be increased by 40 % with double depth of coverage as compared to the other techniques (Poland and Trevor 2012; Sonah et al. 2013).

Biplex Invader Technology

Invader assay (Third Wave Technologies, Madison, WI, now Hologic) was introduced as an alternative to TaqMan Assay. A cleavage (flap endonuclease that cleaves predetermined sequence of nucleic acids) recognizes and cleaves an invader structure which is formed due to hybridization of two overlapping oligonucleotides to the target sequence. The signal thus generated by the initial cleavage of flap is further amplified via second fluorescence resonance energy transfer reaction (FRET), and the fluorescent signals generated are detected using a traditional signal reader (Lyamichev and Neri 2003). This method can use PCR products as well as template DNA for the reaction. PCR amplicons are assayed using 384-well plates which are set in Biomek2000 system (Bechman, CA). Plates are read using fluorescent signals, and data are automatically transferred and analysed using two different clustering algorithms (Olivier et al. 2002; Olivier 2005).

MassCode System

MassCode system is based on reporter tags (Agilent Technologies). Theses tags are covalently conjugated to a photocleavable linker (6-amino-1-hexanol) attached to the DNA primers. These oligonucleotide primers are SNP-based which genotype the allele-specific products and differentiate them through their tag assignments. The MassCode liquid array detection system can be run as a single tube or 96-well plate formats. It is based on a cleavage step that depends on the photolysis of the tags from the amplification product. The signal is based on a single quadrupole mass spectrometry detection system called Agilent 1200 Series HiP-ALS (high performance automated liquid sampler, G1367B) for analysis, and genotyping is determined by the relative proportions of the paired allele tags (Richmond et al. 2011). This system can detect as less as femtomolar range (10^{-15} M). There are about 30 different MassCode tags available which can be multiplexed to provide more than 40,000 SNP genotyping data in one day (Kokoris et al. 2000).

WAVE DNA Fragment Analysis System

The effort and cost involved in the detection of sufficient SNPs across larger physical distances on the genome is one of the limiting steps for establishing SNP markers. One of the effective alternatives to the problem is SNP detection using denaturing high-performance liquid chromatography (DHPLC) performed on pooled DNA. This technique is based on the detection of sequence variation in PCR products. This leads to the formation of mismatched heteroduplexes during reannealing of wild-type and mutant DNA, and the melting temperature differences between heteroduplexes and homoduplexes allow separation by IP-RP-DHPLC (ion-pair reversed-phase HPLC), thus identifying variation (Spiegelman et al. 2000).

DHPLC has the efficiency to detect SNPs with frequency less than 5 %, whereas fluorescent sequencing techniques detect variants in the same pools only if the allele frequency is ≥10 %. Thus, pooling of DNA samples in conjunction with the use of DHPLC presents an effective way to increase efficiency of SNP genotyping (Kuklin et al. 1997; Wolford et al. 2000).

Microarray-Based Techniques

Microarray may be defined as an arrayed set of probes for detecting complementary sequences (nucleic acids) present in the target genomes. It is a high-throughput technique which utilizes hybridization followed by signal detection of the target. Microarray-based techniques can be further classified as below.

Golden Gate and Infinium Assays (Illumina)

Golden Gate and Infinium assay platforms are devised by Illumina for high-density assaying of SNP markers (96–1536 SNPs) (Steemers and Gunderson 2007). Golden Gate assay is based on BeadArray technology which involves a multiplexed SNP genotyping reaction. It uses two oligonucleotides for each SNP: one of which is allele-specific and the second is locus-specific. The locus-specific oligonucleotide contains an anti-tag sequence which is used for detection by the BeadArray. Multiple oligonucleotide pools can also be used at one time in order to increase the number of SNPs genotyped per reaction (Fan et al. 2003; Hyten et al. 2008). A custom Oligo Pool Assay (OPA) format in Golden Gate has been developed in crop plants such as Arabidopsis, barley (3000-SNP-based BeadArray) (Hayes and Szucs 2006), soya bean (384-SNP-based BeadArray), and wheat and maize by Southern California Genotyping Consortium (Rostoks et al. 2006).

Infinium assay is also based on BeadArray technology which allows parallel detection of SNPs in a genome. The multiplexing efficiency of Infinium assay relays on the array feature density (Fan et al. 2006). Total genomic DNA is randomly amplified and hybridized to BeadArrays or BeadChips which are microscopic slides containing 12 sections, each section containing 1.1 million beads holding decoded oligonucleotides. These chips may be used for a single sample multiplexing by loading 12 different bead pools for 720,000 assays or 12 different sample multiplexing by loading a single bead pool 12 times for 60,000 assays. Robotic automated chambers have been developed to analyse 24, 48, or 96 BeadChips assaying multiple genomic DNA samples simultaneously (Syvanen 2005; Gunderson et al. 2005).

MegAllele Assays (Affymetrix)

MegAllele genotyping system is based on ParAllele's molecular inversion probe (MIP) method, multiplexing thousands of genotyping reactions. The MIP oligonucleotide probe has recognition sequences at each terminus which hybridize with the genomic sequences in a way that it forms circular structure leaving a gap at the location of SNP. This probe is now allowed to circularize in four separate reaction tubes by providing one of the four dNTPs in each tube. The covalently circularized probes (suggesting the correct allele) are amplified using a universal primer pair. Each amplified probe contains a unique tag complementary to the sequence on universal tag array that helps in detection (Fan et al. 2006; Hardenbol et al. 2005).

SNPstream Genotyping System (Beckman Coulter)

SNPstream assay is based on two step detection method using base-specific polymerase-dependent extension followed by hybridization capture. Probe for hybridization is represented by a unique DNA sequence attached to a solid phase of the SNPware Tag Array plate (384-well microplate format). DNA samples are amplified using tagged extension primers, the extension step involves a single fluorescence-labelled nucleotide terminator reaction, and the final products are hybridized to the complementary unique tags. These tags are fixed to each well of the microplate and the position of the tag in the well confirms the SNP calling. This technique allows high-throughput genotyping of 384 samples for either 12 or 48 SNPs in one array (Bell et al. 2002; Meirmans et al. 2007).

Tagged Microarray Markers (TAMs)

This is a microarray-based method which can be used for scoring thousands of individuals for SNP markers on a glass slide. SNP alleles are amplified, biotin terminated, and spotted on glass slides coated with streptavidin. The SNP detection is based on the hybridization of fluorescent detector oligonucleotides complementary to tags already attached with SNP allele-specific primers. For each SNP loci, two primer pairs are used and tag is detected by hybridization to a concatameric DNA probe which is labelled with multiple set of fluorochromes (Flavell et al. 2003; Jing et al. 2007).

High-Resolution Melting Curve Analysis (HRM)

High-resolution melting curve (HRM) is one of the most recent advances for genotyping SNP markers. This technique monitors the decreasing fluorescence signal of intercalating dye in the process of dissociation of double-stranded DNA in PCR (Wittwer et al. 2003). The strand separation process of short PCR-amplified fragments based on temperature is detected as variation, and it can detect even one base difference between two samples. This technique is very fast, low cost, and efficient (Wu et al. 2008). HRM has been used recently for SNP genotyping in almond (Wu et al. 2008, 2009), wheat (Dong et al. 2009), rice (Li et al. 2011), barley (Lehmensiek et al. 2008; Hofinger et al. 2009), olive (Muleo et al. 2009), and capsicum (Jeong et al. 2010).

7.3 High-Throughput Genic Markers

Genic markers are based on the transcribed regions of genome, viz. RNA, mRNA, ESTs, and cDNA. These markers can be utilized for diversity, genome fingerprinting, and studying gene expression.

Various types of genic markers can be classified as given below.

7.3.1 Serial Analysis of Gene Expression (SAGE)

SAGE is an efficient technique which can be used for genome-wide analysis of gene expression. This technique was developed by Velculescu et al. in 1995. It is a sequence-based technology and quantifies a 'tag' representing the transcriptome product of a gene. It is a high-throughput technique which generates accurate and non-biased transcript expression profile for a genome. cDNA is cleaved with the help of anchoring enzyme (AE), and short sequence tags of 10–14 bp are obtained from a unique position within each transcript. These tags are concatenated together to form long serial molecules, which are cloned and sequenced using modern high-throughput DNA sequencers. The quantification of a particular tag provides the expression level of the respective transcript. Each read allows information for more than 50 transcripts, and the abundance of specific mRNA is defined by the frequency of each tag. This technology possesses the capability to detect minor differences in gene expression (Trendelenburg et al. 2002). Data produced by SAGE contain a list of tags along with their count values, thus providing a digital representation of expression profile for an organism. SAGE has been used for transcript profiling of plants to analyse their response to various biotic and abiotic stresses, host–pathogen interactions, metabolism of stress-induced compounds, and gene expression data of a particular cell, tissue, or organ. SAGE has been successfully employed to study quantitative gene expression in rice (Matsumura et al. 1999) to profile transcript levels in Arabidopsis roots in response to TNT (2,4,6-trinitrotoluene) exposure (Ekman et al. 2003), low temperature (Robinson and Parkin 2008), and profiling of gene expression of CMV (Cassava mosaic virus) in cassava cultivars (Fregene et al. 2004). SAGE has also been reported in maize (Poroyko et al. 2005), rice (Bao et al. 2005), and soya bean (Moy et al. 2004).

7.3.2 Quantitative Nuclease Protection Assay (qNPA)

qNPA platform can be utilized for high-throughput analysis of plant gene expression by detecting the expected changes in gene expression pattern in response to specific treatments. Gene expression is analysed using the plants grown in 96-well plate format or from plant tissues.

Quantitative nuclease protection assays can directly analyse mRNA from formalin-fixed or paraffin-embedded (FFPE) tissues. Probes are designed for genes of interest which are allowed to hybridize with complementary DNA probe–mRNA heteroduplexes. The remaining unhybridized probes and RNA are digested by nucleases which are specific for single-stranded nucleic acids only, and the mRNA from the heteroduplexes is destroyed using alkaline hydrolysis. The remaining intact probe concentrations which are directly proportional to the amounts of specific mRNA present in the sample are transferred to an Array Plate. An oligonucleotide array and a sandwich hybridization linker are used to capture and label the probes for chemiluminescent detection and quantification. This assay can be used to scan small amounts of preserved tissue from large sets of samples (Roberts et al. 2007).

Quantitative gene expression assays can be utilized to deduct the molecular phenotypes (described as the combination of genes whose expression gives rise to specific cellular state) of the samples (Hughes et al. 2000). This platform has been widely used in mammalian cells as well as plant cells to identify small molecules that initiate the expression of defined targeted genes, viz. *Arabidopsis* (Kris et al. 2007; Martel et al. 2002; Roberts et al. 2007). This technique is cost-efficient when compared with microarrays as standard 96-well plates are used to detect the gene expression.

7.3.3 GeneChips for Transcript-Derived Markers (TDMs)

In GeneChip-based TDMs, RNA is hybridized to chips and used for marker development as well as genome-wide gene expression analysis in crop plants. TDMs can be used to construct genetic linkage maps and genome-wide QTL analysis for large population. (Potokina et al. 2008). About 2000 genetic polymorphisms from an experiment have been developed for two commercial varieties of barley (*Hordeum vulgare*; Steptoe and Morex).

The various types of TDMs, viz. expression level polymorphisms (ELPs) or gene expression markers (GEMs), single-feature polymorphisms (SFPs), copy number variations (CNVs), and presence–absence variations (PAVs), are developed from sequenced transcripts.

7.4 Expression Level Polymorphisms (ELPs) or Gene Expression Markers (GEMs)

GEMs are developed based on differential gene expression or transcript level differences that show bimodal distribution in segregating progeny. These can be developed from any type of DNA-based expression microarray technology (West et al. 2006). Affymetrix GeneChip expression data are used to survey ELPs. ELPs are generally considered as e-traits to establish expression QTL (eQTL) (Jordan et al. 2010; West et al. 2006).

7.4.1 Single-Feature Polymorphisms (SFPs)

SFP is nucleotide variation captured by using expression array. It represents sequence variation between two genotypes within an individual oligonucleotide probe that can be detected by hybridization or affinity differences (Borevitz et al. 2003, 2007). SFPs within a transcribed gene may reflect in phenotypic differences. Sequencing of SFPs confirms that they come into existence as a result of SNPs. However, a SFP is only a prediction of a possible SNP within an oligonucleotide. SFPs have been reported in Arabidopsis (West et al. 2006; Borevitz et al. 2007), rice (Kumar et al. 2007; Kim et al. 2009), wheat (Bernardo et al. 2009; Banks et al. 2009), cowpea (Das et al. 2008), and barley (Rostoks et al. 2005).

7.4.2 Copy Number Variations (CNVs)

CNV describes the genomic rearrangements resulting from gain or loss of DNA segments (larger than 1 kb) (Shaikh et al. 2009). CNVs are found on all chromosomes, more concentrated in regions devoid of genes, although they also represent the regions that contain protein-coding genes or important regulatory elements (Redon et al. 2006). CNVs originate in a genome due to non-allelic homologous recombination (NAHR) events among low copy number, highly similar but non-allelic DNA segments, whereas segments larger than 1 kb originate due to ancient duplication events (Gu et al. 2008). The most efficient approaches to establish high-throughput CNVs are array-based comparative genome hybridization (CGH) and NGS based on reference genome. CGH involves immobilization of DNA probes on an array and hybridization of the target sequences. The detection and resolution of the target signal are determined by the number and type of immobilized probes on the array. The copy number is deduced based on the relative amounts of signal produced by tested sample and reference genome. The capturing of the target sequences is directly monitored by the sequence homology with the probes. DNA segments which are present in the tested genome but absent in the reference genome are not detected because of the lack of complementary probes (Springer et al. 2009). Discovering CNVs using NGS methodology involves bioinformatics approach involving data analysis for calculating the absolute copy number of various genomic segments by finding the relative differences (increase or decrease) in the sequence coverage by mapping the short reads to a reference genome and assembling non-mapping reads for the discovery of new sequences. CNVs have been established in maize (Springer et al. 2009; Belo et al. 2010; Swanson-Wagner et al. 2010), rice (Yu et al. 2011), and soybean (McHale et al. 2012).

7.4.3 Presence–Absence Variations (PAVs)

PAV is used to describe the sequences that are present in one genome but totally absent in the other genomes when compared. The presence/absence variants may also possess the structural variations that could be a result of the presence of segmental DNA (>10-bp) in one genome, which is altogether absent in another genome. PAVs generally affect a larger portion of the genome as compared to other sequence variations, viz. SNPs or insertions or deletions (Zheng et al. 2011; Zhang et al. 2014). PAVs are used to examine the phenotypic variations in a set of genotypes, and the information thus generated can be utilized for fine-mapping the genes of interest (Salathia et al. 2007). Putative PAVs are discovered by aligning paired-end reads with already known insert sizes from the sequenced reference genome, and these alignments consider the assumption that the paired-end reads should not align in regions containing structural variations (Zheng et al. 2011). Two primer pairs are designed for each PAV, and these primers are used for massive genotyping programmes. In plants, PAVs have been discovered in maize (Springer et al. 2009), *Arabidopsis thaliana* (Borevitz et al. 2007; Tan et al. 2012), and *Glycine max* (Lam et al. 2010).

7.4.4 Target Region Amplification Polymorphism (TRAP)

TRAP is an efficient and high-throughput PCR-based marker technique which uses already reported EST database to generate polymorphic markers in close proximity to targeted genes (Hu and Vick 2003). A set of 18 nucleotides long primers are designed (one from the target EST and second being arbitrary with AT- or GC-rich core sequence to anneal with the intron or exon, respectively) to generate markers. Amplified products are separated on polyacrylamide sequencing gel, and data are analysed. TRAP has been used in genotyping valuable germplasm containing desirable gene pools and marker-based tagging of the genes governing agronomically important traits of crop plants (Hu and Vick 2003). This technique has been utilized for QTL-mapping in wheat (Liu et al. 2005; Chu et al. 2008), genetic fingerprinting in lettuce cultivars (Hu et al. 2005), mapping and gene tagging in common bean (Miklas et al. 2006), and genetic diversity analysis in sugar cane (Alwala 2006).

7.4.5 Single-Nucleotide Polymorphisms (SNPs)

Genic SNP markers are discovered from the transcribed regions of the genome. The techniques which can be utilized for genotyping genic SNPs have been discussed earlier in this chapter in the section of genomic high-throughput markers.

Development of high-throughput techniques for gene pool analysis is a dynamic tool which will keep on improving in the future. The older technologies will be replaced by new sophisticated ones, and every development will keep on revolutionizing our understanding of the germplasm and its utilization for crop improvement programmes. Among all the marker types, SNPs seem to have a greater influence on genotyping for crop improvement programmes in future because of its higher abundance in the genomes (Koebner and Summers 2003).

Similarly, among the recent marker detection (validation) assays, DNA chips have revolutionized the germplasm characterization and genome mapping-based plant breeding methodologies. But with the rapid development of highly efficient, low cost, and faster next-generation sequencing (NGS) technologies, GBS will further boost up the molecular marker-based crop improvement programmes (Thompson 2014).

References

Agarwal M, Shrivastava N, Padh H (2008) Advances in molecular marker techniques and their applications in plant sciences. Plant Cell Rep 27(4):617–631

Akbari M, Wenzl P, Caig V, Carling J, Xia L, Yang S, Uszynski G et al (2006) Diversity arrays technology (DArT) for high-throughput profiling of the hexaploid wheat genome. Theor Appl Genet 113(8):1409–1420

Alsop BP, Farre A, Wenzl P, Wang JM, Zhou MX, Romagosa I, Kilian A, Steffenson BJ (2011) Development of wild barley-derived DArT markers and their integration into a barley consensus map. Mol Breed 27(1):77–92

Alwala S, Suman A, Arro JE, Veremis JC, Kimbeng CA (2006) Target region amplification polymorphism (TRAP) for assessing genetic diversity in sugarcane germplasm collections. Crop Sci 46(1):448–455

Banks TW, Jordan MC, Somers DJ (2009) Single-feature polymorphism mapping in bread wheat. The Plant Genome 2(2):167–178

Bao J, Lee S, Chen C, Zhang X, Zhang Y, Liu S, Clark T et al (2005) Serial analysis of gene expression study of a hybrid rice strain (LYP9) and its parental cultivars. Plant Physiol 138(3):1216–1231

Bell PA, Chaturvedi S, Gelfand CA, Huang CY, Kochersperger M, Kopla R, Modica F et al (2002) SNPstream UHT: ultra-high throughput SNP genotyping for pharmacogenomics and drug discovery. Biotechniques 32:70–72

Belo A, Beatty MK, Hondred D, Fengler KA, Li B, Rafalski A (2010) Allelic genome structural variations in maize detected by array comparative genome hybridization. Theor Appl Genet 120(2):355–367

Bennett ST, Barnes C, Cox A, Davies L, Brown C (2005) Toward the $1000 human genome, pp 373–382

Berkman PJ, Lai K, Lorenc MT, Edwards D (2012) Next-generation sequencing applications for wheat crop improvement. Am J Bot 99(2):365–371

Bernardo AN, Bradbury PJ, Ma H, Hu S, Bowden RL, Buckler ES, Bai G (2009) Discovery and mapping of single feature polymorphisms in wheat using Affymetrix arrays. BMC Genom 10(1):251

Borevitz JO, Liang D, Plouffe D, Chang HS, Zhu T, Weigel D, Berry CC, Winzeler E, Chory J (2003) Large-scale identification of single-feature polymorphisms in complex genomes. Genome Res 13:513–523

Borevitz JO, Hazen SP, Michael TP, Morris GP, Baxter IR, Hu TT, Chen H et al (2007) Genome-wide patterns of single-feature polymorphism in *Arabidopsis thaliana*. PNAS 104(29):12057–12062

Brinez B, Blair MW, Kilian A, Carbonell SAM, Chiorato F, Rubiano LB (2012) A whole genome DArT assay to assess germplasm collection diversity in common beans. Mol Breed 30(1): 181–193

Chu CG, Xu SS, Friesen TL, Faris JD (2008) Whole genome mapping in a wheat doubled haploid population using SSRs and TRAPs and the identification of QTL for agronomic traits. Mol Breed 22(2):251–266

Crossa J, Burgueno J, Dreisigacker S, Vargas M, Herrera-Foessel SA, Lillemo M, Singh RP et al (2007) Association analysis of historical bread wheat germplasm using additive genetic covariance of relatives and population structure. Genetics 177(3):1889–1913

Das S, Bhat PR, Sudhakar C, Ehlers JD, Wanamaker S, Roberts PA, Cui X, Close TJ (2008) Detection and validation of single feature polymorphisms in cowpea (*Vigna unguiculata* L. Walp) using a soybean genome array. BMC Genom 9(1):107

Davey JW, Hohenlohe PA, Etter PD, Boone JQ, Catchen JM et al (2011) Genome-wide genetic marker discovery and genotyping using next-generation sequencing. Nat Rev Genet 12(7):499–510

Dong C, Vincent K, Sharp P (2009) Simultaneous mutation detection of three homoeologous genes in wheat by High Resolution Melting analysis and Mutation Surveyor®. BMC Plant Biol 9(1):143

Edenberg HJ, Liu Y (2009) Laboratory methods for high-throughput genotyping. Cold Spring Harb Protoc 2009(11):pdb-top62

Eid J, Fehr A, Gray J, Luong K, Lyle J, Otto G, Peluso P et al (2009) Real-time DNA sequencing from single polymerase molecules. Science 323(5910):133–138

Ekman DR, Lorenz WW, Przybyla AE, Wolfe NL, Dean JFD (2003) SAGE analysis of transcriptome responses in *Arabidopsis* roots exposed to 2,4,6-trinitrotoluene. Plant Physiol 133(3):1397–1406

Ellis TP, Humphrey KE, Smith MJ, Cotton RGH (1998) Chemical cleavage of mismatch: a new look at an established method. Hum Mutat 11:345–353

Etter PD, Bassham S, Hohenlohe PA, Johnson EA, Cresko WA (2011) SNP discovery and genotyping for evolutionary genetics using RAD sequencing. Molecular methods for evolutionary genetics. Humana Press, New York, pp 157–178

Fakhrai-Rad H, Zheng J, Willis TD, Wong K, Suyenaga K, Moorhead M, Eberle J, Thorstenson YR (2004) SNP discovery in pooled samples with mismatch repair detection. Genome Res 14:1404–1412

Fan JB, Oliphant A, Shen R, Kermani BG, Garcia F, Gunderson KL, Hansen M et al (2003) Highly parallel SNP genotyping. Cold Spring Harbor symposia on quantitative biology, vol 68. Cold Spring Harbor Laboratory Press, New York, pp 69–78

Fan JB, Gunderson KL, Bibikova M, Yeakley JM, Chen J, Garcia EW, Lebruska LL, Laurent M, Shen R, Barker D (2006) [3] Illumina universal bead arrays. Methods Enzymol 410:57–73

Flavell AJ, Bolshakov VN, Booth A, Jing R, Russell J, Ellis THN, Isaac P (2003) A microarray-based high throughput molecular marker genotyping method: the tagged microarray marker (TAM) approach. Nucleic Acids Res 31(19):115

Fregene M, Matsumura H, Akano A, Dixon A, Terauchi R (2004) Serial analysis of gene expression (SAGE) of host–plant resistance to the cassava mosaic disease (CMD). Plant Mol Biol 56(4):563–571

Glaszmann JC, Kilian B, Upadhyaya HD, Varshney RK (2010) Accessing genetic diversity for crop improvement. Curr Opin Plant Biol 13(2):167–173

Goldrick MM (2001) RNase cleavage-based methods for mutation/SNP detection, past and present. Hum Mutat 18:190–204

Gu W, Zhang F, Lupski JR (2008) Mechanisms for human genomic rearrangements. Pathogenetics 1(1):4

Gunderson KL, Steemers FJ, Lee G, Mendoza LG, Chee MS (2005) A genome-wide scalable SNP genotyping assay using microarray technology. Nat Genet 37(5):549–554

Gundry CN, Vandersteen JG, Reed GH, Pryor RJ, Chen J, Wittwer CT (2003) Amplicon melting analysis with labelled primers: a closed-tube method for differentiating homozygotes and heterozygotes. Clin Chem 49:396–406

Gupta PK, Roy JK, Prasad M (2001) Single nucleotide polymorphisms (SNPs): a new paradigm in molecular marker technology and DNA polymorphism detection with emphasis on their use in plants. Curr Sci 80(4):524–535

Hajjar R, Hodgkin T (2007) The use of wild relatives in crop improvement: a survey of developments over the last 20 years. Euphytica 156(1–2):1–13

Hamblin MT, Buckler ES, Jannink JL (2011) Population genetics of genomics-based crop improvement methods. Trends Genet 27(3):98–106

Hardenbol P, Yu F, Belmont J, MacKenzie J, Bruckner C, Brundage T, Boudreau A et al (2005) Highly multiplexed molecular inversion probe genotyping: over 10,000 targeted SNPs genotyped in a single tube assay. Genome Res 15(2):269–275

Hayes P, Szucs P (2006) Disequilibrium and association in barley: thinking outside the glass. Proc Natl Acad Sci 103(49):18385–18386

Hearnden PR, Eckermann PJ, McMichael GL, Hayden MJ, Eglinton JK, Chalmers KJ (2007) A genetic map of 1000 SSR and DArT markers in a wide barley cross. Theor Appl Genet 115(3):383–391

Hofinger BJ, Hai-Chun J, Hammond-Kosack KE, Kanyuka K (2009) High-resolution melting analysis of cDNA-derived PCR amplicons for rapid and cost-effective identification of novel alleles in barley. Theor Appl Genet 119(5):851–865

Hu J, Vick BA (2003) Target region amplification polymorphism: a novel marker technique for plant genotyping. PMB Reporter 21(3):289–294

Hu J, Ochoa OE, Truco MJ, Vick BA (2005) Application of the TRAP technique to lettuce (*Lactuca sativa* L.) genotyping. Euphytica 144(3):225–235

Hughes TR, Marton MJ, Jones AR, Roberts CJ, Stoughton R, Armour CD, Bennett HA et al (2000) Functional discovery via a compendium of expression profiles. Cell 102(1):109–126

Hyten DL, Song Q, Choi IY, Yoon MS, Specht JE, Matukumalli LK, Nelson RL, Shoemaker RC, Young ND, Cregan PB (2008) High-throughput genotyping with the GoldenGate assay in the complex genome of soybean. Theor Appl Genet 116(7):945–952

Hyten DL, Song Q, Fickus EW, Quigley CV, Lim JS, Choi IY, Hwang EY, Pastor-Corrales M, Cregan PB (2010) High-throughput SNP discovery and assay development in common bean. BMC Genom 11(1):475

Jaccoud D, Peng K, Feinstein D, Kilian A (2001) Diversity arrays: a solid state technology for sequence information independent genotyping. Nucleic Acids Res 29(4):25

Jeong HJ, Jo YD, Soung-Woo P, Byoung-Cheorl K (2010) Identification of capsicum species using SNP markers based on high resolution melting analysis. Genome 53(12):1029–1040

Jing R, Bolshakov V, Flavell AJ (2007) The tagged microarray marker (TAM) method for high-throughput detection of single nucleotide and indel polymorphisms. Nat Protoc 2(1): 168–177

Jordan DR, Mace ES, Henzell RG, Klein PE, Klein RR (2010) Molecular mapping and candidate gene identification of the Rf2 gene for pollen fertility restoration in sorghum [*Sorghum bicolor* (L.) Moench]. Theor Appl Genet 120(7):1279–1287

Kim S, Misra A (2007) SNP genotyping: technologies and biomedical applications. Annu Rev Biomed Eng 9:289–320

Kim SH, Bhat PR, Cui X, Walia H, Xu J, Wanamaker S, Ismail AM, Wilson C, Close TJ (2009) Detection and validation of single feature polymorphisms using RNA expression data from a rice genome array. BMC Plant Biol 9(1):65

King J, Thomas A, James C, King I, Armstead I (2013) A DArT marker genetic map of perennial ryegrass (*Lolium perenne* L.) integrated with detailed comparative mapping information; comparison with existing DArT marker genetic maps of *Lolium perenne*, L. multiflorum and *Festuca pratensis*. BMC Genom 14(1):437

Koebner R, Summers RW (2003) 21st century wheat breeding: plot selection or plate detection? Trends Biotechnol 21(2):59–63
Kokoris M, Dix K, Moynihan K, Mathis J, Erwin B, Grass P, Hines B, Duesterhoeft A (2000) High-throughput SNP genotyping with the Masscode system. Mol Diagn 5(4):329–340
Kris RM, Felder S, Deyholos M, Lambert GM, Hinton J, Botros I, Martel R, Seligmann B, Galbraith DW (2007) High-throughput, high-sensitivity analysis of gene expression in *Arabidopsis*. Plant Physiol 144(3):1256–1266
Kuklin A, Munson K, Gjerde D, Haefele R, Taylor P (1997) Detection of single-nucleotide polymorphisms with the WAVE™ DNA fragment analysis system. Genet Test 1(3):201–206
Kumar R, Qiu J, Joshi T, Valliyodan B, Xu D, Nguyen HT (2007) Single feature polymorphism discovery in rice. PLoS ONE 2(3):284
Kwok PY (2001) Methods for genotyping single nucleotide polymorphisms. Annu Rev Genomics Hum Genet 2:235–258
Kwok PY, Chen X (2003) Detection of single nucleotide polymorphisms. Curr Issues Mol Biol 5:43–60
Lam HM, Xu X, Liu X, Chen W, Yang G, Wong FL, Li MW et al (2010) Resequencing of 31 wild and cultivated soybean genomes identifies patterns of genetic diversity and selection. Nat Genet 42(12):1053–1059
Lee M (1998) Genome projects and gene pools: new germplasm for plant breeding? Proc Natl Acad Sci 95(5):2001–2004
Lehmensiek A, Sutherland MW, McNamara RB (2008) The use of high resolution melting (HRM) to map single nucleotide polymorphism markers linked to a covered smut resistance gene in barley. Theor Appl Genet 117(5):721–728
Li J, Wang X, Dong R, Yang Y, Zhou J, Yu C, Cheng Y, Yan C, Chen J (2011) Evaluation of high-resolution melting for gene mapping in rice. Plant Mol Biol Reporter 29(4):979–985
Liu ZH, Anderson JA, Hu J, Friesen TL, Rasmussen JB, Faris JD (2005) A wheat intervarietal genetic linkage map based on microsatellite and target region amplified polymorphism markers and its utility for detecting quantitative trait loci. Theor Appl Genet 111(4):782–794
Luca F, Hudson RR, Witonsky DB, Rienzo AD (2011) A reduced representation approach to population genetic analyses and applications to human evolution. Genome Res 21(7):1087–1098
Luikart G, England PR, Tallmon D, Jordan S, Taberlet P (2003) The power and promise of population genomics: from genotyping to genome typing. Nat Rev Genet 4(12):981–994
Lyamichev V, Neri B (2003) Invader assay for SNP genotyping. Single nucleotide polymorphisms. Springer, New York, pp 229–240
Mace ES, Xia L, Jordan DR, Halloran K, Parh DK, Huttner E, Wenzl P, Kilian A (2008) DArT markers: diversity analyses and mapping in *Sorghum bicolor*. BMC Genom 9(1):26
Mace ES, Jean-Francois Rami, Bouchet S, Klein PE, Klein RR, Kilian A, Wenzl P, Xia L, Halloran K, Jordan DR (2009) A consensus genetic map of *Sorghum* that integrates multiple component maps and high-throughput diversity array technology (DArT) markers. BMC Plant Biol 9(1):13
Martel RR, Botros IW, Rounseville MP, Hinton JP, Staples RR, Morales DA, Farmer JB, Seligmann BE (2002) Multiplexed screening assay for mRNA combining nuclease protection with luminescent array detection. Assay Drug Dev Technol 1(1):61–71
Matsumura H, Nirasawa S, Terauchi R (1999) Transcript profiling in rice (*Oryza sativa* L.) seedlings using serial analysis of gene expression (SAGE). Plant J 20(6):719–726
Matukumalli LK, Grefenstette JJ, Hyten DL, Choi IY, Cregan PB, Van Tassell CP (2006) SNP-PHAGE-high throughput SNP discovery pipeline. BMC Bioinf 7:468
McHale LK, Haun WJ, Xu WW, Bhaskar PB, Anderson JE, Hyten DL, Gerhardt DJ, Jeddeloh JA, Stupar JM (2012) Structural variants in the soybean genome localize to clusters of biotic stress-response genes. Plant Physiol 159(4):1295–1308
Meirmans PG, Lamothe M, Perinet P, Isabel N (2007) Species-specific single nucleotide polymorphism markers for detecting hybridization and introgression in poplar (This article is one of a selection of papers published in the special issue on Poplar Research in Canada). Botany 85(11):1082–1091

Miklas PN, Hu J, Grunwald NJ, Larsen KM (2006) Potential application of TRAP (targeted region amplified polymorphism) markers for mapping and tagging disease resistance traits in common bean. Crop Sci 46(2):910–916

Miller TP, Gu Z, Li Q, Hillier L, Kwok PY (1998) Overlapping genomic sequences: a treasure trove of single-nucleotide polymorphisms. Genome Res 8(7):748–754

Minoche AE, Dohm JC, Himmelbauer H (2011) Evaluation of genomic high-throughput sequencing data generated on Illumina HiSeq and genome analyzer systems. Genome Biol 12(11):112

Moy P, Qutob D, Chapman BP, Atkinson I, Gijzen M (2004) Patterns of gene expression upon infection of soybean plants by *Phytophthora sojae*. Mol Plant Microbe Interact 17(10): 1051–1062

Muleo R, Colao MC, Miano D, Cirilli M, Intrieri MC, Baldoni L, Rugini E (2009) Mutation scanning and genotyping by high-resolution DNA melting analysis in olive germplasm. Genome 52(3):252–260

Oleykowski CA, Mullins SRB, Godwin AK, Yeung AT (1998) Mutation detection using a novel plant endonuclease. Nucleic Acids Res 26:4597–4602

Olivier M (2005) The Invader® assay for SNP genotyping. Mutat Res Fundam Mol Mech Mutagen 573(1):103–110

Olivier M, Chuang LM, Chang MS, Chen YT, Pei D, Ranade K, Witte A et al (2002) High-throughput genotyping of single nucleotide polymorphisms using new biplex invader technology. Nucleic Acids Res 30(12):53

Orita M, Iwahana H, Kanazawa H, Hayashi K, Sekiya T (1989) Detection of polymorphisms of human DNA by gel electrophoresis as single-strand conformation polymorphisms. Proc Natl Acad Sci USA 86:2766–2770

Orsini L, Pajunen M, Hanski I, Savilahti H (2007) SNP discovery by mismatch-targeting of Mu transposition. Nucleic Acids Res 35:44

Pareek CS, Smoczynski R, Tretyn A (2011) Sequencing technologies and genome sequencing. J Appl Genet 52(4):413–435

Petroli CD, Carolina PS, Carling J, Dorothy AS, Vaillancourt RE, Myburg AA, da Silva Jr OB, Pappas GJ Jr, Kilian A, Grattapaglia D (2012) Genomic characterization of DArT markers based on high-density linkage analysis and physical mapping to the *Eucalyptus* genome. PLoS ONE 7(9):44684

Poland JA, Trevor WR (2012) Genotyping-by-sequencing for plant breeding and genetics. The Plant Genome 5(3):92–102

Poroyko V, Hejlek LG, Spollen WG, Springer GK, Nguyen HT, Sharp RE, Bohnert HJ (2005) The maize root transcriptome by serial analysis of gene expression. Plant Physiol 138(3):1700–1710

Potokina E, Druka A, Luo Z, Wise R, Waugh R, Kearsey M (2008) Gene expression quantitative trait locus analysis of 16,000 barley genes reveals a complex pattern of genome-wide transcriptional regulation. Plant J 53(1):90–101

Pourmand N, Elahi E, Davis RW, Ronaghi M (2002) Multiplex pyrosequencing. Nucleic Acids Res 30(7):31

Przyborowski JA, Sulima P, Kuszewska A, Załuski D, Kilian A (2013) Phylogenetic relationships between four *Salix* L. species based on DArT markers. Int J Mol Sci 14(12):24113–24125

Redon R, Ishikawa S, Fitch KR, Feuk L, Perry GH, Andrews TD, Fiegler et al (2006) Global variation in copy number in the human genome. Nature 444(7118):444–454

Richmond GS, Khine H, Zhou TT, Ryan DE, Brand T, McBride MT, Killeen K (2011) MassCode liquid arrays as a tool for multiplexed high-throughput genetic profiling. PLoS ONE 6(4):18967

Roberts RA, Sabalos CM, LeBlanc ML, Martel RR, Frutiger YM, Unger JM, Botros IW et al (2007) Quantitative nuclease protection assay in paraffin-embedded tissue replicates prognostic microarray gene expression in diffuse large-B-cell lymphoma. Lab Invest 87(10):979–997

Robinson SJ, Parkin IAP (2008) Differential SAGE analysis in *Arabidopsis* uncovers increased transcriptome complexity in response to low temperature. BMC Genom 9(1):434

Ronaghi M (2003) Pyrosequencing for SNP genotyping. Methods in molecular biology™, vol 212. Springer, New York, pp 189–195

Rostoks N, Borevitz JO, Hedley PE, Russell J, Mudie S, Morris J, Cardle L, Marshall DF, Waugh R (2005) Single-feature polymorphism discovery in the barley transcriptome. Genome Biol 6(6):R54

Rostoks N, Ramsay L, MacKenzie K, Cardle L, Bhat PR, Roose ML, Svensson JT et al (2006) Recent history of artificial outcrossing facilitates whole-genome association mapping in elite inbred crop varieties. Proc Natl Acad Sci 103(49):18656–18661

Rothberg JM, Hinz W, Rearick TM, Schultz J, Mileski W, Davey M, Leamon JH et al (2011) An integrated semiconductor device enabling non-optical genome sequencing. Nature 475(7356): 348–352

Rowe HC, Renaut S, Guggisberg A (2011) RAD in the realm of next-generation sequencing technologies. Mol Ecol 20(17):3499–3502

Royo JL, Galan JJ (2009) Pyrosequencing for SNP genotyping. Single nucleotide polymorphisms. Humana Press, New York, pp 123–133

Salathia N, Lee HN, Sangster TA, Morneau K, Landry CR, Schellenberg K, Behere AS, Gunderson KL, Cavalieri D, Jander G, Queitsch C (2007) Indel arrays: an affordable alternative for genotyping. Plant J 51:727–737

Semagn K, Bjornstad A, Skinnes H, Maroy AG, Tarkegne Y, William M (2006) Distribution of DArT, AFLP, and SSR markers in a genetic linkage map of a doubled-haploid hexaploid wheat population. Genome 49(5):545–555

Shaikh TH, Gai X, Perin JC, Glessner JT, Xie H, Murphy K, O'Hara R et al (2009) High-resolution mapping and analysis of copy number variations in the human genome: a data resource for clinical and research applications. Genome Res 19(9):1682–1690

Sokurenko EV, Tchesnokova V, Yeung AT, Oleykowski CA, Trintchina E, Hughes KT, Rashid RA, Brint JM (2001) Detection of simple mutations and polymorphisms in large genomic regions. Nucleic Acids Res 29:e111

Sonah H, Bastien M, Iquira E, Tardivel A, Legare G et al (2013) An improved genotyping by sequencing (GBS) approach offering increased versatility and efficiency of SNP discovery and genotyping. PLoS ONE 8(1):54603

Spiegelman JI, Mindrinos MN, Oefner PJ (2000) High-accuracy DNA sequence variation screening by DHPLC. Biotechniques 29(5):1084–1093

Springer NM, Ying K, Fu Y, Ji T, Yeh Cheng-Ting J, Yi WuW et al (2009) Maize inbreds exhibit high levels of copy number variation (CNV) and presence/absence variation (PAV) in genome content. PLoS Genet 5(11):1000734

Steemers FJ, Gunderson KL (2007) Whole genome genotyping technologies on the BeadArray™ platform. Biotechnol J 2(1):41–49

Steffenson BJ, Olivera P, Roy JK, Jin Y, Smith KP, Muehlbauer GJ (2007) A walk on the wild side: mining wild wheat and barley collections for rust resistance genes. Crop Pasture Sci 58(6):532–544

Suh Y, Vijg J (2005) SNP discovery in associating genetic variation with human disease phenotypes. Mutat Res 573:41–53

Swanson-Wagner RA, Eichten SR, Kumari S, Tiffin P, Stein JC, Ware D, Springer NM (2010) Pervasive gene content variation and copy number variation in maize and its undomesticated progenitor. Genome Res 20(12):1689–1699

Syvanen AC (2005) Toward genome-wide SNP genotyping. Nat Genet 37:S5–S10

Tabone T, Sallmann G, Webb E, Cotton RG (2006) Detection of 100 % of mutations in 124 individuals using a standard UV/Vis microplate reader: a novel concept for mutation scanning. Nucleic Acids Res 34:45

Tan S, Zhong Y, Hou H, Yang S, Tian D (2012) Variation of presence/absence genes among Arabidopsis populations. BMC Evol Biol 12(1):86

Tassell CPV, Smith TPL, Matukumalli LK, Taylor JF, Schnabel RD, Lawley CT, Haudenschild CD, Moore SS, Warren WC, Sonstegard TS (2008) SNP discovery and allele frequency estimation by deep sequencing of reduced representation libraries. Nat Methods 5:247–252

Thomson MJ (2014) High-throughput SNP genotyping to accelerate crop improvement. Plant Breed Biotech 2:195–212

Till BJ, Burtner C, Comai L, Henikoff S (2004) Mismatch cleavage by single-strand specific nucleases. Nucleic Acids Res 32:2632–2641

Tinker NA, Kilian A, Wight CP, Heller-Uszynska K, Wenzl P, Rines HW, Bjornstad A et al (2009) New DArT markers for oat provide enhanced map coverage and global germplasm characterization. BMC Genom 10(1):39

Trendelenburg G, Prass K, Priller J, Kapinya K, Polley A, Muselmann C, Ruscher K, Kannbley U, Schmitt AO, Castell S, Wiegand F, Meisell A, Rosenthal A, Dirnagl U (2002) Serial analysis of gene expression identifies metallothionein-II as major neuroprotective gene in mouse focal cerebral ischemia. J of Neuroscience 22(14): 5879-5888

Twyman RM (2004) SNP discovery and typing technologies for pharmacogenomics. Curr Top Med Chem 4:1423–1431

Vega FM, Lazaruk KD, Rhodes MD, Wenz MH (2005) Assessment of two flexible and compatible SNP genotyping platforms: TaqMan® SNP genotyping assays and the SNPlex™ genotyping system. Mut Res Fundam Mol Mech Mutagen 573(1):111–135

Velculescu VE, Zhang L, Vogelstein B, Kinzler KW (1995) Serial analysis of gene expression. Science 270:484–487

Wenzl P, Carling J, Kudrna D, Jaccoud D, Huttner E, Kleinhofs A, Kilian A (2004) Diversity arrays technology (DArT) for whole-genome profiling of barley. PNAS 101(26):9915–9920

Wenzl P, Li H, Carling J, Zhou M, Raman H, Paul E, Hearnden P et al (2006) A high-density consensus map of barley linking DArT markers to SSR, RFLP and STS loci and agricultural traits. BMC Genom 7(1):206

West MAL, Leeuwen HV, Kozik A, Kliebenstein DJ, Doerge RW, Clair SDA, Michelmore RW (2006) High-density haplotyping with microarray-based expression and single feature polymorphism markers in *Arabidopsis*. Genome Res 16(6):787–795

White J, Law JR, MacKay I, Chalmers KJ, Smith JSC, Kilian A, Powell W (2008) The genetic diversity of UK, US and Australian cultivars of *Triticum aestivum* measured by DArT markers and considered by genome. Theor Appl Genet 116(3):439–453

Wittwer CT, Reed GH, Gundry CN, Vandersteen JG et al (2003) High-resolution genotyping by amplicon melting analysis using LCGreen. Clin Chem 49:853–860

Wolford J, Blunt D, Ballecer C, Prochazka M (2000) High-throughput SNP detection by using DNA pooling and denaturing high performance liquid chromatography (DHPLC). Hum Genet 107(5):483–487

Wu SB, Wirthensohn MG, Hunt P, Gibson JP, Sedgley M (2008) High resolution melting analysis of almond SNPs derived from ESTs. Theor Appl Genet 118(1):1–14

Wu SB, Tavassolian I, Rabiei G, Hunt P, Wirthensohn M, Gibson JP, Ford CM, Sedgley M (2009) Mapping SNP-anchored genes using high-resolution melting analysis in almond. Mol Genet Genomics 282(3):273–281

Xie Y, McNally K, Li CY, Leung H, Zhu YY (2006) A high-throughput genomic tool: diversity array technology complementary for rice genotyping. J Integr Plant Biol 48(9):1069–1076

Yanagihara K, Mizuuchi K (2002) Mismatch-targeted transposition of Mu: a new strategy to map genetic polymorphism. Proc Natl Acad Sci USA 99:11317–11321

Yang SY, Saxena RK, Kulwal PL, Ash GJ, Dubey A, Harper JD, Upadhyaya HD, Gothalwal R, Kilian A, Varshney RK (2011) The first genetic map of pigeon pea based on diversity arrays technology (DArT) markers. J Genet 90(1):103–109

Yu P, Wang C, Xu Q, Feng Y, Yuan X, Yu H, Wang Y, Tang S, Wei X (2011) Detection of copy number variations in rice using array-based comparative genomic hybridization. BMC genomics 12(10):372

Zhang LM, Luo H, Liu ZQ, Zhao Y, Luo JC, Hao DY, Jing HC (2014) Genome-wide patterns of large-size presence/absence variants in sorghum. J Integr Plant Biol 56(1):24–37

Zheng LY, Guo XS, He B, Sun LJ, Peng Y, Dong SS, Liu TF et al (2011) Genome-wide patterns of genetic variation in sweet and grain sorghum (*Sorghum bicolor*). Genome Biol 12(11):R114

Chapter 8
Introgression and Exploitation of QTL for Yield and Yield Components from Related Wild Species in Rice Cultivars

Kumari Neelam, Kishor Kumar, Harcharan Singh Dhaliwal and Kuldeep Singh

Abstract Green revolution led to the narrowing of the genetic base of cultivated rice gene pool. Genetic diversity is the prerequisite for increasing yield and for stabilizing production under series of biotic and abiotic stresses. The wild *Oryza* species comprising AA, BB, CC, BBCC, CCDD, EE, FF, GG, HH, and JJ genomes are the important reservoir of useful genes. The wild relatives of crop species with hidden potential for useful variability are, however, phenotypically less desirable than the modern cultivars in their overall appearance. They have been utilized extensively for introgression of major genes for disease and insect resistance, but their utilization in enhancing yield and yield-related traits of modern cultivars has remained limited. The related wild species *Oryza rufipogon* (AA genome) has been utilized widely for transferring yield and yield-related traits to the elite rice cultivars followed by reports on *O. glaberrima*, *O. minuta*, *O. nivara,* and *O. glumaepatula*. The availability of advance molecular breeding techniques has enabled the use of alien species with minimum linkage drag. Yield QTLs have been identified on almost all the rice chromosomes though the QTL clusters are confined to only four (1, 2, 3, and 4) chromosomes. Some of the component traits of yield have higher heritability and correlation among themselves. This provides an opportunity for their simultaneous improvement for more than one trait using marker-assisted selection. Many QTLs from different wild species are mapped to the identical chromosomal regions, thus giving an idea of orthologous yield QTLs across the species and populations. This chapter deals with the utilization of wild species for introgression of QTLs for yield and yield-related traits for the improvement of rice productivity.

K. Neelam (✉) · K. Kumar · K. Singh
Punjab Agricultural University, Ludhiana 141004, Punjab, India
e-mail: kneelam@pau.edu

H.S. Dhaliwal
Eternal University, Baru Sahib, Sirmour 173101, Himachal Pradesh, India

V.R. Rajpal et al. (eds.), *Molecular Breeding for Sustainable Crop Improvement*, Sustainable Development and Biodiversity 11,
DOI 10.1007/978-3-319-27090-6_8

Keywords The wild species of rice · Yield component QTL · Grain number · Spikelet number per panicle · Grain weight

8.1 Introduction

Rice (*Oryza sativa* L.) is one of the most important staple food crops serving nearly 21 % and 76 % of the total calorific intake of the world population and of southeast Asia, respectively (Brar and Singh 2011, Miura et al. 2011). The rice production has increased by 130 percent during 1966 to 2000 which needs to be increased further by 25–30 percent by 2035 to feed the ever-growing population (Khush 2013; http://www.gigasciencejournal.com/content/3/1/7; 2014). In order to achieve this goal, continuous work on improving yield potential and other superior agronomic traits is needed. Many important agronomic traits including yield are controlled simultaneously by multiple genes with smaller effects known as quantitative trait loci (QTL) which are strongly influenced by the environment. The identification and understanding of QTL-controlled agronomical traits are difficult because of their complex inheritance. For gaining insight into the mechanism behind the contribution of each of the QTLs, the dissection of complex traits into each QTL (quantitative trait locus) and their isolation and characterization are important. The availability of genomic sequences of the two rice subspecies, *O. sativa* ssp. *japonica* (cv. Nipponbare) and *O. sativa* ssp. *indica* 93–11, paved the way for detailed genetic analysis and isolation of many important QTLs (Goff et al. 2002, Yu et al. 2002a, b, International rice genome sequencing project, 2005). Further, availability of sequences of 3000 rice genomes by the efforts of international rice resequencing group has provided foundation for many more novel alleles and QTLs to be identified and cloned http://www.gigasciencejournal.com/content/3/1/7, 2014). Presently, the information on rice QTLs can be found on Gramene QTL data base (www.gramene.org/qtl). These QTLs are categorized into nine categories. The largest QTL found is of plant height (1011), followed by days to headings (618), spikelet number (353), spikelet fertility, and panicle length (Yamamoto et al. 2009). Only a few QTLs from the wild species of rice for yield-related traits have been reported (Li et al. 2004; Xie et al. 2006, 2008; Luo et al. 2013). Advances in molecular marker techniques, genomics, and statistical methods have facilitated the analysis of QTLs in greater depth (Miah et al. 2013; Huang et al. 2014). Currently, all rice breeding programs aiming at yield improvement are facing a major problem of narrow genetic base of high yielding varieties and hybrids due to severe domestication of cultivated rice (Tanksley and McCouch 1997). Furthermore, the narrow genetic pool reduces the probability of additional selection gains in breeding programs (Rangel et al. 1996). As the QTLs are derived from natural variation, the use of a wider range of variations as that found in wild species is important. The wild rice species represent a magnificent reservoir of genetic variability for many of the traits of agronomic interest (Harlan 1976; Hawkes 1977; Brar and Khush 2002, 2006). They have been exploited vigorously for genes for biotic stress tolerance, but their use remained limited earlier in improving yield and yield components. Later on, it has been discovered that the phenotypically poor wild species can contribute genes for improving yield and yield component traits (Eshed and Zamir 1995; Gur and Zamir 2004; Swamy et al. 2012 ; Gaikwad et al. 2014). The advent of new technologies and dense molecular maps opened up the opportunity for utilizing wild relatives for the improvement of complex traits. In this chapter, we will discuss the current status of the quantitative traits loci which have been introgressed from wild relatives and utilized for improving yield and its component traits in rice.

8.2 The Genus *Oryza*

The genus *Oryza* was first described by Linnaeus (1753) with only one species (*O. sativa*). Today, more than 100 species have been recognized in *Oryza* by various scientists (Vaughan 1989). The cultivated rice (*O. sativa* L.) belongs to the tribe Oryzeae, subfamily Oryzoideae of Poaceae. The tribe Oryzeae has 11 genera, of which genus *Oryza* is the only one with cultivated species. The genus *Oryza* includes two cultivated ($2n$ = 24, AA) and 22 wild species ($2n$ = 24, 48) representing the AA, BB, CC, BBCC, CCDD, EE, FF, GG, KKLL, and HHJJ genome types (Brar and Singh 2011). This has been further divided into four species complexes: (1) sativa complex, (2) officinalis complex, (3) meyeriana complex, and (4) ridleyi complex. Two species, *O. brachyantha* and *O. schlechteri,* could not be placed in any of these groups (Vaughan 1989, 1994). The wild *Oryza* species is known to have a large number of genes for disease and insect resistance such as bacterial blight, blast, brown plant hopper, white backed plant hopper, tungro virus, and abiotic stress tolerance such as to heat, cold, phosphorus deficiency, aluminum toxicity, and others (Heinrich et al. 1985; Amante-Brodeos et al. 1992; Brar and Khush 1997, 2006; Bhasin et al. 2012; Zeliang and Pattanayak 2013). These resistance genes have been transferred to the cultivated rice through wide hybridization and continuous backcrossing with the recurrent parents. This has been facilitated by the use of molecular markers, fluorescence in situ hybridization and other genomics tools. Using this approach, a number of varieties have been released for cultivation in rice-growing countries of Asia, Bangladesh, Philippines, India, China, USA, and many other countries (Sanchez et al. 2013). The different species, their chromosome number, genomic constitutions along with their useful traits are given in Table 8.1 (Inferred from Sanchez et al. 2013; Brar and Singh 2011).

Table 8.1 Species, chromosome number, genomic constitution, and distribution of *Oryza* and related genera along with their useful traits

S.No.	Species	2*n*	Genome	Distribution	Useful traits
***O. sativa* complex**					
1	*O. sativa* L.	24	AA	World wide	Cultigen: high yielding
2	*O. glaberrima* Steud.	24	A^gA^g	West Africa	Cultigen: tolerance to drought, acidity, iron toxicity, P-deficiency, resistance to BB, blast, RYMV, African gall midge, nematodes
3	*O. nivara* Sharma et Shastry	24	AA	Tropical and subtropical Asia	Resistance to grassy stunt virus, BB
4	*O. rufipogon* Griff.	24	AA	Tropical and subtropical Asia, tropical Australia	Resistance to BB, blast, BPH, tungro virus, tolerance to aluminum and soil acidity, increased elongation under deep water, source of CMS and yield-enhancing loci
5	*O. breviligulata* A. Chev. et Roehr. *O. barthii*	24	A^gA^g	Africa	Resistance to GLH, BB, drought avoidance, tolerance to heat, and drought
6	*O. longistaminata* A. Chev et Roehr	24	A^lA^l	Africa	Resistance to BB, nematodes, stem borer, drought avoidance
7	*O. meridionalis* Ng	24	A^mA^m	Tropical Australia	Drought avoidance, heat tolerance
8	*O. glumaepatula* Steud.	24	$A^{gp}A^{gp}$	South and Central America	Elongation ability, source of CMS, tolerance to heat
***O. officinalis* complex**					
9	*O. punctata* Kotschy ex Steud.	24, 48	BB, BBCC	Africa	Resistance to BPH, BB, zigzag leafhopper, tolerance to heat and drought
10	*O. minuta* J.S. Presl. ex C.B. Presl.	48	BBCC	Philippines and Papua New Guinea	Resistance to BB, blast, BPH, GLH
11	*O. officinalis* Wall ex Watt	24	CC	Tropical and subtropical Asia, tropical Australia	Resistance to thrips, BPH, GLH, WPH, BB, stem rot, tolerance to heat
12	*O. rhizomatis* Vaughan	24	CC	Sri Lanka	Drought avoidance, resistance to blast, tolerance to heat
13	*O. eichingeri* A. Peter	24	CC	South Asia and East Africa	Resistance to BPH, WBPH, GLH

(continued)

Table 8.1 (continued)

S.No.	Species	2*n*	Genome	Distribution	Useful traits
14	*O. latifolia* Desv.	48	CCDD	South and Central America	Resistance to BPH, BB, high biomass production
15	*O. alta* Swallen	48	CCDD	South and Central America	Resistance to striped stem borer, high biomass production
16	*O. grandiglumis* (Doell) Prod.	48	CCDD	South and Central America	High biomass production
17	*O. australiensis* Domin.	24	EE	Tropical Australia	Resistance to BPH, BB, blast, drought avoidance, tolerance to heat and drought
	***O. meyeriana* complex**				
18	*O. granulata* Nees et Arn. ex Watt	24	GG	South and South Asia	Shade tolerance, adaptation to aerobic soil
19	*O. meyeriana* (Zoll. et (Mor. ex Steud.) Baill.)	24	GG	Southeast Asia	Shade tolerance, adaptation to aerobic soil
***O. ridleyi* complex**					
20	*O. longiglumis* Jansen	48	HHJJ	Irian Jaya, Indonesia, and Papua New Guinea	Resistance to blast, BB
21	*O. ridleyi Hook . F .*	48	HHJJ	South Asia	Resistance to blast, BB, tungro virus, stem borer, whorl maggot
Unclassified					
1	*O. brachyantha A . Chev . et Roehr*	24	FF	Africa	Resistance to BB, yellow Stem borer, leaf folder, whorl maggot, tolerance to laterite soil
2	*O schlechteri* Pilger	48	KKLL	Papua New Guinea	Stoloniferous
3	*O. coarctata* Tateoka	48	KKLL	Asian Coastal Area	Tolerance to salinity, Stoloniferous
4	*Leersia perrieri A . Camus*	24	unknown	Africa	Shade tolerance, stoloniferous

BPH brown plant hopper, *GLH* green leaf hopper, *WBPH* white backed plant hopper, *BB* bacterial blight, *Shb* sheath blight, *CMS* cytoplasmic male sterility, *RYMV* rice yellow mottle virus. Adopted from Sanchez et al. (2013), Brar and Singh (2011)

8.3 QTL Mapping

Quantitative trait loci have been a major area of study in genetics for over a century. For most of the period up to 1980, the study of genetics of QTLs was based on means, variances, and covariances of relatives and these provided a base for partitioning the total phenotypic variance into genetic and environmental variances. Further studies by Sax (1923) on beans demonstrated that the effects of an individual locus affecting a quantitative trait locus could be isolated. Very little success on utilization of marker–QTL linkage in crop plants during the 1930–1980s could be obtained due to lack of availability of adequate polymorphic markers. The discoveries of series of molecular markers and statistical packages that could help in analyzing variation in quantitative traits changed the scenario during the 1980s. This led to more systematic studies on quantitative trait loci, a term first coined by Gelderman (1975). A QTL is defined as "a region of the genome that contributes to the variation of the traits" (Kearsay 1998). The basic objective of QTL mapping is to (1) identify the regions of genome afof interest an2) analyze the effect of QTL on the trait, *i.e.,* how much variation for the trait is controlled by a specific region, what type of gene actions are involved (additive, dominant and other effects), and which allele is associated with favorable effects. The salient requirement for QTL mapping is as follows: (1) a suitable mapping population generated from phenotypically contrasting parents, (2) a suitable linkage map based on molecular markers, (3) extensive phenotypic data of the mapping population, and (4) appropriate statistical packages to analyze the genotyping information in combination with the phenotypic data for QTL detection. The QTL analysis needs a mapping population such as F_2 plants, recombinant inbred lines (RILs), backcrossed inbred lines (BILs), double haploid lines (DHLs), near-isogeneic lines (NILs), and linkage map constructed with a large number of molecular markers (Yano and Sasaki 1997; Yano et al. 2001). Out of these, use of RILs for QTL mapping is most suitable as the RILs can be tested in different environments to avoid G × E interaction. Near-isogenic lines with only target trait QTL in unique genomic background further facilitates comprehensive analysis of the trait of interest. The desirable loci without any linkage drag can be introgressed and pooled in various genetic backgrounds to study their individual and combined effects. With the completion of rice genome sequencing and availability of dense genetic maps, isolation of QTLs based on NILs has become a routine (Zhang et al. 2006). In NILs, the traits of interest are behaving as single Mendelian factor and thus easy to clone. Over the last five years, several studies utilized NILs for fine mapping and cloning of QTLs in rice and tomato (Frary et al. 2000; Spielmeyer et al. 2002; Li et al. 2004; Fan et al. 2006). Some other mapping populations such as obtained from intersubspecies crosses between *indica* and *japonica* species, backcrossed and advanced backcrossed populations (AB-QTL), and introgression lines made from wild rice were also used for QTL mapping (Li et al. 2006; Song et al. 2007; Nonoue et al. 2008). The effectiveness of AB-QTL was demonstrated by McCouch et al. (2007) in improving yield-related QTLs from weedy rice *O. rufipogon* using different parallel populations and under diverse environments. The favorable alleles from *O. rufipogon* were found to improve recurrent parent performance by 5–20 % for most of the characters studied. Other successful example of AB-QTL includes mapping of yield components in interspecific population derived from crosses between *O. sativa* and the wild species *O. glaberrima* (acc. IRGC#103544 from Mali) (Li et al. 2004), and *O. nivara* (Eizenga et al. 2013). Recently, Gaikwad et al. (2014) reported the presence of yield-enhancing (panicle length, spikelet per panicle, grains per panicle, 1000 grain weight, and grain yield) heterotic QTLs in the introgression lines of *O. glaberrima*, *O. grandiglumis*, *O. glumaepatula,* and *O. longistaminata*. Some other mapping populations such as advanced intercrossed lines (AILs), chromosome segment substitution lines (CSSLs), multiparent advanced generation intercross (MAGIC), association mapping(AM) panel, and nested association mapping (NAM) are also gaining importance for mapping yield QTLs. Jacquemin et al. (2013) summarized different advanced mapping populations derived from wild *Oryza* species for the improvement of elite *O. sativa* cultivars. For QTL mapping, the composite interval mapping (Zeng 1994) is the most commonly used method. Based on the principles of interval mapping and composite interval mapping, various softwares have been designed and are freely available. These include R/qtl (Broman et al. 2003; http://www.rqtl.org/), QTL cartographer (Zeng 1994; http://statgen.ncsu.edu/qtlcart/WQTLCart.htm), and many others.

8.4 Advanced QTL Mapping Approaches

The presence of next-generation sequencing techniques and availability of SNP chips for high-throughput genotyping greatly enhances the precision of QTL mapping (Huang et al. 2009a, b; Bai et al. 2012). Advanced QTL mapping includes various approaches such as whole genome sequencing, QTL-Seq, MutMap, and many others (Fig. 8.1) which are dealt in detail as below.

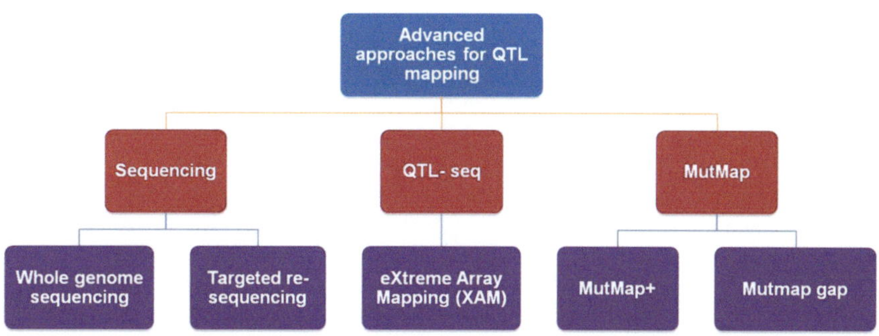

Fig. 8.1 Advanced techniques for QTL mapping

8.4.1 Whole Genome Sequencing

Single-nucleotide polymorphic (SNP) markers are gaining importance in plant breeding and genetics as they are more abundant, amenable to automation, and cost-effective and help in high-throughput profiling of large populations (Edward and Batley 2010, McCouch et al. 2010). SNPs are the most abundant form of genetic variation in eukaryotic genomes, and they occur in both coding and non-coding regions. SNPs markers have wide applicability in marker-assisted selection, QTL mapping, genome-wide association mapping, positional cloning of genes, germplasm characterization, pedigree analysis, and many others. Huang et al. (2009 a, b) reported approximately 20 times faster efficacy of whole genome resequencing methods in data collection and 35 times more precise in recombination breakpoint determination, when used with 287 PCR-based markers along with resequencing with Illumina Genome Analyzer of 150 RILs. This depicts the potential of next-generation sequencing technologies in detecting and accelerating cloning of the genes underlying QTLs. Yu et al. (2011) constructed ultra high-density map from low-coverage sequences of a RIL population of rice and validated positions of previously identified QTL for grain size (*GS3*), grain weight (*GW5*), two major QTL for grain length, and also mapped three yield-related QTL, tillers per plant, number of grains per panicle, and grain length. Gao et al. (2013) resequenced a segregating population of 132 Liang-You-Pei- Jiu (a widely used super hybrid rice) RILs, using Illumina HiSeq 2000 platform. Using 1, 71, 847 high-quality polymorphic SNPs markers, they were able to identify 43 yield-associated loci and also fine-mapped two quantitative trait loci spikelet number per plant (*qSN8*) and secondary panicle branch number (*qSPB1*). High-throughput single-nucleotide polymorphism (Affymetrix SNP chips, 96 Plex set, 384 Plex OPA set) is being actively used for diversity analysis, DNA fingerprinting, QTL mapping, and MAS (McNally et al. 2009; www.oryzasnp.org; Thomson et al. 2011; Huang et al. 2010).

8.4.2 QTL-Seq

This is a rapid method for the identification of QTLs by combining the bulk segregant analysis with high-throughput genotyping technologies. Initially, microarray-based genotyping (eXtreme Array Mapping) was used for the identification and mapping of QTLs in yeast (Ehrenreich et al. 2010). Later, Takagi et al. (2013) developed the QTL-Seq method for rapid identification of QTLs in the progenies obtained from diverse crosses. In this method, two bulk DNA were used. One is H bulk, consisting of 20–50 progenies with higher score of phenotype, and the other one is L bulk with lower phenotypic value. These two bulks along with parents were sequenced using whole genome sequencing approaches. The short reads obtained were aligned and compared with the reference genome and genomic

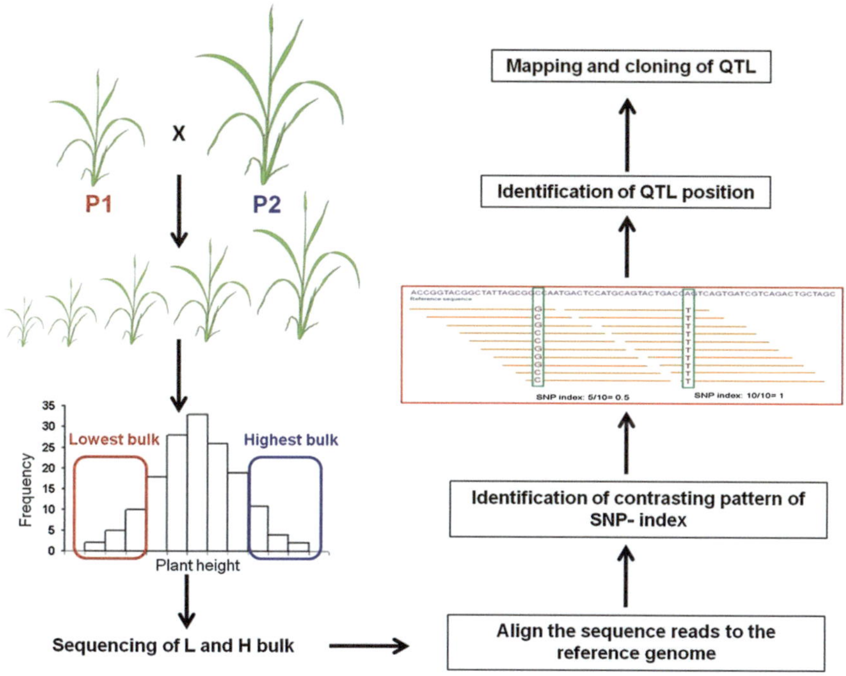

Fig. 8.2 Graphical representation of QTL-Seq: Contrasting bulks are made based on phenotype. The H and L bulks are genotyped and compared with the reference sequence for QTL identification

regions displaying contrasting patterns of SNP-index plots between the two bulks indicated the positions of QTLs (Fig. 8.2). The QTLs for partial resistance to the rice blast disease, seedling vigor, and seedling cold tolerance were mapped using this method (Takagi et al. 2013; Yang et al. 2013). The advantage offered by NGS-assisted BSA over other methods are as follows: (i) It does not require precise phenotyping of each individuals but only identification of individuals with extreme phenotype and (ii) the development of large number SNP-based markers in the desired region helps in fine mapping and cloning of the QTLs. The developed markers could be used for marker-assisted breeding.

8.4.3 MutMap Approach

Mutants are one of the major resources for identification of genes underlying the QTLs responsible for important agronomic traits. The MutMap approach was proposed by Abe et al. (2012). This method involves crossing of the desired mutant to its wild type followed by selfing of F_1 individuals to generate F_2 progeny. The plant

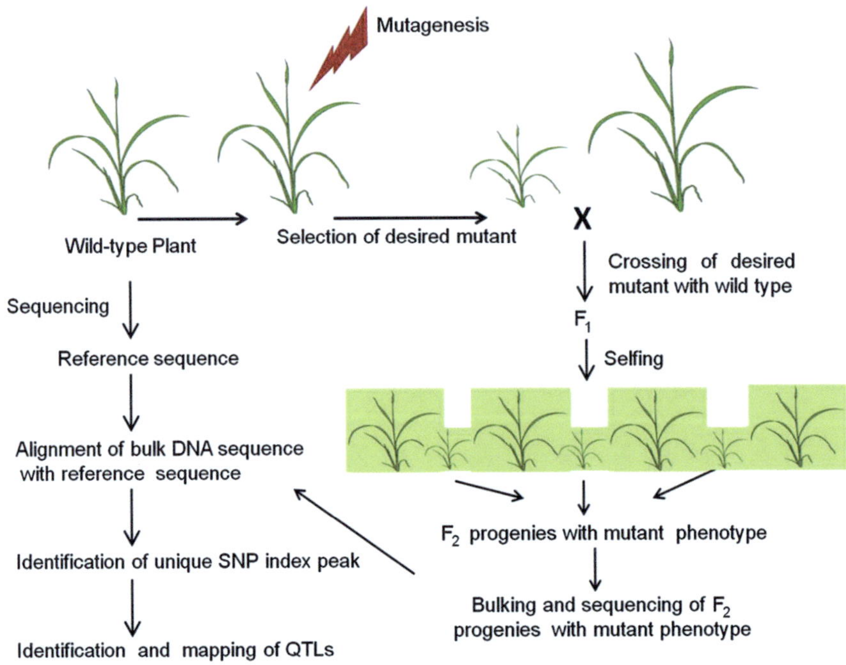

Fig. 8.3 Pictorial representation of MutMap approach for QTL mapping

from approximately 20 F_2 individuals showing mutant phenotype is pooled in equal ratio and subjected to whole genome sequencing. The sequences were aligned to the reference sequence for identifying unique SNP-index peak (SNP-index = 1) which is responsible for the causal phenotype (Fig. 8.3). Further, modification of this approach was given by Fekih et al. (2013) and known as MutMap+. The MutMap+ identifies causal mutation by comparing SNP frequency of bulked DNA of M_3 mutant progenies and wild type. This method has an advantage over other methods as this does not involve artificial crossing and also in mapping of early development genes, sterility genes, and isolation of genes from the plants which are not amenable to crossing.

8.4.4 *TILLING and Eco-TILLING*

Targeting induced local lesions in genomes (TILLING) is a non-transgenic, reverse genetics method that allows rapid screening of thousands of mutagenized lines for mutations in a particular gene (McCallum et al. 2000). In TILLING, mutations are induced using certain mutagenic agents (such as ethyl methyl sulfonate, EMS) to generate variation in a gene of interest. DNA from mutagenized population is then pooled and arrayed in microtiter plates. Pooled DNA is amplified using fluorescently labeled primers designed from the gene of interest. The PCR products are denatured and re-annealed resulted in formation of heteroduplexes if a mutation is present. Heteroduplexes are digested using a crude protein extract of celery juice containing the nuclease CEL I. CEL I enzyme targets point mutation, such as SNP and InDel, present in genome. The digested products are resolved on denaturing polyacrylamide gels and visualized using fluorescence imaging. TILLING combines classical mutagenesis with high-throughput screening of nucleotide polymorphisms using molecular biological techniques in a targeted sequence. Unlike other methods of reverse genetics (i.e., RNA interference, T-DNA mutagenesis), TILLING does not relies on transformation and hence can be used for species that are not transformable or recalcitrant. Since TILLING is non-transgenic method, not any IPR regulation or gene containment regulation is applicable. In rice, Wu et al. (2005) developed more than 38,000 mutant lines through chemical mutagenesis and used them for identifying allelic series of particular genes using TILLING. Serrat et al. (2014) screened *OsACS1* and *OsSGR* genes to identify variation in 6912 mutant population generated from mutagenized mature seed-derived calli. TILLING methodology can also be used to uncover natural nucleotide variation linked to important phenotypic traits using a process termed as EcoTILLING (Comai et al. 2004). Yu et al. (2012) reported allelic series of transcription factors responsible for drought stress tolerance using EcoTILLING. They identified allelic variation within promoter region of 24 transcription factors in indica and japonica rice. EcoTILLING was successfully used by Negrao et al. (2011) for indentifying nucleotide variation in salt-tolerant genes, *SalT* and *OsCPK17,* while screening 375 diverse germplasm of rice. This method can be used for getting allelic variations of yield-related cloned QTL of rice.

8.4.5 Association Mapping

Association mapping holds a great potential in identifying and resolving quantitative variations of complex traits of agronomic and economic importance. Association mapping recognizes QTLs by identifying marker trait associations that can be contributed to the strength of linkage disequilibrium (LD) between markers and functional polymorphisms across a set of diverse germplasm (Zhu et al. 2008). Traditionally, QTL mapping approaches have been based on the analysis of populations derived from biparental crosses that segregated for trait(s) of interest with limited number of recombination events. This leads to the poor resolution of the QTLs and also hampers their direct use in breeding programs. Whereas association mapping utilizes natural populations, landraces, the collection of cultivars released over years and the material within a breeding program and hence harnesses the maximum diversity present in wider genetic pool. However, it is important to critically consider population structure and kinship among individuals, because false associations may be detected due to the confusing effects of population admixture and unequal allele frequency distribution between subgroups. Therefore, it is essential to apply appropriate statistical methods that account for population structure. Various statistical methods, *i.e.,* structured association (Pritchard et al. 2000; Falush et al. 2003), genomic control (Devlin and Roeder 1999), mixed model approach (Yu et al. 2006), and principal component approach (Price 2006) have been proposed to take into account for population structure and family relatedness. The essence of these approaches is to use genotypic information from random and independent molecular markers across the genome for detecting population structure. In addition to population structure, the extent and the distribution of LD across the genome also affect the resolution of association mapping (Remington et al. 2001). LD, or gametic-phase disequilibrium, measures the degree of non-random association between alleles at different loci. The pattern of LD is usually affected by population history, but other factors such as population structure, selection, mutation, relatedness, and genetic drift also have an effect on LD. It is found that in case of self-pollinated crops, LD extends to a much larger distance than that in cross-pollinated species. The detailed reviews on LD in plant species have been given previously by various scientists (Flint-Gracia et al. 2005; Ersoz et al. 2008) and can be referred for better understanding of LD. A range of software packages are available for data analysis in association mapping. TASSEL is the most commonly used software for association mapping in plants (Bradbury et al. 2007). Other software also includes SAS, R, STRUCTURE, SPAGeDi, EINGENSTRAT, and MTDFREML which can be used for analysis. In rice, large number of studies have been conducted for getting insight into population structure, its effect on genetic diversity, LD, and utilization in fine mapping of QTLs (Garris et al. 2005; Olsen et al. 2006; Agrama et al. 2007; Jin et al. 2010, Choudhury et al. 2014). These studies suggest that the extent of LD varies among different genomic regions, different rice accessions studied (Agrama and Eizenga 2008), and different markers used.

8.4.6 Genome Editing Tools

Mapping traits to the gene level remains a daunting task despite the tremendously reduced cost of DNA sequencing and a growing number of success stories. In particular, regions of low recombination such as inversions, centromeres, and telomeres have been found to harbor many alleles of interest but can frustrate fine-mapping efforts. Some of them could be overcome by using association studies as discussed above in outbred populations, but these approaches have additional complications such as low statistical power. New approaches to genomic engineering can be applied to partially alleviate these difficulties. This includes the use of nucleases in genome editing such as zinc finger nucleases (ZFNs) and transcription activator-like effector nucleases (TALENs). Zinc finger nucleases (ZFNs) are a class of engineered DNA-binding proteins that facilitate genome editing by creating a double-stranded break in DNA at predetermined position. These enzymes generate targeted double-strand DNA breaks that are fixed by the cell's own repair pathways, through either non-homologous end joining or homologous recombination. These enzymes lead to a new gene or regulatory element or to knock existing gene out. ZFNs are capable of creating several types of genomic alterations including point mutations, deletions, insertions, inversions, duplications, and translocations, thus providing opportunities to perform genetic manipulations. Transcription activator-like (TAL) effectors are virulent factors from *Xanthomonas* bacteria. TALENs are similar to ZFNs and comprise a non-specific FokI nuclease domain fused to a customizable DNA-binding domain. This DNA-binding domain is composed of highly conserved repeats derived from transcription activator-like effectors (TALEs), which are proteins secreted by *Xanthomonas* bacteria to alter transcription of genes in host plant cells. Several groups have used TALENs to modify endogenous genes in yeast, fruit fly, roundworm, crickets, zebrafish, frog, rat, pig, cow, thale cress, rice, silkworm, human, etc. (Joung and Sander 2013). The next one is CRISPR/Cas9 technology, "clustered regularly interspaced short palindromic repeats," which was recently discovered to be a component of the immune responses of bacteria and archaea (Barrangou et al. 2007, Turner 2014). This technology involves the spacer sequences acquired from foreign DNA to be positioned between host repeats, and transcribed together as CRISPR RNA (crRNA). In the type II CRISPR system, a single nuclease Cas9, guided by a dual-crRNA:tracrRNA, is sufficient to cleave cognate DNA homologous to the spacer (Miao et al. 2013). Comprehensive studies depicting detail methods and use of CRISPR in genome manipulation are provided by many scientists (Cong et al. 2013; Mali et al. 2013; Gratz et al. 2013; Hsu et al. 2014; Wang et al. 2014).

8.5 Yield and Its Component Traits in Rice

Grain yield (GY) of rice is a complex trait consisting of three major yield components, panicles per plant, spikelets per panicle, and grain weight (Yoshida1983, Sakamoto and Matsuoka 2008). Some indirect components such as plant height, tiller number, and heading date also have an impact on yield. Each component traits has varying degree of contribution toward yield increment. It is also known that yield components have higher heritability than GY; therefore, increase in GY could effectively be achieved through the improvement of yield components (Xiong 1992). High broad-sense heritability estimate of 98.89 % for days to maturity, 75.20 % for the number of tillers per plant, 41.74 % for the number of panicles per plant, 98.97 % for 1000 grain weight, and 90.87 % for panicle weight was observed by El-Malky et al. (2008). Akinwale et al., (2011) observed significant positive correlation of grain yield with the number of tillers per plant ($r = 0.58^{**}$), panicle weight ($r = 0.60^{*}$), and number of grains per panicle ($r = 0.52^{*}$). Along with the influence of environment on component traits, the correlation among them also played a role in deciding yield of rice plant (Frankel 1935; Adams 1967). The relationship between rice yield and yield components has been studied extensively.

Subramanian and Rathinam (1984) observed highly significant associations of grain yield with the 1000 grain weight and number of tillers per plant. Deosarkar et al. (1989) and Mehetre et al. (1994) reported significant positive associations between grain yield per plant and number of grains per panicle. In the last decade, a number of studies were conducted to dissect the genetic basis of the interactions between QTL × environment and epistatic interactions among various yield components (Yu et al. 1997; Zhuang et al. 1997; Yamamoto et al. 2000; Xing et al. 2002). However, negative correlation between yield and component traits is also there which ultimately limits the maximum achievable targets for yield improvements. The details of utilization of wild species for the improvement of yield and component traits will be discussed under following subheadings.

8.5.1 QTLs for Grain Number

Among the yield components, grain number showed the largest range of variation and was the major objective of improvement in rice high yield breeding program (Li et al. 1998; Yamagishi et al. 2002). During the course of domestication from wild rice to cultivated rice, profound changes of agronomic traits and genetic diversity occurred (Sun et al. 2001). One of the most important hallmarks of rice domestication is the dramatic increases in grain number, as evidenced by the fact that most cultivated rice showed more grain number than wild rice (Tian et al. 2006). Rice grain number is quantitatively inherited, and a great deal of QTL mapping for grain number has been conducted using various mapping populations derived from interspecific crosses (Xiao et al. 1998; Thomson et al. 2003), *indica–japonica* intersubspecific crosses (Lu et al. 1996; Xing et al. 2002; Bai et al. 2012), *indica–indica* crosses (Lin et al. 1996; Zhuang et al. 1997), and *japonica–Japonica* crosses (Yamagishi et al. 2002). These detected QTLs are distributed in all the 12 rice chromosomes and created a firm basis to investigate the genetic control of grain number. The *O. rufipogon* Griff. is the wild ancestor of cultivated rice (Second 1982, Wang et al. 1992) and has been extensively used for improving yield and related traits of cultivated rice. Tian et al. (2006) mapped a QTL for grain number (*gpa7*) on chromosome 7 by using F_3 population derived from SIL040 (an introgression line derived from *O. rufipogon* in *O. sativa* background) with Guichao 2. However, the contributing alleles from *O. rufipogon* were not favorable for grain number. Using AB-QTL approach, Moncada et al. (2001) identified eight yield-related traits (plant height, panicle length, number of panicles per plant, grains per plant, 1000 grain weight, days to heading, and plant yield) in BC_2F_2 population of Caiapo and *O. rufipogon* (Table 8.2). The AB-QTL approach has also been used by others for mapping QTLs in rice (Marri et al. 2005; Thomson et al. 2003). Ashikari et al. (2005) cloned *Gn1a*, a QTL located on chromosome 1, that increases grain number in rice and elucidated the molecular mechanism of this gene. Table 8.2 summarizes other important QTLs introgressed and mapped from wild species of rice for enhancing grain number.

Table 8.2 The yield and yield-related QTLs introgressed from wild species of rice

	Wild species	Mapping populations	Yield-enhancing QTL	Chromosomes	References
Grain number					
1	*O. rufipogon*	BC_2	*gpl1.1, gpl2.1, gpl4.1, gpl5.1, gpl8.1, gpl8.2*	1, 2, 4, 5, 8	Xiao et al. (1998)
2	*O. rufipogon*	BC_2F_1	*gnp2.1, gnp2.2, gnp5.1, gnp5.1*	2, 5	Marri et al. (2005)
3	*O. rufipogon*	BC_2	*gpl1.1, gpl2.1, gpl11.1*	1, 2, 11	Moncada et al. (2001)
4	*O. rufipogon*	BC_3F_4	*gn9.1*	9	Xie et al. (2008)
5	*O. rufipogon*	BC_2F_2	*gpl1.1*	1	Septiningsih et al. (2003)
6	*O. rufipogon*	$F_{2:3}$	*gpp8*	8	Jin et al. (2009)
7	*O. rufipogon*	BC_2F_4	*gpp1.1, gpp3.1, gpp7.1, gpp 12.1*	1, 3, 7, 12	Fu et al. (2010)
Grain size					
1	*O. rufipogon*	BC_3F_3	*gw*8.1	8	Xie et al. (2006)
2	*O. grandiglumis*	BC_5F_3	*gw2, gw6, gw11, gt2, gt6, gt11, gl6, gl11*	2, 6, 11	Yoon et al. (2006)
3	*O. rufipogon*	BC_2F_2	*gw1.1, gw3.1, gw3.2*	1, 3	Septiningsih et al. (2003)
Panicles per plant					
1	*O. rufipogon*	BC_2F_2	*ppl2.1*	2	Septiningsih et al. (2003)
2	*O. grandiglumis*	BC_5F_3	*pn11*	11	Yoon et al. (2006)
3	*O. rufipogon*	BC_2	*ppl6.1, ppl11.1*	6, 11	Moncada et al. (2001)
4	*O. rufipogon*	BC_2F_1	*np2.1, np2.2*		Marri et al. (2005)
5	*O. rufipogon*	BC_2F_4	*ppl1.1, ppl2.1, ppl7.1, ppl8.1, ppl11.1*	1, 2, 7, 8, 11	Fu et al. (2010)
6	*O. rufipogon*	BC_2	*ppl3.1, ppl7.1*	3, 7	Thomson et al. (2003)
7	*O. minuta*	$F_{2:3}$	*pn4, pn6*	4, 6	Rahman et al. (2007)
8	*O. glumaepatula*	BC_2F_2	*pn5*, pn8, *pn11*	5, 8, 11	Brondani et al. (2002)
Panicle length					
1	*O. rufipogon*	BC_2	*pl1.1, pl8.1*	1, 8	Xiao et al. (1998)
2	*O. grandiglumis*	BC_5F_3	*pl6*	6	Yoon et al. (2006)

(continued)

Table 8.2 (continued)

	Wild species	Mapping populations	Yield-enhancing QTL	Chromosomes	References
3	*O. minuata*	$F_{2:3}$	*pl6, pl7, pl8*	6, 7, 8	Rahman et al. (2007)
4	*O. rufipogon*	BC_2	*pl1.1, pl2.1, pl4.1, pl9.1, pl12.1*	1, 2, 4, 9, 12	Thomson et al. (2003)
5	*O. rufipogon*	BC_2F_1	*pl2.1, pl5.1, pl9.1*	2, 5, 9	Marri et al. (2005)
6	*O. rufipogon*	BC_2F_2	*pl1.1, pl9.1, pl10.1, pl10.2*	1, 9, 10	Septiningsih et al. (2003)
Spikelet number per panicle					
1	*O. rufipogon*	BC_2F_1	*snp2.1, snp5.1, snp5.2*	2, 5	Marri et al. (2005)
2	*O. minuta*	BC_7F_2	*qspp7*	7	Balkunde et al. (2013)
3	*O. rufipogon*	BC_2F_2	*spp2.1, spp3.1, spp9.1*	2, 3, 9	Septiningsih et al. (2003)
4	*O. rufipogon*	ILs	*qspp1, qspp11*	1, 11	Liu et al. (2009)
5	*O. longistaminata*	ILs	*qspp2.1, qspp 2.2*	2	Sidana et al. (2012)
Number of primary and secondary branches NPB					
1	*O. nivara*	BC_2F_2	*qNPB2.1, qNSB1.1, qNSB2.1*	1, 2	Swamy et al. (2012)
2	*O. nivara*	F_2	*pbr1, sbr1, sbr 3, sbr 7*	1, 3, 7	Li et al. (2006)
3	*O. rufipogon*	ILs	*QPbn1, QPbn2, QPbn8, QPbn11, QPbn12, QSbn1, QSbn2, QSbn4, QSbn7*	1, 2, 8, 11, 12, 4, 7	Luo et al. (2008)
1000 grain weight					
1	*O. glaberrima*	CSSL	*TGRWT4, TGRWT6*	4, 6	Gutiérrez et al. (2010)
2	*O. rufipogon*	BC_2F_1	*gy2.1, gy2.2, gy2.3, gy9.2*	2, 9	Marri et al. (2005)
3	*O. minuata*	$F_{2:3}$	*tgw7, tgw11*	7, 11	Rahman et al. (2007)
4	*O. rufipogon*	BC_2F_4	*kgw1.1, kgw2.1, kgw3.1, kgw 4.1, kgw7.1, kgw11.1*	1, 2, 3, 4, 7, 11	Fu et al. (2010)
Yield					
1	*O. glaberrima*	CSSL	*YLD3, YLD4, YLD6, YLD9*	3, 4, 6, 9	Gutiérrez et al. (2010)
2	*O. rufipogon*	BC_2F_2	*yld1.1, yld1.2, yld2.1*	1, 2	Septiningsih et al. (2003)

(continued)

Table 8.2 (continued)

	Wild species	Mapping populations	Yield-enhancing QTL	Chromosomes	References
3	*O. rufipogon*	BC_2F_4	*yld1.1, yld2.1, yld8.1, yld12.1*	1, 2, 8	Fu et al. (2010)
4	*O. rufipogon*	BC_2F_1	*Yldp2.1, yldp2.2, yldp9.1*	2, 9	Marri et al. (2005)

Abbreviations: *gpl* grains per panicle, *gn* grain number, *gpp* grains per panicle, *gw* grain weight, *ppl* panicle per plant, *pn* panicle number, *np* number of panicles, *pl* panicle length, *snp* spikelet number per panicle, *spp* spikelets per plant, *NPB* number of primary branches, *NSB* number of secondary branches, *pbr* primary branches, *sbr* secondary branches, *QPbn* QTL for primary branch number, *QSbn* QTL for secondary branch number, *TGRWT* thousand grain weight, *gy* grain yield, *tgw* thousand grain weight, *YLD* yield per plant, *ILs* introgression lines, and CSSL chromosome segment substitution line

8.5.2 QTLs for Panicle Architecture

Grain number per panicle is basically determined by the panicle architecture, *i.e.,* the number and length of primary branches of a panicle, the number and length of secondary branches on each primary branch, and the number of branches on secondary and higher order branches and panicle length. Panicle-related traits are important agronomic traits which are directly associated with grain yield. A total of 39 QTLs were found to be associated with panicle-related traits including panicle length (PN), primary branch number (PBN), secondary branch number (SBN), spikelet number per panicle (SPP), and spikelet density (SD) and were detected using a set of 265 ILs of common wild rice (*O. rufipogon* Griff.) in the background of *indica* high yielding cultivar Guichao 2 (*O. sativa* L.) with single-point analysis. The alleles of 20 QTLs derived from wild rice showed positive effects, and some QTLs such as *QPl1b* for PL, *QPbn8* for PBN, *QSd4* and *QSd11b* for SD, and *QSpp4* for SPP showed larger positive effects, providing good candidates and useful information for marker-aided improvement of yield potential of rice (Luo et al. 2008). Brondani et al. (2002) studied 11 agronomic traits in BC_2F_2 families of the interspecific cross *O. sativa* × *O. glumaepatula* and reported positive effect of *O. glumaepatula* alleles on panicle number and tiller number. Further, Rangel et al. (2013) conducted CIM analysis and confined the region to the marker interval 4752-RM82 on chromosome 7 with LOD 4.2.

8.5.3 QTLs for Spikelet Number

The number of spikelets per rice panicle is very important in determining yield, which is defined as the product of spikelet yield (or sink) and ripening ability (or source) (Hua et al. 2002). Luo et al. (2013) has demonstrated that 2 QTLs, *qSPP5* for spikelets per panicle (SPP), and *qTGW5* for grain weight (TGW) are tightly linked on chromosome 5 in a BC_5F_4 (NILs) populations that were derived from a cross between the Korean japonica cultivar *Hwayeongbyeo* and *O. rufipogon*. Sidana (2012) extensively studied the yield-contributing characters from the wild species *O. longistaminata*. She reported the colocalization of QTLs for number of spikelet per panicle, fertile grain per panicle, and plant height on chromosome 2 between the marker intervals RM13742-RM13750 and RM13750-RM13781, and the favorable allele is contributed by *O. longistaminata*.

8.5.4 QTLs for Grain Size

Grain size in rice is a major determinant of grain yield and market value (Huang et al. 2012). A large diversity in grain size has been observed within and between different subpopulations of *O. sativa*. The four genes (*GS3*, *GW2*, *G1F1,* and *GS5*) contributing to seed or grain size have been identified and characterized (Fan et al. 2006; Song et al. 2007; Wang et al. 2008; Shomura et al. 2008; Weng et al. 2008). The wild species of rice also possess positive effects in enhancing grain size either by improving grain length and grain width. Li et al. (2004) fine mapped a grain weight QTL, *gw*3.1, using a set of near-isogenic lines (NILs) developed from *O. sativa*, and cv. Jefferson X *O. rufipogon* (IRGC105491) population based on five generations of backcrossing and seven generations of selfing. Yoon et al. (2006) identified five QTLs for grain width located on chromosome 2, 3, 6, 8, and 11; four QTLs for grain thickness located on chromosome 2, 6, 7, and 11; and two QTLs for grain weight, using a $F_{2:3}$ families derived from a cross between introgression line of *O. grandiglumis* and *Hwaseong Yeo*. Oh et al. (2011) further fine mapped the grain weight QTL *tgw*11 between the two SSR markers RM224 and RM27358 on chromosome 11. Yield-enhancing QTLs cluster for grain weight on chromosome 8 has been fine mapped using NILs population of *O. sativa* and *O. rufipogon* (Xie et al. 2006, 2008).

The summary of QTLs mapped and utilized from wild species of rice for improving yield and component traits is given in Table 8.2.

8.5.5 Colocalization of QTLs for Yield and Yield-Related Traits

The QTLs with major effect on yield and yield-related component traits are usually clustered in a few chromosomal segments. As obvious from the Table 8.2, each chromosome of rice was found harboring a minimum of one or two QTLs for yield and yield-related traits although more QTLs were identified on chromosomes 1, 2, 3, and 4 of rice. Xie et al. (2008) reported colocalization of *sn9.1*, *gn9.1*, *dn9.1*,

Table 8.3 Chromosomal region harboring QTL cluster for yield and yield-related traits across population and species in rice

Chromosome	Marker (s)	QTL clusters*	References
1	RM220-RM272	*ph1.1, spp1, gyp1, fgp1*	Brondani et al. (2002), Li et al. (2004)
	RM272-RM259	*ph1.2, dth1.1, spp1.1, gpp1.1*	Thomson et al. (2003), Xiao et al. (1996), Marri et al. (2005)
	RM212-RM315	*pl1.1, pss1.1, dth1.1, sf1.1, ph1.1*	Thomson et al. (2003), Septiningsih et al. (2003), Marri et al. (2005)
2	RM250-RM208	*pl2.1, sn2.1, yld2.1, gpp2.1, gw2.1*	Marri et al. (2005), Septiningsih et al. (2003), Li et al. (2004), Moncada et al. (2001)
	RM262-RM263	*gnp2.1, yld2.1, yld2.2*	Li et al. (2004), Marri et al. (2005)
5	RM194-RM249	*qspp5, qtgw5, qsb5, qgw5, snp5.1, gn5.1*	Li et al. (2004), Marri et al. (2005), Luo et al. (2013)
8	RM350-RM210	*yld8.1, snp8.1, gpp8.1, ph8.1*	Xiao et al. (1998), Thomson et al. (2003), Marri et al. (2005), Li et al. (2004)
9	RM242-RM250	*pl9.1,* gw9.2, *spp9.1, pp9.1, yld9.1*	Marri et al. (2005), Li et al. (2004)

pl9.1, and *yl9.1* QTLs with *gw9, hd9.1,* and *ph9.1* in the 37.4 kb interval flanked by markers RM24718-RM30005. In BC_5F_3 population derived from a cross involving *O. grandiglumis,* six QTLs (panicle length, panicle number, spikelets per panicle, grain width, grain thickness, and 1000 grain weight) were found between the same marker interval RM224-RM144 on chromosome 11. Yield QTLs on chromosome 1 were colocalized with QTLs for grain yield per panicle, grains per plant, filled grains, and spikelets per panicle (Septiningsih et al. 2003; Brondani et al. 2002). The presence of the QTLs for yield and yield-related traits at the similar marker interval of the chromosome has been reported in several other studies (Table 8.3) which indicated non-random distribution of yield-related traits in rice genome. These QTLs are also having some common genomic regions shared across the species and populations. The colocalization of QTLs for different yield-related traits might be due to independent but closely related genes or might be because of single gene with pleiotropic effects. Cloning of colocalized QTLs for multiple traits has unequivocally shown that the colocalized QTLs is due to the same major gene with pleiotropic effect (Xue et al. 2008).

Such hot spot QTL regions are important for manipulating or improving more than one trait at a time. These regions can be further dissected, fine mapped, and cloned for knowing the biology of the genes controlling underlying QTLs (Ashikari et al. 2005; Eshed and Zamir 1995). Till date, a large number of QTLs related with yield and yield components have been cloned though they are not from wild

Table 8.4 Summary of QTLs cloned by map-based cloning techniques in rice

Chromosome	Trait*	QTL/Genes**	Encoded protein	References
1	NGP	*Gn1a*	Cytokinin oxidase/dehydrogenase	Ashikari et al. (2005)
1	NGP	*LOG*	Cytokinin-activating enzyme	Kurakawa et al. (2007)
1	NGP, NPP	*LAX1*	A bHLH transcription factor	Komatsu et al. (2001)
2	NPP, NGP	*qGY2-1*	Leucine-rich repeat receptor-like kinase	Zha et al. (2009)
2	NGP	*LP*	Kelch repeat-containing F-box protein	Li et al. (2011a, b)
2	GW, GS	*GW2*	RING-type E3 ubiquitin ligase	Song et al. (2007)
2	GW, GS	*PGL2*	Atypical bHLH protein	Heang and Sassa (2012a, b)
3	GW, GS	*GS3*	Transmembrane protein	Mao et al. (2010)
3	GW, GS	*BRD1*	Brassinosteroid-6-oxidase	Mori et al. (2002)
3	GW, GS	*GL3.1/qGL3*	Phosphatase with Kelch-like repeat domain	Zhang et al. (2012b), Qi et al. (2012)
4	NPP	*D17/HTD1*	Carotenoid cleavage dioxygenase	Zou et al. (2006)
4	NGP, NPP	*APO2*	Plant-specific transcription factor	Ikeda-Kawakatsu et al. (2012)
4	GW, GF	*GIF1*	Cell wall invertase	Wang et al. (2008)
4	GW, GF	*FLO2*	Protein with a tetratricopeptide repeat motif	She et al. (2010)
4	TSN	*Nal1/GPS*	Trypsin-like serine and cysteine proteases	Fujita et al. (2013), Takai et al. (2013)
5	GW, GS	*APG*	Typical bHLH protein	Heang and Sassa (2012a, b)
5	GW, GS	*SRS3*	Kinesin 13 protein	Kitagawa et al. (2010)
5	GW, GS	*GS5*	Putative serine carboxypeptidase	Li et al. (2011a, b)
5	GW, GS	*qSW5/GW5*	Nuclear protein	Weng et al. (2008), Shomura et al. (2008)
6	NPP	*D3*	F-box leucine-rich repeat protein	Ishikawa et al. (2005)
6	NPP	*MOC1*	GRAS family nuclear protein	Li et al. (2003)

(continued)

Table 8.4 (continued)

Chromosome	Trait*	QTL/Genes**	Encoded protein	References
6	NGP	*SCM2/APO1*	F-box protein	Ookawa et al. (2010), Ikeda et al. (2007)
6	NGP	*Hd1*	Protein with a zinc finger domain	Zhang et al. (2012a)
6	GW, GS	*TGW6*	Iindole-3-acetic acid (IAA)-glucose hydrolase	Ishimaru et al. (2013)
6	GW, GF	*HGW*	ubiquitin-associated domain protein	Wang et al. (2008)
7	NPP, NGP	*PROG1*	Zinc finger nuclear transcription factor	Jin et al. (2008)
7	NGP, NPP	*EP2/DEP2/SRS1*	Plant-specific protein	Zhu et al. (2010), Li et al. (2010)
7	NGP, NPP	*DEP3*	Patatin-like phospholipase A2 (PLA2) superfamily	Quiao et al. (2011)
7	NGP, NPP	*FZP*	Ethylene-responsive element-binding factor	Komatsu et al. (2003)
7	NGP	*Ghd7*	CCT domain protein	Xue et al. (2008)
8	NPP, NGP	*WFP/IPA1* (*OsSPL14*)	Squamosa promoter-binding protein-like 14	Miura et al. (2010), Jiao et al. (2010)
8	NGP	*Ghd8/DTH8*	OsHAP3 subunit of a CCAAT-box-binding protein	Wei et al. (2010), Yan et al. (2011)
8	GW, GS	*GW8/OsSPL16*	Squamosa promoter-binding protein-like 16	Wang et al. (2012)
9	NGP	*DEP1*	PEBP-like domain protein	Huang et al. (2009a, b)
9	GW, GS	*SG1*	Novel protein	Nakagawa et al. (2012)
11	NGP	*SP1*	Transporter of the peptide transporter family	Li et al. (2009)
11	GW, GS	*SRS5*	Alpha-tubulin protein	Segami et al. (2012)

*Traits: *NPP* number of panicles per plant, *NGP* number of grains per panicle, *GW* grain weight, *GS* grain size, *GF* grain filling, and *TSN* total spikelet number per panicle

** Gene(s): *Gn1* grain number, *LOG* lonely guy, *LAX qGY2* grain yield, *LP* large panicle, *GW* grain weight, *PGL* positive regulator of grain length, *GS* grain size, *BRD* brassinosteroid-6-oxidase, *GL* grain length, *D17/HTD* DWARF17 high-tillering dwarf, *APO* aberrant panicle organization, *GIF1* grain incomplete filling, *FLO2* floury endosperm2, *NGL* narrow leaf, *GPS* green for photosynthesis, *APG* antagonist of PGL, *SRS3* small and round seed, *GS* grain size, *SW* seed weight, *MOC* monoculm, *FZP* frizzy panicle, *PROG* prostrate growth, *Ghd7* heading date, *DTH* days to heading, *WFP* wealthy farmers panicle, *SG* short grain, *OsSPL* Squamosa promoter-binding protein-like, *DEP* dense and erect panicle, *TWG* thousand grain weight, *SCM* strong culm, and *HGW* heading and grain weight species. Their sequences can, however, be used for discovering and mining their superior alleles among the wild-related species for introgression and exploitation. These QTLs controls several rice traits, including grain productivity (*Gn1a*, *OsSPL14*, *DEP1*), grain size and weight (*GW2*, *qSW5*, *GS3*), yield potential (*Ghd7*), number of panicles per plant (NPP), and heading date (*Hd1*, *Hd3a*, *Hd6*, *Ehd1*). A summary of cloned QTLs is given below (Table 8.4).

8.6 Future Perspective

The wild relatives of rice harbor many useful genes particularly for resistance to biotic and abiotic stresses along with many other agronomically useful traits. With the progress made in plant breeding, biotechnology, bioinformatics, and other related technologies, it has become easier to reduce the linkage drag and transfer only desired variability from wild relatives. The New Rice for Africa (NERICA), which was developed from a cross between African rice (*O. glabrrima* Steud.) and Asian rice (*O. sativa* L.), is a successful example of utilization of wild rice towards sustainable agricultural development. NERICA varieties have high yield potential, short growth cycle, weed competitiveness, early vigor trait, and resistant to African insect and pests along with higher protein content and amino acid balance. Tables 8.2 and 8.3 revealed the presence of useful alleles for yield improvement in wild species of rice, and further, their colocalization provides an opportunity to pool and introgress major effect QTLs together. The progress made in genomics enables researchers to enrich the QTLs region with more number of SNPs markers, also construct high-density haplotype map, and perform genome-wide association studies and the dissection of complex traits. Further, systematic integration of the developments made in functional genomics, proteomics, and metabolomics is required for understanding the functions of underlying genes.

References

3K RGP (2014) The 3,000 rice genomes project. GigaScience 3:7

Abe A, Kosugi S, Yoshida K, Satsume S, Takagi H, Kanzaki H, Matsumura H, Yoshida K, Mitsuoka C, Tamiru M, Innan H, Canon L, Kamoun S, Terauchi R (2012) Genome sequencing reveals agronomically important loci in rice using MutMap. Nat Biotech 30:174–179

Adams MW (1967) Basis of yield components compensation in crop plants with special reference to the field bean (*Phaseolus vulgaris*). Crop Sci 7(5):505–510

Agrama HA, Eizenga GC, Yan W (2007) Association mapping of yield and its components in rice cultivars. Mol Breed 19:341–356

Agrama HA, Eizenga GC (2008) Molecular diversity and genome-wide linkage disequilibrium patterns in a worldwide collection of *Oryza sativa* and its wild relatives. Euphytica 160:339–355

Akinwale MG, Gregorio G, Nwilene F, Akinyele BO, Ogunbayo SA, Odiyi AC (2011) Heritability and correlation coefficient analysis for yield and its components in rice (*Oryza sativa* L.) *African*. J Plant Sci 5(3):207–212

Amante-Bordeos A, Sitch LA, Nelson R, Dalmacio RD, Oliva NP, Aswidinnoor H, Leung H (1992) Transfer of bacterial blight and blast resistance from the tetraploid wild rice *Oryza minuta* to cultivated rice, *Oryza sativa*. Theor Appl Genet 84(3–4):345–354

Ashikari M, Sakakibara H, Lin S, Yamamoto T, Takashi T, Nishimura A, Angeles ER, Qian Q, Kitano H, Matsuoka M (2005) Cytokinin oxidase regulates rice grain production. Science 309 (5735):741–745

Bai XF, Wu B, Xing YZ (2012) Yield-related QTL and their applications in rice genetic improvement. J Integr Plant Biol 54(5):300–311

Balkunde S, Le HL, Lee HS, Kim DM, Kang JW, Ahn SN (2013) Fine mapping of a QTL for the number of spikelets per panicle by using near-isogenic lines derived from an interspecific cross between *Oryza sativa* and *Oryza minuta*. Plant Breed 132(1):70–76

Barrangou R, Fremaux C, Deveau H, Richards M, Boyaval P, Moineau S, Romero DA, Horvath P (2007) CRISPR provides acquired resistance against viruses in prokaryotes. Science 2315:1709–1712

Bhasin H, Bhatia D, Raghuvanshi S, Lore JS, Sahi GK, Kaur B, Vikal Y, Singh K (2012) New PCR-based sequence-tagged site marker for bacterial blight resistance gene *Xa38* of rice. Mol Breed 30(1):607–611

Bradbury PJ, Zhang Z, Kroon DE, Casstevens TM, Ramdoss Y, Buckler ES (2007) TASSEL: software for association mapping of complex traits in diverse samples. Bioinformatics 23:2633–2635

Brar DS, Khush GS (1997) Alien introgression in rice. Plant Mol Biol 35(1–2):35–47

Brar DS, Khush GS (2002) Transferring genes from wild species into rice. In: Kang MS (ed) Quantitative genetics, genomics and plant breeding. CABI, Wallingford, pp 197–217

Brar DS, Khush GS (2006) Cytogenetic Manipulation and Germplasm Enhancement of Rice (*Oryza sativa* L.). In: Genetic resources, chromosome engineering and crop improvement. Genetic Resources Chromosome Engineering & Crop Improvement. CRC Press, New York, pp 115–158

Brar DS, Singh K (2011) Oryza. In: Kole C (ed) Wild crop relatives: genomic and breeding resources, cereals. Springer, Berlin, pp 321–365

Broman KW, Wu H, Sen S, Churchill GA (2003) R/qtl: QTL mapping in experimental crosses. Bioinformatics 19:889–890

Brondani C, Rangel PHN, Brondani RPV, Ferreira ME (2002) QTL mapping and introgression of yield-related traits from *Oryza glumaepatula* to cultivated rice (*Oryza sativa*) using microsatellite markers. Theor Appl Genet 104(6–7):1192–1203

Choudhury DR, Singh N, Singh AK, Kumar S, Srinivasan K, Tyagi RK, Ahmad A, Singh NK, Singh R (2014) Analysis of genetic diversity and population structure of rice germplasm from North-Eastern region of India and development of a core germplasm set. PLoS ONE 9:e113094

Comai L, Young K, Till BJ, Reynolds SH, Greene EA, Codomo CA, Enns LC, Johnson J, Burtner C, Oden AR, Henikoff S (2004) Efficient discovery of DNA polymorphisms in natural populations by ecotilling. Plant J 37:778–786

Cong L, Ran FA, Cox D, Lin S, Barretto R, Habib N, Hsu PD, Wu X, Jiang W, Marraffini LA, Zhang F (2013) Multiplex genome engineering using CRISPR/Cas systems. Science 339:819–823

Deosarkar DB, Misal MB, Nerkar YS (1989) Variability and correlation studies in breeding lines of upland rice. J Maharashtra Agric Univ 14(1):20–28

Devlin B, Roeder K (1999) Genomic control for association studies. Biometrics 55:997–1004

Edward D, Batley J (2010) Plant genome sequencing: applications for crop improvement. Plant Biotech J 8(1):2–9

Ehrenreich IM, Torabi N, Jia Y, Kent J, Martis S, Shapiro JA, Gresham D, Caudy AA, Kruglyak L (2010) Dissection of genetically complex traits with extremely large pools of yeast segregants. Nature 464:1039–1042

Eizenga GC, Prasad B, Jackson AK, Jia MH (2013) Identification of rice sheath blight and blast quantitative trait loci in two different *O. sativa*/*O. nivara* advanced backcross populations. Mol Breed 31(4):889–907

El-Malky M, El-Habashy M, Abdelkhalik AF (2008) Rice germplasm evaluation for agronomic traits and their influence on stem borer (*Chilo agamemnon* bles.) resistance. J Agric Res 46 (3):206

Ersoz ES, Yu J, Buckler ES (2008) Applications of linkage disequilibrium and association mapping in crop plants. In: Varshney R, Tuberosa R (eds) Genomic assisted crop improvement: Vol. I: genomics approaches and platforms. Springer, Germany, pp 97–120

Eshed Y, Zamir D (1995) An introgression line population of *Lycopersicon Pennellii* in the cultivated tomato enables the identification and fine mapping of yield-associated QTL. Genetics 141(3):1147–1162

Falush D, Stephens M, Pritchard JK (2003) Inference of population structure using multilocus genotype data: linked loci and correlated allele frequencies. Genetics 164:1567–1587

Fan CH, Xing YZ, Mao HL, Lu TT, Han B, Xu CG, Li XH, Zhang QF (2006) GS3, a major QTL for grain length and weight and minor QTL for grain width and thickness in rice, encodes a putative transmembrane protein. Theor Appl Genet 112(6):1164–1171

Fekih R, Takagi H, Tamiru M, Abe A, Natsume S et al (2013) MutMap+: genetic mapping and mutant identification without crossing in rice. PLoS ONE 8(7):e68529

Flint-Garcia SA, Thuillet AC, Yu J, Pressoir G, Romero SM, Mitchell S, Doebley J, Kresovich S, Goodman MM, Buckler ES (2005) Maize association population: a high-resolution platform for quantitative trait locus dissection. Plant J 44:1054–1064

Frankel OH (1935) Analytical yield investigations on New Zealand wheat: II. Five years' analytical variety trials. J Agricul Sci 25(04):466–509

Frary A, Nesbitt TC, Grandillo S, Knaap E, Cong B, Liu J, Meller J, Elber R, Alpert KB, Tanksley SD (2000) fw2.2: a quantitative trait locus key to the evolution of tomato fruit size. Science 289(5476):85–88

Fu Q, Zhang P, Tan L, Zhu Z, Ma D, Fu Y, Zhan X, Cai H, Sun C (2010) Analysis of QTL for yield-related traits in Yuanjiang common wild rice (*Oryza rufipogon* Griff.). J Genet Genomics 37(2):147–157

Fujita D, Trijatmiko KR, Tagle AG, Sapasap MV, Koide Y, Sasaki K, Tsakirpaloglou N, Gannaban RB, Nishimura T, Yanagihara S, Fukuta Y, Koshiba T, Slamet-Loedin IH, Ishimaru T, Kobayashi N (2013) NAL1allele from a rice landrace greatly increases yield in modern indicacultivars. Proc Natl Acad Sci USA 110:20431–20436

Gaikwad KB, Singh N, Bhatia D, Kaur R, Bains NS, Tajinder SB, Singh K (2014) Yield-enhancing heterotic QTL transferred from wild species to cultivated rice *Oryza sativa* L. PLoS ONE 9(6):e96939

Gao ZY, Zhao SC, He WM, Guo LB, Peng YL, Wang JJ, Guo XS, Zhang XM, Rao YC, Zhang C, Dong GJ, Zheng FY, Lu CX, Hu J, Zhou Q, Liu HJ, Wu HY, Xu J, Ni PX, Zeng DL, Liu DH, Tian P, Gong LH, Ye C, Zhang GH, Wang J, Tian FK, Xue DW, Liao Y, Zhu L, Chen MS, Li JY, Cheng SH, Zhang GY, Wang J, Qian Q (2013) Dissecting yield-associated loci in super hybrid rice by resequencing recombinant inbred lines and improving parental genome sequences. Proc Natl Acad Sci USA 110(35):14492–14497

Garris A, Tai T, Coburn J, Kresovich S, McCouch S (2005) Genetic structure and diversity in *Oryza sativa* L. Genetics 169:1631–1638

Geldermann H (1975) Investigations on inheritance of quantitative characters in animals by gene markers I. Methods. Theor Appl Genet 46(7):319–330

Goff SA, Ricke D, Lan TH, Presting G, Wang R, Dunn M, Glazebrook J, Sessions A, Oeller P, Varma H, Hadley D, Hutchison D, Martin C, Katagiri F, Lange BM, Moughamer T, Xia Y, Budworth P, Zhong J, Miguel T, Paszkowski U, Zhang S, Colbert M, Sun WL, Chen L, Cooper B, Park S, Wood TC, Mao L, Quail P, Wing R, Dean R, Yu Y, Zharkikh A, Shen R, Sahasrabudhe S, Thomas A, Cannings R, Gutin A, Pruss D, Reid J, Tavtigian S, Mitchell J, Eldredge G, Scholl T, Miller RM, Bhatnagar S, Adey N, Rubano T, Tusneem N, Robinson R, Feldhaus J, Macalma T, Oliphant A, Briggs S (2002) A draft sequence of the rice genome (*Oryza sativa* L. ssp. *japonica*). Science 296(5565):92–100

Gratz SJ, Cummings AM, Nguyen JM, Hamm DC, Donohue LK, Harrison MM, Wildonger J, O'Connor-Giles KM (2013) Genome engineering of *Drosophila* with the CRISPR RNA-Guided Cas9 nuclease. Genetics 194:1029–1035

Gur A, Zamir D (2004) Unused natural variation can lift yield barriers in plant breeding. PLoS Biol 2(10):1610–1615

Gutierrez A, Carabalí S, Giraldo O, Martínez C, Correa F, Prado G, Tohme J, Lorieux M (2010) Identification of a Rice stripe necrosis virus resistance locus and yield component QTL using *Oryza sativa* x *O.* glaberrima introgression lines. BMC Plant Biol 10(1):6. doi:10.1186/1471-2229-10-6

Harlan JR (1976) Genetic resources in wild relatives of crops. Crop Sci 16(3):329–333

Hawkes JG (1977) The importance of wild germplasm in plant breeding. Euphytica 26(3):615–621

Heang D, Sassa H (2012a) An atypical bHLH protein encoded by POSITIVE REGULATOR OF GRAIN LENGTH 2 is involved in controlling grain length and weight of rice through interaction with a typical bHLH protein APG. Breed Sci 62:133–141

Heang D, Sassa H (2012b) Antagonistic actions of HLH/bHLH proteins are involved in grain length and weight in rice. PLoS ONE 7:e31325

Heinrichs EA, Medrano FG, Rapusas HR, International Rice Research Institute (1985) Genetic evaluation for insect resistance in rice. International Rice Research Institute, Manila, Philippines

Hsu PD, Lander ES, Zhang F (2014) Development and applications of CRISPR-Cas9 for genome engineering. Cell 157:1262–1278

Hua JP, Xing YZ, Xu CG, Sun XL, Yu SB, Zhang Q (2002) Genetic dissection of an elite rice hybrid revealed that heterozygotes are not always advantageous for performance. Genetics 162 (4):1885–1895

Huang A, Xu S, Cai X (2014) Whole-genome quantitative trait locus mapping reveals major role of epistasis on yield of rice. PLoS ONE 9(1):e87330

Huang XH, Wei XH, Sang T, Zhao Q, Feng Q, Zhao Y, Li CY, Zhu CR, Lu TT, Zhang ZW, Li M, Fan DL, Guo YL, Wang AH, Wang L, Deng LW, Li WJ, Lu YQ, Weng QJ, Liu KY, Huang T, Zhou TY, Jing YF (2010) Genome-wide association studies of 14 agronomic traits in rice landraces. Nat Genet 42:961–967

Huang X, Feng Q, Quian Q, Zhao Q, Wang L, Wang A, Guan J, Fan D, Weng Q, Huang T, Dong G, Sang T, Han B (2009a) High throughput genotyping by whole genome resequencing. Genome Res 19:1068–1076

Huang X, Kurata N, Wei X, Wang ZX, Wang A, Zhao Q, Zhao Y, Liu K, Lu H, Li W, Guo Y, Lu Y, Zhou C, Fan D, Weng Q, Zhu C, Huang T, Zhang L, Wang Y, Feng L, Furuumi H, Kubo T, Miyabayashi T, Yuan X, Xu Q, Dong G, Zhan Q, Li C, Fujiyama A, Toyoda A, Lu T, Feng Q, Qian Q, Li J, Han B (2012) A map of rice genome variation reveals the origin of cultivated rice. Nature 490(7421):497–501

Huang X, Qian Q, Liu Z, Sun H, He S, Luo D, Xia G, Chu C, Li J, Fu X (2009b) Natural variation at the *DEP1*locus enhances grain yield in rice. Nat Genet 41:494–497

Ikeda K, Ito M, Nagasawa N, Kyozuka J, Nagato Y (2007) Rice *ABERRANT PANICLE ORGANIZATION 1,* encoding an F-box protein, regulates meristem fate. Plant J 51:1030–1040

Ikeda-Kawakatsu K, Maekawa M, Izawa T, Itoh J, Nagato Y (2012) *ABERRANT PANICLE ORGANIZATION 2/RFL*, the rice ortholog of Arabidopsis LEAFY, suppresses the transition from inflorescence meristem to floral meristem through interaction with *APO1*. Plant J 69:168–180

International Rice Genome Sequencing Project (2005) The map-based sequence of the rice genome. Nature 436(7052):793–800

Ishikawa S, Maekawa M, Arite T, Onishi K, Takamure I, Kyozuka J (2005) Suppression of tiller bud activity in tillering dwarf mutants of rice. Plant Cell Physiol 46:79–86

Ishimaru K, Hirotsu N, Madoka Y, Murakami N, Hara N, Onodera H, Kashiwagi T, Ujiie K, Shimizu B, Onishi A, Miyagawa H, Katoh E (2013) Loss of function of the IAA-glucose hydrolase gene *TGW6* enhances rice grain weight and increases yield. Nat Genet 45:707–711

Jacquemin J, Bhatia D, Singh K, Wing RA (2013) The international *Oryza* map alignment project: development of a genus-wide comparative genomics platform to help solve the 9 billion-people question. Curr Opin Plant Biol 16:147–156

Jiao Y, Wang Y, Xue D, Wang J, Yan M, Liu G, Dong G, Zeng D, Lu Z, Zhu X, Qian Q, Li J (2010) Regulation of *OsSPL14* by OsmiR156 defines ideal plant architecture in rice. Nat Genet 42:541–544

Jin FX, Kim DM, Ju HG, Ahn SN (2009) Mapping quantitative trait loci for awnness and yield component traits in isogenic lines derived from an *Oryza sativa/O. rufipogon* cross. J Crop Sci Biotech 12:9–15

Jin J, Huang W, Gao JP, Yang J, Shi M, Zhu MZ, Luo D, Lin HX (2008) Genetic control of rice plant architecture under domestication. Nat Genet 40:1365–1369

Jin L, Lu Y, Xiao P, Sun M, Corke H, Bao J (2010) Genetic diversity and population structure of a diverse set of rice germplasm for association mapping. Theor Appl Genet 121:475–487

Joung KJ, Sander JD (2013) TALENs: a widely applicable technology for targeted genome editing. Nat Rev Mol Cell Biol 14(1):49–55

Kearsey MJ (1998) The principles of QTL analysis (a minimal mathematics approach). J Exp Bot 49(327):1619–1623

Khush GS (2013) Strategies for increasing the yield potential of cereals: case of rice as an example. Plant Breed 132:433–436

Kitagawa K, Kurinami S, Oki K, Abe Y, Ando T, Kono I, Yano M, Kitano H, Iwasaki Y (2010) A novel kinesin 13 protein regulating rice seed length. Plant Cell Physiol 51:1315–1329

Komatsu M, Chujo A, Nagato Y, Shimamoto K, Kyozuka J (2003) *FRIZZY PANICLE* is required to prevent the formation of axillary meristems and to establish floral meristem identity in rice spikelets. Development 130:3841–3850

Komatsu M, Maekawa M, Shimamoto K, Kyozuka J (2001) *The LAX1*and *FRIZZY PANICLE 2* genes determine the inflorescence architecture of rice by controlling rachis branch and spikelet development. Dev Biol 231:364–373

Kurakawa T, Ueda N, Maekawa M, Kobayashi K, Kojima M, Nagato Y, Sakakibara H, Kyozuka J (2007) Direct control of shoot meristem activity by a cytokinin-activating enzyme. Nature 445:652–655

Li C, Zhou A, Sang T (2006) Genetic analysis of rice domestication syndrome with the wild annual species, *Oryza nivara*. New Phytol 170(1):185–193

Li F, Liu W, Tang J, Chen J, Tong H, Hu B, Li C, Fang J, Chen M, Chu C (2010) Rice *DENSE AND ERECT PANICLE 2* is essential for determining panicle outgrowth and elongation. Cell Res 20:838–849

Li J, Thomson M, McCouch SR (2004) Fine mapping of a grain-weight quantitative trait locus in the pericentromeric region of rice chromosome 3. Genetics 168(4):2187–2195

Li M, Tang D, Wang K, Wu X, Lu L, Yu H, Gu M, Yan C, Cheng Z (2011a) Mutations in the F-box gene LARGER PANICLE improve the panicle architecture and enhance the grain yield in rice. Plant Biotechnol J 9:1002–1013

Li S, Qian Q, Fu Z, Zeng D, Meng X, Kyozuka J, Maekawa M, Zhu X, Zhang J, Li J, Wang Y (2009) Short panicle1 encodes a putative PTR family transporter and determines rice panicle size. Plant J 58:592–605

Li X, Qian Q, Fu Z, Wang Y, Xiong G, Zeng D, Wang X, Liu X, Teng S, Hiroshi F, Yuan M, Luo D, Han B, Li J (2003) Control of tillering in rice. Nature 422:618–621

Li Y, Fan C, Xing Y, Jiang Y, Luo L, Sun L, Shao D, Xu C, Li X, Xiao J, He Y, Zhang Q (2011b) Natural variation in *GS5* plays an important role in regulating grain size and yield in rice. Nat Genet 43:1266–1269

Li ZK, Pinson SRM, Stansel JW, Paterson AH (1998) Genetic dissection of the source-sink relationship affecting fecundity and yield in rice (*Oryza sativa* L.). Mol Breed 4:419–426

Lin HX, Qian HR, Zhuang JY, Lu J, Min SK, Xiong ZM, Huang N, Zheng KL (1996) RFLP mapping of QTL for yield and related characters in rice (*Oryza sativa* L.). Theor Appl Genet 92(8):920–927

Linnaeus C (1753) Species plantarum 1: 333. Stock holm

Liu TM, Mao DH, Zhang SP, Xing YZ (2009) Fine mapping *SPP1*, a QTL controlling the number of spikelets per panicle, to a BAC clone in rice (*Oryza sativa*). Theor Appl Genet 118:1509–1517

Lu C, Shen L, Tan Z, Xu Y, He P, Chen Y, Zhu L (1996) Comparative mapping of QTL for agronomic traits of rice across environments using a doubled haploid population. Theor Appl Genet 93(8):1211–1217

Luo JH, Liu H, Zhou TY, Gu BG, Huang XH, Shangguan YY, Zhu JJ, Li Y, Zhao Y, Wang YC, Zhao Q, Wang AH, Wang ZQ, Sang T, Wang ZX, Han B (2013) An-1 encodes a basic helix-loop-helix protein that regulates awn development, grain size, and grain number in rice. Plant Cell 25(9):3360–3376

Luo JJ, Hao W, Jin J, Gao JP, Lin HX (2008) Fine mapping of Spr3, a locus for spreading panicle from African cultivated rice (*Oryza glaberrima* steud.). Mol Plant 1(5):830–838

Mali P, Yang L, Esvelt KM, Aach J, Guell M, DiCarlo JE, Norville JE, Church GM (2013) RNA-Guided human genome engineering via Cas9. Science 339:823–829

Mao H, Sun S, Yao J, Wang C, Yu S, Xu C, Li X, Zhang Q (2010) Linking differential domain functions of the GS3 protein to natural variation of grain size in rice. Proc Natl Acad Sci USA 107:19579–19584

Marri PR, Sarla N, Reddy LV, Siddiq EA (2005) Identification and mapping of yield and yield related QTL from an Indian accession of *Oryza rufipogon*. BMC Genet 6:33

McCallum CM, Comai L, Greene EA, Henikoff S (2000) Targeted screening for induced mutations. Nat Biotechnol 18:455–457

McCouch SR, Sweeney M, Li J, Jiang H, Thomson M, Septiningsih E, Edwards J, Moncada P, Xiao J, Garris A, Tai T, Martinez C, Tohme J, Sugiono M, Mcclung A, Yuan LP, Ahn SN (2007) Through the genetic bottleneck: *O. rufipogon* as a source of trait-enhancing alleles for *O. sativa*. Euphytica 154:317–339

McCouch SR, Zhao K, Wright M, Tung CW, Ebana K, Thomson M, Reynolds A, Wang D, DeClerck G, Ali ML, McClung A, Eizenga G, Bustamante C (2010) Development of genome-wide SNP assays for rice. Breed Sci 60:524–535

McNally K, Childs K, Bohnert R, Davidson R, Zhao K, Ulat V, Zeller G, Clark R, Hoen D, Bureau T, Stokowski R, Ballinger D, Frazer K, Cox D, Padhukasahasram B, Bustamante C, Weigel D, Mackill D, Bruskiewich R, Rätsch G, Buell C, Leung H and Leach J (2009) Genome wide SNP variation reveals relationships among landraces and modern varieties of rice. Proc Natl Acad Sci USA 106:12 273–12 278

Mehetre SS, Mahajan CR, Patil PA, Lad SK, Dhumal PM (1994) Variability, heritability, correlation, path analysis, and genetic divergence studies in upland rice. Int Rice Res Notes 19(1):8–10

Miah G, Rafii MY, Ismail MR, Puteh AB, Rahim HA, Islam KN, Latif MA (2013) A review of microsatellite markers and their applications in rice breeding programs to improve blast disease resistance. Int J Mol Sci 14(11):22499–22528

Miao J, Guol D, Zhang J, Huang Q, Qin G, Zhang X, Wan J, Gu H, Qu LJ (2013) Targeted mutagenesis in rice using CRISPR-Cas system. Cell Res 23:1233–1236

Miura K, Ashikari M, Matsuoka M (2011) The role of QTL in the breeding of high-yielding rice. Trends Plant Sci 16(6):319–326

Miura K, Ikeda M, Matsubara A, Song XJ, Ito M, Asano K, Matsuoka M, Kitano H, Ashikari M (2010) *OsSPL14* promotes panicle branching and higher grain productivity in rice. Nat Genet 42:545–549

Moncada P, Martinez CP, Borrero J, Chatel M, Gauch H, Guimaraes E, Tohme J, McCouch SR (2001) Quantitative trait loci for yield and yield components in an *Oryza sativa* x *Oryza rufipogon* BC2F2 population evaluated in an upland environment. Theor Appl Genet 102(1):41–52

Mori M, Nomura T, Ooka H, Ishizaka M, Yokota T, Sugimoto K, Okabe K, Kajiwara H, Satoh K, Yamamoto K, Hirochika H, Kikuchi S (2002) Isolation and characterization of a rice dwarf mutant with a defect in brassinosteroid biosynthesis. Plant Physiol 130:1152–1161

Nakagawa H, Tanaka A, Tanabata T, Ohtake M, Fujioka S, Nakamura H, Ichikawa H, Mori M (2012) Short grain1decreases organ elongation and brassinosteroid response in rice. Plant Physiol 158:1208–1219

Negrão S, Almadanim C, Pires McNally KL, Oliveira MM (2011) Use of EcoTILLING to identify natural allelic variants of rice candidate genes involved in salinity tolerance. Plant Genet Resour 9:300–304

Nonoue Y, Fujino K, Hirayama Y, Yamanouchi U, Lin SY, Yano M (2008) Detection of quantitative trait loci controlling extremely early heading in rice. Theor Appl Genet 116(5): 715–722

Oh JM, Balkunde S, Yang P, Yoon DB, Ahn SN (2011) Fine mapping of grain weight QTL, tgw11 using near isogenic lines from a cross between *Oryza sativa* and *O. grandiglumis*. *Genes*. Genome 33(3):259–265

Olsen KM, Caicedo AL, Polato N, McClung A, McCouch S, Purugganan D (2006) Selection under domestication: evidence for a sweep in the rice Waxy genomic region. Genetics 173:975–983

Ookawa T, Hobo T, Yano M, Murata K, Ando T, Miura H, Asano K, Ochiai Y, Ikeda M, Nishitani R, Ebitani T, Ozaki H, Angeles ER, Hirasawa T (2010) New approach for rice improvement using a pleiotropic QTL gene for lodging resistance and yield. Nat Commun 1:132

Price AH (2006) Believe it or not. QTL are accurate. Trends Plant Sci 11:213–216

Pritchard JK, Stephens M, Donnelly P (2000) Inference of population structure using multilocus genotype data. Genetics 155:945–959

Qi P, Lin YS, Song XJ, Shen JB, Huang W, Shan JX, Zhu MZ, Jiang L, Gao JP, Lin HX (2012) The novel quantitative trait locus *GL3.1*controls rice grain size and yield by regulating Cyclin-T1. Cell Res 22:1666–1680

Qiao Y, Piao R, Shi J, Lee SI, Jiang W, Kim BK, Lee J, Han L, Ma W, Koh HJ (2011) Fine mapping and candidate gene analysis of dense and erect panicle 3, *DEP3*, which confers high grain yield in rice (*Oryza sativa* L.). Theor Appl Genet 122:1439–1449

Rahman ML, Chu SH, Choi MS, Qiao YL, Jiang W, Piao R, Khanam S, Cho YI, Jeung JU, Jena KK, Koh HJ (2007) Identification of QTL for some agronomic traits in rice using an introgression line from *Oryza minuta*. Mol Cells 24(1):16–26

Rangel PH, Guimaraes EP, Neves PCF (1996) Base gene´tica das cultivares de arroz (*Oryza sativa* L.) irrigado do Brasil. Pesqui Agropec Bras 31:349–357

Rangel PN, Vianello RP, Melo ATO, Rangel PHN, Mendonça JA, Brondani C (2013) Yield QTL analysis of *Oryza sativa* x *O. glumaepatula* introgression lines. Pesq agropec Bras 48(3):280–286

Remington DL, Thornsberry JM, Matsuola Y, Wilson LM, Whitt SR, Doebley J, Kresovich S, Goodman MM, Buckler ES IV (2001) Structure of linkage disequilibrium and phenotypic associations in the maize genome. Proc Natl Acad Sci USA 98:11479–11484

Sakamoto T, Matsuoka M (2008) Identifying and exploiting grain yield genes in rice. Curr Opin Plant Biol 11(2):209–214

Sanchez P, Wing R, Brar D (2013) The wild relative of rice: genomes and genomics: crops and models. In: Zhang Q, Wing RA (eds) Genetics and genomics of rice, vol 5. Plant genetics and genomics. Springer, New York, pp 9–25

Sax K (1923) The association of size differences with seed-coat pattern and pigmentation in *Phaseolus vulgaris*. Genetics 8(6):552–560

Second G (1982) Origin of the genic diversity of cultivated rice (*Oryza* spp.). Jap J Genet 57:25–58

Segami S, Kono I, Ando T, Yano M, Kitano H, Miura K, Iwasaki Y (2012) Small and round seed 5 gene encodes alpha-tubulin regulating seed cell elongation in rice. Rice 5:4

Septiningsih EM, Prasetiyono J, Lubis E, Tai TH, Tjubaryat T, Moeljopawiro S, McCouch SR (2003) Identification of quantitative trait loci for yield and yield components in an advanced backcross population derived from the *Oryza sativa* variety IR64 and the wild relative *O. rufipogon*. Theor Appl Genet 107(8):1419–1432

Serrat X, Esteban R, Guibourt N, Moysset L, Nogués S, Lalanne E (2014) EMS mutagenesis in mature seed-derived rice calli as a new method for rapidly obtaining TILLING mutant populations. Plant Methods 10:5

She KC, Kusano H, Koizumi K, Yamakawa H, Hakata M, Imamura T, Fukuda M, Naito N, Tsurumaki Y, Yaeshima M et al (2010) A novel factor *FLOURY ENDOSPERM 2* is involved in regulation of rice grain size and starch quality. Plant Cell 22:3280–3294

Shomura A, Izawa T, Ebana K, Ebitani T, Kanegae H, Konishi S, Yano M (2008) Deletion in a gene associated with grain size increased yields during rice domestication. Nat Genet 40:1023–1028

Sidana K (2012) Molecular mapping of qtl for yield and yield components transferred from *Oryza longistaminata* (a.chev. et roehr.) to *Oryza sativa* (L.). M.Sc. thesis, Punjab Agricultural University, Ludhiana, Punjab

Song XJ, Huang W, Shi M, Zhu MZ, Lin HX (2007) A QTL for rice grain width and weight encodes a previously unknown RING-type E3 ubiquitin ligase. Nat Genet 39:623–630

Spielmeyer W, Ellis MH, Chandler PM (2002) Semidwarf (*sd-1*), "green revolution" rice, contains a defective gibberellin 20-oxidase gene. Proc Natl Acad Sci USA 99(13):9043–9048

Subramanian S, Rathinam M (1984) Association of grain yield attributes in the hybrids of crosses between tall and semi-dwarf varieties of rice. Madras Agri J 71(8):531–538

Sun CQ, Wang XK, Li ZC, Yoshimura A, Iwata N (2001) Comparison of the genetic diversity of common wild rice (*Oryza rufipogon* Griff.) and cultivated rice (*O. sativa* L.) using RFLP markers. Theor Appl Genet 102(1):157–162

Swamy BM, Kaladhar K, Rani NS, Prasad G, Viraktamath B, Reddy GA, Sarla N (2012) QTL analysis for grain quality traits in 2 BC_2F_2 populations derived from crosses between *Oryza sativa* cv Swarna and 2 accessions of *O. nivara*. J Hered 103(3):442–452

Swamy BPM, Kaladhar K, Ramesha MS, Viraktamath BC, Sarla N (2011) Molecular Mapping of QTL for Yield and Yield-Related Traits in *Oryza sativa* cv Swarna × *O. nivara* (IRGC81848) Backcross Population. Rice Sci 18(3):178–186

Takagi H, Abe A, Yoshida K, Kosugi S, Natsume S, Mitsuoka C, Uemura A, Utsushi H, Tamiru M, Takuno S, Innan H, Cano LM, Kamoun S, Terauchi R (2013) QTL-seq: rapid mapping of quantitative trait loci in rice by whole genome resequencing of DNA from two bulked populations. Plant J 74(1):174–183

Tanksley SD, McCouch SR (1997) Seed banks and molecular maps: unlocking genetic potential from the wild. Science 277(5329):1063–1066

Thomson MJ, Zhao K, Wright M, McNally KL, Leung H, McCouch SR (2011) Development and application of 96- and 384-plex single nucleotide polymorphism (SNP) marker sets for diversity analysis, mapping and marker-assisted selection in rice. In: Second Africa Rice Congress, Bamako, Mali, 22–26 Mar 2010: Innovation and Partnerships to Realize Africa's Rice Potential

Thomson MJ, Tai TH, McClung AM, Lai XH, Hinga ME, Lobos KB, Xu Y, Martinez CP, McCouch SR (2003) Mapping quantitative trait loci for yield, yield components and morphological traits in an advanced backcross population between *Oryza rufipogon* and the *Oryza sativa* cultivar Jefferson. Theor Appl Genet 107(3):479–493

Tian F, Li DJ, Fu Q, Zhu ZF, Fu YC, Wang XK, Sun CQ (2006) Construction of introgression lines carrying wild rice (*Oryza rufipogon* Griff.) segments in cultivated rice (*Oryza sativa* L.) background and characterization of introgressed segments associated with yield-related traits. Theor Appl Genet 112(3):570–580

Turner TL (2014) Fine-mapping natural alleles: quantitative complementation to the rescue. Mol Ecol 23:2377–2382

Vaughan DA (1989) The genus *Oryza* L.: current status of taxonomy, IRRl Res Paper Sr 138. International Rice Research Institute, Manila, Philippines, p 21

Vaughan DA (1994) The wild relatives of rice: a genetic resources handbook. International Rice Research Institute, Manila, Philippines 137 p

Wang E, Wang J, Zhu XD, Hao W, Wang LY, Li Q, Zhang LX, He W, Lu BR, Lin HX, Ma H, Zhang GQ, He ZH (2008) Control of rice grain-filling and yield by a gene with a potential signature of domestication. Nat Genet 40(11):1370–1374

Wang S, Wu K, Yuan Q, Liu X, Liu Z, Lin X, Zeng R, Zhu H, Dong G, Qian Q, Zhang G, Fu X (2012) Control of grain size, shape and quality by *OsSPL16* in rice. Nat Genet 44:950–954
Wang Y, Xi Cheng, Shan Q, Yi Zhang, Liu J, Gao C, Qiu JL (2014) Simultaneous editing of three homoeoalleles in hexaploid bread wheat confers heritable resistance to powdery mildew. Nature Biotech 32:947–951
Wang ZY, Second G, Tanksley SD (1992) Polymorphism and phylogenetic-relationships among species in the genus *Oryza* as determined by analysis of nuclear RFLPs. Theor Appl Genet 83 (5):565–581
Wei X, Xu J, Guo H, Jiang L, Chen S, Yu C, Zhou Z, Hu P, Zhai H, Wan J (2010) *DTH8* suppresses flowering in rice, influencing plant height and yield potential simultaneously. Plant Physiol 153:1747–1758
Weng J, Gu S, Wan X, Gao H, Guo T, Su N, Lei C, Zhang X, Cheng Z, Guo X, Wang J, Jiang L, Zhai H, Wan J (2008) Isolation and initial characterization of *GW5*, a major QTL associated with rice grain width and weight. Cell Res 18:1199–1209
Wu JL, Wu C, Lei C, Baraoidan M, Bordeos A, Madamba MR, Ramos-Pamplona M, Mauleon R, Portugal A, Ulat VJ, Bruskiewich R, Wang G, Leach J, Khush G, Leung H (2005) Chemical- and irradiation-induced mutants of *indica* rice IR64 for forward and reverse genetics. Plant Mol Biol 59(1):85–97
Xiao JH, Li JM, Grandillo S, Ahn SN, Yuan LP, Tanksley SD, McCouch SR (1998) Identification of trait-improving quantitative trait loci alleles from a wild rice relative, *Oryza rufipogon*. Genetics 150(2):899–909
Xiao J, Li J, Yuan L, Tanksley SD (1996) Identification of QTL affecting traits of agronomic importance in a recombinant inbred population derived from a subspecific rice cross. Theor Appl Genet 92:230−244
Xie X, Jin F, Song MH, Suh JP, Hwang HG, Kim YG, McCouch SR, Ahn SN (2008) Fine mapping of a yield-enhancing QTL cluster associated with transgressive variation in an *Oryza sativa* x *O. rufipogon* cross. Theor Appl Genet 116(5):613–622
Xie X, Song MH, Jin F, Ahn SN, Suh JP, Hwang HG, McCouch SR (2006) Fine mapping of a grain weight quantitative trait locus on rice chromosome 8 using near-isogenic lines derived from a cross between *Oryza sativa* and *Oryza rufipogon*. Theor Appl Genet 113(5):885–894
Xing Z, Tan F, Hua P, Sun L, Xu G, Zhang Q (2002) Characterization of the main effects, epistatic effects and their environmental interactions of QTL on the genetic basis of yield traits in rice. Theor Appl Genet 105(2–3):248–257
Xiong ZM (1992) Research outline on rice genetics in China. In: Xiong ZM, Cai HF (eds) Rice in China. Chinese Agricultural Science Press, Beijing, pp 40–57
Xue W, Xing Y, Weng X, Zhao Y, Tang W, Wang L, Zhou H, Yu S, Xu C, Li X, Zhang Q (2008) Natural variation in *Ghd7* is an important regulator of heading date and yield potential in rice. Nat Genet 40:761–767
Yamagishi M, Takeuchi Y, Kono I, Yano M (2002) QTL analysis for panicle characteristics in temperate japonica rice. Euphytica 128(2):219–224
Yamamoto T, Lin H, Sasaki T, Yano M (2000) Identification of heading date quantitative trait locus Hd6 and characterization of its epistatic interactions with Hd2 in rice using advanced backcross progeny. Genetics 154(2):885–891
Yamamoto T, Yonemaru J, Yano M (2009) Towards the understanding of complex traits in rice: substantially or superficially? DNA Res 16(3):141–154
Yan WH, Wang P, Chen HX, Zhou HJ, Li QP, Wang CR, Ding ZH, Zhang YS, Yu SB, Xing YZ, Zhang Q (2011) A major QTL, *Ghd8*, plays pleiotropic roles in regulating grain productivity, plant height, and heading date in rice. Mol Plant 4:319–330
Yang ZM, Huang DQ, Tang WQ, Zheng Y, Liang KJ, Cutle AJ, Wu WR (2013) Mapping of quantitative trait loci underlying cold tolerance in rice seedlings via high-throughput sequencing of pooled extremes. PLoS ONE 87(7):e88433
Yano M, Kojima S, Takahashi Y, Lin HX, Sasaki T (2001) Genetic control of flowering time in rice, a short-day plant. Plant Physiol 127(4):1425–1429

Yano M, Sasaki T (1997) Genetic and molecular dissection of quantitative traits in rice. Plant Mol Biol 35(1–2):145–153

Yoon DB, Kang KH, Kim HJ, Ju HG, Kwon SJ, Suh JP, Jeong OY, Ahn SN (2006) Mapping quantitative trait loci for yield components and morphological traits in an advanced backcross population between *Oryza grandiglumis* and the *O. sativa* japonica cultivar Hwaseongbyeo. Theor Appl Genet 112(6):1052–1062

Yoshida S (1983) Rice. In: Smith WH, Banta SJ (eds) Potential productivity of field crops under different environments. International Rice Research Institute Publishing, Los Bãnos, pp 103–127

Yu HH, Xie WB, Wang J, Xing YZ, Xu CG, Li XH, Xiao JH, Zhang QF (2011) Gains in QTL detection using an ultra-high density SNP map based on population sequencing relative to traditional RFLP/SSR markers. PLoS ONE 6(3):e17595

Yu J, Buckler ES (2006) Genetic association mapping and genome organization of maize. Curr Opin Biotech 17:155–160

Yu J, Hu SN, Wang J, Wong GKS, Li SG, Liu B, Deng YJ, Dai L, Zhou Y, Zhang XQ, Cao ML, Liu J, Sun JD, Tang JB, Chen YJ, Huang XB, Lin W, Ye C, Tong W, Cong LJ, Geng JN, Han YJ, Li L, Li W, Hu GQ, Huang XG, Li WJ, Li J, Liu ZW, Li L, Liu JP, Qi QH, Liu JS, Li L, Li T, Wang XG, Lu H, Wu TT, Zhu M, Ni PX, Han H, Dong W, Ren XY, Feng XL, Cui P, Li XR, Wang H, Xu X, Zhai WX, Xu Z, Zhang JS, He SJ, Zhang JG, Xu JC, Zhang KL, Zheng XW, Dong JH, Zeng WY, Tao L, Ye J, Tan J, Ren XD, Chen XW, He J, Liu DF, Tian W, Tian CG, Xia HG, Bao QY, Li G, Gao H, Cao T, Wang J, Zhao WM, Li P, Chen W, Wang XD, Zhang Y, Hu JF, Wang J, Liu S, Yang J, Zhang GY, Xiong YQ, Li ZJ, Mao L, Zhou CS, Zhu Z, Chen RS, Hao BL, Zheng WM, Chen SY, Guo W, Li GJ, Liu SQ, Tao M, Wang J, Zhu LH, Yuan LP, Yang HM (2002a) A draft sequence of the rice genome (*Oryza sativa* L. ssp *indica*). Science 296(5565):79–92

Yu S, Liao F, Wang F, Wen W, Li J, Mei H, Luo L (2012) Identification of rice transcription factors associated with drought tolerance using the ecotilling method. PLoS ONE 7(2):e30765. doi:10.1371/journal.pone.0030765

Yu SB, Li JX, Tan YF, Gao YJ, Li XH, Zhang QF, Maroof MAS (1997) Importance of epistasis as the genetic basis of heterosis in an elite rice hybrid. Proc Natl Acad Sci USA 94(17):9226–9231

Yu SB, Li JX, Xu CG, Tan YF, Li XH, Zhang QF (2002b) Identification of quantitative trait loci and epistatic interactions for plant height and heading date in rice. Theor Appl Genet 104(4):619–625

Zeliang PK, Pattanayak A (2013) Wide hybridization in the genus *Oryza*: aspects and prospects. Indian J Hill Farm 26(2):71–77

Zeng ZB (1994) Precision mapping of quantitative trait loci. Genetics 136(4):1457–1468

Zha X, Luo X, Qian X, He G, Yang M, Li Y, Yang J (2009) Over-expression of the rice *LRK1* gene improves quantitative yield components. Plant Biotechnol J 7:611–620

Zhang X, Wang J, Huang J, Lan H, Wang C, Yin C, Wu Y, Tang H, Qian Q, Li J, Zhang H (2012a) Rare allele of *OsPPKL1* associated with grain length causes extra-large grain and a significant yield increase in rice. Proc Natl Acad Sci USA 109:21534–21539

Zhang Y, Luo L, Xu C, Zhang Q, Xing Y (2006) Quantitative trait loci for panicle size, heading date and plant height co-segregating in trait-performance derived near-isogenic lines of rice (*Oryza sativa*). Theor Appl Genet 113(2):361–368

Zhang ZH, Wang K, Guo L, Zhu YJ, Fan YY, Cheng SH, Zhuang JY (2012b) Pleiotropism of the photoperiod-insensitive allele of *Hd1* on heading date, plant height and yield traits in rice. PLoS ONE 7:e52538

Zhu C, Gore M, Buckler ES, Yu J (2008) Status and prospects of association mapping in plants. Plant Genome 1:2–16

Zhuang JY, Lin HX, Lu J, Qian HR, Hittalmani S, Huang N, Zheng KL (1997) Analysis of QTL × environment interaction for yield components and plant height in rice. Theor Appl Genet 95:799–808

Zou J, Zhang S, Zhang W, Li G, Chen Z, Zhai W, Zhao X, Pan X, Xie Q, Zhu L (2006) The rice *HIGH-TILLERING DWARF1* encoding an ortholog of Arabidopsis *MAX3* is required for negative regulation of the outgrowth of axillary buds. Plant J 48:687–698

Zhu KM, Tang D, Yan CJ, Chi ZC, Yu HX, Chen JM, Liang JS, Gu MH, Cheng ZK (2010) ERECT PANICLE2 encodes a novel protein that regulates panicle erectness in indica rice. Genetics 184:343−350

Chapter 9
A Molecular Phylogenetic Framework for Timothy (*Phleum pratense* L.) Improvement

Alan V. Stewart and Nicholas W. Ellison

Abstract The recent molecular phylogenetic analysis of *Phleum* germplasm has provided a clear evolutionary history of the genus from which modern hexaploid germplasm and their cultivars have evolved. This framework will allow us to effectively use the full set of germplasm from all ploidies (2x, 4x, 6x and 8x) for a more systematic improvement of Timothy. The many new molecular forms of *Phleum* now known offer a huge potential to expand the gene pool of commercial hexaploid Timothy. The opportunity exists to hybridize and incorporate many new forms of molecular diversity into Timothy. However, genebank samples of many of these new forms are absent and urgently require collection. In addition, many of the novel diploid, tetraploid, hexaploid and octoploid forms are under serious threat from habitat degradation and climate warming in situ. It is also critical that core collections are developed and maintained using molecular phylogenetic and genetic diversity information as a basic framework. In order to apply molecular resources in an effective and balanced manner, it is important to ensure pragmatic field breeding programmes are continued in all major regions. This is a concern for Timothy, as it is a species with limited international breeding investment.

Keywords Timothy · *Phleum pratense* · Germplasm · Molecular breeding · Phylogeny

9.1 Introduction

The genus *Phleum* contains one important commercial hexaploid species, Timothy, *Phleum pratense* and two minor commercial species, turf Timothy, *P. pratense* subsp. *bertolonii* and Alpine Timothy, *P. alpinum*. Timothy, *P. pratense*

A.V. Stewart (✉)
PGG Wrightson Seeds, 175, Lincoln 7640, New Zealand
e-mail: astewart@pggwrightsonseeds.co.nz

N.W. Ellison
AgResearch Grasslands, Private Bag 11008, Palmerston North, New Zealand

V.R. Rajpal et al. (eds.), *Molecular Breeding for Sustainable Crop Improvement*, Sustainable Development and Biodiversity 11,
DOI 10.1007/978-3-319-27090-6_9

ssp. *pratense,* is commonly used in cold winter pastures for high-quality hay. It is particularly widely used in Scandinavia, northern Europe, northern Japan, Canada and northern USA, but there is also a very small amount used in Patagonia and in the south of New Zealand. Cultivars are all hexaploid and frequently divided into early- and late-flowering types.

The diploid turf Timothy, *P. pratense* ssp. *bertolonii,* is used occasionally in the northern Europe for turf purposes, often in mixtures with other turf grass species.

Alpine Timothy, *P. alpinum,* is used to a limited extent in Europe for high-altitude revegetation plantings.

Each year approximately 34,000 tons of Timothy seed are harvested, a few hundred tons of turf Timothy and only a few tons of alpine Timothy. This makes *Phleum* the third most widely sown grass genus with 8 % of the world's temperate grass seed, following *Lolium* and *Festuca* (Bondesen 2007).

9.2 Taxonomy

The genus *Phleum* contains 14 species in four sections over a polyploid series from diploid to octoploid as outlined by Joachimiak (2005) and Stewart et al. (2010).

The three commercial species are in section Phleum. In this paper, we use the widely accepted nomenclature of Humphries (1978, 1980) for European species and of Barkworth (2007) for American species, as followed by Stewart et al. (2010).

1. *P. alpinum* L. or alpine Timothy is an alpine species differentiated into three very different diploid or tetraploid cytotypes with ciliate or glabrous awns:
   (a) A glabrous awned allotetraploid; *P. alpinum* L. ≡ *P. commutatum* Gaudin with a circumpolar northern hemisphere and South American distribution.
   (b) A ciliate awned diploid form, known from the Rhaetic Alps of Italy and the Balkans; *P. alpinum* ssp. *rhaeticum* Humphries, ≡ *P. rhaeticum* (Humphries) Rauschert.
   (c) A glabrous awned diploid form also known as *P. alpinum* L., currently referred to here by the informal name "commutatum" following Joachimiak and Kula (1993). This form grows among the snow-bed vegetation at high altitudes (Zernig 2005) and occurs in the mountains of central Europe from the Northern Alps to the Carpathian Mountains.
2. *P. pratense* L. is a lowland species represented by a diploid to octoploid polyploid series:
   (a) The diploids occur throughout much of Europe and parts of North Africa; *P. pratense* ssp. *bertolonii* (DC.) Bornm., ≡ *P. bertolonii* DC., ≡ *P. nodosum* L.
   (b) Less common tetraploid forms in southern Europe; *P. pratense* ssp. *pratense.*

(c) Widespread agricultural hexaploid forms; *P. pratense* ssp. *pratense.*
(d) An octoploid form restricted to southern Italy; *P. pratense* ssp. *pratense*

3. *P. echinatum* Host is a winter active annual grass of eastern Mediterranean mountains which is not used commercially.

The other three sections, *Chilochloa, Achnodon and Maillea,* contain 11 species but none of these is commercial so will not be described here (Stewart et al. 2010).

9.3 Botanical Origin and Evolution

The molecular results of Stewart et al. (2010) show an Asian origin for the section *Phleum* and at least two separate migrations into Europe have been identified.

The first migration into Europe was of an ancestor of diploid *P. alpinum* subsp. *rhaeticum*. The penultimate Riss glaciation 130,000–150,000 year B.P. provided ample opportunity for this subalpine species to migrate vast distances through lowland areas to eventually become isolated on the Alps during the subsequent warmer interglacial period. Subsequent migration along mountain ranges has occurred so that today *rhaeticum* occurs in the Alps, Pyrenees, Apennines and the Balkans (with differing molecular signatures).

Migration of the "*rhaeticum*" also occurred onto the colder mountain ranges to the north into Germany and to the Carpathian Mountains of Poland and Romania. This was also associated with micro-evolutionary changes in morphology and cytology to develop into diploid "commutatum". The overlap of the range of *rhaeticum* and "commutatum" has since allowed considerable hybridization so that a swarm of hybrids overlaps the range of "commutatum" and the part of the *rhaeticum* range. Occasional tetraploid hybrids have developed from these, and these have migrated back east again, at least as far as Kazakhstan.

Migration of *rhaeticum* populations back into the lowlands as a result of climate cooling eventually resulted in the first lowland species of this group, *P. pratense* subsp. *bertolonii*. This was also accompanied by micro-evolutionary changes in cytology, morphology and adaptation. As the climate cooled during the last glaciation (the Würm 22,000–13,000 years B.P.), this lowland species retreated into southern European glacial refugia. Upon warming, these subsequently reinvaded northern Europe from the Balkan/Italy refugia. Those in the Spanish/Portuguese glacial refuge remained restricted to that region. Hybridization occurred when these 2 forms met at the interface in France resulting in the generation of a tetraploid (probably best termed an autotetraploid).

Hybrids formed in the Italian Alps where subsp. *bertolonii* and the Balkans *rhaeticum* overlapped resulting in an allotetraploid *pratense*. It is probable that a further hybridization with the adjacent northern European subsp. *bertolonii* leads to the formation of the agricultural hexaploid *pratense*. Upon warming in the Holocene, these subsequently reinvaded northern Europe from the Balkan/Italy

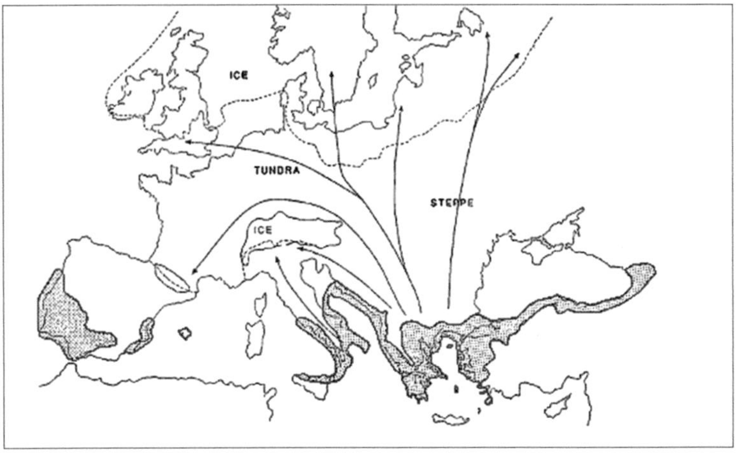

Fig. 9.1 Generic glacial refugia of southern Europe (*shaded areas*) (after Hewitt 1999) and potential post-glacial migration route of diploid ssp. *bertolonii* and agricultural hexaploid *pratense*

refugium, a refugium common to a wide range of European biota (Hewitt 1996, 1999) (Fig. 9.1).

Hexaploid and octoploid forms occur today within known glacial refugia, with two different hexaploids in southern Italy and Morocco and an octoploid in the mountains of southern Italy.

The very widespread allotetraploid *P. alpinum* formed over 300,000 years B.P. in Asia from the hybridization of an ancestral *rhaeticum* with another unknown genome (possibly from a taxonomic different section of *Phleum*). This form remained in Asia until eventually migrating into Europe during the last glaciation (the Würm 22,000–13,000 years B.P.), when conditions became suitable. At the same time, many species, including this one, were able to migrate into the Americas via the Bering/Aleutian route, although probably not completing their entry into the America until this route became open around 8000 years ago (Hong et al. 1999; Weber 2003). This divergent migration has led to a divergence in molecular forms, one in northern Europe and the other in Japan and the Americas. This circumpolar migration was completed in Iceland where derivatives of both the forms now occur.

Dogan (1991) describes the Mediterranean and western Asia as the centre of origin for the genus *Phleum*. This remains true for *P. pratense* today, although it may be considered surprising that a species so common in high-latitude cold temperate zones has originated within Southern European glacial refugia and that these areas still retain high genetic diversity today. Although it has been suggested that northern Europe is a centre of diversity (Guo et al. 2003), this appears unlikely as it has only been free of glaciation in the last 12,000 years and any diversity must be recent, or of migratory origin.

9.4 Germplasm and Gene pools

Genetic resources for commercial Timothy breeding at the hexaploid level can be divided into four gene pools on the following basis (Stewart et al. 2010).

9.4.1 Primary Gene pool

These can be defined as cultivars and elite breeding lines adapted to the region of agricultural use. This is largely northern Europe, northern Asia and North America and to a much lesser extent southern hemisphere regions such as New Zealand. This is the primary source of material for Timothy breeders and consists almost exclusively of hexaploids, although the diploids subsp. *bertolonii* are used occasionally for turf. In general, these resources are well used by breeders and their "working" collections largely represent this gene pool.

Interestingly, all the commercial materials explored to date and the vast bulk of hexaploid germplasm lines in genebanks represent a single molecular form with a common ITS sequence of the genomic formula $\mathbf{B_N B_N B_N B_N R_G R_G}$ as assigned by Stewart et al. (2010). The uniform molecular genomic constitution suggests that this pool has originated from a very narrow series of hybridization events. Events most probably occur in the Balkans/Italy glacial refugia during the last ice age (the Würm 22,000–13,000 years B.P.). This hexaploid gene pool most likely results from a limited series of hybridizations of diploid and tetraploid plants suggesting that the genetic basis for this gene pool is quite limited, possibly even as narrow as a single cross of two plants and probably no more than a few at most. Furthermore, the tetraploid parent would also most likely have resulted from limited hybridization events. Although the diversity of each genome may be limited, genetic exchange between the 3 genomes has likely contributed greatly to the diversity available today within this gene pool, as would micro-mutations and any introgression which may have occurred from other forms.

9.4.2 Secondary Gene pool

This pool includes hexaploid germplasm of *P. pratense* from the centre of origin in the Balkans/Italy region of Southern Europe, the Mediterranean mountains and North Africa, a region largely outside the region of commercial use. These represent hexaploid forms with different molecular ITS sequences than the primary gene pool. They will represent different hybridization events between different molecular forms of diploid and tetraploid ancestors. To date, it appears that breeders have not used this material due to their poor winter hardiness in northern Europe.

The genomic formula assigned by Stewart et al. (2010) to a unique hexaploid found in Basilicata in Southern Italy is $\mathbf{R_G R_G XXXX}$ (represented by one unique accession, line 6091 in the Kew collection collected in 1977). This form never expanded beyond the glacial refuge and could potentially be a reciprocal cross of the forms which contributed to the primary gene pool but further molecular and cytological characterization would be required to confirm this. The author has hybridized this with commercial hexaploids to form fertile hybrids.

The genomic formula assigned by Stewart et al. (2010) to a unique hexaploid found in Morocco $\mathbf{B_S B_S XXXX}$ (line CGN10072 in the Netherlands genebank). Interestingly, this form was labelled *bertolonii* and has many features in common with diploid *bertolonii* including much shorter reproductive stems than the other hexaploid forms. It could potentially be an auto-hexaploid of *bertolonii* but further molecular and cytological characterization would be required to confirm this. The author has hybridized this with commercial hexaploids to form fertile hybrids.

To date, it appears that other breeders have not used these two unique forms of material due to their poor winter hardiness in the northern regions. However, further collections are urgently required as these southern regions have almost never been targeted for genebank collection because of this lack of winter hardiness. However, such material deserves special attention as it represents unique hexaploid genomic constitutions. Sadly, genetic erosion in these southern regions is occurring at an alarming rate as many in situ populations are under threat from climate warming and human-induced habitat degradation.

9.4.3 Tertiary Gene pool

This consists of germplasm of ploidy levels other than hexaploid. Ploidy remains the major barriers to hybridization between *Phleum* species but apart from ploidy difficulties most forms cross readily (Nath 1967). The one exception to this is the widespread and more ancient palaeo-allotetraploid *P. alpinum* (genomic formula of the eastern form from Europe to Iceland $\mathbf{R_E R_E XX}$ and the western form from East Asia, the Americas, to Iceland $\mathbf{R_W R_W XX}$) which is difficult to cross with other forms and here we would classify that into the quaternary gene pool.

With the latest knowledge of the genomic constitution of *P. pratense* (Stewart et al. 2010), it should now be possible to either resynthesize *P. pratense* from different forms of the same genomes as listed in the "Genomic Formula" table, or use genomes from other forms and ploidy levels of section *Phleum* for introgression into *P. pratense* hexaploids.

Resynthesis of new and novel hexaploid forms from a full range of diploid and tetraploid genomic forms should be possible. For example, crosses involving the following diploids (with examples of germplasm lines, note forms lacking examples were determined from herbarium specimens and no germplasm lines are known to occur in collections):

Diploid subsp. *bertolonii* in Spain and Portugal	$\mathbf{B_S B_S}$	e.g. PI319076, RBG43968
Diploid subsp. *bertolonii* in Greece and the Balkans	$\mathbf{B_G B_G}$	
Diploid *rhaeticum* in the Alps	$\mathbf{R_S R_S}$	
Diploid *rhaeticum* in the Pyrenees	$\mathbf{R_P R_P}$	
Diploid *rhaeticum* in Greece	$\mathbf{R_G R_G}$	e.g. H729
Diploid *rhaeticum* in Italy	$\mathbf{R_I R_I}$	
Diploid "commutatum" in the Carpathian Mountains	**CC**	e.g. 2189
Diploid hybrids of *rhaeticum* and "commutatum"	**RC**	e.g. 13G2304001

with any of the following tetraploids (or their reciprocals)

Tetraploid *rhaeticum* "commutatum" hybrids, Italy	$\mathbf{R_S R_S CC}$	
Tetraploid *rhaeticum* "commutatum" hybrids, Caucasus	$\mathbf{CCR_G R_G}$	eg. PI619539, PI619567
Autotetraploid *pratense* in France	$\mathbf{B_S B_S B_N B_N}$	e.g. 58702
Allotetraploid *pratense* in the Italian Alps	$\mathbf{B_N B_N R_G R_G}$	e.g. H677
Tetraploid hybrid of *bertolonii* and hexaploid *pratense*	$\mathbf{B_N B_N B_N R_G}$	

Similarly, crosses of any of the tetraploid forms above with the octoploid from southern Italy $\mathbf{R_8 R_8 XXXXXX}$ (represented by one unique accession, line RBG20633 from Kew collection and duplicated in the Margot Forde Genebank in New Zealand as accession H736).

These will provide a huge diversity of new forms for hexaploid Timothy improvement.

Crosses between diploid *bertolonii* and hexaploid *P. pratense* are easy to produce artificially (Nordenskiold 1945), and although they are not always tetraploid (Løhde 1978), they are found in nature (Müntzing 1935; Foerster 1968, 2005). These readily backcrossed to hexaploid *P. pratense* or could be crossed to the octoploid forms.

Hybrids between tetraploid *P. pratense* forms and the Southern Italian octoploids have been made by the senior author and these form semi-fertile hexaploid plants. These plants have been crossed with regular agricultural hexaploids to generate fertile progeny.

Hybrids of tetraploid and hexaploid forms are pentaploid as expected (Nielsen and Nath 1961) and these may be backcrossed to hexaploids quite readily.

This tertiary germplasm pool represents an enormous untapped pool of unexplored material for breeders, but collections and molecular characterization will be necessary.

9.4.4 Quaternary Gene pool

By definition, these would consist of the more difficult to cross material such as widespread allotetraploid *P. alpinum* (genomic formula of the eastern form from Europe to Iceland $\mathbf{R_E R_E XX}$ and the western form from East Asia, the Americas, to Iceland $\mathbf{R_W R_W XX}$), as well as species in other sections. Other crosses which appear to be possible with hexaploid *P. pratense* include *P. phleoides* in section *Chilochloa*, of which there is a herbarium sample in the Leiden University National Herbarium in the Netherlands; *P. hirsutum* in section *Chilochloa,* which the senior author has crossed to obtain tetraploids; *P. subulatum* in section *Achnodon*, which Myers (1941) crossed to obtain a male sterile tetraploid progeny.

There is also a report of a sterile cross between tetraploid *Dactylis glomerata* and hexaploid *P. pratense* (Nakazumi et al. 1997).

At this stage, the quaternary gene pool offers much less potential for breeders than the secondary and tertiary gene pools and resources would be much better targeted at the resources which are easiest to utilize.

9.5 Recommendations for Future Actions

The current understanding of the genomic constitution within *Phleum* should allow breeders to utilize the genetic resources more effectively than previously. It should now be possible to resynthesize or introgress a much wider range of diverse *P. pratense* than has occurred naturally.

However, many of the genetic resources of wild relatives are under threat from climate warming and human-induced habitat degradation (‘t Mannetje 2007).

There is an urgent need to collect hexaploid *P. pratense* germplasm from Mediterranean mountain glacial refuge areas as well as a wide range of genetically diverse diploid, tetraploid and octoploid *P. pratense* and the readily crossable forms of *P. alpinum.* These include diploid subspecies *rhaeticum* and “commutatum” as well as their diploid and tetraploid hybrids.

These collections should be integrated into core collections to maximize molecular diversity of the available genomes.

It is also important that each of the major regions where Timothy is used maintains strong functional field breeding programs to allow adequate cultivar development, germplasm collection, introgressions of wild germplasm and exploration of molecular resources.

References

Barkworth ME, Capels KM, Long S, Anderton LK (2007) Flora of North America north of Mexico; v.24: Magnoliophyta; Commelinidae (in part); Poaceae, part 1. Oxford University Press, Oxford

Bondesen OB (2007) Seed production and seed trade in a globalised world. Seed production in the northern light. In: Proceedings of 6th international herbage seed conference, Norway, pp 9–12

Dogan M (1991) A taxonomical revision of the genus *Phleum* L. (Gramineae). Karaca Arbor Mag 1:53–70

Foerster E (1968) Ein beitrag zur untercheidung von *Phleum pratense* und *Phleum nodosum.* Göttinger Flor Rundbreife 1:9

Foerster E (2005) Natürliche Hybriden zwischen *Phleum pratense* und *Phleum bertolonii* (Natural hybrids between *Phleum pratense* and *Phleum bertolonii*). Gründungstagung der Vereinigung zur Erforschung der Flora Deutschlands, 29 und 30 Oktober 2005 in Vechta

Guo Y-D, Yli-Matilla T, Pulli S (2003) Assessment of genetic variation in timothy (*Phleum pratense* L.) using RAPD and UP-PCR. Hereditas 138:101–113

Hewitt GM (1996) Some genetic consequences of ice ages, and their role in divergence and speciation. Biol J Linn Soc 58:247–276

Hewitt GM (1999) Post glacial recolonisation of European biota. Biol J Linn Soc 68:87–112

Hong Q, White P, Klinka K, Chourmouzis C (1999) Phytogeographical and community similarities of alpine tundras of Changbaishan Summit, and Indian Peaks, USA. J Veg Sci 10:869–882

Humphries CJ (1978) Notes on the genus *Phleum* L. Bot J Linn Soc 76:337–340

Humphries CJ (1980) *Phleum.* In: Tutin TC et al (eds) Flora Europeaea 5. *Alismataceae* to *Orchidaceae* (*Monocotyledones*). Cambridge University Press, Cambridge, pp 239–241

Joachimiak A (2005) Heterochromatin and microevolution in *Phleum.* In: Sharma AK, Sharma A (eds) Plant genome: biodiversity and evolution. Vol. 1, Part B: phanerogams. Science Publishers Inc., Enfield, New Hampshire, pp 89–117

Joachimiak A, Kula A (1993) Cytotaxonomy and karyotype evolution in *Phleum* sect. *Phleum* (Poaceae) in Poland. Plant Syst Evol 188:11–25

Løhde JJH (1978) *Phleum pratense* and *Phleum bertolonii* hybridisation, morphology and ecology in Denmark. Thesis, Kongelige veterinaer- og Landbohoejskole, Denmark (1977) 80 pp From Årsskrift, Kongelige veterinaer- og Landbohoejskole 1978, p 167

Müntzing A (1935) Cyto-genetic studies on hybrids between two *Phleum*-species. Hereditas 20:103–136

Myers WM (1941) Meiotic behaviour of *Phleum pratense, Phleum subulatum* and their F1 hybrid. J Ag Res 63:649–655

Nakazumi H, Furuya M, Shimokouji H, Fujii H (1997) Wide hybridization between timothy (*Phleum pratense* L.) and orchardgrass (*Dactylis glomerata* L.). Bull Hokkaido Prefectural Agric Expt Stat (Japan) 72:11–16

Nath J (1967) Cytogenetical and related studies in the genus *Phleum* L. Euphytica 16:267–282

Nielsen EL, Nath J (1961) Cytogenetics of a tetraploid form of *Phleum pratense* L. Euphytica 10:343–350

Nördenskiold H (1945) Cyto-genetic studies in the genus *Phleum.* Acta Agr Suec 1:1–138

Stewart AV, Ellison N, Joachimiak A (2010) The genus *Phleum*; wealth of wild species: role in plant genome elucidation and improvement. New York, Springer

’t Mannetje L (2007) Climate change and grasslands through the ages: an overview. Grass Forage Sci 62:113–117

Weber WA (2003) The Middle Asian element in the Southern Rocky Mountain flora of the western United States: a critical biogeographical review. J Biogeog 30:649–688

Zernig K (2005) *Phleum commutatum* and *Phleum rhaeticum* (*Poaceae*) in the Eastern Alps: characteristics and distribution. Phyton 45:65–79

Chapter 10
Genetic Improvement of Basmati Rice—The Journey from Conventional to Molecular Breeding

A.K. Singh and S. Gopala Krishnan

Abstract Marker-assisted breeding provides a great opportunity to the present-day researchers for breeding new crop varieties by design through precise transfer of desirable gene(s). Among the molecular breeding approaches, marker-assisted backcross breeding (MABB) is an attractive proposition for breeders as it can help in improving an already popular variety for specific target traits such as resistance to disease/pest/improvement in quality. MABB has been successfully employed in transferring genes (*xa13* and *Xa21*) governing resistance to diseases such as bacterial blight (BB), blast (*Pi54, Pita, Pi1, Pi9, Pib, Piz5* and *Pi5*), sheath blight (*qSBR 11-1*), and brown plant hopper (BPH; *Bph18, Bph20* and *Bph21*) into a number of Basmati rice varieties, namely Pusa Basmati 1, Pusa Basmati 1121, and Pusa Basmati 6 as well as long slender grain aromatic rice varieties such as Pusa Sugandh 5. Further, a major QTL for salt tolerance (*Saltol*) has been transferred to Pusa Basmati 1121 and Pusa Basmati 1, which are widely grown in northwestern India. Genetically enhanced donor sources in the form of near-isogenic lines (NILs) carrying major gene(s)/QTLs for resistance to biotic (BB, blast, sheath blight, and BPH) and abiotic (salt tolerance) stresses in the background Pusa Basmati 1, the first semi-dwarf, high-yielding Basmati rice variety, have been developed. QTL mapping using recombinant inbred line (RIL) population has unveiled several novel QTLs for different agronomic, grain and cooking quality traits. Besides their effective use in Basmati rice improvement, molecular markers are also utilized in basic studies as well as in maintenance breeding of Basmati rice varieties, which is discussed in the present chapter.

Keywords Basmati rice · Disease resistance · Gene pyramiding · Molecular marker-assisted selection · Near-isogenic lines

---

A.K. Singh (✉) · S. Gopala Krishnan
Division of Genetics, ICAR-Indian Agricultural Research Institute, New Delhi 110012, India
e-mail: aks_gene@yahoo.com

S. Gopala Krishnan
e-mail: gopal_icar@yahoo.co.in; gopalakrish@iari.res.in

V.R. Rajpal et al. (eds.), *Molecular Breeding for Sustainable Crop Improvement*, Sustainable Development and Biodiversity 11,
DOI 10.1007/978-3-319-27090-6_10

10.1 Introduction

Basmati rice from Indian subcontinent has unique quality characteristics, because of which it is highly sought-after rice in the international market. The traditional Basmati cultivars are poor yielders owing to their tall stature, which makes them prone to lodging, photoperiod and sensitive to temperature and have low input response. Genetic improvement of Basmati rice at ICAR Indian Agricultural Research Institute (ICAR-IARI) led to the development of number of high-yielding Basmati rice varieties, namely Pusa Basmati 1, Pusa Basmati 1121, Improved Pusa Basmati 1, Pusa Basmati 6, and Pusa Basmati 1509 (Singh et al. 2013). Through concerted research, progressive improvement has been brought about in cooked kernel length (Fig. 10.1) and the duration of traditional Basmati rice varieties has been reduced from 160 to 115–140 days with the enhancement of productivity from 2.5 to 6–8 tons/ha in improved dwarf Basmati rice varieties (Fig. 10.2). As a result, India's forex earning from the export of Basmati rice has gone up from a mere 294 crores in 1990–1991 to 29,300 crores in 2013–2014 (www.apeda.gov.in) (Fig. 10.3). The ICAR-IARI-bred Basmati rice varieties alone contribute to more than 75 % of the Basmati rice export. Although, these high-yielding Basmati rice varieties are widely grown and liked by the farmers, millers, and traders, their productivity and quality suffered on account of their susceptibility to biotic stresses such as bacterial blight (BB) caused by *Xanthomonas oryzae pv. oryzae*, blast caused by *Magnoporthae oryzae*, sheath blight caused by *Rhizoctonia solani* and brown plant hopper (BPH) infection by *Nilaparvata lugens*. Therefore, it was imperative to improve them for resistance to biotic and abiotic stresses.

Utilizing host plant resistance is one the most cost-effective and environment-friendly strategies for managing these stresses. Availability of major genes imparting resistance to these stresses along with gene-based/gene-linked markers for 39 BB resistance genes, 105 blast resistance genes (Sharma et al. 2010), and 21 BPH resistance genes (Jena et al. 2010) provided an excellent opportunity for targeted

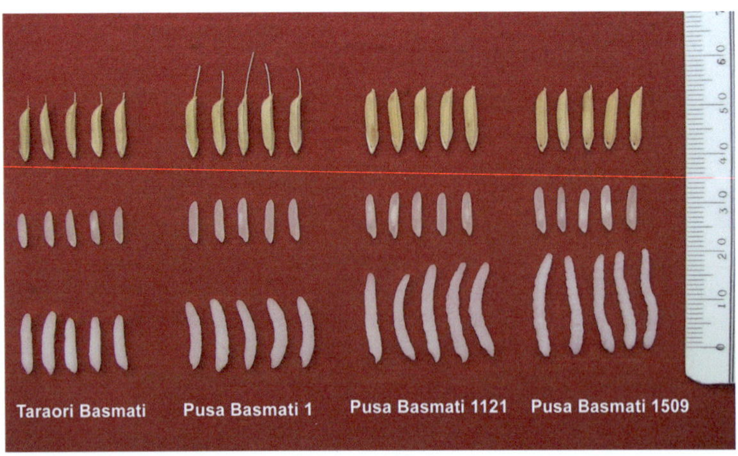

Fig. 10.1 Improvement in milled rice length and kernel length after cooking in the Basmati rice varieties released by ICAR-IARI, New Delhi

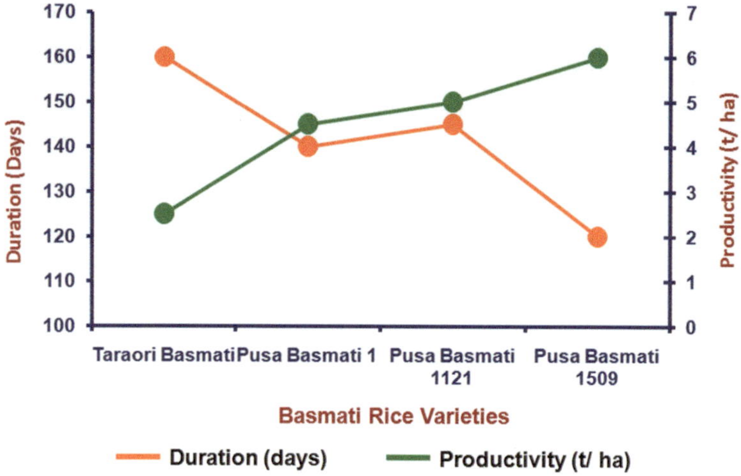

Fig. 10.2 Graph representing the change brought about by combining high yield with short duration in Basmati rice through genetic improvement at ICAR-IARI, New Delhi

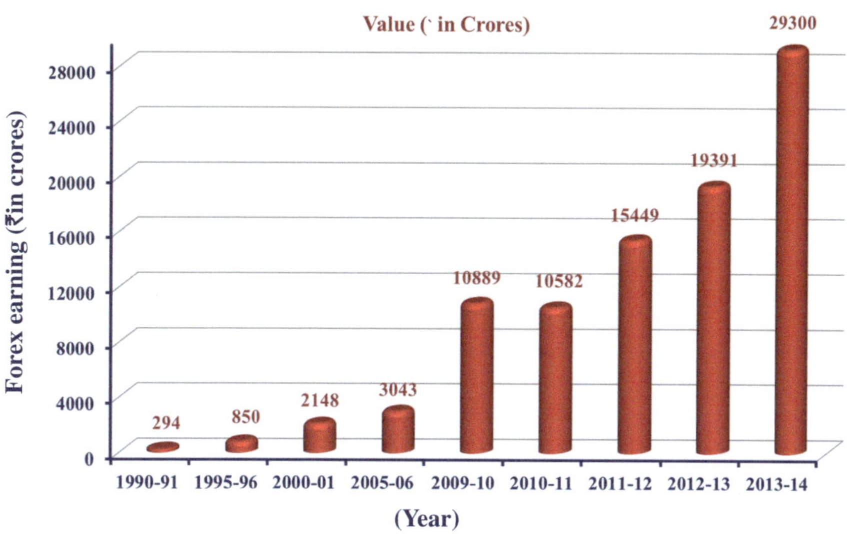

Fig. 10.3 Trend showing the increase in foreign exchange (forex) earnings through the export of Basmati rice from India since 1990–1991

transfer of these genes into popular Basmati varieties. However, all these genes were available in non-Basmati sources, and their transfer to Basmati background could have adverse effects on the superior grain and cooking quality traits of Basmati rice varieties. Under such situations, marker-assisted backcross breeding (MABB) offers a unique opportunity for transferring desirable genes from unadapted donors to otherwise agronomically superior cultivars having specific weakness (Singh et al.

2011). The wealth of genomic resources available in the case of rice eases the effectiveness of marker-aided transfer of not only the traits controlled by major genes but also the QTLs (Gopala Krishnan et al. 2012). Molecular markers have been deployed in MABB aiding foreground selection, the selection of target locus using a gene-linked/gene-based marker and background selection, the selection for recovery of recurrent parent genome using markers uniformly distributed across the genome (Singh et al. 2013). MABB has been employed for the incorporation of genes governing BB, blast, BPH resistance, and salt-tolerant cultivars in Basmati rice improvement, through precise marker-aided transfer of a number of resistance genes/QTLs (Singh et al. 2011). A schematic outline for MABB is presented in Fig. 10.4, which involves three steps: (1) foreground selection, (2) recombinant selection and (3) background selection.

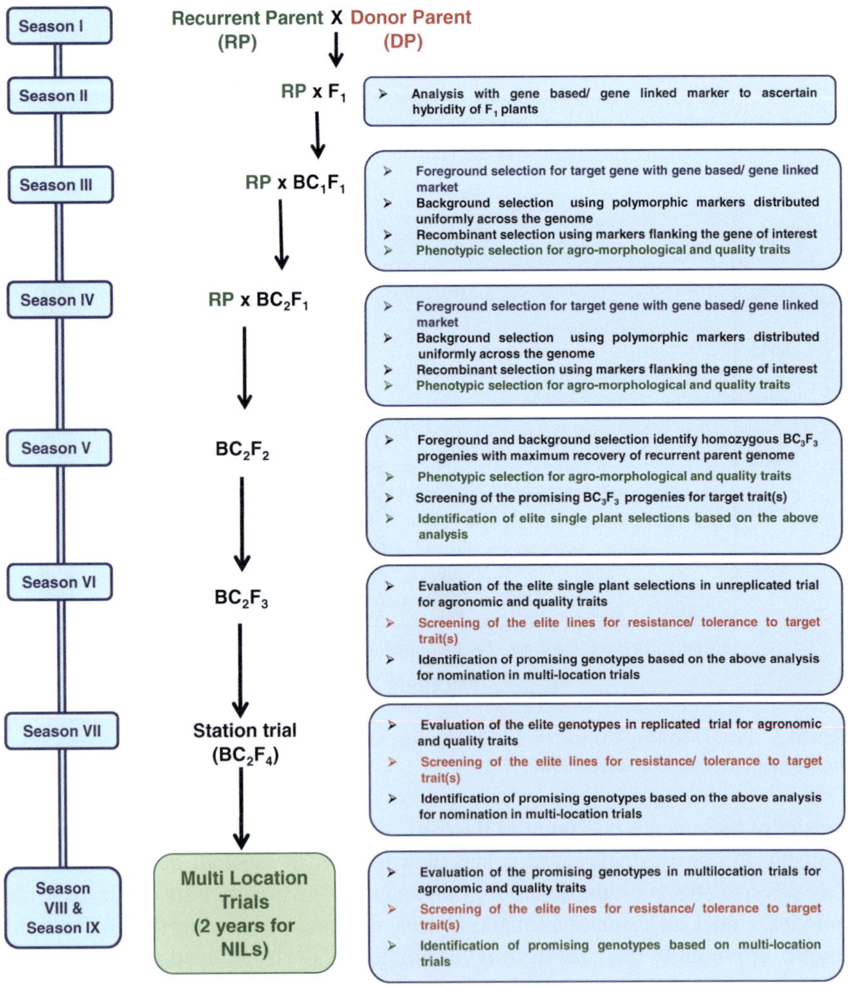

Fig. 10.4 A general scheme for marker-assisted backcross breeding

1. Foreground selection: It refers to the selection for the target gene/locus using a marker tightly linked to the gene/locus under selection. The distance between marker and the gene/locus should preferably be less than one cM. However, the most ideal situation would be to use a marker based on functional polymorphism within the gene. Using gene-based markers for foreground selection increases the selection efficiency for the target locus to 100 %.
2. Recombinant selection: In a backcross breeding program, when we transfer a desirable gene from an unadapted donor to an otherwise agronomically superior variety, which lacks the desirable gene under question, the donor segment in the flanking genomic regions of the target locus is also introduced, which may be associated with some undesirable characteristics, often described as "linkage drag." It is therefore important to eliminate undesirable flanking genomic region from donor during the backcross generations. In order to accomplish this goal, a set of markers polymorphic between donor and recurrent parents in the genomic regions flanking the target locus are selected, and in backcross generations, say in BC_1F_1, plants having recurrent parent allele at the nearest flanking marker on one side of the target locus are selected and used for generating BC_2F_1 seeds. In BC_2F_1, the same exercise is repeated on the other side of the target locus. The recombinant selection reduces the donor segment in the flanking genomic region of the target locus to a minimum size possible and thus the associated undesirable introgressions. However, the extent to which donor segment can be reduced will depend upon the fact that how closely the polymorphic markers between donor and recurrent parents are identified in the genomic regions flanking of the target locus.
3. Background selection: Recovery of recurrent parent genome in the plants positive for target locus based on foreground selection and having minimum linkage drag based on recombinant selection is an important activity in a MABB program. This involves the identification of a set of markers polymorphic between donor and recurrent parents providing genome-wide coverage. Usually, the polymorphic markers should be spaced 10–20 cM apart. In backcross generations, the plants selected for target locus and having minimum linkage drag are analyzed using polymorphic markers in the background genome to identify the plants showing homozygosity for recurrent parent alleles at maximum number of loci and thus the higher recovery of recurrent parent genome. This exercise reduces the number of backcrosses required for reconstituting the recurrent parent genome and thus reduces the time required for product development.

In addition to the three-step selection as described above, stringent phenotypic selection for recurrent parent phenotype is also practiced as done in conventional backcross breeding to expedite the recovery of recurrent parent genome and phenome.

In the ongoing discussion, we present the achievements of MABB in Basmati rice. The target genes, their location, markers used for foreground selection, and donors and the recurrent parents are presented in Table 10.1.

Table 10.1 Genes, their chromosomal location, donors, recurrent parents and markers used for foreground selection employed in marker-assisted breeding for resistance to biotic and abiotic stress tolerance in Basmati rice

Disease	Gene	Donor parent	Recurrent parents	Chr' location	Linked marker	Linkage distance	References
Bacterial blight	*xa13*	IRBB55, Pusa 1460	Pusa Basmati 1, Pusa Basmati 1121, Pusa Basmati 6	8	*xa13*prom (STS)	Promoter based	Singh et al. (2011)
	Xa21			11	pTA248 (STS)	Gene based	Ronald et al. (1992)
Blast	*Piz-5*	C101A51, Pusa 1602, IRBLZ5-a	Pusa Basmati 1, Pusa Basmati 1121, Pusa Basmati 6	6	AP5930 (STMS)	0.05 cM	Fjellstrom et al. (2006)
				6	AP4007	<0.01 cM	
	Pi-54	Tetep, Pusa 1603, DHMAS-70Q164-2a		11	RM206 (STMS)	0.6 cM	Sharma et al. (2005)
	Pi-1	DHMAS70Q164-2a	Pusa Basmati 1	11	RM224 (STMS)	0.0 cM	Fuentes et al. (2008)
	Pi-ta	DHMAS70Q164-2a	Pusa Basmati 1	12	RM247 (STMS)	3.0 cM	Eizenga et al. (2006)
					YL155/YL87	Gene based	Eizenga et al. (2006)
					YL153/YL154	Gene based	Eizenga et al. (2006)
	Pi-b	IRBLB-b	Pusa Basmati 1	2	RM208 (STMS)	1.2 cM	Fjellstrom et al. (2004)
					Pibdom	Gene based	Fjellstrom et al. (2004)
	Pi-5	IRBL5-M	Pusa Basmati 1	9	S04G03 (STS)	0.8 cM	Jeon et al. (2003)
					C1454	0.06 cM	Lee et al. (2009)

(continued)

Table 10.1 (continued)

Disease	Gene	Donor parent	Recurrent parents	Chr' location	Linked marker	Linkage distance	References
	Pi-9	IRBL-9-W	Pusa Basmati 1	6	AP5930 (STMS)	0.05 cM	Fjellstrom et al. (2004)
					AP5659-5 (STMS)	0.05 cM	Fjellstrom et al. (2006)
					NBS2-Pi9 195-1 (STS)	Gene based	Qu et al. (2006)
Brown plant hopper	*Bph18*	IR65482	Pusa Basmati 1121, Pusa Basmati 6	12	RM6217 (STMS)	Gene based	Jena et al. (2006)
	Bph20	IR71033	Pusa Basmati 1121, Pusa Basmati 6	4	B120 (STS)	Gene based	Singh et al. (2011)
	Bph21	IR71033	Pusa Basmati 1121, Pusa Basmati 6	12	B121 (STS)	Gene based	Rahman et al. (2009)
Salinity tolerance	*Saltol*	FL478	Pusa Basmati 1121	1	RM3412 (STMS)	QTL region	Thomson et al. (2010)
	Saltol	Pusa 1734	Pusa Basmati 1	1	RM10793 (STMS)	QTL region	Thomson et al. (2010)

10.2 Marker-Assisted Improvement of Pusa Basmati 1 for BB Resistance

Pusa Basmati 1 (PB1), the first semi-dwarf, high-yielding Basmati quality rice variety, was released in 1989. It became highly susceptible to BB disease caused by *Xanthomonas oryzae pv. oryzae*. MABB approach was used to incorporate two genes for BB resistance, namely *xa13* and *Xa21* into PB1 using IRBB55 as a donor. The CAPS marker RG136 linked to *xa13* and STS marker *pTA248* linked to *Xa21* were used for foreground selection. Foreground selection for *xa13* and *Xa21* coupled with phenotypic selection for agronomic, grain, and cooking quality traits in BC_1F_1, BC_1F_2, and BC_1F_3 generations made it possible to recover as high as 86.7 % recurrent parent genome as assessed with 252 polymorphic AFLP markers (Joseph et al. 2004). Further, marker-assisted background analysis using simple sequence repeat (SSR) markers was effectively integrated with foreground selection to identify superior BB-resistant recombinants with minimal linkage drag (Gopalakrishnan et al. 2008), which led to the release of Improved Pusa Basmati 1 (Pusa 1460), the first product of molecular breeding in rice in India (Singh et al. 2007).

10.3 MABB for Incorporating BB and Blast Resistance in Pusa Basmati 1121 and Pusa Basmati 6

Pusa Basmati 1121 is one of the most widely grown Basmati rice varieties in India, occupying an area of 1.2 m ha out of the total Basmati-cultivating area of 2 m ha. Pusa Basmati 6 (Pusa 1401), another Basmati rice variety, surpasses Pusa Basmati 1121 in several attributes such as non-lodging, non-shattering habit, response to input use, dwarf stature, higher yield, non-chalky grains, strong aroma, and better cooking quality. In spite of these advantages, both these varieties are also susceptible to BB and blast disease. In order to incorporate BB resistance in both these varieties, the improved Basmati quality donor, "Pusa 1460," was used as the donor parent for marker-assisted transfer of BB-resistant genes *xa13* and *Xa21*. As shown in Fig. 10.4, selection of plants in each backcross generation was carried out for respective genes using foreground markers, followed by the rigorous phenotypic and background selection to hasten the recovery of both the recurrent parent phenome and genome. A large number of genotypes, with Pusa 1718 (PB1121 + *xa13* + *Xa21*) (Pusa Basmati 1121/Improved Pusa Basmati 1/Pusa Basmati 1121^{*3}) with recurrent parent genome (RPG) recovery of more than 85.0 %, were developed possessing grain and cooking quality in the BB-resistant PB1121 genotypes, which were on par with the recurrent parent for their grain length, cooked kernel length, elongation ratio, and aroma (Ellur et al. 2012, 2013a, b, 2016). Similarly, MABB in combination with phenotypic selection helped in the development of Pusa 1728 (*xa13* + *Xa21*), the BB resistant improved PB6 genotypes with recurrent RPG recovery of as high as 97.7 % estimated using a minimum of 60 polymorphic SSR markers providing genome-wide coverage. The rapid recovery of Basmati quality traits in both the improved BB-resistant PB1121 and PB6 genotypes is attributed to the utilization of Basmati quality donor parent, Improved Pusa Basmati 1 (Ellur et al. 2016). Pusa 1728 (PB6 + *xa13* + *Xa21*) was found resistant against the virulent BB isolates collected from Basmati-growing areas of the country (Ellur et al. 2013a, b). The improved BB-resistant lines showed performance on par with PB6 for agronomic and cooking quality traits. Additionally, MABB in combination with stringent phenotypic selection helped in the development of PB1121 + *Xa38*, using *O. nivara*-derived novel BB-resistant gene, *Xa38* from PR114-*Xa38* to PB1121.

10.4 Marker-Assisted Improvement of Basmati Rice Varieties PB1121, Pusa Basmati 6, and Pusa Basmati 1 for Resistance to Rice Blast

Basmati rice is severely affected by rice blast caused by *Magnaporthe grisea*, and till date, there is no resistant source available in the Basmati rice germplasm. Until now, 105 major blast resistance genes have been documented in rice. However, all these genes are present in non-Basmati background, and their transfer to Basmati varieties poses the problem of impairment of grain and cooking quality traits.

The blast resistant donors developed by marker-assisted breeding Pusa 1602 (PRR78 + *Piz-5*) and Pusa 1603 (PRR78 + *Pi54*) were used to transfer the respective genes into Pusa Basmati 1121 and Pusa Basmati 6 (Singh et al. 2012a). Marker-assisted simultaneous but stepwise backcross breeding (MASS-BB) was adopted for the transfer of two blast resistance genes, *Piz5* and *Pi54* from Pusa 1602 and Pusa 1603, respectively, into PB1121 and PB6 using gene-linked markers, which was followed by intercrossing the BC_3F_1 plants to pyramid both the blast resistant genes. In each backcross generation, the plants were selected for respective genes using foreground markers and were subjected to rigorous phenotypic and background selection to accelerate the recovery of both the recurrent parent phenome and genome. Marker-aided selection in combination with phenotypic selection helped in the development of Pusa 1716 (PB1121 + *Piz5*), Pusa 1717 (PB1121 + *Pi54*), Pusa 1726 (PB6 + *Piz5*), and Pusa 1727 (PB6 + *Pi54*) with RPG recovery of 93.8, 94.7, 94.2, and 93.5 % estimated using 56, 58, 54, and 59 STMS markers representing genome-wide coverage, respectively (Ellur et al. 2013a, b, 2016).

Additionally, two gene pyramided lines for blast resistance have been developed including Pusa 1883 (PB1121 + *Piz5* + *Pi54*) and Pusa 1884 (PB6 + *Piz5* + *Pi54*), which exhibited resistance to blast disease under artificial inoculation with respective diagnostic isolates. The improved lines showed performance on par with the recurrent parents, PB1121 and PB6, for agronomic and cooking quality traits. Further, these lines have been evaluated and found to be resistant in natural epiphytotics at uniform blast nurseries located at 3 hotspot locations across India (Ellur et al. 2013a, b, 2016). The pyramided lines will be further evaluated under multilocation trials for release as improved cultivar and as an invaluable donor for BB and blast resistance in Basmati rice improvement.

In a unique attempt, for the first time a set of seven major blast resistance genes have been transferred into the genetic background of PB1. The development of near-isogenic lines (NILs) was carried out using the series of IRBL lines (IRBL5-M: *Pi5*; IRBLb-B: *Pib*; IRBL9-W: *Pi9*; and IRBLz5-CA: *Piz-5*) and a doubled haploid line carrying three blast resistant genes (*Pi1*, *Pi54*, *Pita*) as the donors and PB1 as the recurrent parent in independent backcross breeding programs (Khanna et al. 2012, Shikar et al. 2012). Marker-assisted backcross breeding was carried out to transfer seven blast genes governing resistance to blast disease of rice, namely *Pi1*, *Pi54*, *Pita*, *Pi*-5, *Pib*, *Pi9, and Piz*-5 with DHMAS70Q164-2a, IRBL5-M, IRBLb-B, IRBL9-W, and IRBLz5-CA as donors, respectively (Khanna et al. 2012, 2013). Foreground selection using gene-linked markers along with simultaneous selection for plant phenotype, agronomic performance, and grain and cooking quality was adapted to recover the recurrent parent phenotype as well as disease resistance. The lines with higher recovery of RPG namely Pusa 1634-19 (*Piz-5*) with 92.7 %, Pusa 1635-12 (*Pib*) with 93.6 %, Pusa 1636-2 (*Pi5*) with 94.6 % and Pusa 1637-6 (*Pi9*) with 97.2 % recovery of RPG have been developed (Khanna et al. 2015). Artificial screening with virulent isolates and screening in the uniform blast nursery at two hotspot locations confirmed the resistance in the NILs. The monogenic lines developed in the background of Pusa Basmati 1 will serve as useful donors for blast resistance genes in the Basmati breeding program (Khanna et al. 2012).

The monogenic lines developed in the background of Pusa Basmati 1 were then used for pyramiding these genes to build in durable resistance for blast resistance in Basmati rice. Intercrosses of the NILs were effected to develop gene pyramided NILs with 2, 3, and 4 genes in different combinations in the background of Pusa Basmati 1. Three and two gene pyramided lines homozygous for the genes, *Pi54 + Pi1 + Pita, Pi54 + Pi1, Pi54 + Pita, Pi1 + Pita, Pi9 + Pi5, Pi9 + Pib, Pib + Pi5,* etc., were developed with RPG recovery ranging from 93.6 to 98.6 %, and the improved lines, *viz.* Pusa-1633-4-2, Pusa-1633-170-6, Pusa-1633-30-8, Pusa-1633-162-3, and Pusa-1633-101-4, expressed resistance reaction to diagnostic isolates from Basmati-growing regions (Shikari et al. 2013, Khanna et al. 2015b). These monogenic and pyramided NILs will not only serve as an excellent resource for functional genomic analyses for blast resistance but also serve as valuable donors for blast resistance genes in Basmati rice improvement. The improved lines possessed excellent grain and cooking quality besides good agronomic performance, *at par* with check Pusa Basmati 1. Further, a set of improved blast resistant genotypes are being tested in the national Basmati trials for their deployment as improved cultivars, which will help to achieve durable resistance to rice blast in Basmati-growing regions of India (Khanna et al. 2015a, b).

10.5 Development of Multiple Biotic Stress-Resistant Basmati Rice

Besides BB and blast diseases, sheath blight (ShB) caused by *Rhizoctonia solani* not only causes severe yield losses but also impairs the quality of the rice grain (Singh et al. 2011). ShB resistance is quantitatively inherited (Pinson et al. 2005), and 16 QTLs for ShB resistance have been identified (Srinivasachary et al. 2011), out of which *qSBR11-1* is a major QTL that has been found stable and effective against the ShB (Channamallikarjuna et al. 2010). "Tetep," an *indica* rice cultivar from Vietnam, is the source of resistance to both blast (*Pi54*), and ShB (*qSBR11-1*) disease was used a donor for developing Basmati rice genotypes with resistance to multiple diseases, namely BB, blast, and ShB, in the Basmati rice background using Improved Pusa Basmati 1 as recipient parent in the MABB. Development of Pusa 1608 is the first successful example, where marker-assisted selection has been utilized for transferring of genes conferring resistance to three different diseases in rice namely, *xa13* and *Xa21* for BB resistance; *Pi54* for blast resistance and a major QTL *qSBR11-1* (Singh et al. 2012).

10.6 Marker-Aided Improvement of Pusa Basmati 1121 and Pusa Basmati 6 for BPH Resistance

The donors for BPH resistance Rathu Heenati (*Bph3, Bph17*), IR68542 (*Bph18*), and IR71033 (*Bph20, Bph21*) were screened for their resistance level in the greenhouse using the standard protocol (Pathak et al. 1969). The genotype, Rathu Heenati, was found to be highly resistant, followed by IR68542 and IR71033. The donors IR68542 and IR71033 were used for marker-assisted introgression of three genes, *Bph18*, *Bph20*, and *Bph21* into Pusa Basmati 1121 and Pusa Basmati 6 through MASS-BB. Advanced backcross-derived lines have been developed with *Bph18*, *Bph20*, and *Bph21* in the genetic background of Pusa Basmati 1121 and Pusa Basmati 6, which are in advanced stages of evaluation.

10.7 Marker-Assisted Improvement of Pusa Basmati 1121 and Pusa Basmati 1 for Salt Tolerance

Increasing soil salinity is a serious threat to rice production worldwide in twenty-first century. Rice is highly sensitive to salt stress at seedling and reproductive stages, and development of salinity tolerance at seedling stage is important for crop establishment (Vinod et al. 2013). A major QTL among these was *Saltol* (for *salt tolerance*) mapped on the short arm of chromosome 1 by using an F_8 recombinant inbred lines (RILs) developed from the cross of a salt-tolerant land race, Pokkali, from Kerala and IR29, a salt-sensitive rice variety (Gregario 1997). A set of 23 diverse germplasm including land races, wild germplasm, and improved varieties were characterized for the haplotype diversity in the *Saltol* region using 20 SSR markers, in order to find associations with seedling stage salt tolerance under controlled condition (ECe of 12 dSm^{-1}). It was observed that haplotypes possessing Pokkali alleles at both RM8094 and RM3412 markers could be useful for marker-assisted selection (Babu et al. 2014).

Pusa Basmati 1121 is widely cultivated in Haryana, occupying almost 0.8 m ha (80 %) of the Basmati rice area in the state, which is severely affected by inland salinity owing to brackish underground water used for irrigation. Since Pusa Basmati 1121 is sensitive to soil salinity, its full potential is not realized in saline soil. *Saltol* is a major QTL for seedling stage salinity tolerance, which explains up to 70 % phenotypic variance for seedling stage salinity stress (Gregario 1997). The introgression of *Saltol* in Pusa Basmati 1121 was undertaken using FL478 as donor parent through MABB. The foreground selection for *Saltol* was carried out with linked molecular marker RM3412. Recombinant selection on the carrier chromosome was carried out with 21 polymorphic markers flanking/including the *Saltol* region, and 58 polymorphic markers having genome-wide coverage were used for background analysis. Background analysis with the three best advanced backcross-derived Pusa 1734 genotypes revealed the RPG recovery ranging from 92.8 to 96.4 %. The advanced backcross-derived *Saltol*-positive PB1121 genotypes (Pusa 1734) were evaluated for agronomic performance as well as grain and cooking quality analysis and were found to be on par with the Pusa Basmati 1121 for grain length, kernel elongation, elongation ratio, and aroma (Babu et al. 2012). A set of these lines is being tested in AICRP trials for further evaluation and variety release.

Pusa Basmati 1, another popular Basmati rice variety, is also susceptible to salinity. MABB approach was employed to incorporate "*Saltol*" into the genetic background of Pusa Basmati 1 by using backcross-derived genotype, Pusa 1734, developed from the cross between Pusa Basmati 1121 and FL478. Foreground selection for *Saltol* QTL was carried out using linked marker RM10793 and background selection done using SSR markers spanning across the rice genome. Further, foreground selection was coupled with rigorous phenotypic selection for agronomic, grain, and other cooking quality traits, to accelerate the recurrent parent phenome recovery. Eighteen superior BC_4F_2 homozygous plants were selected and advanced to next generation through pedigree selection to develop improved version of Pusa Basmati 1 with salt tolerance. The agronomic performance and grain and cooking quality of the improved lines were on par with the original Pusa Basmati 1. These improved lines along with susceptible check and parents were artificially screened for salinity tolerance at seedling stage using hydroponics under controlled condition. Salt stress was imposed 14 days after germination by adding 60 mM NaCl (ECe of 6.9 dSm^{-1}), and salt concentration was increased to 120 mM (ECe of 13.9 dSm^{-1}). These lines are further being evaluated for their performance and will help in the development of Improved Pusa Basmati 1 with salinity tolerance (Chaudhary et al. 2014).

10.8 Haplotype Analysis of *Pup1* Gene in Basmati Rice Varieties

Phosphorus use efficiency of rice is an important trait in rice improvement, and it would aid in the development of rice with higher P fertilizer use efficiency. *Pup1* (*Phosphorous uptake* 1) is a major QTL conferring tolerance to the P deficiency under field conditions identified in rice (Wissuwa et al. 1998), which was further fine-mapped to 130 kb region (Chin et al. 2010). *PSTOL1*, a gene governing phosphorus starvation tolerance, have been cloned from an *Aus* rice variety Kasalath (Gamuyao et al. 2012). It was also shown that modern rice varieties lack *PSTOL1*. Based on the screening of rice germplasm including Basmati rice varieties using gene-based markers, namely K29, K46, K59, and K41, spanning the *Pup1* locus, along with checks, Vandana and Anjali, it was demonstrated for the first time that showed that all Basmati/aromatic rice varieties were *Pup1* positive, while most of the non-aromatic varieties were devoid of *Pup1* (Singh et al. 2011). Further, validation of "P" use efficiency of Basmati varieties vis-à-vis presence of *Pup1* locus is being undertaken in P sick plot, and the findings have great relevance in management and use of phosphorous in Basmati rice (Singh et al. 2011).

10.9 Mapping Grain Quality Traits in Basmati Rice

Mapping of milled rice grain length was carried out in a mapping population developed from a cross between a short grain aromatic rice landrace, Sonasal, and long slender grain Basmati rice variety, Pusa Basmati 1121. Based on the phenotyping of 300 F_2 plants for grain dimensions before and after cooking for different grain dimension traits (milled rice length and breadth, L/B ratio, cooked kernel length, and elongation ratio) and genotypic data of 141 SSR markers across rice genome, a major QTL responsible for milled rice length was mapped on chromosome 3, which explained phenotypic variance as high as 74 % (Singh et al. 2011).

10.10 Validation and Use of Fragrance Gene-Linked Markers

Aroma in rice is mainly due to the production of 2-acetyl-1-pyrroline (2-AP) even though more than 100 volatile aroma compounds have been identified in cooked rice (Buttery et al. 1988). Selection for aroma in Basmati rice has been mainly based on sensory evaluation, which is labor-intensive and time-consuming and varies with individual perception. Further, with each successive analysis, the ability to perceive aroma in the samples diminishes due to saturation of sensory organ and/or physical abrasions to the tongue.

A recessive gene governing aroma in rice was mapped on chromosome 8 (Ahn et al. 1992). The gene, *badh2*, has been cloned, and it has been demonstrated that aroma is due to an eight-base-pair deletion in aromatic varieties compared to non-aromatic varieties (Bradbury et al. 2005). However, the efficacy of the 8-bp deletion marker to differentiate between Basmati and non-Basmati rice genotypes was low. Therefore, a new marker *nksdel* based on *BADH2* gene sequence information from Basmati and non-Basmati rice varieties was developed (Amaravathi et al. 2008) and validated in a set of aromatic and non-aromatic rice genotypes with 100 % efficacy (Singh et al. 2011).

10.11 Conclusion

Marker-assisted breeding has provided an unprecedented opportunity for precise transfer of genes responsible for biotic and abiotic stress-tolerant genes/QTLs into various popular Basmati rice varieties. Conventional breeding, essentially based on phenotypic selection, was the mainstay of Basmati rice improvement which had helped in making significant impacts on the development of improved cultivars. However, with the evolution of marker technology in rice, it has been possible to refine Basmati rice improvement through mapping important Basmati quality traits in rice. Further, marker-assisted selection has enabled pyramiding of genes governing resistance/tolerance for different biotic and abiotic stresses, respectively. Marker-assisted breeding has been successfully employed for the development of Improved Pusa Basmati 1 and the improved versions of PB1121, Pusa Basmati 6 and Pusa Basmati 1 with resistance to BB, blast, sheath blight, BPH and tolerance to seedling stage salinity, which are in different stages of testing for release as improved varieties. This has been possible through the adoption of cost-effective MAS strategy complemented by phenotypic selection aiding in precise gene transfer for the improvement of Basmati rice varieties.

Acknowledgement The research work on Basmati rice improvement is funded by Indian Council of Agricultural Research, New Delhi; ICAR National Agricultural Innovation Project (ICAR-NAIP) and by the "Accelerated Crop Improvement Programme" of Department of Biotechnology, India.

References

Ahn SN, Bollich CN, Tanksley SD (1992) RFLP tagging of a gene for aroma in rice. Theor Appl Genet 84:825–828

Amarawathi Y, Singh R, Singh AK, Singh VP, Mohapatra T, Sharma TR, Singh NK (2008) Mapping of quantitative trait loci for basmati quality traits in rice (*Oryza sativa* L.). Mol Breed 21:49–65

Babu NN, Sharma SK, Ellur RK, Singh VK, Pal M, Pathania S, Singh VK, Singh D, Gopalakrishnan S, Bhowmick PK, Nagarajan M, Vinod KK, Singh NK, Prabhu KV, Singh AK (2012) Marker assisted improvement of Pusa Basmati 1121 for salinity tolerance. In: Abstracts of poster presentation (MB52) international conference on plant biotechnology for food security: new frontiers, National Agricultural Science Centre, New Delhi, India, 21–24 Feb 2012, pp 72–73

Babu NN, Vinod KK, Gopalakrishnan S, Bhowmick PK, Vanaja T, Krishnamurthy SL, Nagarajan M, Singh NK, Prabhu KV, Singh AK (2014) Marker based haplotype diversity of *Saltol* QTL in relation to seedling stage salinity tolerance in selected genotypes of rice. Indian J Genet 74(1):16–25

Bradbury LM, Fitzgerald TL, Henry RJ, Jin Q, Waters DL (2005) The gene for fragrance in rice. Plant Biotech J 3:363–370

Buttery RG, Turnbaugh JG, Ling LC (1988) Contribution of volatiles to rice aroma. J Agric Food Chem 36(5):1006–1009

Channamallikarjuna V, Sonah H, Prasad M, Rao GJN, Chand S, Upreti HC, Singh NK, Sharma TR (2010) Identification of major quantitative trait loci *qSBR11-1* for sheath blight resistance in rice. Mol Breed 25:155–166

Chaudhary S, Yadav AK, Singh V, Singh BD, Ellur RK· Babu N, Bhowmick PK, Gopala Krishnan S, Nagarajan M, Vinod KK, Singh AK (2014) Development of salt tolerant "Pusa Basmati 1" through marker assisted transfer of "*Saltol*", a major QTL for seedling stage salinity tolerance. In: National conference on "perspectives and trends in plant sciences and biotechnology (PTPB-2014) held at Panjab University, Chandigarh, 21–23 Feb 2014, C11

Chin JH, Lu X, Haefele SM, Gamuyao R, Ismail A, Wissuwa M, Heuer S (2010) Development and application of gene-based markers for the major rice QTL *Phosphorus uptake 1*. Theor Appl Genet 120(6):1073–1086

Eizenga GC, Agrama HA, Lee FN, Yan WG, Jia Y (2006) Identifying novel resistance genes in newly introduced blast resistant rice germplasm. Crop Sci 46:1870–1878

Ellur RK, Pathania S, Singh VK, Singh D, Gopala Krishnan S, Bhowmick PK, Nagarajan M, Vinod KK, Singh UD, Mondal KK, Chander S, Mohapatra T, Sharma TR, Prabhu KV, Singh AK (2012) Marker assisted breeding for incorporation of bacterial blight, blast and brown plant hopper resistance in Pusa Basmati 1121. In: Abstracts of poster presentation (*MB48*) international conference on plant biotechnology for food security: new frontiers, National Agricultural Science Centre, New Delhi, India, 21–24 Feb 2012, pp 70–71

Ellur RK, Pathania S, Gopala Krishnan S, Bhowmick PK, Nagarajan M, Vinod KK, Singh UD, Rathour R, Prabhu KV, Singh AK (2013a) Pyramiding of blast resistance genes in the Basmati rice variety Pusa Basmati 1121 (P0049). In: Abstract book: 7th international rice genetics symposium, Manila, Philippines, 5–8 Nov 2013, pp. 363–364

Ellur RK, Pathania S, Gopala Krishnan S, Bhowmick PK, Vinod KK, Singh UD, Mondal KK, Rathour R, Prabhu KV, Singh AK (2013b). Pyramiding of bacterial blight and blast resistance genes in the Basmati rice variety Pusa Basmati 6 through marker assisted backcross breeding (PVII-192). In: 11th international symposium on rice functional genomics—sustaining food and nutritional security, New Delhi, India, 20–23 Nov 2013, pp 146

Ellur RK, Khanna A, Yadav A, Pathania S, Rajashekara H, Singh VK, Gopala Krishnan S, Bhowmick PK, Nagarajan M, Vinod KK, Prakash G, Mondal KK, Singh NK, Singh AK (2016) Improvement of Basmati rice varieties for resistance to blast and bacterial blight diseases using marker assisted backcross breeding. Plant Sci 242:330–341

Fjellstrom R, Conaway-Bormans CA, McClung AM, Marchetti MA, Shank AR, Park WD (2004) Development of DNA markers suitable for marker-assisted selection of three *Pi* genes conferring resistance to multiple *Pyricularia grisea* pathotypes. Crop Sci 44:1790–1798

Fjellstrom R, McClung AM, Shank AR (2006) SSR markers closely linked to the *Pi-z* locus are useful for selection of blast resistance in a broad array of rice germplasm. Mol Breed 17:149–157

Fuentes JL, Correa-Victoria FJ, Escobar F, Prado G, Aricapa G, Duque MC, Tohme J (2008) Identification of microsatellite markers linked to the blast resistance gene *Pi-1*(*t*) in rice. Euphytica 160:295–304

Gamuyao R, Chin JH, Tanaka JP, Pesaresi P, Catausan S, Dalid C, Slamet-Loedin IS, Mendoza EMT, Wissuwa M, Heuer S (2012) The protein kinase *Pstol1* from traditional rice confers tolerance of phosphorus deficiency. Nature 488:535–539

Gopala Krishnan S, Waters DLE, Katiyar SK, Sadananda AR, Satyadev V, Henry R (2012) Genome-wide DNA polymorphisms in elite *indica* rice inbreds discovered by whole-genome sequencing. Plant Biotech J 10(6):623–634

Gopalakrishnan S, Sharma RK, Rajkumar KA, Joseph M, Singh VP, Singh AK, Bhat KV, Singh NK, Mohapatra T (2008) Integrating marker assisted background analysis with foreground selection for identification of superior bacterial blight resistant recombinants in Basmati rice. Plant Breed 127:131–139

Gregorio GB (1997) Tagging salinity tolerance genes in rice using amplified fragment length polymorphism (AFLP). Ph.D. thesis, University of the Philippines, Los Baños, p 118

Jena KK, Kim S (2010) Current status of brown planthopper (BPH) resistance and genetics. Rice 3:161–171

Jena KK, Jeung JU, Lee JH, Choi HC, Brar DS (2006) High-resolution mapping of a new brown plant hopper (BPH) resistance gene, *Bph18*(t), and marker-assisted selection for BPH resistance in rice (*Oryza sativa* L.). Theor Appl Genet 112:288–297

Jeon JS, Chen D, Yi GH, Wang GL, Ronald PC (2003) Genetic and physical mapping of *Pi5*(*t*), a locus associated with broad-spectrum resistance to rice blast. Mol Gen Genomics 269:280–289

Joseph M, Gopalakrishnan S, Sharma RK, Singh AK, Singh VP, Singh NK, Mohapatra T (2004) Combining bacterial blight resistance and Basmati quality characteristics by phenotypic and molecular marker assisted selection in rice. Mol Breed 13:377–387

Khanna A, Sharma V, Shikari AB, Gopala Krishnan S, Sharma TR, Nagarajan M, Vinod KK, Singh D, Singh UD, Ellur RK, Pathania S, Prabhu KV, Singh AK (2012) Development of monogenic lines for blast resistant genes in Basmati rice. In: Abstracts of poster presentation (*MB50*) international conference on plant biotechnology for food security: new frontiers, National Agricultural Science Centre, New Delhi, India, Feb 21–24 2012, pp 71–72

Khanna A, Shikari AB, Gopala Krishnan S, Singh UD, Sharma TR, Rathour R, Variar M, Prashanthi SK, Nagarajan M, Singh AK (2013) Development of near isogenic lines for functional genomics of blast resistance in Basmati rice (PVII-197). In: 11th International symposium on rice functional genomics—sustaining food and nutritional security, New Delhi, India, 20–23 Nov 2013, pp 149

Khanna A, Sharma V, Ellur RK, Shikari AB, Gopala Krishnan S, Singh UD, Prakash G, Sharma TR, Rathour R, Variar M, Prashanthi SK, Nagarajan M, Vinod KK, Bhowmick PK, Singh NK, Prabhu KV, Singh BD, Singh AK (2015a) Development and evaluation of near isogenic lines for major blast resistance gene(s) in Basmati rice. Theor Appl Genet 128:1243–1259

Khanna A, Sharma V, Ellur RK, Shikari AB, Gopala Krishnan S, Singh UD, Prakash G, Sharma TR, Rathour R, Variar M, Prashanthi SK, Nagarajan M, Vinod KK, Bhowmick PK, Rajashekhara H, Singh NK, Prabhu KV, Singh AK (2015b) Marker assisted pyramiding of major blast resistance genes *Pi9* and *Pita* in the genetic background of an elite Basmati rice variety, Pusa Basmati 1. Indian J Genet 75(4):417–425

Lee SK, Song MY, Seo YS, Kim HK, Ko S, Cao PJ, Suh JP, Yi G, Roh JH, Lee S, An G, Hahn TR, Wang GL, Ronald P, Jeon JS (2009) Rice *Pi5*-mediated resistance to *Magnaporthe oryzae* requires the presence of two coiled-coil–nucleotide-binding–leucine-rich repeat genes. Genetics 181(4):1627–1638

Pathak MD, Cheng CH, Fortuno ME (1969) Resistance to *Nephotettix impictiveps* and *Nilaparvata lugens* in varieties of rice. Nature 223:502–504

Pinson SRM, Capdevielle FM, Oard JH (2005) Confirming QTLs and finding additional loci conditioning sheath blight resistance in rice (*Oryza sativa* L.) using recombinant inbred lines. Crop Sci 45:503–510

Qu S, Liu G, Zhou B, Bellizzi M, Zeng L, Dai L, Han B, Wang GL (2006) The broad-spectrum blast resistance gene *Pi9* encodes a nucleotide-binding site-leucine-rich repeat protein and is a member of a multigene family in rice. Genetics 172:1901–1914

Rahman ML, Jiang W, Chu SH, Qiao Y, Ham TH, Woo MO, Lee J, Khanam MS, Chin JH, Jeung JU, Brar DS, Jena KK, Koh HJ (2009) High-resolution mapping of two rice brown planthopper resistance genes, *Bph20(t)* and *Bph21(t)*, originating from *Oryza minuta*. Theor Appl Genet 119:1237–1246

Ronald PC, Albano B, Tabien R, Abenes L, Wu KS, McCouch SR, Tanksley SD (1992) Genetic and physical analysis of the rice bacterial blight disease resistance locus, *Xa21*. Mol Gen Genet 236:113–120

Sharma TR, Shanker P, Singh BK, Jana TK, Madhav MS, Gaikwad K, Singh NK, Plaha P, Rathour R (2005) Molecular mapping of rice blast resistance gene *Pi-kh* in the rice variety, tetep. J Plant Biochem Biotechnol 14:127–133

Sharma TR, Rai AK, Gupta GK, Singh NK (2010) Broad-spectrum blast resistance gene *Pikh* cloned from the rice line tetep designated as *Pi54*. J Plant Biochem Biotechnol 19:87–89

Shikari AB, Khanna A, Gopala Krishnan S, Sharma TR, Nagarajan M, Vinod KK, Singh D, Singh UD, Ellur RK, Pathania S, Prabhu KV, Singh AK (2012) Marker assisted development of near isogenic lines for major blast resistance genes (*Pi54, Pi1* and *Pita*) in Pusa Basmati 1. In: Abstracts of poster presentation (*MB51*) international conference on plant biotechnology for food security: new frontiers, National Agricultural Science Centre, New Delhi, India, 21–24 Feb 2012, pp 72

Shikari AB, Khanna A, Gopala Krishnan S, Singh UD, Sharma TR, Rathour R, Variar M, Prashanthi SK, Prabhu KV, Singh AK (2013) Marker assisted pyramiding of major blast resistance genes in Basmati rice variety 'Pusa Basmati 1' (P0048). In: Abstract book: 7th international rice genetics symposium, Manila, Philippines, 5–8 Nov 2013, pp 362–363

Singh VP, Singh AK, Mohapatra T, Joseph M, Gopala Krishnan S, Atwal SS, Nagarajan M, Pandey KR, Gopalakrishnan J, Sinha SN, Chopra NK, Singh VK, Singh R, Tyagi JP, Singh D, Singh J, Singh J, Ravindran G (2007) Notification of Basmati rice variety Pusa 1460 (improved Pusa Basmati 1). Indian J Genet 67(3):304

Singh AK, Gopala Krishnan S, Singh VP, Mohapatra T, Prabhu KV, Singh NK, Sharma TR, Nagarajan M, Vinod KK, Singh D, Singh UD, Chander S, Atwal SS, Seth R, Singh VK, Ellur RK, Singh A, Anand D, Khanna A, Yadav S, Goel N, Singh A, Shikari AB, Singh A, Marathi B (2011) Marker assisted selection: a paradigm shift in Basmati breeding. Indian J Genet 71(2):1–9

Singh AK, Gopala Krishnan S, Bhowmick PK, Prabhu KV, VP Singh (2013) Pusa Basmati 1509–a new dawn in climate-smart cultivation. Rice India 2:6–11

Singh A, Singh VK, Singh SP, Pandian RTP, Ellur RK, Singh D, Bhowmick PK, Gopala Krishnan S, Nagarajan M, Vinod KK, Singh UD, Prabhu KV, Sharma TR, Mohapatra T, Singh AK (2012) Molecular breeding for the development of multiple disease resistant Basmati rice. AoB Plants pls029. doi:10.1093/aobpla/pls029

Singh VK, Singh A, Singh SP, Ellur RK, Choudhary V, Sarkel S, Singh D, Gopala Krishnan S, Nagarajan M, Vinod KK, Singh UD, Rathore R, Prashanthi SK, Aggrawal PK, Bhatt JC, Mohapatra T, Prabhu KV, Singh AK (2012b) Incorporation of blast resistance into "PRR78", an elite Basmati rice restorer line, through marker assisted backcross breeding. Field Crops Res 128:8–16

Singh VK, Singh A, Singh SP, Ellur RK, Singh D, Gopala Krishnan S, Bhowmick PK, Nagarajan M, Vinod KK, Singh UD, Mohapatra T, Prabhu KV, Singh AK (2013) Marker-assisted simultaneous but stepwise backcross breeding for pyramiding blast resistance genes *Piz5* and *Pi54* into an elite Basmati rice restorer line 'PRR78'. Plant Breed 132 (5):489–495

Srinivasachary Willocquet L, Savary S (2011) Resistance to rice sheath blight (*Rhizoctonia solani* Kuhn) [(teleomorph: *Thanatephorus cucumeris* (A.B. Frank) Donk.] disease: current status and perspectives. Euphytica 178:1–22

Thomson MJ, Ocampo MD, Egdane J, Rahman MA, Sajise AG, Adorada DL, Raiz ET, Blumwald E, Seraj ZI, Singh RK, Gregorio GB, Ismail AM (2010) Characterizing the *Saltol* quantitative trait locus for salinity tolerance in rice. Rice 3:148–160

Vinod KK, Gopala Krishnan S, Naresh BN, Nagarajan M, Singh AK (2013) Improving salt tolerance in rice: looking beyond the conventional. In: Ahmad P, Azooz MM, Prasad MNV (eds) Salt stress in plants: signalling, omics and adaptations. Springer, New York, pp 219–260

Wissuwa M, Yano M, Ae N (1998) Mapping of QTLs for phosphorus-deficiency tolerance in rice (*Oryza sativa* L.). Theor Appl Genet 97:777–783

Chapter 11
Genetic Diversity and Coffee Improvement in India

N. Surya Prakash, Jeena Devasia, Y. Raghuramulu and Ramesh K. Aggarwal

Abstract Coffee, often referred to as 'Brown Gold', is a popular beverage crop produced in more than 80 countries including India, having an annual turnover of ca. US$ 70 billion globally. This economically important plantation crop of tropics and subtropics comprises only two commercial species, *Coffea arabica* L. and *Coffea canephora* Pierre ex Froehner (popularly known as arabica and robusta coffee, respectively) although there exists large number of *Coffea* species worldwide. *C. arabica* L. is the only allotetraploid species ($2n = 4x = 44$) in the genus and self-fertile while all other species including *C. canephora* are diploids ($2n = 22$) and self-incompatible. Arabica and robusta coffee types differ significantly in terms of phenotype, agronomic behaviour, bean and liquor quality, and breeding behaviour as well as genetic variability in their extant germplasm. Arabicas produce superior quality coffee but are susceptible to major diseases and pests while robustas are more tolerant to the diseases and pests but the bean and the liquor qualities are inferior to arabicas. Therefore, the major focus of coffee improvement in India is on the development of high-yielding hybrids having durable host resistance in arabica and evolving drought-tolerant robusta genotypes to cope with the changing climate, more efficiently. Till date, the Central Coffee Research Institute (CCRI), India, has developed 13 improved arabica and three robusta selections for commercial cultivation, by employing conventional breeding approaches and utilizing coffee germplasm that was introduced in the nineteenth century through multiple international expeditions. Occurrence of spontaneous tetraploid inter-specific hybrids between tetraploid arabica and other diploid species has also facilitated arabica coffee improvement through introgressive breeding. More recently, the DNA marker tools and technologies that provide new opportunities to overcome some of the limitations of the conventional breeding strategies are being integrated for more

N. Surya Prakash · J. Devasia · Y. Raghuramulu
Central Coffee Research Institute, Coffee Research Station (P.O), Chikmagalur 577 117, Karnataka, India

R.K. Aggarwal (✉)
Centre for Cellular & Molecular Biology (CSIR-CCMB), Habsiguda, Uppal Road, Hyderabad 500 007, Telangana, India
e-mail: rameshka@ccmb.res.in

V.R. Rajpal et al. (eds.), *Molecular Breeding for Sustainable Crop Improvement*, Sustainable Development and Biodiversity 11,
DOI 10.1007/978-3-319-27090-6_11

precise characterization of primary as well as secondary gene pools for exploitation in breeding through marker-assisted selection. This chapter highlights the key aspects of Indian coffee sector *vis-a-vis* world scenario and unique features of Indian coffee cultivation, genetic resources, analysis and exploitation of genetic diversity for crop improvement through conventional breeding.

Keywords Coffee germplasm · Genetic resources · Molecular analysis · Genetic diversity · Coffee breeding · Indian coffee varieties

List of Abbreviations

%	Per cent
AD	Anno Domini
AFLP	Amplified fragment length polymorphism
BC	Backcross
CxR	*Coffea congensis* × Robusta (*Coffea canephora)*
C	*Coffea*
CBD	Coffee berry disease
CCRI	Central Coffee Research Institute
CIFC	Centro Investigacao das Ferrugens do Cafeeiro
CLR	Coffee leaf rust
DMS	Dimethyl sulphate
DNA	Deoxyribonucleic acid
Dt.	District
EST-SSRs	Expressed sequence tags–simple sequence repeats
etc.	Etcetera
EMS	Ethyl methane sulphonate
f-AFLP	Fluorescent-amplified fragment length polymorphism
F_1	First filial generation
FAO	Food and agricultural organization
FAQ	Fair average quality
Fig	Figure
GDP	Gross domestic product
ha	Hectares
HdT	Hibrido de Timor
ICO	International coffee organization
INEAC	Institut National pour l'Etude Agronomique du Congo Belge
IPR	Intellectual property rights
ISSR	Inter-simple sequence repeats
ITS	Internal transcribed spacer
kg	Kilogram
m	Metres
MSL	Mean sea level
MT	Metric tonnes
mt-DNA	Mitochondrial DNA

P.	*Psilanthus*
PLS	Pure line selection
RAPD	Random-amplified polymorphic DNA
rDNA	Ribosomal DNA
RFLP	Restricted fragment length polymorphism
Sln.	Selection
sp.	Species
SSR	Simple sequence repeats
US$	US dollars
viz.	Namely

11.1 Introduction

Coffee is a popular tropical plantation crop, produced in more than 80 countries spread over in both tropical and subtropical regions of the world. A beverage of choice in most part of the world, coffee ranks second in International trade next to petroleum products and is often referred to as 'Brown gold' with an annual turnover touching US$ 70 billion.

Coffee is mostly produced in developing countries while is consumed mainly in developed countries and hence forms an important source of export earnings for the producer countries. The coffee industry also provides employment to an estimated 100 million people in the areas of cultivation, processing, trading, transportation and marketing. Coffee cultivation is predominantly a small holder enterprise with 26 million farmers belonging to this category across the globe. Coffee in India occupies a place of pride among the plantation crops with a planted area of ca. 0.40 million ha. India ranks sixth in global coffee production after Brazil, Vietnam, Columbia, Indonesia and Ethiopia, with a total production of 5.3 million bags contributing to 3.7 % of world production (ICO Ann Rev 12/13). The coffee sector is dominated by small holdings, with 99 % of coffee farmers having holdings of less than 10 ha and coffee farming forms their main livelihood activity.

Coffee belongs to family Rubiaceae and the genus *Coffea* that includes over 100 species. Of these, only two species, namely *C. arabica* L. or arabica coffee and the *C. canephora* Pierre ex A. Froehner or robusta coffee, are commercially grown which together constitute 99 % of global coffee production. Another species, *C. liberica* Bull ex Hiern or liberica coffee, is grown to a smaller extent in East Africa and Asia and accounts for about 1 % of global production.

11.1.1 Brief Overview of Indian Coffee Industry

Coffee in India is an export-oriented commodity with the exports showing steady increase over the past 15 years and for the year 2012–13 about 95 % of the produce was exported, realizing Rs. 4552.7 crores to state exchequer. The area and production of coffee in India has shown a steady increase over the years (Fig. 11.1), from 92,523 ha during 1950 to over 4.09 lakh ha by 2011–12. There has been a phenomenal increase in production from 0.19 lakh MT to 3.14 lakh MT during the corresponding period (Anonymous 2014a). The major coffee-growing areas are distributed in the Southern states such as, Karnataka (56 %), Kerala (20 %) and Tamil Nadu (8 %), which forms the traditional coffee-growing areas. The non-traditional growing region comprises 58,000 ha spread across Andhra Pradesh, Odisha and North-Eastern states of India (Fig. 11.2).

The share of planted area under coffee cultivation in different coffee-growing states of India is presented in Fig. 11.3, while the distribution pattern of the two types of coffees, arabica and robusta, is depicted in Fig. 11.4.

For many decades, there was stagnation in domestic consumption of coffee in India at 60,000 MT. With Café culture coming to the forefront in most metropolitan cities, the domestic consumption of coffee has seen a huge surge with upward trend from the year 2000 that touched over a lakh tonnes by the year 2010 with an average growth rate of about 5–6 %. Notably, there seems to be a concurrence between the growing areas under coffee plantation and the major part of domestic consumption (78 %), which is concentrated in the Southern states of the country. Of the total consumption of coffee in India, urban share is 73 % while the remaining 27 % accounts for rural sector (Reddy 2009).

11.1.2 Unique Features of Indian Coffee

Coffee cultivation in India is unique as it is cultivated in eco-friendly manner under two-tier shade canopy, compared to open cultivation practices followed in most of

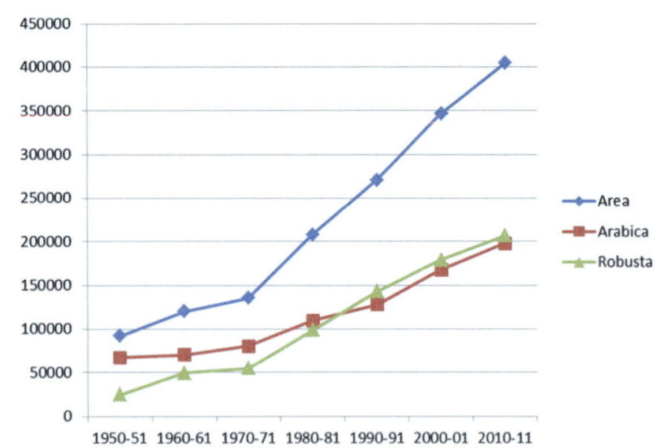

Fig. 11.1 Area under coffee in India from 1950–51 to 2010–11

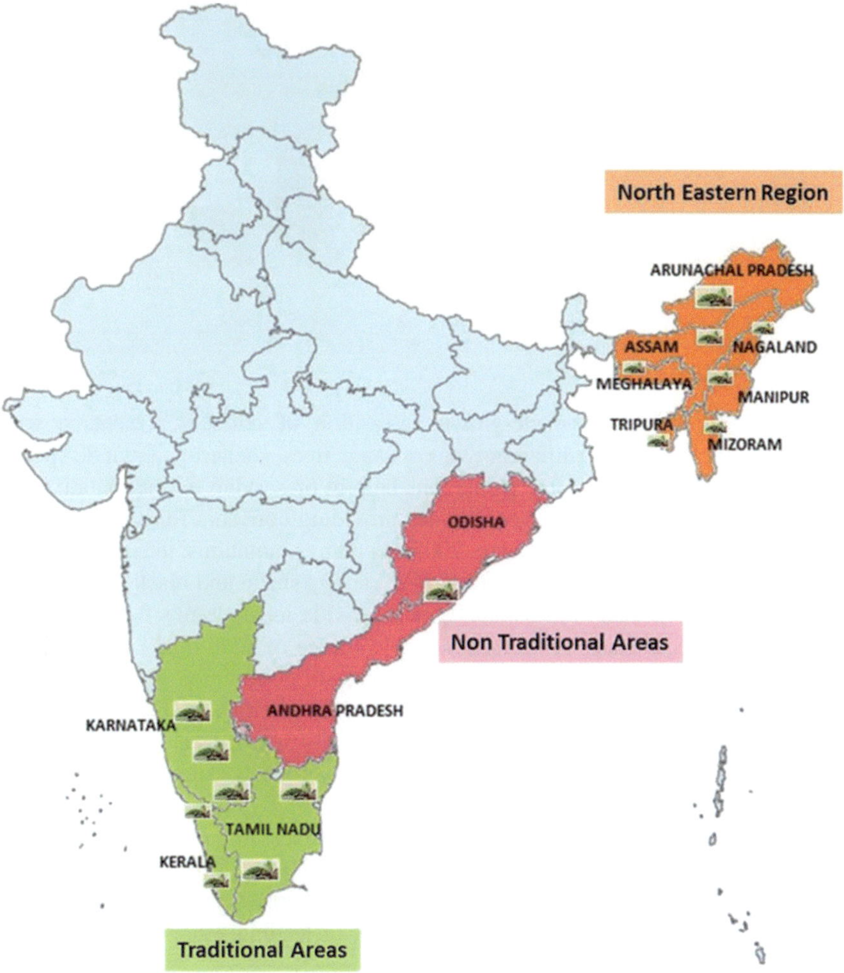

Fig. 11.2 Map showing coffee-growing regions of India

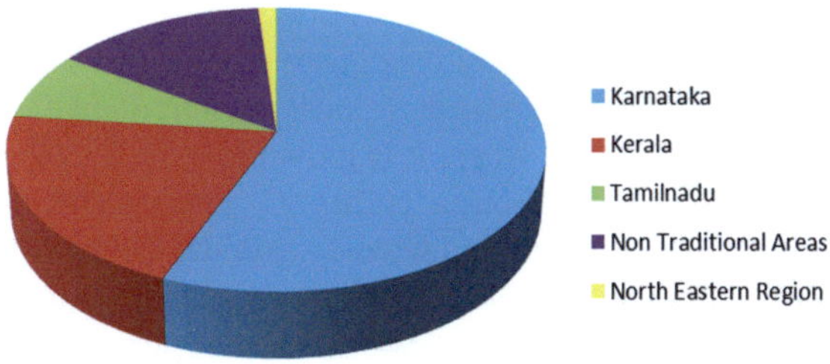

Fig. 11.3 Area under coffee plantation in different states of India

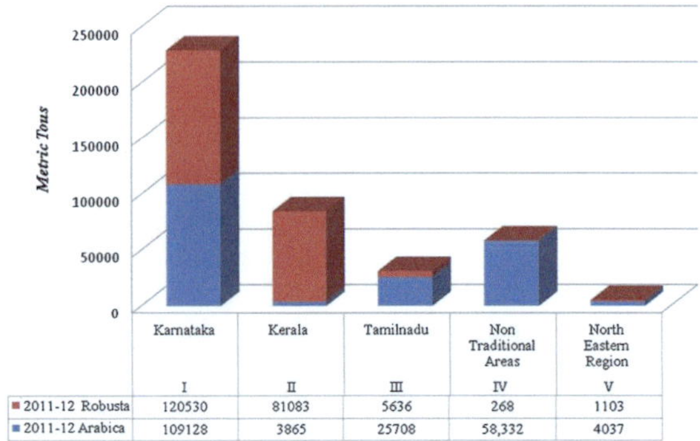

Fig. 11.4 Distribution of area under arabica and robusta cultivation in major coffee-growing states of India

the other countries. The shade-grown cultivation of coffee represents a wide diversity of more than 250 different types of shade trees, sheltering about 80 species of flora and fauna. Thus, coffee plantations help in preserving the rich biodiversity of western and eastern hill ranges, besides providing corridors for movements for several wild animals (Venkatachalam 2005). In many plantations, intercrops such as areca, orange and banana, serve as the lower canopy shade and black pepper is the most suitable associated crop as shade trees provide ideal strands for training the pepper vines. These crops serve as a valuable source of additional income for the planters. The Indian arabicas are rated as good with good body, good acidity and fairly good flavour, especially the high-grown coffees that show distinct flavour in cup. On the robusta front, India has been globally recognized for meticulously prepared shade-grown washed robustas. The Indian robusta coffees are often rated as neutral in taste with light to fair acidity with liquor quality rating as FAQ to Good. Because of the sustainable shade-grown practices, Indian robusta coffees are acclaimed world over for their unique quality attributes. Some of the specialty coffee grades such as washed robusta (Robusta Kaapi Royale) and Monsooned Malabar have become an important constituent of leading espresso blends in the Global market and earn substantial premiums.

11.2 Origin and Distribution

Arabica coffee (*C. arabica*) is reported to have originated in the high lands of Abyssinia in south-west Ethiopia where it is grown at altitudes ranging from 1300 to 2000 m above MSL, while robusta coffee (*C. canephora*) has its origin in western Africa and the species is distributed at altitudes below 1000 m MSL. *C. arabica* is adapted to highlands while *C. canephora* is adapted to lowland tropical areas. The two commercially grown species also differ significantly in terms of their morphology, vegetative vigour, ploidy level, breeding behaviour, genetic diversity, yielding potential, bean quality traits and also in genes conditioning resistance for major diseases and pests (Herrera et al. 2012). Robustas grow up to 10 m in untrained conditions and is more vigorous and productive than arabicas. Further, robusta is also an important source of disease-resistant genes but the bean and beverage quality is inferior to arabica. Among all the identified species of coffee, only *C. arabica* is self-fertile allotetraploid ($2n = 4x = 44$) while all other species, including *C. canephora,* are diploids ($2n = 22$) and are generally self-incompatible (Charrier and Berthaud, 1985). Molecular–cytogenetic analysis established *C. eugenioides* and *C. canephora* or ecotypes related to these diploid species as the likely progenitors of *C. arabica* (Lashermes et al. 1999).

11.3 Taxonomy of the Genus *Coffea*

Initially, the taxonomy of coffee species was described mainly based on morphological characters of the specimens preserved in different herbaria. Subsequently, advanced techniques including the DNA sequence information were used to describe the species relationships. The first tropical flora published by Kew Botanical Gardens in 1877 contained only 10 coffee species. In the taxonomic classification of Chevalier (1947), the genus *Coffea* included four sections, viz. Eucoffea, Mascarocoffea, Paracoffea and Argocoffea, among which the first two sections are presently grouped under the genus *Coffea* and the other two are considered as distinct genera as *Psilanthus* and *Argocoffeopsis*, respectively. Eucoffea was further divided into five subsections, viz. Erythrocoffea, Pachycoffea, Nanocoffea, Melanocoffea and Mozambicoffea. All the cultivated species were placed under subsection Erythrocoffea. Leroy (1980) described three genera under tribe Coffeeae: *Coffea*, *Psilanthus* and *Nostalachma.* The first two genera are reported to have close affinity and most of the species including the commercially cultivated species belong to the subgenera *Coffea.* The classification of genus *Coffea* into two subgenera *Coffea* and *Psilanthus* was mainly based on differences in their floral structure. The flowers in *Coffea* are characterized by a long style, medium corolla tube with anthers protruding from the corolla tube, while those under genus *Psilanthus* with short style, long corolla tube with anthers encased in the corolla tube. The genus *Psilanthus* with around 20 species has wider distribution in the tropical humid regions of Africa, India, South-East Asia and Pacific while the geographic distribution of the genus *Coffea* is confined to tropical humid regions of Africa and islands of West Indian Ocean (Charrier and Eskes 2004).

The East African species of *Coffea* were revised by Bridson (1982) and Berthaud (1986), who proposed some revisions to the classification of Chevalier (1947). However, the discovery of several new species during twentieth century in East, West and Central Africa and in Madagascar (Charrier and Eskes 2004) made the taxonomic classification increasingly complex. Majority of the coffee species are found to occur naturally in the humid evergreen forests of Africa, Madagascar and Mascarenes, but some species are also found in seasonally dry deciduous forest and/or bush land (Maurin et al. 2007).

11.4 Genetic Resources and Diversity of Coffee Gene pool

The coffee genetic resources mainly consist of primary gene pool comprising the wild and cultivated varieties of *C. arabica* and *C. canephora* and secondary gene pool comprising over 100 related diploid species of *Coffea* and *Psilanthus*. Among cultivated species, *C. arabica* is characterized by low genetic diversity (Lashermes et al. 1996a) which has been attributed to the allotetraploid origin, reproductive biology and evolutionary process of this species. However, diploid species harbors considerable variability. Some of the diploid species form valuable gene reservoir for various breeding programmes (Berthaud and Charrier 1988). The diploid *Coffea* varieties are known to interbreed freely with each other and produce relatively fertile progeny (Anthony 1992; Louarn 1993). Thus, the primary as well as secondary gene pool is of great significance in coffee improvement; however, there is a need for thorough characterization of the variability in this extant germplasm using high-resolution genetic approaches. To this end, a number of studies have been carried out in recent years using different types of DNA markers to ascertain the genetic diversity in the available coffee germplasm. Genetic diversity within *C. arabica* cultivars and wild collections has been extensively analysed using various DNA marker approaches such as RAPD (Lashermes et al. 1996b; Zezlina et al. 1999; Anthony et al. 2001; Aga et al. 2003; Sera et al. 2003; Cristancho et al. 2004), AFLP (Anthony et al. 2002a; Steiger et al. 2002; Prakash et al. 2002; Aggarwal et al. 2004) and SSR markers (Anthony et al. 2002a, b; Aggarwal et al. 2004). All these studies have largely confirmed that there exists a low genetic diversity within the arabica germplasm.

In the case of the second commercial species *C. canephora* (robusta), initial studies to ascertain genetic diversity were based on isozyme variability (Berthaud, 1986; Montagnon et al. 1992). These studies, although limited to few samples, revealed genetic structuring of the species in two clearly distinct groups. The first one was the 'Guinean Group' consisting of wild populations of Ivory Coast and the second one was 'Congolese group' comprising wild and cultivated germplasm from Central African Republic and Cameroon. Subsequent studies using RFLP polymorphism (Dussert et al. 1999) grouped the wild and cultivated forms of robusta coffee into five diversity groups (A, B, C, D, and E). All the studies established high genetic diversity in robustas compared to arabicas.

Among the other *Coffea* species, DNA sequence data were used to establish the molecular phylogeny and phylogenetic relationships. Further, the internal transcribed spacer ITS-2 region of the nuclear ribosomal DNA (Lashermes et al. 1997) and the chloroplast DNA variation (Lashermes et al. 1996a; Cros et al. 1998) were successfully used to infer phylogenetic relationships of *Coffea* species. The results suggest a radial mode of speciation and a recent origin in Africa for the genus *Coffea* (Etienne et al. 2002). In addition, several major clads were also identified, which reveal a strong geographical, correspondence i.e. coffee clads specific to West Africa, Central Africa, East Africa and Madagascar.

11.5 Coffee Genetic Resources in India

India is one of the countries that started systematic research on coffee in early nineteenth century with the primary mandate of addressing the leaf-rust problem and to evolve rust-resistant varieties. Initially, the Mysore Coffee Experiment Station was established near Balehonnur in Chikmagalur district of Karnataka, during 1925. This station was later expanded as Central Coffee Research Institute (CCRI) under administrative control of Coffee Board of India. To start with, concerted efforts were made to collect and conserve the available variability through systematic surveys of existing coffee plantations. This process undertaken during 1925–1940 resulted in the establishment of indigenous germplasm comprising more than 250 collections of arabica and robusta. This massive achievement however was a collective effort of a number of researchers/workers, but became a reality due to the singular efforts of Dr. M.K. Venkat Rao, the first Research Officer in-charge of Mysore Coffee Experimental Station, Balehonnur, for which he needs to be acknowledged here. These collections also formed the base material for the development of early Indian selections of arabica and robusta in India.

Subsequently, with the support from International agencies during 1954–1955, several world collections were introduced to India and an exotic germplasm bank was established at CCRI. Later during 1964, Dr. R.L. Narasimhaswamy, Botanist from CCRI, participated in an FAO-sponsored expedition to Ethiopia, the original habitat of arabica coffee, and collected 80 wild collections from different provinces of Ethiopia and established the same in CCRI (Meyer et al. 1968). At present, the Coffee Genebank at CCRI, Chikmagalur, has about 300 surviving collections of arabica, 73 types of robusta and 17 different species of *Coffea* (Figs. 11.5 and 11.6). In addition, two field genebanks with 76 world collections of arabica and 73 collections of robusta were established and maintained at Coffee Research Sub-Station, Chettalli, Coorg Dt., Karnataka, and Regional Coffee Research Station, Chundale, Kerala, respectively. These germplasm resources have been thoroughly characterized and the collections with useful agronomic traits were exploited in breeding programmes towards the development of improved genotypes for commercial cultivation.

11.5.1 Distribution of Coffee (Psilanthus) Species in India

Although coffee is considered as an introduced plant to India, some of the *Coffea* species (currently regrouped under the genus *Psilanthus)* are reported to be distributed in India (Narasimhaswamy and Vishweshwara 1963). Popularly called as indigenous species, these belong to two distinct phyto-geographical groups: the first group comprising *P. bababudanii, P. bengalensis, P. khasiana* and *P. fragrans* is distributed predominantly in the Eastern Himalayas, while the second group with *P. malabaricus, P. travancorensis* and *P. wightianus* is confined to Western and Eastern Ghats in the Southern parts of peninsular India. These indigenous species are of special interest because their beans contain either low levels of caffeine or free from caffeine.

Fig. 11.5 A view of coffee genebank at CCRI

Fig. 11.6 S.1587 (Rume Sudan) an exotic collection from Kenya

Five indigenous coffee species such as *P. bengalensis*, *P. travancorensis, P. wightianus, P. khasiana* and *P. bababudanii* were collected from their natural habitats in the forests of North East India and Tamil Nadu (South India) and established in the coffee genebanks maintained at CCRI and its Regional stations.

11.5.2 Molecular Characterization of Representative Coffee Germplasm Available in India

The genetic diversity of representative coffee germplasm collections available in India was analysed using various DNA marker approaches. The collections included a set of 25 accessions representing all the Ethiopian provinces, 33 rust differential clones, 16 different species of *Coffea* and 16 superior selections (14 arabica and 2 robusta) developed by the CCRI for commercial cultivation. Over all, four fragment size-based marker approaches (RAPD, ISSR, f-AFLP and SSRs), as well as DNA sequencing-based nucleotide variations in nuclear and organelle genomic domains, were employed for molecular characterization (Aggarwal 2005), as briefly described below.

11.5.2.1 Diversity Among Ethiopian Collections

The 80 Ethiopian arabica collections introduced into coffee genebank in India during 1964–65 belongs to different provinces of Ethiopia, viz. Shoa, Illubabor, Gojam, Kaffa, Erytrea, Sidamo and Harar.

Molecular analysis of a set of 25 accessions representing the collections from all these provinces using 25 RAPD and 15 ISSR markers revealed very low variation among different Ethiopian arabica accessions. Interestingly, the data generated using both the multilocus markers, viz. RAPD and ISSR, suggested a very narrow genetic base of the collection. Further, the analysis revealed no distinct generic affinities/clusters between Ethiopian arabicas and their provincial origin, thereby indicating that the geographical isolation did not result into variation among them. Interestingly, almost close correspondence was noticed between the grouping patterns of Ethiopian arabicas based on the two marker approaches, viz. RAPD and ISSR (Aggarwal 2005).

11.5.2.2 Molecular Characterization of *Coffea* Species Present in the Coffee Genebank at CCRI

Molecular characterization of 14 *Coffea* species and four species belongs to *Psilanthus* present in genebank was analysed using mobility-based DNA markers such as RAPD, ISSR and SSRs (Aggarwal 2005). Further, direct sequencing of three phylogenetically informative domains of nuclear and organelle genomes, viz. internal transcribed spacer regions ITS1-5SrDNA-ITS2 of the nuclear ribosomal DNA, 16S rDNA domain of mitochondrial genome and intergenic 'trnL' region of chloroplast genome, was also undertaken. Both individual plant sample and pooled samples were used for analysis to define the sampling strategy for DNA marker-based analysis of inherently heterozygous study material. The intra-species variation was also analysed using founder genotypes for four of the coffee species, viz. *C. eugenioides*, *C. stenophylla*, *C. dewevrei* and *C. salvatrix,* using RAPDs/ISSRs and ITS1-5SrDNA-ITS2 sequencing. The analysis revealed considerable variability in the species gene pool as 96 % of the 625 reproducible amplicons generated using 20 RAPD primers were polymorphic. Similarly, all the 162 reproducible amplicons generated using nine selected ISSR primers were polymorphic and informative for the purpose of species relationships. Comparative analysis of pooled or individual samples using different DNA marker systems established that multiple individual samples are not necessary to draw valid inferences to define the generic affinities between coffee species. Sequencing of around 30 kb of the three genomic domains of the 19 representative genotypes covering all the coffee species and related genera revealed significant variation in the form of both indel and base substitutions across species. Internal Transcribed Spacer regions ITS 1 and ITS 2 of the ITS1-5SrDNA-ITS2 nuclear ribosomal domain had average sequence sizes of 247 and 231 bp, respectively. Similarly, average sequence sizes of the sequenced partial Mt 16S rDNA conserved domain and the intergenic 'trnL' region of chloroplast genome were 756 bp and 532 bp for all the species. Among the diploid species, *C. canephora* was found to be phylogenetically most close to *C. arabica* followed by the cluster comprising *C. congensis, C. liberica* and *C. dewevrei*. A strong geographical correspondence was observed for the six Pachycoffea species, namely *C. arnoldiana, C. abeokutae, C. arwensiensis, C. exelsa, C. liberica* and *C. dewevreii.*

The molecular data validated the placement of the indigenous species, *Psilanthus bengalensis*, *P. travencorensis*, *P. khasiana* and *P. wightiana,* under the related Paracoffea genus *Psilanthus*. Both the dominant DNA markers (RAPD and ISSRs) were found to be effective and promising for detecting genetic variation and phylogenetic inferences. Moreover, ISSRs were found to be more informative than RAPD in terms of polymorphic bands detected per primer, and reliable in terms of reproducibility.

The analysis of ITS sequences defined the utility of nuclear ITS domain to derive the realistic affinities between different *Coffea* species, and the need to analyse the whole domain rather than only one of the variable segments, i.e. ITS-1 or ITS-2, for reliable inference in phylogenetic reconstruction (Hendre and Aggarwal 2007). Furthermore, the sequence-based genomic analysis of the three organellar compartments, viz. nucleus, chloroplast, and mitochondrion, suggests that organelle DNA may not be the ideal candidate for phylogenetic analysis of coffee species (Aggarwal 2005).

11.5.2.3 Molecular Characterization of Coffee Rust Differential Clones

A set of 33 rust differential clones, comprising of 29 different clones of *C. arabica* and one differential host each of *C. racemosa*, *C. excelsa*, *C. canephora* and *C. congensis* maintained in the coffee germplasm bank at CCRI, was subjected to DNA marker assays using 35 RAPD primers, 7 AFLP primer combinations and >150 in-house developed SSR markers. The analysis indicated presence of only subtle variation among the arabica-based rust differentials while it was substantial between arabicas and diploid rust differentials. Among the rust differentials, all the arabica genotypes formed one cluster and the four diploid rust differentials clustered out as per their genetic origin. The grouping of the arabica-based rust differentials was however not in accordance with rust-resistant factors (Aggarwal 2005).

11.5.2.4 Molecular Characterization of CCRI Station-Bred Selections

Genetic improvement of coffee undertaken at CCRI since 1925, using both indigenous and exotic collections, resulted in the development of 13 improved arabica coffee selections and three superior robusta selections. Released for commercial cultivation from time to time, these improved selections have been cultivated across the Indian coffee tracts depending on their agro-climatic suitability.

All these selections were fingerprinted using high-resolution DNA marker techniques involving nuclear genomic markers such as RAPD, f-AFLP, SSR and IRAP. The DNA profiles generated using all the marker systems showed clear distinction between tetraploid arabicas and diploid robustas, with more polymorphism among the later. The arabica selections indicated limited variability while it was substantial for robustas (Fig. 11.7). Average per cent polymorphic markers using the different marker approaches were found to range from 51 to 62 % for diploid selections and only 18–23 % for tetraploid arabicas (Aggarwal 2005). All the selections were grouped into two clusters, representing arabica genotypes and robustas as per their genetic origin which also confirms to their pedigree (Fig. 11.8).

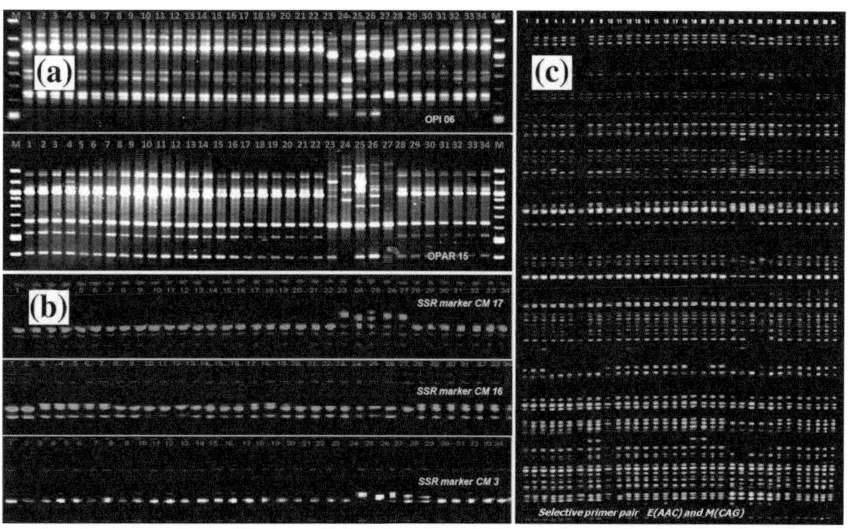

Fig. 11.7 Representative DNA profiles of *Coffea arabica* selections generated using: **a** RAPD markers; **b** in-house developed coffee-specific SSR markers; and **c** AFLP markers

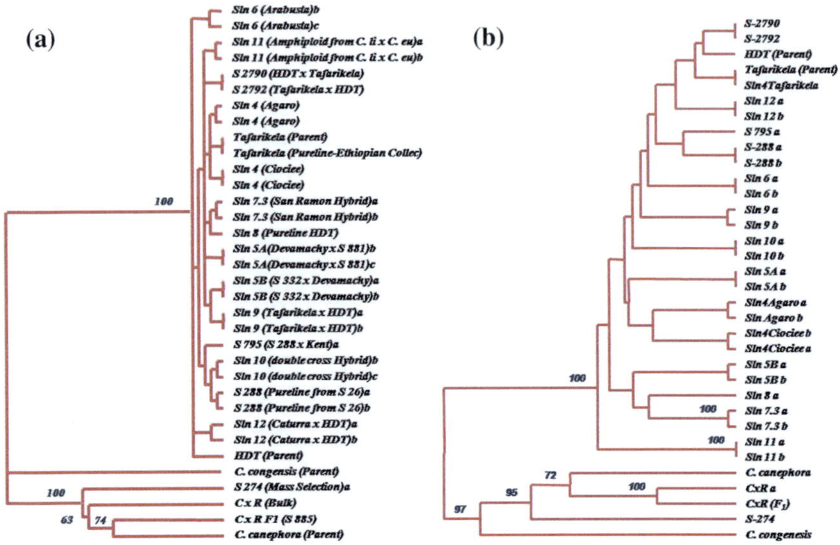

Fig. 11.8 UPGMA trees generated using: **a** RAPD markers, and **b** SSR markers, showing genetic affinities between different arabica selections

The low variability among the arabica-based selections confirms to the narrow genetic base of arabica germplasm.

11.5.3 Reference Polymorphism Panels for Individualization of Coffee Genotypes

Despite the narrow genetic base and very low detectable genetic variability, the cultivated arabica selections could be uniquely discriminated from each other but required the use of large number of DNA markers. Based on fingerprint data generated by use of the various marker systems, a reference DNA polymorphism panel was prepared for distinguishing the CCRI selections that are expected to be useful for varietal registration/IPR protection (Aggarwal 2005).

11.5.4 Molecular Characterization of Robusta Coffee Germplasm

The genetic diversity in 40 germplasm collections of robusta, including both indigenous and exotics, maintained in Genebank at Regional Coffee Research Station, Chundale, was assessed in comparison with 14 representative samples of core collections of *C. canephora* and three accessions of *C. congensis* using AFLP and SSR marker approaches (Prakash et al. 2005a, b). The accessions of Indian gene pool grouped together with the robusta types identified as diversity group ‘E’. The study clearly revealed high amount of diversity present in core samples, which was not represented in Indian gene pool. Further, the vast diversity found in the accessions representing a core collection from Africa, the centre of genetic diversity of robusta coffee, was however not present in the cultivated lines of robusta genotypes.

11.6 Brief Overview of Coffee Genetic Improvement Worldwide

Availability of genetic variability, especially for characters of agronomic importance, is an important prerequisite for the success of genetic improvement of any crop. In the case of coffee, the two commercially cultivated species, *C. arabica* popularly known as arabica coffee and *C. canephora* called as robusta coffee, differ to a great extent in morphology, vegetative vigour, genetic diversity, yield potential, bean quality traits and breeding behaviour. Arabica coffee grows into a small tree under natural growth but attains bushy growth and looks like a shrub when regulated by training. On the other hand, robusta coffee is characterized by more erect and robust growth as the name denotes and grows into large vigorous bushes.

Arabica produces superior quality coffee but the crop yields are often affected by major diseases and pests as arabica coffee plants manifest susceptibility to several diseases and pests such as leaf rust (*Hemileia vastatrix Berk & Br*), coffee berry disease (CBD) (*Colletotrichum kahawae* Waller et Bridge), stem borer (*Xylotrechus quadripes* Chevrolat) and nematodes (*Meloidogyne* sp. and *Pratylenchus* sp.). Robusta is more tolerant to these diseases and pests and has potential to give consistently high yields but the bean and liquor qualities of robusta are inferior to arabica.

Therefore, the main objectives of genetic improvement in coffee have been largely focussed to improve production coupled with resistance in arabica and production with bean and liquor qualities in robusta.

11.6.1 Genetic Improvement of Arabica Coffee

The main focus of early breeding programmes for arabica improvement undertaken world over during 1920s–1950s was on yield, quality and adaptability to local conditions. Simple selection and crossing within genetically homogeneous base populations were the strategies adopted. In general, breeding for disease resistance was given low priority except in India where the primary focus of arabica coffee breeding was on leaf-rust resistance since the initiation of organized research in the 1920s with the inception of Mysore Coffee Experiment Station, located at Balehonnur, Chikmagalur district, Karnataka, India, currently known as CCRI. The early programmes resulted in considerable success in the development of vigorous and productive cultivars such as 'Mundo Novo', 'Caturra' and 'Catuai' from Brazil, 'Kents', 'S.288' and 'S.795' from India, 'Blue Mountain' from Jamaica and several others. Consequent to the gradual spread of leaf-rust disease to various other coffee-growing countries across Africa, Central and South America, there was a shift in breeding focus towards rust resistance. During the second phase of arabica coffee improvement, from the 1950s to the 1990s, the priority was on disease resistance, especially coffee leaf rust (CLR) and CBD. This phase of coffee improvement was very productive, as several high-yielding varieties with broad spectrum of resistance were developed in a relatively short period. The success could be largely attributed to the coordinated efforts of different coffee research groups across the continents in enriching the valuable genetic resources (Meyer et al. 1968) through germplasm exchange and exploitation of new genetic diversity by the application of advanced selection and breeding methods (Van der Vossen 1985). Generation of basic information on coffee genetics in Brazil (Sybenga 1960; Carvalho et al. 1969) and also the establishment of Coffee Rust Research Centre (CIFC) in Oeiras, Portugal, to work exclusively on various aspects of coffee rust pathogen, *Hemileia vastatrix,* were instrumental in driving forward the rust-resistant breeding programmes in many countries. Further, identification of high-yielding mutants of arabica such as 'Caturra', 'San Ramon' and 'Villasarchi' (having dwarf/compact bush stature) and spontaneous hybrids of robusta and arabica, such as 'Hibrido de Timor (HdT)' and 'Devamachy' (with high levels of disease resistance), contributed to a great extent towards the development of several high-yielding and disease-tolerant cultivars with compact bush stature, viz. 'Catimor', 'Ruiru-11', 'Sarchimor', that have been extensively used for commercial cultivation.

Further, based on the objective, appropriate breeding strategy was employed for evolving improved varieties. The progress of selection and breeding until 2000 has been excellently reviewed by Van der Vossen (1985), Carvalho (1988), Berthaud and Charrier (1988), Wrigly (1988) and Van der Vossen (2001). An exclusive review on rust-resistant breeding was made by Bettencourt and Rodrigues (1988). Recently, the status of arabica coffee breeding with respect to durable resistance to CLR in different countries has been reviewed during the first Workshop on durable resistance to coffee rust, held at University of Vicosa, Brazil, in September 2005 (Alvarado 2005; Fazouli et al. 2005; Matiello et al. 2005; Pereira et al. 2005; Prakash et al. 2005b; Sera et al. 2005). More recently, a comprehensive account of coffee breeding was published by Herrera et al. (2012) while Montagnon et al. (2012) reviewed the status of breeding coffee for quality.

11.6.1.1 Arabica Coffee Improvement in India

Arabica coffee was introduced in India sometime during 1600 AD by a Muslim pilgrim, Baba Budan, who is believed to have brought seven seeds from Yemen, presumably *Mokka* coffee and planted on his hermitage located in Chandragiri hills near Chikmagalur in Karnataka which later became popular as Baba Budan Giris. However, the arabica coffee plants remained as backyard plants for long time and it is only during the late 1820s, when the British entrepreneurs started coffee cultivation on commercial scale in South India. The cultivation of coffee rapidly progressed during the next 40 years, till the outbreak of CLR disease caused by an obligate parasitic fungus, *Hemileia vastatrix* (orange rust) affecting arabica coffee plants. Soon, by the later part of the eighteenth century, CLR posed a serious threat for arabica coffee plantations in South-East Asia including India, as the disease was relatively unknown and no fungicides were invented. In India, some of the enterprising planters made efforts to select the disease-tolerant plants from existing populations. The early cultivars such as 'Coorgs', 'Chicks' and 'Kents' belong to this category of selections, of which 'Kents' became popular and was largely used for cultivation in the 1920s. Soon, the 'Kents' variety also succumbed to rust disease. To tide over the situation, the rust-resistant diploid *Coffea* species such as *C. liberica* and *C. canephora* were also introduced. Some efforts were also made to develop rust-tolerant hybrids by crossing arabica with rust hardy diploid species. But, fertile hybrids could not be generated due to variation in ploidy level (tetraploid vs. diploid); nevertheless, a few of the hybrids such as 'Hamiltons', 'Jacksons', 'Netrakonda' and 'Chandrapore' were used for cultivation. Despite these efforts, the leaf-rust problem of arabica coffee could not be effectively tackled. Consequently, the coffee paved the way for tea in neighbouring Sri Lanka, while in Indonesia arabica coffee was replaced with robusta coffee. In India, the need for systematic research on coffee was felt and as a result the Mysore Coffee Experiment Station was established in 1925 with the primary mandate of developing rust-resistant varieties and to address the problems in coffee cultivation (Anonymous 2014b). Thus, unlike other coffee-growing countries, the main focus of arabica coffee improvement in India has been on breeding for rust resistance coupled with high production, improved quality and wide adaptability.

For genetic improvement in any crop, genetic variability and heritability are the two important prerequisites. In the early breeding programmes, the indigenous germplasm collections established during 1925–1940 were exploited. Subsequently, the exotic germplasm and the spontaneous tetraploid inter-specific hybrids have been used in breeding programmes. Similar to the observation elsewhere that arabica coffee has a narrow genetic base, the arabica coffee gene pool in India, collected from indigenous and exotic sources, represent limited variability in relation to the various agronomic traits such as plant size (dwarf, semi-dwarf, tall), branching habit (erect, spreading, drooping), fruit size (small, medium, bold), fruit ripening (early, late), quality traits, yield potential and a wide spectrum of resistance from complete susceptibility to mild resistance to leaf-rust pathogen. However, some of the tetraploid inter-specific arabica hybrids of spontaneous origin manifest complete resistance to leaf-rust pathogen due to introgression of diploid genomes, and these provided additional variability for resistance breeding. In line with the scope and objectives of breeding, CCRI has successfully utilized the available variability and developed 13 improved arabica genotypes (selections) for commercial cultivation by employing proven breeding methods (Anonymous 2014b; Srinivasan and Narasimhaswamy 1975). The station-bred selections have been designated as Sln.1 to Sln.13 and depending on the location-specific adaptability, the improved arabica genotypes have been recommended for commercial cultivation across the arabica-growing tracts in India. The CCRI selections, viz. Sln.3 (S.795), Sln.5A, Sln.5B, Sln.6, Sln.9 and Sln.13, are very popular, and occupy sizeable area (ca. 90 %) under arabica coffee.

The pedigree and characteristic features of some popular arabica selections are summarized in Table 11.1.

11.6.1.2 Breeding Strategies Employed for Arabica Coffee Improvement

The classical genetic studies undertaken in Brazil established that *C. arabica* is the only allotetraploid and self-fertile species under the genus *Coffea*, and it exhibits diploid mode of inheritance for all the characters (Krug and Carvalho 1951). Hence, some of the breeding methods relevant to self-pollinated diploid crops, such as pure line selection, inter-varietal hybridization followed by pedigree selection, inter-specific hybridization followed by backcross breeding, multiple crosses and introgressive breeding using spontaneous tetraploid inter-specific hybrids, have been successfully employed for arabica improvement, world over. The resultant hybrids were sufficiently homogeneous to permit propagation by seed and practically majority of the arabica cultivars grown in the world today have been established from seedlings (Van der Vossen 1985). There are some reports on success achieved in large-scale multiplication of F_1 hybrids using somatic embryogenesis technology and their superior field performance (Betrand et al. 2010). The breeding strategies adopted and arabica improvement in Indian context are detailed below.

Pure Line Selection

In arabica coffee improvement programmes taken up initially, pure line selection was the main strategy applied wherein the elite individual plants with desired agronomical traits were identified in the populations and selected individual plants were advanced by recurrent selfing to derive commercial lines. The early Indian coffee varieties, ‘Kents’ and S.288 were developed by this strategy. Similarly, pure line selection strategy was used to exploit exotic Ethiopian collections from the genebank for race-specific resistance to leaf-rust pathogen. The major constraint in arabica coffee breeding has been its narrow genetic base, which is a bottleneck to achieve a selection efficiency sufficient enough for quick progress besides maintaining adequate variability in subsequent generations. In advanced generations, the selection efficiency from individual and even family selection within same

Table 11.1 Details and distinguishing features of arabica (*Coffea arabica*) selections released by the Central Coffee Research Institute, India

Selection	Breeding method	Distinguishing characters	Agronomical features/adaptability
Sln.1 (S.288)	– Pure line selection (PLS) from S.26, a putative natural hybrid between *C. arabica* and *C. liberica*. – Given for cultivation during in 1936–37	– Tall phenotype with vigorous growth – Leaves are dark green, thick, elliptic in shape; young leaves (tip) bronze coloured – Fruits round with broad disc and orange yellow to red in colour popularly known as 'Golden drops' – Fruits show relatively high percentage of multilocular condition	– Moderate yielder (800–1000 kg/ha) – Manifest resistance to leaf-rust races I and II due to presence of S_H3 rust-resistant gene introgressed from *C. liberica* – Found superior to other varieties such as 'Chicks' and 'Kents' under cultivation during that time – Liquor quality is FAQ (fair average quality) – Adaptable to all coffee-growing regions
Sln.3 (S.795)	– Developed from the cross between S.288 (*C. liberica* introgressed line) × 'Kents' – Given for cultivation during 1945–46	– Vigorous and widespreading bush with profuse growth – Leaves are oblong, broad and elliptic in shape – Young leaves (tip) bronze coloured – Fruits round to oblong with broad honey disc, red in colour – Number of fruits 12–16 per cluster. – Beans are bold in size, bluish-grey colour with good visual appearance when wet processed	– Yield potential 1500–2000 kg/ha – Manifest resistance to leaf-rust races I and II, prevalent at the time of release due to presence of S_H3 rust-resistant gene introgressed from *C. liberica* – Later fell susceptible to new virulent races VII, VIII, XII, XIV and XVI – 'A' grade beans 60–65 % – Liquor balanced with good body, good acidity and fairly good flavour – Most widely adaptable and cultivated tall arabica variety in India

(continued)

Table 11.1 (continued)

Selection	Breeding method	Distinguishing characters	Agronomical features/adaptability
Sln.4	– Composite selection of 3 Ethiopian arabica collections, viz. Cioccie, Agaro and Tafarikela – Given for cultivation during the 1960s	– Tall bush types, Cioccie and Agaro, exhibit semi-erect branching while Tafarikela shows drooping growth habit – The young leaf (leaf tip) colour is green (Cioccie)/bronze (Agaro) and dark bronze (Tafarikela) – Fruits are long, bold, flat orange red with projected honey disc, occasionally with persistent calyx in Agaro and Cioccie – Fruits relatively small in Tafarikela and ripen early compared to other arabicas – Beans are long and bold in size	– Moderate yielders with yield potential of 1000 kg/ha – Shows specific resistance to leaf-rust race VIII. In addition, Tafarikela manifests horizontal resistance – Released for commercial cultivation as a strategy to check the devastation of race VIII on S.795 plantations – 65 % of the beans belong to 'A' grade with excellent liquor quality – performs well in high altitudes and under well-maintained two-tier shade canopy
Sln.5A	– Derivative of the cross between Devamachy (a spontaneous Robusta × Arabica hybrid) x S.881 (a wild arabica collection from Rume Sudan) – Given for cultivation during the early 1970s	– Plants are tall, widespreading and exhibits vigorous vegetative growth – leaves are narrow and elliptic; young leaves (leaf tip) are generally green in colour – Fruits round to oblong, variable in size with long peduncles and 8–12 per cluster, medium size beans – Relatively late ripener	– Consistent yielder (1200–1500 kg/ha) – Manifest high field resistance to leaf rust; normally defoliation is not seen even in rust susceptible plants – High percentage of B grade beans (~40 %), with liquor quality rating of FAQ to FAQ+ – Adaptable to hot and humid climate and performs well across the arabica-growing areas – Largely cultivated variety in non-traditional coffee areas especially the tribal holdings of Andhra Pradesh

(continued)

Table 11.1 (continued)

Selection	Breeding method	Distinguishing characters	Agronomical features/adaptability
Sln.5B (S.2931)	– Derivative of the cross between Devamachy × S.333 – Given for cultivation during the early 1970s	– Tall bushes with vigorous growth and semi-erect branches that droop on bearing – Leaves are elliptic and young leaves (tip) are bronze coloured – Fruits are round, bold with round naval, 12–16 per cluster, normal in ripening – Beans are round, bold in size	– Yield potential of 1500 kg/ha. – The plants exhibit high field tolerance to leaf rust – 60 % of the beans represent 'A' grade and liquor quality is FAQ to FAQ+ – Shows wider adaptability, suitable for cultivation in different arabica-growing regions and performs well in medium-to-high altitudes
Sln.6 (S.2828)	– Developed by crossing robusta cv. S.274 and 'Kents Arabica' followed by recurrent backcrosses to Kents parent – Given for cultivation during the 1970s	– Tall and wide spreading bushes – Leaves are broad, elliptic and young leaves (tip) are bronze coloured – Fruits are bold, round in tight clusters with 16–20 fruits per cluster	– Yield ranges between 1200 and 1500 kg/ha – Shows mixed type of rust reaction with about 80 % plants manifesting high field tolerance to rust – Beans bluish-grey, round and bold with 60–65 % 'A' grade – Adaptable to medium altitudes
Sln.7.3	– Derived from multiple crosses involving 'San Ramon', a dwarf mutant and other tall arabicas such as S.795, Agaro and Hibrido de Timor (HdT) – Given for cultivation during the 1970s	– Majority of the plants show dwarf bush stature with few medium and to tall segregants – Branches are compact with close internodes – Leaves are broad elliptic and wrinkled; young leaves (tip) are bronze coloured – Fruits are round, red in colour and late ripener	– Yield potential of over 1500 kg/ha, annual variations for production with alternate bearing behaviour – Susceptible to virulent races of rust – Withstand drought conditions because of deep-rooted system, suitable for marginal areas – Around 60 % of the beans belong to 'A' grade with FAQ to above FAQ cup quality

(continued)

Table 11.1 (continued)

Selection	Breeding method	Distinguishing characters	Agronomical features/adaptability
Sln.8	– Developed by pure line selection from 'Hibrido de Timor' popularly known as HdT, a spontaneous Robusta x Arabica hybrid, identified in Timor island – Given for cultivation during the late 1970s	– Plants are tall and vigorous – Plant phenotype closely resembles other arabica varieties with semi-erect to drooping growth habit – Leaves are dark green, elliptic in shape; young leaves (tip) are bronze coloured – Fruits are round, red in colour and normal in ripening	– Moderate yielder, yield potential of 1000–1200 kg/ha – Manifest high resistance to leaf rust, majority of the plants show resistance to all known races of rust – Beans are round and medium in size with over 60 % 'A' grade. Cup quality is FAQ to FAQ+ – Suitable for cultivation in medium altitudes
Sln.9 (S.2790)	– Derived from HdT × Tafarikela – Given for cultivation during the late 1970s	– Plants tall, vigorous with semi-erect to drooping branches – Inter-node length is medium to long – Leaves broad, elliptic, and young leaves are dark bronze coloured – Fruits are bold, dark red, flat, oblong and 12–16 per cluster and ripen early compared to other arabica varieties	– Yield potential of 1200–1600 kg/ha – Manifests high tolerance to leaf rust under field conditions and drought hardy – Beans are bluish green, round, bold with 60–65 % 'A' grade – Liquor quality is excellent with strong body and possessing distinct flavour in cup – Widely adaptable across all the coffee-growing regions – Suitable for gap filling and plants establishes well in supply positions
Sln.10	– Double cross-hybrid developed by crossing two F_1s, Caturra (a high-yielding dwarf mutant) × Cioccie and Caturra × S.795	– Plants are compact and semi-dwarf in bush stature with profuse branching habit – Leaves are broad and elliptic, and young	– Moderate Yielder (1000–1200 kg/ha) – Exhibits resistance to common races of rust due to presence of S_H3 rust-resistant gene, but susceptible

(continued)

Table 11.1 (continued)

Selection	Breeding method	Distinguishing characters	Agronomical features/adaptability
	– Given for cultivation during the 1980s	leaves are green to light bronze coloured – Fruits bold, 12–15 per cluster, normal ripening	to virulent races with V_3 gene combinations – Cup quality is similar to S.795. Fruits are bold and produce nearly 65 % 'A' grade beans – Shows location-specific adaptability and performs well in higher altitudes
Sln.11	– Amphiploid of the diploid inter-specific hybrid between *C. arabica* and *C. eugenioides* – Given for cultivation during the 1980s	– Tall, vigorous bushes with thin branches – Leaves are small in size, narrow, linear and elliptic, and young leaves are green in colour – Fruits are small, oblong, 10–15 per cluster and late ripening	– Moderate yielder (1000–1200 kg/ha) – Plants show high field tolerance to leaf rust and moderately tolerant to drought conditions – Produces more B grade beans (30–40 %) with FAQ and FAQ + cup quality – Exhibits better adaptability to hot and humid climate
Sln.12 (Cauvery/Catimor)	– Derived from cross between Caturra and HdT – Given for commercial cultivation during 1986	– Semi-dwarf bushes, vigorous growth with profuse primaries and secondaries and close internodes – Leaves are medium in size and elliptic, and young leaves are green in colour – Fruits are dark red, round, bold, 12–18 in tight clusters – Early bearing habit	– High-yield potential of 2000 kg/ha – In initial years, manifested high resistance to leaf rust, later broke down due to appearance of seven new rust races – Beans are round, bold with over 60 % 'A' grade and liquor quality is FAQ to FAQ + with pleasant aroma and taste – Widely adaptable across all the coffee regions and suitable for cultivation in high elevations

(continued)

Table 11.1 (continued)

Selection	Breeding method	Distinguishing characters	Agronomical features/adaptability
			(above 1000 m MSL) – Suitable for high-density planting
Sln.13 (Chandragiri)	– Derived from the cross Villasarchi × HdT – Given for commercial cultivation during 2007	– Semi-dwarf bushes with vigorous growth and widespreading branches and closer nodes – Leaves are broad, thick, dark green in colour and elliptic in shape, and young leaves (tip) are green in colour – Fruits oblong, flat, elongated, 12–18, in number and borne in loose clusters	– Yield potential of 1500–1800 kg/ha. – Manifests high field tolerance to leaf rust – Suitable for cultivation at higher elevations (1000 m and above MSL) – Beans are long and bold with over 70 % 'A' grade of which 20–25 % belong to extrabold ('AA' grade). – Cup quality is FAQ to FAQ+

population based on phenotypic performance becomes uncertain and slow. In situations where the heritability is low, it is more reliable to select parents on the performance of their offspring, but this method has several disadvantages including long time frame.

Inter-varietal Hybridization Followed by Pedigree Selection
Predominantly, this strategy has been used to improve the agronomically desirable characters such as yield, tolerance to diseases and quality. In general, a proven genotype is selected as recipient parent that is crossed with a donor parent for any specific trait. The hybrid progenies are subjected to recurrent selection by pedigree method. The famous and well-known variety of India, S.795 (Fig. 11.9), was developed during the late 1940s, by following this strategy.

Subsequently, several other genotypes such as Sln.5A, Sln.5B (Fig. 11.10) and Sln.9 (Fig. 11.11) were developed by crossing the selected arabica genotypes with diverse sources of rust resistance. In majority of these cases, spontaneous tetraploid inter-specific hybrids such as Devamachy and Hibrido de Timor were used as donors of resistance. Both these donor sources are the natural hybrids between *C. arabica* and *C. canephora*, very much resembling arabica coffee, with high levels of resistance to major coffee diseases and pests due to the genes introgressed from diploid species. Thus, this strategy helped to transfer the genes conferring disease/pest resistance into selected arabica genotypes. The introgressive breeding strategy has been the most successful strategy for arabica improvement especially for host resistance and the Timor hybrid (HdeT) has been extensively used as the

Fig. 11.9 S.795, the most popular arabica variety of India

donor for broad spectrum of resistance to diseases and pests, worldwide. This strategy assumed greater significance for arabica coffee breeding as it helps in minimizing the time frame compared to inter-specific hybridization strategy.

Inter-specific Hybridization

The diploid species of *Coffea* are more heterogeneous because of cross-pollinating nature compared to tetraploid arabica, and thus have been the important source of variability for arabica improvement. Among several diploid species, *C. canephora* is the major source for resistance genes for several important diseases, such as CLR (*Hemileia vastatrix*), CBD (*Colletotrichum kahawae*) and root knot nematode (*Meloidogyne* spp.). Similarly, other diploid species such as *C. liberica* have been successfully used as source of rust-resistant breeding in India (Srinivasan and Narasimhaswamy 1975) while *C. racemosa* was used as source of resistance to coffee leaf minor (Guerreiro Filho et al. 1999). Hence, inter-specific hybridization with an objective to transfer the desirable genes (in particular) for disease resistance from diploid species such as *C. canephora* and *C. liberica* into tetraploid arabica cultivars without affecting quality traits has remained the main objective of arabica coffee

Fig. 11.10 Sln.5B, a popular inter-varietal hybrid having wider adaptability

Fig. 11.11 Sln.9, a popular tall arabica selection known for its superior bean and liquor quality

breeding (Van der Vossen 2001). However, the ploidy difference between *C. arabica* (tetraploid) and other species (diploids) is the main limitation for developing fertile hybrids from direct crosses. To overcome this limitation, two different methods, the tetraploid breeding method and triploid breeding method, were followed to obtain fertile tetraploid inter-specific hybrids. In tetraploid breeding method followed in Brazil, the chromosome number of the diploid species (*C. canephora*) was doubled using colchicine and such induced tetraploid robusta types were crossed with normal tetraploid arabicas. The tetraploid breeding strategy helped in realizing fertile F_1 hybrids due to gametic balance. The resultant hybrids were backcrossed to arabicas and selection was exercised in the progeny for desirable types.

The triploid breeding strategy, which was followed in India and Colombia, direct crosses were made between tetraploid *C. arabica* and diploid *C. canephora*. The resultant triploids were recurrently backcrossed to arabica. One of the arabica selections, Sln.6 (S.2828), developed in India by using triploid breeding strategy resembles arabica phenotype with vigorous growth and compact fruit clusters as well as resistance of robusta coupled with superior bean and liquor quality of arabica (Fig. 11.12).

Exploitation of Spontaneous Mutants

As arabica coffee is known for narrow genetic base, efforts to induce variability by using physical mutagens such as gamma irradiation and chemical mutagens such as EMS and DMS were made, but in general, most such efforts were unsuccessful except for generation of a few mutants such as 'purpurascens' that were found to be of no agronomical value. Nevertheless, screening of large populations in Brazil resulted in the identification of few spontaneous mutants with dwarf (San Ramon) and semi-dwarf bush stature (Caturra, Villasarchi). These mutants with high-yield potential and precocious bearing nature were exploited for breeding purposes and

Fig. 11.12 Sln.6, a popular inter-specific hybrid performing well at medium altitudes

Fig. 11.13 Sln.7.3, a dwarf arabica variety

several outstanding compact arabica varieties such as Catuai, Catimor and Sarchimor, suitable for high-density planting were developed. In India also, several such coffee varieties were developed, a few prominent being: Sln.7.3 (Fig. 11.13) derived from multiple crosses involving dwarf mutant San Ramon; Sln.12 (Cauvery-Catimor) developed from a cross between high yielding 'Caturra' mutant and resistant donor Hibrido de Timor; and the semi-dwarf variety Chandragiri (Sln.13—Sarchimor; Fig. 11. 14) developed from a cross of Villasarchi, a vigorous dwarf mutant with Hibrido de Timor.

Heterosis Breeding and Development of F_1 Hybrids

In order to enhance the chances of exploiting transgressive hybrid vigour, creation of hybrids between genetically diverse subpopulations, such as crosses between common cultivars and Ethiopian accessions, was suggested by Lashermes et al. (1996b) and Van der Vossen (2001). Further, Lashermes et al. (2008) envisaged the development of F_1 hybrids between diverse genetic groups, such as wild Sudan and Ethiopian origins, as one of the promising strategies in arabica breeding to obtain high percentage of heterosis. This strategy was successfully applied in Kenya and Ethiopia to develop commercial cultivars such as, variety 'Ababuna' in Ethiopia and Ruiru-II, a composite F_1 hybrid variety in Kenya. In Central America, a programme for development of F_1 hybrids between high-yielding and cultivated varieties of arabica (Caturra, Catuai, Catimor, Sarchimor) and semi-wild trees of

Fig. 11.14 Chandragiri, a semi-dwarf arabica variety

Ethiopia or Sudan origin was initiated during 1992 (Etienne et al. 2002). Field trials of these F_1 hybrids recorded 30 % higher productivity, along with resistance to leaf rust and nematodes, besides maintaining an excellent cup quality (Bertrand et al. 2005). Because of promising performance, the selected F_1 hybrids have been multiplied in large numbers using somatic embryogenesis technique for commercial cultivation (Bertrand et al. 2010). Availability of male sterile lines provides great practical advantages for development of heterotic F_1 hybrids and Georget et al. (2014) reported the successful exploitation of male sterility for development of F_1 hybrids that recorded superior performance over the pollinator parent in field trials conducted at Nicaragua. In India, four male sterile plants have been identified recently from exotic germplasm collections and efforts are being made to identify the best pollinators, as well as for exploitation in heterosis breeding programmes.

11.6.2 Genetic Improvement of Robusta Coffee

The epidemics of *H. vastatrix* (orange rust) in South-East Asia between 1870 and 1900 threatened the arabica coffee cultivation in several coffee-growing countries of the region. Consequently, other tolerant species such as *Coffea canephora* (Robusta) and *Coffea liberica* were introduced in Indonesia and India where as in Sri Lanka, there was a shift towards tea cultivation. These diploid species, especially *C. canephora,* adopted well to the low-altitude regions. As a result, robusta breeding programmes were initiated even earlier to arabica with the objective of improving bean quality. The pioneering work on coffee biology and selection carried out in East Java in the early years of nineteenth century was reviewed by Cramer (1957) that formed the basis for subsequent breeding programmes of robusta coffee not only in India but also in Africa (Van der Vossen 1985).

11.6.2.1 Robusta Coffee Improvement in India

According to the earlier reports, the nucleus robusta stock introduced into Java in 1901 came from trees already under cultivation in Zaire in 1895, originating in Lomani River region. The material selected in Java was reintroduced in the Belgian Congo around 1916 at INEAC (Institut National pour l'Etude Agronomique du Congo Belge), which has become the major selection centre of *C. canephora* from 1930 to 1960 (Montagnon et al. 1998). It was also believed that the improved seed from Java was used to establish robusta plantations in India and also in African countries such as Uganda, Ivory Coast and Zaire, from where robusta coffee originated (Van der Vossen 1985). Apart from Java, Ceylon was reported to be the other source of robusta introduced into India during early nineteenth century which is popularly known as Peradeniya Robusta. At present, both these robusta types still occupy larger area under robusta and broadly known as 'Old robusta'. The 'Old robustas' show vigorous growth compared to arabica and grows into moderately large bushes. Their fruits are red, round to oblong with pronounced navel, small to medium in size, 30–40 per node and borne in tight clusters. Beans of old robustas are greenish grey, small in size, comprises less than 50 % AB grade and the liquor quality rated as Fair Average Quality (FAQ) to Good.

Therefore, the main objective of robusta coffee improvement in India was bean and liquor quality improvement. As *C. canephora* is strictly allogamous species, breeding methods suitable to cross-pollinated crops such as mass selection and intra-specific, as well as inter-specific hybridization, have been followed. The efforts at CCRI for robusta improvement have resulted in the development of three superior robusta selections (Table 11.2), of which S.274 (Sln.1R) and C×R (Sln. 3R) are very popular among the coffee growers and occupy sizeable area under robusta coffee.

Table 11.2 Details and distinguishing features of robusta (*Coffea canephohra*) selections released by the Central Coffee Research Institute, India

Selection	Breeding method	Distinguishing characters	Agronomical features/adaptability
Sln.1R (S.274)	– Seedling progenies of two individual high-yielding mother plants – Given for cultivation during the 1940s	– The plants are vigorous and grow into moderately large trees – Leaves are large, broader, light green in colour – Fruits dark red, medium to bold, round to oblong with pronounced navel and borne in tight clusters of 30–50 fruits each	– Yield potential is 2000–2500 kg/ha under irrigated and 1000–1500 kg/ha under unirrigated conditions – Beans are medium to bold in size, round in shape and about 45 % represent 'A' grade – The cup quality is neutral with rating of FAQ to good – Adaptable to almost all robusta-growing regions
Sln.2R (BR Series 9,10,11)	– High-yielding clonal progenies (BR 9, 10, 11) of 12 single plant progenies (S.267–S.278) that yielded twice the family mean yield – Given for cultivation during the 1960s	– This plants resemble S.274 in growth habit, yield potential and bean/cup characteristics	– Yield potential similar to S.274 – High stability for production and 'A' grade beans – The cup quality is neutral with rating of FAQ to good – Adaptable to all robusta-growing regions
Sln.3R (C×R)	– Inter-specific hybridization involving *C. congensis* and *C. canephora* (Robusta) followed by back cross to robusta – Given for cultivation during the 1970s	– Plants are compact in bush stature compared to S.274 – Branches show semi-erect to drooping growth habit; nodes are close – Early bearer, fruits arranged in tight clusters of 30–40 per node with prominent and projected naval – Fruits contain high mucilage content that enable easy pulping	– Yield potential is 2000–2500 kg/ha under irrigated and 1000–1500 kg/ha under unirrigated conditions – Early and uniform ripener – Suitable for planting at closer spacing, 2.7 × 2.7 m or 2.4 × 2.4 m (9 × 9 ft or 8 × 8 ft) – Beans are medium to bold, 50–55 % are 'A' grade – Liquor is soft, neutral with light to fair acidity and good cup quality – Suitable to all robusta-growing regions of India

11.6.2.2 Breeding Methods Adopted for Robusta Coffee Improvement

Mass Selection

In mass selection strategy, the plants with outstanding vigour, yield and bean quality characters were selected in the base populations and progenies were advanced through open pollinated seed. The varieties, Apoata of Brazil, S.274 of India, Nemaya of Central America were derived by using this strategy.

Bi-clonal and Poly-clonal Gardens

In the second strategy, superior plants which yielded higher than the family mean yields in single plant progenies were selected and used for establishing bi- and poly-clonal gardens depending on the combining ability. The seed mixture of these clones or the mixture of clones was released for commercial cultivation in different countries. The Balehonnur Robusta clones (BR series) of India, SA and BP selections of Indonesia, IF clones of Ivory Coast are some of the varieties developed by using this strategy.

Hybridization

Intra-specific Breeding

In robusta, both intra-specific and inter-specific hybridizations were employed. In intraspecific hybridization strategy, emphasis has been given to exploit the available diversity within the species. Initial studies on genetic diversity of *C. canephora* using isozyme profiles (Berthaud 1986) distinguished two major diversity groups, the 'Guinean' group from West Africa (Ivory Coast and Guinea) and 'Congolese' group from Central African countries. Among these two groups, the Congolese coffee types generally showed better agronomic value than Guinean types, and majority of the cultivated canephora populations in the world constitute Congolese genotypes. The Guinean types are limited to Ivory Coast and Guinea, both as wild and as commercial cultivations. Dussert et al. (1999) have carried out extensive studies on the *C. canephora* populations and grouped the wild and cultivated forms of robusta coffee into five diversity groups (A, B, C, D, and E) based on the analysis of RFLP polymorphism. Berthaud (1986) emphasized the importance of Guinean genotypes for *C. canephora* breeding as the most productive clones obtained in Ivory Coast during the 1960s were the hybrids between Congolese and Guinean types. Montagnon et al. (1998) carried out a reciprocal recurrent selection programme among Congolese and Guinean hybrids developed in Ivory Coast. High amount of heterosis for vigour and yield in inter-group hybrids compared to intra-group hybrids was achieved as reported by Leroy et al. (1993). In India, the main limitation for robusta coffee breeding is the non-availability of adequate genetic resources especially the wild genotypes belonging to diversity groups (A, B, C, D) from the centre of diversity.

Inter-specific Breeding

Inter-specific hybridization has also been tried for robusta improvement, beginning in Java during nineteenth century with the discovery of a spontaneous diploid

Fig. 11.15 C×R, a popular robusta variety suitable for high-density planting and known for its intrinsic quality

inter-specific hybrid between *C. canephora* var *ugandae* and *C. congensis* called Congusta or Conuga that proved to be of considerable commercial value (Cramer 1957). The hybrids were fertile and also exhibited several important features, such as good vigour, production, adaptation to sandy soils, tolerance to temporary water logging, and good bean size and cup quality. In India, systematic breeding programme was undertaken between *C. canephora* and *C. congensis* that led to the development of a highly fertile Congensis × Robusta hybrid (C×R). The strategy included the development of F_1 hybrid between *C. congensis* and robusta which was backcrossed to robusta and BC progeny was subjected to mass selection followed by sib-mating that resulted in a highly fertile and compact C×R hybrid variety. The C×R variety shows good vegetative vigour and compact bush stature (Fig. 11.15), thus suitable for planting at closer spacing than other robusta varieties. There was a remarkable improvement for bean size in C×R, besides superior liquor characteristics than that from other robustas.

The distinguishing features of robusta varieties released from CCRI for commercial cultivation are detailed in Table 11.2.

11.7 Current Focus of Coffee Improvement in India

The current focus of coffee improvement programmes in India is aimed at the development of high-yielding hybrids coupled with durable host resistance in arabica for maximizing productivity and evolving drought-tolerant robusta genotypes, to cope with the changing climate and demands of consumer markets, more efficiently. Systematic breeding programme is being pursued with emphasis on development of heterotic F_1 hybrids in both arabica and robusta using genetically distant genotypes and male sterile lines identified in wild gene pool/land races from Ethiopia. In breeding for resistance, pyramiding of the resistance genes in selected arabica cultivars by using marker-assisted selection for durability of rust resistance search for new sources of tolerance/resistance to white stem borer (*Xylotrechus quadripes*) in coffee gene pool for breeding purposes, and integration of genomic information for improving the efficiency of conventional breeding are some of the major priorities of genetic improvement programmes of coffee in India.

11.8 Conclusion

Despite limited genetic diversity in arabica germplasm, remarkable successes have been achieved with respect to the development of new varieties that contributed significantly for sustainable growth of the world coffee industry. Further, availability of suitable varieties prompted the farmers to adopt innovative farming approaches such as high-density planting and new planting designs that suit for mechanization of farm operations. These developments supported by the technical advances in coffee agronomy facilitated the coffee farmers' world over towards realizing high yields and improved product quality both in robusta and in arabica that enabled to maximize the economic returns. However, considering the growing importance of coffee as a popular drink in the new emerging markets and rising growth of coffee consumption in domestic sector of many producing countries, there exists a great scope for assured market of the produce and to realize better prices. Coffee being an important crop for the developing countries with significant contribution to their GDP, efforts to achieve sustainable coffee economy would have profound socioeconomic implications on livelihood of millions of people worldwide including India. Further, the changes in climatic conditions such as rise in temperatures and erratic rainfall patterns are posing new challenges for coffee cultivation. In this realm, there is a continuous need for development and deployment of genetically improved coffee varieties to meet the ever-changing demands of markets and environment. However, the available opportunities are limited for conventional genetic improvement of coffee. The problem is more pronounced in arabica coffee because of its tetraploid status and very narrow genetic base, making genetic improvement more difficult to be realized.

The recent advances in DNA markers/technologies and coffee genomics provide new possibilities to overcome some of the limitations of conventional coffee improvement with requisite speed and efficiency. Thus, the situation warrants serious efforts to integrate the newer molecular genomic tools/technologies for genetic research and crop improvement of coffee. DNA analysis of existing coffee germplasm resources to date have indicated very low genetic diversity in cultivated gene pool, and thus enriching the genetic resources through exchange programmes and exploitation of wild arabica genotypes should be the important components of coffee improvement efforts. Similarly, considering the quick success achieved in resistance breeding programmes of arabica through introgressive breeding involving spontaneous tetraploid inter-specific hybrids between arabica and diploid species, search for new spontaneous hybrids assumes significance and should be aggressively pursued. Furthermore, exploitation of secondary gene pool consisting of diploid species for desired variability/gene(s) remains an important approach as inter-species gene flow is widespread in coffee. Thus, enriching the genetic resources through new explorations, integration of conventional breeding approaches with advanced genotyping methods based on DNA markers, marker-assisted selection breeding and genomics are some of the available options that have great promise for accelerated genetic improvement of coffee in India and elsewhere.

Acknowledgements The authors wish to express deep sense of tribute to all the coffee breeders who contributed immensely for collection, maintenance and exploitation of the coffee genetic resources and varietal improvement in India, starting from erstwhile Mysore Coffee Experimental Station to the present CCRI, facing several difficulties both at personal and at professional levels by working in remote plantation regions. The authors also gratefully acknowledge the administrative support from the authorities of Centre for Cellular and Molecular Biology (CSIR-CCMB), Hyderabad, and financial support (project grant: BT/PR/6292/Agr/16/575/2005) from the Department of Biotechnology, Govt. of India, New Delhi, for the project on molecular characterization of coffee germplasm of India.

References

Aga E, Bryngelsson T, Bekele E, Salomon B (2003) Genetic diversity of forest arabica coffee (*Coffea arabica* L.) in Ethiopia as revealed by random amplified polymorphic DNA (RAPD) analysis. Hereditas 138:36–38

Aggarwal RK (2005) DNA fingerprinting of coffee germplasm for conservation and better management and development of molecular map of coffee: molecular characterization and linkage analysis. Completion report, submitted to Department of Biotechnology, New Delhi, India, Project No. BT/PR1302/AGR/08/095/98, p 48

Aggarwal RK, Rajkumar R, Rajendrakumar P, Hendre PS, Baruah A, Phanindranath R, Annapurna,V Prakash NS, Santaram A, Srinivasan CS, Singh L (2004) Fingerprinting of Indian coffee selections and development of reference DNA polymorphism panels for creating molecular IDs for variety identification. In: Proceedings of the 20th international science colloquium coffee. ASIC, Bangalore and Paris, pp 751–755

Alvarado GA (2005) Evolution of *Hemileia vastatrix* virulence in Colombia. In: Zambolim L, Zambolim EM, Várzea VMP (eds) Durable resistance to coffee leaf rust. Universidade Federal de Viçosa, Viçosa, pp 99–115

Anonymous (2014a) Database of Coffee, September 2014. Coffee Board of India, Bangalore
Anonymous (2014b) Coffee guide. Coffee Board of India, Bangalore
Anthony F (1992) Les ressources génétiques des caféiers: collecte, gestion d'un conservatoire et evaluation de la diversité génétique. Collection TDM (81), ORSTOM (ed), Paris
Anthony F, Bertrand B, Quiros O, Wilches A, Lashermes P, Berthaud J, Charrier A (2001) Genetic diversity of wild coffee (*Coffea arabica* L.) using molecular markers. Euphytica 118:53–65
Anthony F, Combes MC, Astorga C, Bertrand B, Graziosi G, Lashermes P (2002a) The origin of cultivated *Coffea arabica* L. varieties revealed by AFLP and SSR markers. Theor Appl Genet 104:894–900
Anthony F, Quiros O, Topart P, Bertrand B, Lasherms P (2002b) Detection by simple sequence repeat markers of introgression from *Coffea canephora* in *Coffea arabica* cultivars. Plant Breed 121:542–544
Berthaud J (1986) Les resources g_en_etiques pour l' am_elioration des caf_eiers africains diploides. ORSTOM, collection Travaux et documents. Montpellier, p 379
Berthaud J, Charrier A (1988) Genetic resources of *Coffea*. In: Clarke RJ, Macrae R (eds) Coffee. Agronomy, vol 4. Elsevier Applied Science, London, pp 1–42
Bertrand B, Etienne H, Cilas C, Charrier A, Baradat P (2005) *Coffea arabica* hybrid performance for yield, fertility and bean weight. Euphytica 141:255–262
Bertrand B, Alpizar E, Llara L, Santacreo R, Hidalgo M, Quijano JM, Charmetant P, Montagnon C, Georget F, Etienne H (2010) Performance of *Coffea arabica* F1Hybrids in comparison with american pure line varieties. In: Proceedings of the 23rd international conference on coffee science. ASIC, Bali, pp 484–495
Bettencourt AJ, Rodriguez CJ Jr (1988) Principles and practice of coffee breeding for resistance to rust and other diseases. In: Clarke RJ, Macrae R (eds) coffee, vol 4. Agronomy, Elseviers Applied Science, pp 199–235
Bridson DM (1982) Studies in coffee and Psilanthus (Rubiaceae subfam. Cinchonoideae) for part 2 of flora of Tropical East Arica: Rubiaceae. Kew Bull 36:817–859
Carvalho A (1988) Principles and practice of coffee plant breeding for productivity and quality factors: *Coffea arabica*. In: Clarke RJ, Macrae R (eds) coffee, vol 4. AgronomyElsevier Applied Science, London, pp 129–165
Carvalho A, Ferwerda FP, Frahm-Leviveld JA, Medina PM, Mendes AJT, Monaco LC (1969) Coffee. In: Ferwerda FP, Wit F (eds) Outlines of perennial crop breeding in the tropics. Wageningen, Veenman & Zonen NV, pp 189–241
Charrier A, Berthaud J (1985) Botanical classification of coffee. In: Clifford MN, Wilson KC (eds) *Coffee:* botany, biochemistry and production of beans and beverage. Croom Helm, London, pp 13–47
Charrier A, Eskes AB (2004) Botany and genetics of coffee. In: Wintgens JN (ed) *Coffee:* growing, processing, sustainable production, a guidebook for growers, processors, traders, and researchers. Wiley-VCH, Weinhelm, pp 25–56
Chevalier A (1947) Les Vrais Les faux cafeirs Nomcnclature ct systematique Les Cafeirs du globe. Encycl biologique no 28(3):117–228 (P. Lechevalier, Paris)
Cramer PJS (1957) In: Wellman FL (ed) Review of literature of coffee research in Indonesia. Turrialba, SIC International American Institute of Agricultural Sciences
Cristancho MA, Chaparro AP, Cortina HA, Gaitan AL (2004) Genetic variability of*C. arabica* accessions from Ethiopia evaluated with RAPDs. In: Proceedings of the 20th international scientific colloquium on *Coffee*, Bangalore and Paris, pp 672–675
Cros J, Combes M-C, Trouslot P, Anthony F, Hamon S, Charrier A, Lashermes P (1998) Phylogenetic analysis of chloroplast DNA variation in *Coffea* L. Mol Phylogenet Evol 9:109–117
Dussert S, Lashermes P, Anthony F, Montagnon C, Trouslot P, Combes MC (1999) Le caféier, *Coffea canephora*. In: Hamon P, Seguin M, Perrier X, Glaszmann JC (eds) Diversité génétique des plantes tropicales cultivées. CIRAD, Montpellier, pp 175–194

Etienne H, Anthony F, Dussert S, Fernandez D, Lashermes P, Bertrand B (2002) Biotechnological applications for the improvement of coffee (*Coffea arabica* L). In Vitro Cell Dev Biol-Plant 38:129–138

Fazuoli LC, Oliveira ACB, Toma-Braguini M, Silvarolla MB (2005) Identification and use of sources of durable resistance to coffee leaf rust at the IAC. In: Zambolim L et al (eds) Durable resistance to coffee leaf rust. UFV, Viçosa, pp 137–185

Georget F, Alpizar E, Courtel P, Hidalgo JM, Dechamp E, Poncon C, Etienne H, Bertrand B (2014) Arabica F_1 Hybrid seed production based on genetic male- sterility. In: Proceedings of the 25th International conference on coffee science (ASIC), Colombia, Sept 8–13

Guerreiro-Filho O, Silvarolla MB, Eskes AB (1999) Expression and mode of inheritance of resistance to leaf miner. Euphytica 105:7–15

Hendre P, Aggarwal RK (2007) DNA markers: development and application for genetic improvement of coffee. In: Varshney RK, Tuberosa R (eds) Genomics assisted crop improvement, vol 2. Springer, Berlin, Genomics Applications in Crops, pp 399–434

Herrera JC, Cortina HA, Anthony F, Prakash NS, Lashermes P, Gaitan AL, Cristancho MA, Acuna R, Lima DR (2012) Coffee (*Coffea* spp.). In: Singh RJ (ed) Genetic resources, chromosome engineering and crop improvement—medicinal crops, vol 6. CRS press, pp 589–640

Krug CA, Carvalho A (1951) The genetics of coffee. Adv Genet 4:127–158

Lashermes P, Combes M-C, Trouslot P, Anthony F, Charrier A (1996a) Molecular analysis of the origin and genetic diversity of *Coffea arabica* L. Implications for coffee improvement. In: Proceedings of EUCARPIA meeting on tropical plants, Montpellier, pp 23–29

Lashermes P, Cros J, Combes M-C, Trouslot P, Anthony F, Hamon S, Charrier A (1996b) Inheritance and restriction fragment length polymorphism of chloroplast DNA in the genus *Coffea* L. Theor Appl Genet 93:626–632

Lashermes P, Combes M-C, Trouslot P, Charrier A (1997) Phylogenetic relationship of coffee-tree species (*Coffea* L.) as inferred from ITS sequences of nuclear ribosomal DNA. Theor Appl Genet 94:947–955

Lashermes P, Combes M-C, Robert J, Trouslot P, D'Hont A, Anthony F, Charrier A (1999) Molecular characterisation and origin of the *Coffea arabica* L. genome. Mol Gen Genet 261:159–266

Lashermes P, Andrade AC, Etienne H (2008) Genomics of coffee, one of the world's largest traded commodities. In: Moore PH, Ming R (eds) Genomics of tropical crop plants. Springer, New York, pp 203–225

Leroy JF (1980) Les grandes lignées de Caféiers. Association Scientifique Internationale du Café (ASIC) 9th Colloque, pp 473–477

Leroy T, Montagnon C, Eskes EB, Charrier A (1993) Reciprocal recurrent selection applied to *Coffea canephora* Pierre. Characterisation and evaluation of breeding populations and value of intergroup hybrids. Euphytica 67:113–125

Louarn J (1993) La fertilitédes hybrides intrspécifique setles relations génomiques entreles caféiers diplod' origines africaines (Genre *Coffea* L. sous-genre *Coffea*). Thèse del. Université Paris XI, Orsay

Matiello JB, Almeida SR, Carvalho CHS (2005) Resistant cultivars to coffee leaf rust. In: Zambolim L, Zambolim E, Várzea VMP (eds) Durable resistance to coffee leaf rust. Universidade Federal de Viçosa, Brasil, pp 443–450

Maurin O, Davis AP, Chester M, Mvungi EF, Jaufeerally-Fakim Y, Fay MF (2007) Towards a phylogeny for *Coffea* (Rubiaceae): identifying well supported lineages based on nuclear and plastid DNA sequences. Ann Bot 100:1565–1583

Meyer FG, Fernie LM, Narasimhaswamy L, Monaco LC, Greathead DJ (1968) FAO coffee mission to Ethiopia. Food and Agricultural organisation of the United Nations, Rome

Montagnon C, Leroy T, Yapo A (1992) Etude complémentaire de la diversité génotypique et phénotypique des caféiers de l' espéce *C. canephora* en collection en Cote d' Ivorie. In: Proceedings of the 14th International science colloquim on coffee. ASIC, Paris, pp 444–450

Montagnon C, Leroy T, Eskes EB (1998) Amelioration varietale de Coffea canephora. 2. Les programmes de selection et leurs resultas. Plantation, Recherche, Developpement 5:89–98

Montagnon C, Marraccini P, Bertrand B (2012) Breeding for coffee quality. In: Oberthür T, Läderach P, Cock James H (eds), Specialty coffee: managing quality. IPNI [Southeast Asia], Penang, pp 89–117

Narasimhaswamy RL, Vishweswara S (1963) A note on the occurrence of three species of coffee indigenous to India. Coffee 5(19):65–71

Pereira AA, Sakiyama NS, Zambolim L, Moura WM, Maciel Zambolim E, Caixeta ET (2005) Identification and use of sources of durable resistance to coffee leaf rust in the UFV/EPAMIG breeding program. In: Zambolim L, Maciel-Zambolim E, Várzea VMP (eds) Durable resistance to coffee leaf rust. UFV, Viçosa, pp 215–232

Prakash NS, Combes M-C, Somanna N, Lashermes P (2002) AFLP analysis of introgression in coffee cultivars (*Coffea arabica* L.) derived from a natural interspecific hybrid. Euphytica 124:265–271

Prakash NS, Combes MC, Dussert S, Naveen KS, Lashermes P (2005a) Analysis of genetic diversity in Indian robusta coffee gene pool (*Coffea canephora*) in comparison with a representative core collection using SSRs and AFLPs. Genet Resour Crop Evol 52:333–343

Prakash NS, Ganesh D, Bhat SS (2005b) Population dynamics of coffee leaf rust (*Hemileia vastatrix*) and recent advances in India. In: Zambolim L, Zambolim E, Várzea VMP (eds) Durable resistance to coffee leaf rust. Universidade Federal de Viçosa, Brasil, pp 411–442

Reddy B (2009) Domestic coffee consumption in India—recent trends. Indian Coffee LXXIII (11):4–9

Sera T, Ruas PM, Ruas CF, Diniz LEC, Carvalho VP, Rampim L (2003) Genetic polymorphism among 14 elite *Coffea arabica* L. cultivars using RAPD markers associated with restriction digestion. Genet Mol Biol 26:59–64

Sera T, Sera GH, Ito DS, Doi DS (2005) Coffee breeding for durable resistant to leaf rust disease at Instituto Agronomico do Parana. In: Zambolim L, Zambolim E, Várzea VMP (eds) Durable resistance to coffee leaf rust. Universidade Federal de Vic¸osa, Brasil, pp 187–214

Srinivasan KH, Narasimhaswamy RL (1975) A review of coffee breeding work done at the Government coffee experiment station, Balehonnur. Indian Coffee 34:311–321

Steiger DL, Nagai C, Moore PH, Morden CW, Osgood RV, Ming R (2002) AFLP analysis of genetic diversity within and among Coffea arabica cultivars. Theor Appl Genet 105:209–215

Sybenga J (1960) Genetics and cytology of coffee: a literature review. Biblio Genet 19:217–316

Van der Vossen HAM (1985) Coffee selection and breeding. In: Clifford MN, Wilson KC (eds) Coffee: botany, biochemistry and production of beans and beverage. Croom Helm, London, pp 48–96

Van der Vossen HAM (2001) Agronomy I: coffee breeding practices. In: Clarke RJ, Vitzthum OG (eds) Coffee recent developments. Blackwell Science Ltd, London, pp 184–201

Venkatachalam L (2005) Coffee: need to achieve market balance. In: The Hindu Survey of Indian Agriculture, pp 113–115

Wrigley G (1988) Coffee. Longman Scientific and Technical, Essex

Zezlina S, Soranzio M, Rovelli P, Krieger MA, Sodhal MR, Graziosi G (1999) Molecular characterization of the cultivar Bourbon LC. In: Proceedings of the 18th International science colloquim on coffee. ASIC, Paris, pp 314–29

Chapter 12
Introgression and Exploitation of Biotic Stress Tolerance from Related Wild Species in Wheat Cultivars

Parveen Chhuneja, Satinder Kaur and H.S. Dhaliwal

Abstract Bread wheat is one of the three most important cereal crops which has major role in feeding the population globally. Biotic stresses, mainly the fungal diseases, pose major constraint to wheat production. To combat against these diseases, continuous efforts have been made to mine genes from wide variety of sources including primary, secondary, and tertiary gene pools of cultivated wheat which are rich sources of genes against different biotic stresses. More than 100 resistance genes against leaf rust, stripe rust, stem rust, and powdery mildew have been identified from these gene pools and successfully transferred to cultivated wheat. The transfers from primary gene pool are achieved through homologous pairing while transfer from secondary and territory gene pool requires special chromosome engineering techniques for affecting transfers through induced homoeologous pairing or translocations. The introgressions have been reported as small cryptic alien segments or complete chromosome arms or chromosomes such as chromosome addition and substitution lines. Molecular cytogenetic techniques such as genomic in situ hybridization (GISH) have proved to be a highly efficient technique to directly and precisely detect the alien segments in wheat while molecular marker technologies now combined with next-generation sequencing techniques have facilitated the mapping as well as marker-based mobilization of alien genes to cultivated wheat background. Present review gives a brief description of the contributions of different gene pools of wheat toward the biotic stress resistance, methodologies of gene transfer, characterization of these transfers, and use of the molecular marker technologies for precisely mapping the alien genes for resistance to various biotic stresses in wheat.

P. Chhuneja (✉) · S. Kaur
School of Agricultural Biotechnology, Punjab Agricultural University, Ludhiana, India
e-mail: pchhuneja@pau.edu

S. Kaur
e-mail: satinder.biotech@pau.edu

H.S. Dhaliwal
Eternal University, Baru Sahib, Himachal Pradesh, India
e-mail: hsdhaliwal07@gmail.com

V.R. Rajpal et al. (eds.), *Molecular Breeding for Sustainable Crop Improvement*, Sustainable Development and Biodiversity 11,
DOI 10.1007/978-3-319-27090-6_12

Keywords Alien introgression · *Aegilops* · Disease resistance · *Triticum* · Molecular characterization

12.1 Introduction

Wheat is one of the most important staple foods of the world, occupying 17 % of the total crop acreage worldwide, feeding nearly half of the world population. A steadily growing population and limited areas available for growing crops make it necessary to secure and extend yield potential. Exploitation of the basic genetic principles has resulted in the development of high-yielding varieties and this productivity is maintaining its rate of progress. Genetic manipulation of dwarfing genes coupled with improved production technology led to the green revolution during the mid-1960s. This led to quantum jump in yield of two major cereals: wheat and rice. Since then increase in yield has been consistent but slow. Global population has doubled during last 45 years and expected to reach 9 billion by 2050. Feeding a population of 9 billion people would require raising overall food production by 70 %. Production in the developing countries would need to be almost double. Quantity of food produced per capita has been declining for last more than 20 years estimated based on available cereal grains, which make up about 80 % of the world's food supply. The present rate of increase in the production of three major cereals may not keep pace with the growing world population.

The population of India will be 1.4 billion by 2020 and will need ~109 million tons (MT) of wheat to meet its food demands. Wheat production in India has shown an upward trend for the last five years producing a record of 95.91 MT during 2013–2014 harvest (http://agricoop.nic.in/imagedefault/trade/wheatnew.pdf). However, most of the existing wheat varieties, which were released more than a decade ago, are showing signs of fatigue and have succumbed to the new races of the stripe rust. The future techniques will require newer wheat-breeding strategies including quicker and reliable selection methods to have designer plants combining high yield and disease resistance. Breeding for resistance against diseases is an important objective of wheat-breeding programs globally. It is also widely recognized that in the absence of diverse genetic input, the breeding approach may not prove fruitful. The genetic variability for resistance to major diseases, viz., yellow rust, leaf rust, stem rust, Karnal bunt, powdery mildew and leaf blight within bread wheat germplasm needs to be supplemented with identification and mobilization of new genes from untapped germplasm collections. Wild relatives of wheat provide a rich reservoir of genes for resistance to various wheat diseases and can provide a valuable source of genetic variation for the improvement of biotic stress tolerance in cultivated wheat (Sharma and Gill 1983; Jiang et al. 1994; Friebe et al. 1996; Singh et al. 1998). A thorough knowledge of phylogenetic relationships of cultivated wheats with its wild progenitors and related species is absolutely essential for the successful introgression and exploitation of useful variability in the wheat germplasm.

12.2 Major Diseases of Wheat

The rust diseases of wheat have historically been one of the major productivity constraints for wheat cultivation globally. Stem (or black) rust caused by *Puccinia graminis* is one of the most significant threats to global wheat production (Singh et al. 2008) with the emergence of Ug99 group of stem rust races. Leaf rust caused by *Puccinia triticina* and stripe rust caused by *Puccinia striiformis* continue to pose a major threat to wheat production over a large area. Leaf rust and stripe rust could affect production on approximately 60 (63 %) and 43 (46 %) m ha, respectively, in Asia, if susceptible cultivars were grown there (Singh et al. 2004a). Powdery mildew, caused by *Blumeria graminis* f. sp. *tritici*, is another important foliar disease of wheat occurring worldwide. It competes for nutrients and reduces the photosynthetic capacity of the leaves. Severe epidemics of this disease often occur in areas with cool and humid climates, causing significant yield losses (Bennett 1984). Breeding and deployment of powdery mildew resistant cultivars is the economical and environmentally friendly method to avoid fungicide applications and reduction in the yield due to diseases. The discovery and utilization of new powdery mildew-resistant genes has been a long-term objective for wheat geneticists and breeders worldwide.

Karnal bunt (KB) of wheat, caused by *Tilletia indica,* was first reported in Karnal, India (Mitra 1931). The disease was soon detected in numerous other regions throughout Northern and Central India. Later, the disease was found to occur in several other countries such as Nepal, Pakistan, Afghanistan, Iran, Iraq, South Africa, Mexico, and USA (Rush et al. 2005). Wheat grains infected by *T. indica* produce trimethylamine and flour from grains with over 3 % bunted kernels imparts an off-color and unpleasant odour (Mehdi et al. 1973). The disease has become important worldwide due to the strict international quarantine measures imposed by a number of countries (Rush et al. 2005). The pathogen is soil, seed, and airborne in nature and hence difficult to control once introduced and established in an area. The host genetic resistance is the most effective, economical, and eco-friendly method of KB management. However, development of KB resistant varieties is difficult due to limited variability for KB resistance in hexaploid wheat (Dhaliwal et al. 1993), quantitative nature of inheritance, and considerable influence of environment on screening for disease resistance (Dhaliwal and Singh 1997). Also, our knowledge of genetics of this host–pathogen system is limited. Inheritance studies have indicated that KB resistance is governed by two or more genes which act additively (Morgunov et al. 1994; Fuentes-Davila et al. 1995; Singh et al. 1995; Villareal et al. 1995; Singh et al. 1999).

12.3 Phylogeny of Polyploid Wheats

Hexaploid wheat originated in two steps of natural hybridization and chromosome doubling, thus comprising genomes of three diploid species (Fig. 12.1). One wild diploid *Triticum* species and two species of the closely related genus *Aegilops* are the wild progenitors of bread wheat. Kihara (1919) and Sax (1922) based on the cytological data on chromosome pairing in interspecific hybrids among species of different ploidy levels indicated that *T. monococcum* and *T. turgidum* have one genome in common while *T. turgidum* and *T. aestivum* share two genomes in common. A diploid wheat species *T. urartu* was found to be a distinct biological species, reproductively isolated from *T. monococcum* (Johnson and Dhaliwal 1976) and on the basis of molecular data it was found that *T. urartu* and not *T. monococcum* contributed the A genome to polyploid wheats (Dvorak et al. 1993). There has been a lot of controversy regarding *Ae. speltoides* as the donor of the B and G genomes to polyploid wheats (Sarkar and Stebbins 1956; Riley et al. 1958). Evidences from diverse sources indicate that *Ae. speltoides* contributed the G genome to *timopheevii* wheats, whereas *Ae. speltoides* with a different cytotype or a species closely related to it contributed the B genome to *turgidum* wheats (Jiang and Gill 1994).

McFaden and Sears (1946) and Kihara (1944) unequivocally demonstrated that *Ae. tauschii* was the D genome donor of bread wheat which arose from a hybridization between *T. turgidum* and *Ae. tauschii* var. *strangulata* about 7000 years ago (Dvorak et al. 1998). The tetraploid parent probably was cultivated emmer, ssp. *dicoccum* because the range of wild progenitor, ssp. *dicoccoides* does not overlap with that of *Ae. tauschii*. Vast cytological, molecular cytogenetic, and

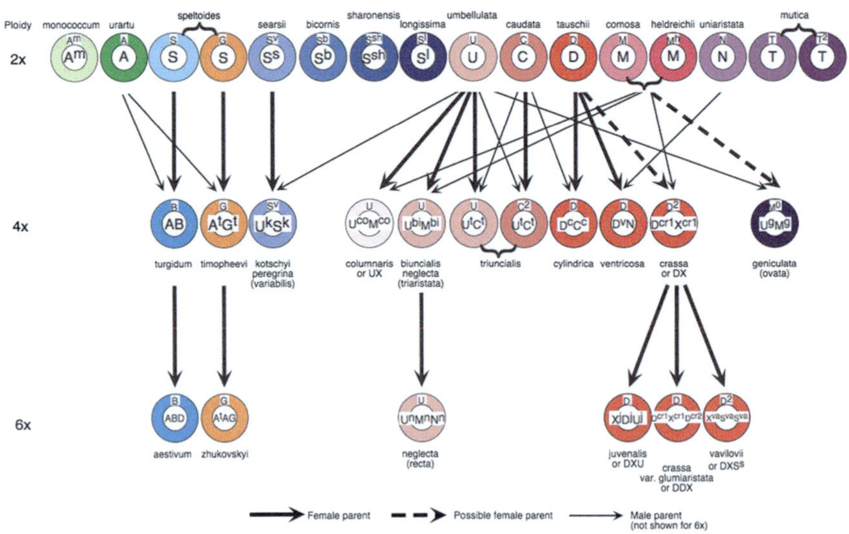

Fig. 12.1 Phylogeny of *Triticum* and *Aegilops* species (*Source* BS Gill, Wheat Genetic and Genomic Resource Centre)

molecular mapping data in Triticeae and related tribes indicate that a very high level of gene content and synteny is maintained among different species thus making it possible to substitute complete chromosome or chromosome segment carrying useful traits from progenitor and non-progenitor genomes into any of the wheat A, B, and D genomes without drastic effects. The ease of transfer of useful variability and its subsequent commercial exploitation would, however, depend on the evolutionary relationship and differentiation between the donor and the recipient genomes.

12.4 Gene Pools of Wheat

The most recent taxonomic status of *Triticum* and *Aegilops* genera to which the cultivated wheats and their progenitors belong as established by van Slageren (1994) is given in Table 12.1 with some modifications. There are three ploidy levels in both the genera with 2n chromosomes 14, 28, 42 and the basic chromosome $x = 7$ in all the species. Other genera of Poaceae such as *Secale*, *Hordeum*, *Dasopyrum, Agropyron, Elymus, Leymus, Elytrigia,* and *Thinopyrum* are also important for introgression of useful variability into cultivated wheats.

On the basis of their genomic constitution, the wild relatives of wheat can be classified into primary, secondary, and tertiary gene pools (Jiang et al. 1994; Friebe et al. 1996). Species belonging to the primary gene pool share homologous genomes with cultivated wheat (Fig. 12.2). This group includes land races of *T. aestivum*, the wild and cultivated forms of *T. turgidum*, and donor species of the A and D genomes of bread wheat, *T. monococcum, T. urartu*, *T. boeoticum,* and *Ae. tauschii*. Gene transfer from these species can be achieved by direct hybridization, backcrossing, and selection (Friebe et al. 1996). No special cytogenetic manipulation except embryo rescue in certain cases is necessary to produce F_1 hybrid (Jiang et al. 1994). Many genes conferring resistance to diseases and insect pests have been transferred using this method and several of them are still being exploited in cultivar improvement (McIntosh et al. 1995a, b).

The secondary gene pool of common wheat includes the polyploid *Triticum* and *Aegilops* species that have at least one genome in common with wheat. Gene transfer from these species by homologous recombination is possible, if the target gene is located on a homologous chromosome. However, if the genes are present in a non-homologous genome, special cytogenetic manipulations are required. These species have contributed several resistance genes that are being used in cultivar development (Jiang et al. 1994).

Species belonging to the tertiary gene pool are more distantly related. Their chromosomes are not homologous to those of wheat. Gene transfer from these species cannot be achieved by homologous recombination, chromosome pairing, and recombination between wheat chromosome and alien chromosomes (Jiang et al. 1994; Friebe et al. 1996). Special cytogenetic techniques are required to ensure compensating transfers with least linkage drag for commercial exploitation of

Table 12.1 Species of genus *Triticum* and *Aegilops* and their genomic constitution

Species	Genome	Synonyms
Triticum		
Diploids		
T.monococcum L. (Einkorn)	A^m	*T. boeoticum*
var. boeoticum		*T. aegilopoides*
var. aegilopoides(wild)		
T. urartu Tumanian ex Gandilyan(wild)	*A*	
Tetraploids		
T. turgidum L.	*AB*	*T. durum*
var durum, (macaroni wheat)		
var. dicoccum (emmer wheat)		*T. dicoccum*
var. polonicum (polish wheat)		*T. polonicum*
var. carthlicum (persian wheat)		*T. carthlicum*
var. dicoccoides (wild emmer)		*T. dicoccoides*
T. timopheevii (cultivated)	A^tG	*T. araraticum*
var. araraticum (wild *timopheevi*)		
Hexaploids		
T. aestivum (common or bread wheat)	*ABD*	*T. vulgare*
var. spelta (spelta or dinkel wheat)		*T. spelta*
var. campactum (club wheat)		*T. campactum*
var. sphaerococcum (Indian dwarf wheat)		*T. sphaerococcum*
var. vavilovii		*T. vavilovii*
var. macha		*T. macha*
T. zhukouskyi	A^tA^mG	
Aegilops		
Diploid		
Ae. speltoides Tausch	*S*	*T. speltoides var. aucheriVar. ligustica*
Ae. longissima Schweinfx Maschl	S^l	*T. longissimum*
Ae. sharonensis Eig	S^{sh}	*T. sharonense*
Ae.searsei Feldman & Kislv ex Harmmer	S^s	*T. searsii*
Ae. bicornis (Forsk) Jaub & spach	S^s	*T. bicorne*
Ae. tauschii (Coss) Schmalh	*D*	*T. tauschii*
		var. strangulata
		var. tauschii
Ae. mutica Boiss	*T*	*T. tripsacoides*
Ae. comosa (Sm.& Sibth)Richter	*M*	*Ae. heldreiohii, T. comosum*
Ae. caudata L.	*C*	*Ae. markgrafii, T. dichasian*
Ae. umbellulata Zhuk	*U*	*T. umbellulatum*
Ae. uniaristata Vis	*N*	*T. uniaristatum*

(continued)

Table 12.1 (continued)

Species	Genome	Synonyms
Tetraploid		
Ae. cylindrica host	D^cC^c	*T. cylindricum*
Ae. ventricosa Tausch	D^VN^V	*T. ventricosum*
Ae. crassa L.	$D^{CL}M^c$	*T. crissum*
Ae. triuncialis L.	UC^t	*T. triunciale*
Ae. geniculata Roth	UM^o	*T. ovatum*
Ae. neglecta	*UM*	*Ae. triaristata*
Ae. columnaris Zhuk	*UM*	*T. columnare*
Ae. biuncialis Vis	*UM*	*Ae. lorentii, T. machrochaetum*
Ae. kotschyi Boiss	US^l	*T. kotschyi*
Ae. peregrina	US^l	*Ae. variabilis*
Hexaploids		
Ae. juvenalis (Thell) eig	*DMU*	*T. juvenile*
Ae. vavilovii (Zhuk) Chennav	*DMS*	*T. syriacum*
Ae. crassa var. *glumiaristata*	*DDM*	*T. crassum(6x)*
Ae. neglecta var. *recta*	*UMN*	*Ae. triaristata(6x)*

Fig. 12.2 Gene pools of wheat

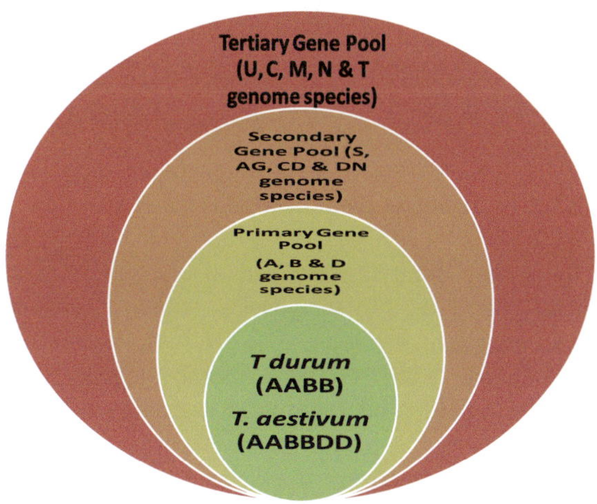

introgressed derivatives. Even though such transfers may include an entire chromosome arm or part of an arm, these have been successfully bred into commercial wheat cultivars because the alien chromosome arm or segment genetically compensates for the missing wheat chromatin.

Though gene transfers from distant relatives is more difficult, but it has been established that the more distant from wheat the relative is, the more likely it is to have genes that are not present in any of the wheat cultivars themselves. Some of the genes may be of great value to wheat growers (Sears 1981).

12.5 Steps and Techniques for Alien Gene Introgression

Various steps for successful hybridization, introgression of useful variability, and characterization of introgressed derivatives are listed below:

12.5.1 Steps of Introgression

i. Synchronization of flowering
ii. Embryo rescue
iii. Overcoming hybrid dysgenesis
iv. Synthetic amphiploid
v. Circumventing gene suppression
vi. Development of alien addition and homoeologous substitutions
vii. Induction of homoeologous pairing and translocations

Introgression of useful variability from related cultivars and wild species into cultivated bread and durum wheats has been reviewed by Cox (1998). In this chapter, we will describe only the techniques for promoting alien gene transfer and characterization of interspecific derivatives.

12.5.2 Techniques for Promoting Alien Gene Transfer

12.5.2.1 Induced Homoeologous Pairing

The pairing affinity among homoeologous chromosomes of the three wheat genomes is suppressed by the genetic activity of the pairing homoeologous gene, *Ph1,* on the long arm of chromosome 5B. In the presence of *Ph1,* the alien chromosomes will not pair with their wheat homoeologues. Therefore, different strategies for promoting pairing between wheat and alien chromosomes must be used.

12.5.2.2 Use of 5B—Deficient Stocks

If an alien species can be crossed directly with bread wheat, plants monosomic for chromosome 5B can be used as female parents in crosses with it. About 75 % of the offspring will be deficient for 5B. Progeny lacking 5B will show considerable homoeologous pairing and desired recombinants may be recovered by backcrossing them with the wheat parent.

12.5.2.3 Crosses with Species Carrying a Suppressor of *Ph1*

An effective method of inducing homoeologous pairing is by crossing wheat with certain genotypes of alien species that inactivate the homoeologous pairing suppressors. *Aegilops speltoides*, for example, is known to suppress the activity of *Ph1* in hybrids with wheat, resulting in high homoeologous pairing (Riley 1960). These hybrids are then backcrossed to wheat and direct transfer of genetic material from *Ae. speltoides* to wheat may be obtained. Riley et al. (1968a) used this technique to transfer stripe rust resistance from *Ae. comosa* (MM) to wheat. Chen et al. (1994) transferred a dominant homoeologous pairing inducer Ph^{I} gene from *Ae. speltoides* into hexaploid wheat cultivar Chinese Spring (CS). Aghaee-Sarbarzeh (2000) has successfully induced homoeologous pairing between alien chromosomes and their wheat homoeologues using Chinese Spring stock with suppressor for *Ph* locus. The availability of this system in *T. aestivum* would allow the exploitation of this system for reducing the linkage drag during the transfer of alien genes.

12.5.2.4 Crosses Involving *Ph* Mutants

A high-pairing mutation involving a small, intercalary deficiency for *Ph1* was produced by Sears (1977) and is designated *ph1b*. Subsequently, another mutation (a terminal deficiency) that conditions an intermediate level of homoeologous pairing was induced. This mutant is designated as *ph2a* and is located on the short arm of 3D. It seems that the level of homoeologous pairing obtained in the absence of *Ph1* is about the highest obtainable. Another mutant induced in the durum wheat cultivar Cappelli has been designated *ph1c* and is being used in promoting homoeologous pairing in intergeneric hybrids. These mutants, when crossed with alien species, may induce wheat–alien chromosome pairing and thus facilitate desired gene transfer into wheat.

12.5.2.5 Chemical Agents

Knight et al. (2010) showed in detached wheat tillers of wheat–rye hybrids that okadaic acid (OA), a drug known to induce chromosome condensation, can be introduced into wheat interspecific hybrids prior to meiosis to induce homoeologous chromosome pairing. This pairing occurs in the presence of the *Ph1* locus, which usually suppresses pairing of related chromosomes through delayed condensation. The timing of chromosome condensation during the onset of meiosis is an important factor in controlling chromosome pairing. This indicated that with the correct concentration of OA, chromosome pairing can be induced in wheat–rye interspecific hybrid plants even in the presence of the *Ph1* locus, mimicking the pairing observed in the absence of *Ph1*. Thus, this approach of treating detached wheat tillers with a drug can in principal provide a powerful method to enhance genetic exchange between chromosomes once standardized for in vivo system.

12.5.2.6 Radiation-Induced Translocations

Sears (1956) pioneered a method using ionizing radiation to produce translocation between homoeologous chromosomes. He developed synthetic amphiploids by crossing *Ae. umbellulata* (UU) with *T. dicoccoides* (AABB) which was further crossed with Chinese Spring wheat to transfer leaf rust resistance from *Ae. umbellulata* to bread wheat. After two backcrosses of the F_1 plants (pollen parent) to Chinese Spring leaf rust-resistant plants with 21 bivalents plus an added *Ae. umbellulata* (UU) chromosome were obtained, including one carrying an isochromosome for the long arm of the *umbellulata* chromosome, the arm carries the resistance gene. Plants carrying the isochromosome were irradiated with X-rays when the first spikes were entering meiosis. Pollen from these plants was used to pollinate Chinese Spring. Since the U chromosome was deleterious and showed low transmission through the pollen, it was expected that most resistant progeny would carry a translocation. Of the 6091 plants, 132 were rust-resistant and 40 proved to carry translocation. One line was named Transfer and its gene for resistance, *Lr9*, was used in several cultivars in the USA.

12.5.2.7 Spontaneous Translocations

During meiosis in wide crosses, chromosomes often occur as univalents which may divide incorrectly to give rise to telocentrics; reunions are known to occur between different telocentrics resulting in novel chromosomes with desired genes. Based on this phenomenon, Sears (1972) proposed a method of transferring genes involving the exchange of whole chromosome arms. In a wheat–alien hybrid having wheat as well as an alien monosome, both univalents may occasionally divide incorrectly in the same sporocyte and thus a wheat chromosome arm may rejoin an alien chromosome arm. Although, frequency of such unions is low, Zeller (1973) could produce two wheat–rye exchanges in crosses between Chinese Spring monosomics and Chinese Spring–Rye addition lines. Several European cultivars carry a spontaneous translocation between chromosome 1B of wheat and 1R of rye. The Veery lines developed in the CIMMYT program in Mexico also carry a 1BL/1RS translocation derived from the winter wheat, Kavkaz. They have proven to be very high yielding as well as having good resistance to several diseases.

12.5.2.8 Introgression via Direct Hybrid

Gene transfer by direct backcrossing is possible in those combinations where there are one or more genomes homologous between recipient and donor species (Table 12.2). Hybrids between durum or common wheat and related donor species are generally male sterile which set seed when backcrossed. Gerechter-Amitai and

Table 12.2 Gene transfer into wheat by direct backcrossing

Recipient parent	Donor parent	Trait	References
T. durum	*T. boeoticum*	Stem rust	Gerechter-Amitai et al. (1971)
T. aestivum	*Ae. speltoides*	Leaf rust	Dvorak (1977)
T. aestivum	*T. monococcum*	Leaf rust	Cox et al. (1994)
T. aestivum	*Ae. squarrosa*	Leaf rust	Cox et al. (1994)
T. durum	*T. monococcum*	Hessian fly	Cox and Hatchett (1994)
		Herbicide tolerance	Gill et al. (1987)
T. durum	*T. timopheevii*	Glume blotch	Ma et al. (1995)
		Stem rust	
		Powdery mildew	
T. aestivum	*T. araraticum*	Powdery mildew	Dhaliwal et al. (2002)
		Stripe rust	
		Leaf rust	
T. aestivum	*Ae. triuncialis*	Leaf rust, powdery mildew	Harjit Singh et al. (2000)
		Cereal cyst nematode	
T. aestivum	*Ae. ovata*	Leaf rust, stripe rust	Dhaliwal et al. (2002)

Grama (1974) transferred stem rust-resistant gene from *T. monococcum spp. aegilopoides* into *T. durum* by simple backcrossing. Sterile triploid hybrid (A^mAB) produces viable female gamete with only 14 chromosomes (A^m/AB) and on backcrossing with recurrent durum parents as male, fertile durum derivatives are recovered in BC_1 generation with complete recovery of B genome and recombination and assortment of A^m/A. During gene transfer via direct hybrids, embryos have to be rescued in wider ploidy level differences between the recipient and the donor species such as between *T. aestivum* and *Ae. tauschii* or *T. monococcum*. Moreover, there is very little seed set on backcrossing the sterile F_1 hybrids as the chances of formation of viable female gametes are very rare unless and until there is unreduced gamete formation in certain combinations. It is desirable in certain cases with partial male fertility to use early backcross generation derivatives as male parent for rapid recovery of euploid and elimination of unwanted chromosomes and translocations. Repeated backcrossing and selfing of recovered euploid with alien introgression accompanied with stringent selection may be required for rapid recovery of cultivated background without any linkage drag. At PAU, we have successfully transferred genes for disease resistance and HMW glutenin subunits from several *Triticum* and *Aegilops* species into wheat and durum cultivars via direct hybridization and backcrossing.

12.5.2.9 Gene Transfer via Synthetic Amphiploids

Synthetic amphiploids, being fertile and true breeding, have been used for transfer of genes for disease resistance from diploid to tetraploid or hexaploid wheat (Table 12.3). In majority of cases of gene transfer from diploid species via synthetic amphiploids, *T. durum* has been used as one of the buffering or bridging species for making synthetic amphiploids for their ultimate hybridization with bread wheat cultivars. During development of synthetic amphiploid between *T. durum* and the donor diploid species such as *T. monococcum, Ae. umbellulata, Ae. caudata*, doubling of chromosomes of sterile triploid hybrids is required, whereas no colchicine treatment is needed in *T. durum* × *Ae. tauschii* and *T. durum x Ae. longissima* hybrids as there is a high degree of seed set on selfing due to unreduced female and male gamete formation. It is, however, very important to use *T. durum* parent susceptible to a particular disease or with low expression of a trait that is intended to be transferred from a particular diploid species so that concomitant transfer for the same trait for *T. durum* parent does not get confounded or interfere with the monitoring of genes transferred from the diploid donor species.

Series of *durum–Ae. tauschii* synthetic amphiploids have been developed and very extensively used at CIMMYT, Mexico, for transfer of disease and insect resistance and quality traits from *Ae. tauschii* into bread wheat. The suppression of disease resistance in certain amphiploids due to the presence of gene suppressors in *T. durum* and *T. aestivum* has to be avoided through careful selection of *T. durum* and *T. aestivum* parents. Due to the presence of *Ne1* gene in most of the *T. durum* cultivars and *Ne2* in most of the CIMMYT-derived *T. aestivum* lines, hybrids between wheat and synthetic *durum-tauschii* often end up with hybrid necrosis. To avoid hybrid necrosis, we have developed a WL711 version without *Ne* alleles, which is being exhaustively used for gene transfer. In case of intended gene transfer from a non-progenitor species with non-homologous genome, it will be desirable to

Table 12.3 Gene transfer into common wheat via synthetic amphiploids

Amphiploid	Trait	References
Ae. speltoides–T. monococcum	Leaf rust	Kerber and Dyck (1990)
	Stem rust	
T. durum–T. monococcum	Leaf rust	Valkoun et al. (1986)
	Stripe rust	
	Karnal bunt	Kuraparthy et al. (2001)
		Dhaliwal et al. (2002)
Tetra Canthatch–*Ae. tauschii*	Leaf rust	Kerber and Dyck (1969)
T. durum–Ae. tauschii	Septorial leaf blotch	May and Lagudah (1992)
	Karnal bunt	Villareal et al. (1994a,b)
T. durum–Ae. umbellulata/T aestivum	Leaf rust	Chhuneja et al. (2008b)
	Stripe rust	

cross the amphiploid with Ph^I stock for inducing homoeologous pairing before backcrossing with the elite wheat variety.

Polyploid wheats have been successfully hybridized with several distantly related species and genera of Poaceae including *Aegilops, Agropyron, Leymus, Elymus, Secale, Hordeum, Haynaldia, Thinopyrum, Sorghum, Pennisetum,* and *Zea mays* due to their higher ploidy levels, buffering genomes, and genes controlling crossability which has been extensively reviewed (Sharma and Gill 1983; Jiang et al. 1994).

12.6 Molecular Cytogenetic Characterization of Alien Introgressions

Characterization of a wheat–alien chromosome translocation includes the identification of the translocated chromosome, localization of the break points, and estimation of the amount of the transferred alien chromatin. Molecular cytogenetic techniques such as genomic in situ hybridization (GISH) have proved to be the most efficient techniques to directly and precisely detect the alien segments in wheat. It allows rapid identification of individual chromosomes in situ (on a glass slide). Non-isotopic methods of mapping DNA sequences in situ on chromosomes on a glass slide were used to construct a molecular karyotype of wheat (Rayburn and Gill 1985). These molecular cytogenetic methods of genome analysis have greatly facilitated cytogenetic analysis in wheat and related species, especially the analysis of alien transfers (Friebe et al. 1991, 1996).

A number of wheat–alien translocations conferring resistances to diseases and pests have been successfully characterized (Table 12.4) through strenuous and collaborative efforts at the international level, some of which have been included in an excellent review by Friebe et al. (1996). Eleven of the 58 wheat–alien translocations analyzed by C-banding and GISH were whole arm translocations with break points within the centromere, whereas 45 translocations were terminal. There were only two intercalary translocations with an alien segment inserted into wheat chromosome arm. The majority of the translocations obtained through irradiation were of non-compensating type involving transfers between non-homoeologous chromosome arms, whereas most of the wheat–alien translocations produced by induced homoeologous recombinations were of compensating type with greater agronomic potential.

In situ hybridization was initially developed, independently, by Gall and Pardue (1969) and John et al. (1969). Genomic in situ hybridization (Pinkel et al. 1986) is a special type of fluorescence in situ hybridization, which uses genomic DNA of donor species as a probe in combination with an excess amount of unlabeled blocking DNA, to monitor alien chromatin introgressions. Genomics in situ hybridization using genomic DNA of the donor species as probe offers advantages as compared to other methods as it leads to the ‘painting’ of all alien chromatin

Table 12.4 Characterization of wheat–alien translocations

S.No	Alien species	Alien target genes	Description	Mode of transfer	Type*	Germplasm
1	*Ae. umbellulata*	*Lr9*[a]	T6BS.6BL-6U#1L	Irradiation	C	Transfer T47
2	*Ae. umbellulata*	*Lr9*	T4BL.4BS-6U#1L	Irradiation	N	T41
3	*Ae. speltoides*	*Lr28*[a]	T4AS.4AL-7S#2S	Homoeol. rec.	C	2A/2 M#4/2
4	*Ae. speltoides*	*Lr32*	T2DL-2S#1L.2S#1S	Homoeol. rec.	C	C82.2
5	*Ae. speltoides*	*Lr35/Sr39*	T2B/2S#2	Homoeol. rec.	C	RL5711
6	*Ae. longissima*	*Pm13*	T3BL.3BS-3S1#1S	Homoeol. rec.	C	RIA
7	*Ae. comosa*	*Yr8/Sr34*	*T2DS-2 M#1L.2 M#1S*	Homoeol. rec.	*C*	*2D-2M#3/8*
8	*Ae. geniculata*	*Lr57/Yr40*	*T5DL·5DS-5MgS*	Homoeol. rec.	*C*	*TA5602*
9	*Ae. triuncilais*	*Lr58*	*T2BS·2BL-2tL*	Spontaneous	*C*	*TA5605*
10	*Ae. umbellulata*	*Lt76/Yr70*	*T5DL·5DS-5US*	Homoeol. rec.	*C*	*T393-4*
11	*T. timopheevii*	*Sr36/Pm6*[a]	*T2B/2G#1*	Homoeol. rec.	*C*	*C747*
12	*T. araraticum*	*Sr40*	*T2BL/2G#2S*	Homoeol. rec.	*C*	*RL6087*
13	*S. cereale*	*Pm8/Sr31/Lr26/Yr9*[a]	*Tl1BL.1R#1S*	Spontaneous	*C*	*WGRC14*
14	*S. cereale*	*Lr25/Pm7*	*T4BS.4BL-2R#1L*	Irradiation	*N*	*Transec*
15	*S. cereale*	*H21*	*T2BS.2R*	Tissue culture	*C*	*KS85HF011*
16	*A. elongatum*	*Lr19/Sr25*[a]	*T7DS.7DL-7Ae#1L*	Irradiation	*C*	*Agatha*
17	*A. elongatum*	*Sr24/Lr24*[a]	*T3DS.3DL-3Ae#1L*	Spontaneous	*C*	*Agent*
18	*A. elongatum*	*Sr24/Lr24*	*T1BL.1BS-3Ae#1L*	Irradiation	*N*	*Teewon*
19	*A. elongatum*	*WSMR*	*T4DS.4DL-1Ae#1L*	Irradiation	*N*	*CI15322*
20	*A. intermedium*	*Wsm1*	*T4DL.4Ai#2S*	Irradiation	*C*	*WGRC27*
21	*A. intermedium*	*BYDR*	*T7DS-7Ai#1S.7Ai#1L*	Tissue culture	*C*	*TC/e*

[a]*Commercially exploited*, **C* -Compensating, *N* -Non-compensating

located in the nucleus. GISH has been widely used to investigate the origin of genomes, chromosomes, and parental genomes in hybrids (Schwarzacher et al. 1989) and to analyze derived introgressed lines from interspecific crosses (Murata et al. 1992; Schwarzacher et al. 1992; Taketa et al. 1997). Technical advances in DNA probe labeling, in situ hybridization and microscopy, allow repeated hybridization, mapping, and processing of chromosome images for multiprobe mapping on a single metaphase.

C-banding and GISH patterns detected alien introgression carrying *Lr9*, the first alien-resistant gene transferred from *Ae. umbellulata* into wheat, on chromosome 6B (Friebe et al. 1996). Alien introgressions carrying leaf and stripe rust resistance genes *Lr57* and *Yr40* from *Ae. geniculata* have been characterized by GISH (Kuraparthy et al. 2007a). One of the introgression lines (IL) had an alien translocation covering complete short arm and half of the long arm of chromosome 5D. Another IL had a small translocation spanning 1/4th of the short arm of 5D while the third IL did not show any GISH signal indicating that the alien segment was very small. Schwarzacher et al. (1992) used fluorescent in situ hybridization (FISH) technique to identify alien chromatin from *H. vulgare*, *Th. bessarabicum*, *Leymus muticaulis,* and *S. cereale* in chromosome spreads of wheat. Radiation-induced wheat–rye translocation lines resistant to Hessian fly were analyzed by the total genomic and highly repetitive rye DNA probes (Mukai et al. 1993). FISH analysis revealed the exact locations of the break points and allowed the estimation of the sizes of the transferred rye segments. Wheat–rye 1B-1R translocation has also been characterized by GISH (Heslop-Harrison et al. 1990). Using biotin labeled total genomic DNA of rye as probe for in situ hybridization, the sizes and 1B-1R translocation points in five wheat varieties were determined. All translocation break points were found to be at or near to centromere.

12.7 Molecular Markers for Characterization of Alien Introgressions

Molecular markers are useful tools for assaying genetic variation and provide an efficient means to link phenotypic and genotypic variations (Varshney et al. 2005). Using molecular markers, high-density genetic linkage maps in various crops have been established which in turn provide a basis for marker-assisted selection (MAS) of agronomically useful traits, for the pyramiding of the genes of interest and their isolation by map-based cloning.

Restriction fragment length polymorphism (RFLP) was one of the first DNA marker techniques used to characterize wheat cultivars (Vaccino et al. 1993). The polymcrase chain reaction (PCR) technique facilitated the development of simpler and low-cost molecular markers, called SSR (also called microsatellites, Tautz and

Renz 1984). PCR-based markers, RAPD (Random Amplified Polymorphic DNA), STS (Sequence Tagged Sites), and SSR (Simple Sequence Repeat), etc. have been found to be invaluable tools for the monitoring of introgressed genes from different wild relatives into cultivated species (Autrique et al. 1995; Brown et al. 1996). Autrique et al. (1995) used RFLP markers to mark the resistance genes *Lr9* from *Ae. umbellulata* and *Lr32* from *Ae. tauschii*. RAPD markers were used to identify addition lines of *Ae. searsii* (Diaz-Salazar and Orellana 1995) and *Ae. caudata* (Peil et al. 1998).

Molecular markers are powerful tools for identifying quantitative traits and dissecting these complex traits into Mendelian factors in the form of quantitative trait loci (QTL) as well as for establishing the genomic locations of such genetic loci. Bulked segregant analysis (BSA), which involves pooling of entries at the two extremes for a segregating trait (Michelmore et al. 1991), has been effectively used for identifying molecular markers associated with disease-resistant genes in a number of species.

Various molecular markers have been widely used to tag and map resistance genes in wheat; however, SSRs have emerged as the choice of marker in gene-mapping studies. This type of molecular marker is genome-specific, appears to be evenly distributed over the wheat genome, and shows a higher level of polymorphism compared to any other marker system (Röder et al. 1998).

Microsatellite markers have been developed and incorporated in already-existing RFLP linkage maps in crops such as wheat (Roder et al. 1998). Wheat has more than 3000 SSR markers mapped so far (Song et al. 2005). Molecular markers can be used for alien gene transfers and understanding the mechanism of gene transfer. The size of the smallest translocation with a particular trait can be revealed by molecular mapping using physically and/or genetically mapped markers. Several DNA markers closely linked with rust-resistant genes have also been developed. Such markers ensure selection of a target gene based on the presence of the linked genotype. The success of selection depends on the close genetic association and robustness of a given marker across different genetic backgrounds. The markers found to be closely linked with the rust-resistant genes transferred from wild species are listed in Tables 12.5 and 12.6.

Schachermayr et al. (1994) developed near isogenic lines (NILs) for the *Ae. umbellulata* leaf rust-resistant gene *Lr9* and used two linked RFLP markers (cMWG 684, PSR 546) to locate it on 6BL of wheat. Gold et al. (1999) converted an ISSR (inter-simple-sequence repeat) marker to a SCAR (sequence characterized amplified region) marker linked to the chromosome segment carrying *Lr35* and *Sr39*. Robert et al. (1999) identified one RAPD and one RFLP marker closely linked to the stripe rust-resistant gene *Yr17*. Gupta et al. (2006) reported tagging of leaf rust-resistant gene *Lr19* (7DL) of wheat derived from *Ag. elongatum* using RAPD and microsatellite markers. Sixteen RAPD markers were identified as linked to the alien gene *Lr19*. Feuillet et al. (1995) screened Thatcher NILs for *Lr1* (5DL)

Table 12.5 List of leaf rust and stripe rust resistance genes, transferred from wild progenitor species and tagged with molecular markers

Gene	Source	Chromosome	Marker	References
Leaf rust resistance genes				
Lr63	*T. monococcum*	3AS	SSR	Kolmer et al. (2010)
Lr21	*Ae. tauschii*	1DS	RFLP/RGA-STS	Rowland and Kerber (1974)
				Huang and Gill (2001)
Lr22a	*Ae. tauschii*	2DS	SSR	Hiebert et al. (2007)
Lr32	*Ae. tauschii*	3D	SSR	Thomas et (2010)
Lr39/Lr41	*Ae. tauschii*	2DS	SSR	Raup et al. (2001)
				Singh et al. (2004)
Lr40	*Ae. tauschii*	1DS	SSR	Spielmeyer et al. (2000)
Lr42	*Ae. tauschii*	1DS	SSR	Liu et al. (2013)
Lr43	*Ae. tauschii*	7DS	SSR	Hussien et al. (1997)
Lr28	*Ae. speltoides*	4AL	RAPD/TPSCAR/SSR	Cherukuri et al. (2005); Vikal et al. (2004)
Lr35	*Ae. speltoides*	2B	RAPD/TPSCAR	Seyfarth et al. (1999)
Lr36	*Ae. speltoides*	6BS	–	Gold et al. (1999)
Lr47	*Ae. speltoides*	7AS	CAPS/SSR	Helguera et al. (2003)
Lr51	*Ae. speltoides*	1BL	CAPS	Helguera et al. (2005)
Lr66	*Ae. speltoides*	3A	SCAR	Marais et al. (2009a, b)
Lr53	*T. dicoccoides*	6BS	SSR	Dadkhodaie et al. (2011)
Lr61	*T. turgidum*	6BS	AFLP	Herrera-Fossel et al. (2008)
Lr64	*T. dicoccoides*	6AL	SSR	Kolmer (2008)
Stripe rust resistance genes				
Yr28	*Ae. tauschii*	4DS	SSR	Singh et al. (2000a, b)
Yr15	*T. dicoccoides*	6BS		Sun et al. (1997)
Yr35	*T. dicoccoides*	6BS	SSR	Dadkhodaie et al. (2011)
Stem rust resistance genes				
Sr2	*T. turgidum*	3BS	SSR/STS	Mago et al. (2011)
Sr13	*T. turgidum*	1DL/1RS	SSR	Knott (1962), Simons et al. (2011)
Sr21	*T. monococcum*	2AL	Sequence-based markers	Chen et al. (2015)

(continued)

Table 12.5 (continued)

Gene	Source	Chromosome	Marker	References
Sr22	*T. monococcum*	7AL	SSR	Oslon et al. (2010)
Sr35	*T. monococcum*	3AL	SSR	Saintenac et al. (2013)
Sr33	*Ae. tauschii*	1DL	SSR	Periyannan et al. (2013)
Sr45	*Ae. tauschii*	1DS	EST/SSR/AFLP	Periyannan et al. (2014)
Sr46	*Ae. tauschii*	2DS	–	Yu et al. (2011)
				Singh et al. (2011)
Sr32	*Ae. speltoides*	2AL, 2BL	SSR/EST	Mago et al. (2013a, b)
Sr39	*Ae. speltoides*	2B	SCAR	Gold et al. (2002)
				Mago et al. (2009)
Sr47	*Ae. speltoides*	2BL	SSR	Faris et al. (2008)
Powdery mildew resistance genes				
Pm1b	*T. monococcum*	*7AL*	STS	Hsam et al. (1998)
Pm1c	*T. monococcum*	*7AL*	RFLP/RAPD	Sears and Briggle (1969)
				Hartl et al. (1995)
Pm4d	*T. monococcum*	*2AL*	SSR/STS	Schmolke et al. (2012)
Pm25	*T. monococcum*	*1A*	RAPD	Shi et al. (1998)
Pm2	*Ae. tauschii*	*5DS*	SSR	Qiu et al. (2006)
Pm19	*Ae. tauschii*	*7D*	–	Lutz et al. (1995a, b)
Pm34	*Ae. tauschii*	*5DL*	SSR	Miranda et al. (2006)
Pm35	*Ae. tauschii*	*5DL*	SSR	Miranda et al. (2007)
Pm1d	*T. spelta var duhamelianum*	*7AL*	STS	Hsam et al. (1998)
Pm12	*Ae. speltoides*	*6BS*	RFLP	Jia et al. (1996)
Pm32	*Ae. speltoides*	*1BL*	Monosomic	Hsam et al. (2003)
Pm3 k	*T. dicoccoides*	*1AS*		Yahiaoui et al. (2009)
Pm16	*T. dicoccoides*	*4A*	SSR	Chen et al. (2005)
Pm26	*T. turgidum var. dicoccoides*	*2BS*	RFLP	Rong et al. (2000)
Pm30	*T. dicoccoides*	*5BS*	–	Liu et al. (2002)
Pm31	*T. dicoccoides*	*6AL*	–	Xie et al. (2003)
Pm36	*T. dicoccoides*	*5BL*	EST	Blanco et al. (2008)
Pm41	*T. dicoccoides*	*3BL*	SSR/RFLP	Li et al. (2009)
Pm42	*T. dicoccoides*	*2BS*	SSR/RFLP	Hua et al. (2009a)

(continued)

Table 12.5 (continued)

Gene	Source	Chromosome	Marker	References
Pm5a	*T. dicoccum*	*7BL*	SSR	Law and Wolfe (1966)
Pm49	*T. dicoccum*	*2BS*		Piarulli et al. (2012)
Pm50	*T. dicoccum*	*2AL*		Mohler et al. (2013)

Table 12.6 List of leaf rust and stripe rust resistance genes, transferred from wild non-progenitor species and tagged with molecular markers

Gene	Source	Chromosome	Marker	References
Leaf rust resistance genes				
Lr9	*Ae. umbellulata*	6BL	STS/RFLP/RAPD/SCAR	Schachermayr et al. (1994)
Lr76	*Ae. umbellulata*	5DS	STS	Bansal et al. (2015)
Lr18	*T. timopheevi*	5BL	N-band	Yamamori (1994)
Lr50	*T. timopheevi subsp. armeniacum*	2BL	SSR	Brown-Guerdira et al. (2003)
Lr52	*T. timopheevi subsp. viticulosum*	2A	SSR	Tar et al. (2008)
Lr54	*Ae. kotschyi*	2DL	SCAR	Marais et al. (2005)
Lr59	*Ae. peregrina*	1AL	SSR	Marais et al. (2008, 2010)
Lr57	*Ae. geniculata*	5DS	CAPS	Kuraparthy et al. (2009)
Lr58	*Ae. triuncialis*	2BL	SSR	Kuraparthy et al. (2011)
Lr62	*Ae. neglecta*	6AS	SSR	Marais et al. (2009a, b)
Lr56	*Ae. sharonensis*	6A	SSR	Marais et al. (2010a, b)
Lr25	*Secale cereale*	4BL	RAPD/SSR	Procunier et al. (1995,)
				Singh et al. (2011)
Lr26	*Secale cereale*	1BL	RFLP	Mago et al. (2005a, b)
Lr44	*T. spelta*	1B	SSR	Dyck and Sykes (1994)
Lr37	*Ae. ventricosa*	2AS	SCAR/CAPS	Helguera et al. (2003)
Lr38	*Ag. intermedium*	2AL	SSR	Mebrate et al. (2008)

(continued)

Table 12.6 (continued)

Gene	Source	Chromosome	Marker	References
Lr19	*Ag. elongatum*	7DL	RFLP/STS/RAPD	Prins et al. (2001) Gupta et al. (2006)
Lr24	*Ag. elongatum*	3DL	STS/SCAR	Dedryver et al. (1996)
				Gupta et al. (2006)
Lr29	*Ag. elongatum*	7DS	RAPD/SCAR	Procunier et al. (1995)
Lr55	*Elymus trachycaulis*	1B	Dart	Friebe et al. (2005)
Stripe rust resistance genes				
Yr37	*Ae. kotschyi*	2DL	SCAR	Marais et al. (2005)
Yr38	*Ae. sharonensis*	6AL	SSR	Marais et al. (2010a, b)
Yr40	*Ae. geniculata*	5DS	CAPS	Kuraparthy et al. (2009)
Yr42	*Ae. neglecta*	6AS	SSR	Marais et al. (2009a, b)
Yr8	*Ae. comosa*	2A, 2D	–	Riley et al. (1968a, b)
Yr5	*T. spelta*	2BL	STS	McGrann et al. (2014)
Yr17	*Ae. ventricosa*	2AS	SCAR/CAPS	Robert et al. (1999)
Yr70	*Ae. umbellulata*	5DS	STS	Bansal et al. (2015)
Yr9	*S. cereal*	1BL/1RS	RFLP	Mago et al. (2005a, b)
Yr50	*Th. intermedium*	4BL	SSR	Liu et al. (2013)
Stem rust resistance genes				
Sr34	*Ae. comosa*	2A, 2D		Friebe et al. (1996)
Sr36	*T. timopheevi*	2BS	SSR	Tsilo et al. (2008)
Sr37	*T. timopheevi*	4BL	SSR	Zhang et al. (2012)
Sr40	*T. timopheevi*	2BS	SSR	Wu et al. (2009)
Sr38	*Ae. ventricosa*	2AS		Helguera et al. (2003)
Sr53	*Ae. geniculata*	5DL	RFLP	Liu et al. (2011)
Sr27	*Secale cereal*	3AS		Singh et al. (2011)
Sr31	*Secale cereal*	1BL		Mago et al. (2002)
Sr50	*Secale cereal*	1DL/1RS	–	Anugrawati et al. (2008)
Sr24	*Th. elongatum*	3DL	SSR	Mago et al. (2005a, b)
Sr25	*Th. elongatum*	7DL	STS	Liu et al. (2010)

(continued)

Table 12.6 (continued)

Gene	Source	Chromosome	Marker	References
Sr26	*Th. elongatum*	6AL	STS	Mago et al. (2005a, b)
Sr43	*Th. elongatum*	7DL	SSR/EST	Xu et al. (2009)
Sr44	*Th. intermedium*	7DS		Liu et al. (2013)
Sr52	*Dasypyrum villosum*	6AL	SSR/RFLP	Qi et al. (2011)
Powdery mildew resistance genes				
Pm1d	*T. spelta var duhamelianum*	7AL	AFLP	Hsam et al. (1998)
Pm4b	*T. carthlicum*	2AL	SSR/STS	The et al. (1979)
Pm33	*T. carthlicum*	2BL	SSR/STS	Zhu et al. (2005)
Pm3b	*T. sphaerococcum*	1AS	SSR/RFLP	Yahiaoui et al. (2004)
Pm7	*S. cereale*	4BL	–	Friebe et al. (1994)
Pm20	*S. cereale*	6BS	–	Heun et al. (1990)
Pm8	*S. cereale*	1BL/1RS	STS	Mohler et al. (2001)
Pm17	*S. cereale*	1BL/1RS	RFLP	Mohler et al. (2001)
Pm29	*Ae. ovata*	*7DL*	*RFLP/AFLP*	Zeller et al. (2002)
Pm13	*Ae. longissima*	*3B*	*STS/RFLP/RAPD*	Ceoloni et al. (1992), Cenci et al. (1999)
Pm6	*T. timopheevii*	*2B*	*RFLP*	Jørgensen (1973)
Pm27	*T. timopheevii*	*6B*	*SSR/RFLP*	Jarve et al. (2000)
Pm37	*T. timopheevii*	*7AL*	*SSR*	Perugini et al. (2008)
Pm40	*Th. intermedium*	*7BS*	*SSR*	Luo et al. (2009)
Pm43	*Th. intermedium*	*2DL*	*SSR*	He et al. (2009)
Pm21	*Haynaldia villosum*	*6AS*	*SCAR/RFLP*	Qi et al. (1996)

with 37 RFLP probes and found three to be linked to the gene after testing on F_2 populations between Thatcher and *Lr1*/Thatcher. Helguera et al. (2003) developed PCR primers based on existing RFLP markers for the *Lr37–Yr17–Sr38* gene cluster.

Lr57 and *Yr40*, a leaf rust and stripe rust resistance gene, respectively, introgressed from *Ae. geniculata* have been mapped on chromosome 5DS of wheat using 11 RFLP probes (Kuraparthy et al. 2007a). PCR-based CAPS markers were later developed by Kuraparthy et al. (2009) which are being used for marker-assisted transfer of these genes to other backgrounds. The leaf rust-resistant gene *Lr21* has been located onto chromosome 1D and successfully cloned by Huang et al. (2003) using RFLP and STS markers. The rust-resistant genes *Lr20–Sr15* (7AL) and *Lr47* (7AS) were mapped using STS, SSR, CAPS, and SCAR

Table 12.7 Stripe rust QTLs on different chromosomes

QTL	*Markers*	References
QPst.jic-1BL	*Xgwm259-Xgwm818*	Melichar et al. (2008)
QYr.uga-2AS	*Xbarc124-Xgwm359*	Hao et al. (2011)
QYr.sgi.2B.1	*Xgwm148*	Ramburan et al. (2004)
QYrl.cau-2BS.2	*Xgwm148-Xbarc167*	Guo et al. (2008)
QYrcaas-5AL	*Xwmc410-Xbarc261*	Lan et al. (2010)
QYrtm.pau-2A	*Xwmc407-Xwmc170*	Chhuneja et al. (2008a)
QYrtb.pau-5A	*Xbarc151-Xcfd12*	Chhuneja et al. (2008a)
QYr-tem-5B.2	*Xwmc235-Xgwm604*	Feng et al. (2011)
QYrst.wgp-6BS.2	*Xgwm132-Xgdm113*	Santra et al. (2008)

markers (Purnhauser et al. 2000; Neu et al. 2002; Stepien et al. 2003, Khan et al. 2005). A leaf rust-resistant gene *Lr58* has been transferred from *Ae. triuncialis* L. into common wheat (*Triticum aestivum* L.). Using RFLP markers *XksuF11*, *XksuH16*, and *Xbg123,* the gene was mapped on distal region of chromosome arm 2BL (Kuraparthy et al. 2007b).

Stripe rust-resistant genes in diploid A genome were mapped and transferred to bread wheat. A linkage map with 169 SSR and RFLP loci generated from a set of 93 RILs from a cross involving *T. monococcum* (acc. pau14087) and *T. boeoticum* (acc. pau5088) was used for mapping stripe rust-resistant genes. The QTL-controlling stripe rust resistance in *T. monococcum* was mapped on chromosome 2A (*QYrtm.pau-2A*), whereas the QTL from *T. boeoticum* was mapped on 5A (*QYrtm.pau-5A*). One stripe rust-resistant gene from *T. boeoticum* acc. pau5088 was confirmed to be introgressed in cultivated wheat which was indicated by co-introgression of *T. boeoticum* sequences linked to stripe rust-resistant QTL, *QYrtb.pau-5A* (Chhuneja et al. 2008a). Some of the reported stripe rust QTLs are listed in Table 12.7.

12.8 Progenitor Gene Pool of Wheat—A Source for Disease Resistance

12.8.1 Diploid 'A' Genome Species

Triticum monococcum L., generally known as einkorn wheat, is an ancient diploid A genome wheat that was domesticated about 10,000 years ago in the southwest Turkey in the Karaca Dag mountains (Heun et al. 1997). *T. monococcum ssp. monococcum* L. is domesticated and *T. monococcum* ssp. *aegilopoides* (Link) Thell. *T. boeoticum* Boiss. ($2n = 2x = 14$) is a wild form of *T. monococcum* ssp. *monococcum.* Primary gene pool of wheat, *T. monococcum* s.l., comprising three closely related species *T. monococcum*, *T. boeoticum,* and *T. urartu*, harbors useful variability for many economically important genes, including resistance to diseases, which can be used for hexaploid wheat improvement (Feldman and Sears 1981; Dhaliwal et al. 1993; Hussien et al. 1997; Yao et al. 2007) but they have not been exploited to the level of the D genome or other *Aegilops* species have been exploited. *T. urartu* has been the A genome donor of the most important polyploid wheat species including the durum or macaroni wheat *T. turgidum* (AABB), *T. timopheevii* (AAGG), and common wheat *T. aestivum* (AABBDD). In contrast, *T. monococcum* has only been used for the generation of *T. zhukovskyi* (A^mA^mAAGG) (Dvorak et al. 1993; Dubcovsky et al. 1995). Thus, the A^m genome is under-represented in hexaploid wheat, and the exploitation of genetic diversity in *T. monococcum* and discovery of novel variant alleles may provide opportunities for further wheat genetic improvement. There is advantage of introgressing traits from *T. monococcum* than other wheat relatives as introgressed chromatin from *T. monococcum* readily recombines with *T. aestivum* chromatin thus facilitating the transfer of traits into wheat germplasm with ease. This provides an opportunity to reduce the size of the alien chromatin and to eliminate linkage to unwanted genes.

T. monococcum has high levels of resistance to the wheat leaf rust. Though some reports suggested the low level of genetic variation in *T. monococcum,* Bai et al. (1998) studied 49 *T. monococcum* accessions for leaf rust and all were found to possess the same gene for leaf rust resistance. Anker and Niks (2001) reported most of the *T. monococcum* accessions (84 %) to be resistant, whereas all *T. urartu* accessions were found susceptible to leaf rust. *T. monococcum* and *T. boeoticum* were found to be closely related but their host status for the wheat leaf rust fungus clearly differed. These three diploid wheat species thus differed in their reaction to wheat leaf rust. This is more likely based on a high allele frequency of one or more effective major genes in *T. monococcum* which are absent in *T. boeoticum* hence, *T. monococcum* has almost a non-host status to the wheat leaf rust (Anker et al. 2001).

Several disease resistance genes have been transferred from *T. monococcum, T. boeoticum,* and *T. urartu* to cultivated wheat. A summary of the resistance genes for leaf rust, stripe rust, stem rust and powdery mildew transferred from A genome species and cataloged is summarized in Table 12.5. Valkoun et al. (1986) reported leaf rust resistance in three accessions of *T. boeoticum* and its transfer into hexaploid wheat. A single gene giving partial resistance to leaf rust was introduced from *T. monococcum* to Thatcher isogenic line RL6137 mapped on chromosome 3A. This gene in RL6137 on chromosome 3AS was designated as *Lr63* (McIntosh et al. 2009). Also, Hussien et al. (1997) mapped leaf rust-resistant genes in winter wheat lines derived from *T. boeoticum* on chromosomes 6A, 1A, and 5A.

Rouse and Jin (2011) screened 1061 accessions of *T. monococcum* and 205 accessions of *T. urartu* against five *P. graminis* f. sp. *tritici* races and found 78.7 % of *T. monococcum* and 93.0 % of *T. urartu* to be resistant to race TTKSK, however, only 6.4 % *T. monococcum* accessions were resistant to all the five races. Infection-type patterns of accessions of both species indicated previously uncharacterized genes for resistance to race TTKSK exist in both *T. monococcum* and *T. urartu.* Stem rust resistance genes, namely *Sr21* (2A), *Sr22* (7A) and *Sr35* (3A), have been successfully transferred and mapped in common wheat from *T. monococcum* (Table 12.5; Paull et al. 1994; McIntosh et al. 1984).

Apart from rust resistance, powdery mildew resistance has also been introduced from *T. monococcum*. *Pm1* and its two alleles *Pm1b* and *Pm1c* have been transferred from *T. monococcum* into cultivated wheat (Sears and Briggle 1969; Hsam et al. 1998; McIntosh et al. 2008). *Pm*25 a major gene for powdery mildew resistance has been transferred to common wheat germplasm NC96BGTA5 from wild einkorn accession PI427662 and mapped on chromosome 1A in close association with gene *Pm*3A (Shi et al. 1998). Plamenov et al. (2009) produced amphiploid between two durum cultivars and the *T. monococcum* ssp. *aegilopoides* which showed resistance to powdery mildew at seedling and adult plant stage and to four leaf rust races at seedling stage showing that amphiploids of A genome and durum are valuable resource of fungal resistance. *Pm*4d from *T. monococcum* is another gene located on *Pm*4 locus, which was found in *Tm*27d2 cultivar (Schmolke et al. 2012). Monneveux et al. (2001) reviewed *T. monococcum* as a useful resource of several resistance genes for root rot (Yamaleev et al. 1989), scab (Saur 1991), Septoria *tritici* avenae (Yu and Sun 1995), nodorum (Ma and Hughes 1993), Hessian fly (Bouhssini et al. 1997) and aphids (Pietro et al. 1998).

At Punjab Agricultural University, Ludhiana, India, about 200 accessions of *T. monococcum* and *T. boeoticum* were screened for leaf rust and stripe rust resistance for several years and we found that all the *T. monococcum* accessions, most of the *T. boeoticum* and a few *T. urartu* accessions, were completely resistant to leaf rust. However, a lot of variation was observed for stripe rust resistance. Most of the *T. monococcum* accessions have shown moderate to complete resistance; most of the *T. boeoticum* accessions showed complete resistance and the majority of the *T. urartu* accessions were highly susceptible. The stripe rust data recorded during 2012–2013, 2013–2014, and 2014–2015 crop seasons and leaf rust data for 2013–2014 are summarized in Table 12.8.

A recombinant inbred line (RIL) population developed by crossing *T. boeoticum* acc. pau5088 with *T. monococcum* acc. pau14087 showed segregation for resistance to several diseases including stripe rust, powdery mildew, Karnal bunt, and cereal cyst nematode (Dhaliwal et al. 2003; Singh et al. 2007a). This population was used for generating a linkage map of the diploid A genome of wheat (Singh et al. 2007a, b) consisting of 179 SSR, RFLP, and bin-mapped EST markers (Fig. 12.3). Using this inter-subspecific map, we have mapped genes/QTLs for a number of wheat diseases including stripe rust, cereal cyst nematode, and Karnal bunt. Two QTLs, one each in *T. monococcum* acc. pau14087, and *T. boeoticum* acc. pau5088, were detected for resistance in the RIL population. The QTL in *T. monococcum* mapped on 2A in a 3.6 cM interval between *Xwmc407* and *Xwmc170*, whereas the QTL from *T. boeoticum* mapped on 5A in 8.3 cM interval between *Xbarc151* and *Xcfd12* (Chhuneja et al. 2008a, b, c).

With the objective of transferring these genes into hexaploid wheat, *T. monococcum* acc. pau14087 and one resistant RIL were crossed to hexaploid wheat, using *T. durum* as a bridging species. The F_1 triploid plants were crossed to

Table 12.8 Rust reaction of different accessions of progenitor species of wheat recorded during 2012–2013 and 2013–2014 crop seasons under field conditions at Punjab Agricultural University, Ludhiana, India

Species	Year	Number of accessions			Total accessions
		Highly resistant[a]	Moderately resistant	Susceptible	
Stripe rust					
T. monococcum	2013	52	6	3	61
	2014	46	2	4	52
	2015	58	1	1	60
T. boeoticum	2013	58	6	9	73
	2014	157	3	14	174
	2015	153	4	14	171
T. urartu	2013	–	–	–	–
	2014	15	8	6	29
	2015	8	3	18	29
Ae. tauschii	2013	67	52	84	203
	2014	72	11	198	281
	2015	22	53	208	283
Ae. speltoides	2013	88	24	3	115
	2014	66	24	9	99
	2015	168	5	2	175
T. diccocoides	2013	3	10	15	28
	2014	57	35	81	173
	2015	72	39	45	156
Leaf rust					
T. monococcum	2014	52	0	0	52
T. boeoticum	2014	166	0	8	174
T. urartu	2014	20	4	6	30
Ae. tauschii	2014	51	0	18	69
Ae. speltoides	2014	50	0	2	52
T. diccocoides	2014	122	0	51	173

[a]The accessions with rust scores of 0, TR, and 5MR were categorized as highly resistant and those with 10MR–20MR and ≥10S were categorized as moderately resistant and susceptible, respectively

susceptible hexaploid wheat cvs. WL711 and PBW343. In the F_1 triploid, only those gametes were viable to those have a full complement of A and B genomes (Gill et al. 1986). However, no resistant plants could be recovered in the backcross generations from this cross indicating that the resistance in diploid wheats was either recessive in nature or was being suppressed by the A and/or B genome of *T. durum*. The A genome of diploid wheat was expected to segregate in the F_1 of the cross *T. durum*/Tm14087//WL711 but not the B genome. The B genome of *T. durum*, however, was expected to segregate in the BC_1F_1 generation of the cross *T.*

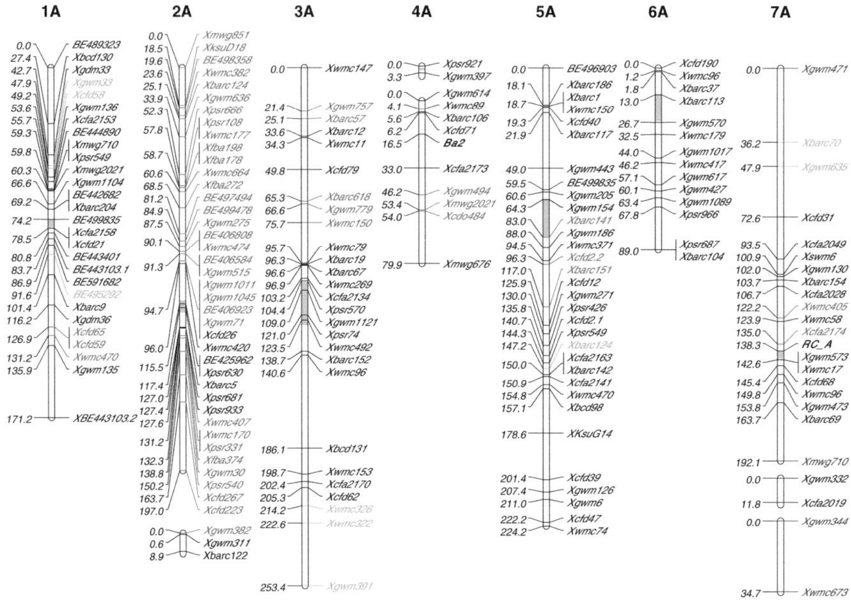

Fig. 12.3 A-genome linkage map based on *T. boeoticum/T. monococcum* RIL population (Singh et al. 2007a, b)

durum/Tm14087//2*WL711. If the B genome suppressed the resistance, then resistant plants were expected in this generation, which turned out to be true. Suppression of leaf and stripe rust-resistant genes in amphiploids generated by crossing susceptible *T. durum* and several accessions of *Ae. umbellulata* and *Ae. caudata* was also observed (Aghaee et al. 2001). The leaf rust- and stripe rust-resistant BC_1F_1 plants were then allowed to self and stripe rust-resistant homozygous progenies with good plant type were developed.

T. boeoticum/T. monococcum RIL population showed digenic segregation for cereal cyst nematode resistance also. Composite interval mapping identified two QTLs, one each on chromosome 1AS and 2AS, conferring CCN resistance to *T. monococcum*. 1A QTL, designated as *Qcre.pau*-1A, appeared to be a major gene with 34 % contribution to the overall phenotypic variance (Singh et al. 2010). *Qcre.pau*-1A is a putative novel CCN resistance gene since this is the only CCN resistance gene mapped in any 'A' genome species and none of other known genes have been mapped on chromosome 1A. The other QTL mapped on 2A, designated as *Qcre.pau*-2A, might be allelic to *Cre5*, a CCN resistance gene transferred from *Ae. ventricosa* and mapped on 2AS almost in the same chromosomal region. *Qcre.pau*-1A was transferred to cultivated wheat using *T. durum* as bridging species. Selected CCN-resistant F_8 homozygous introgression lines with 2n = 28 and 2n = 42 also showed the co-introgression of the molecular markers identified to be linked with CCN resistance gene *Qcre.pau*-1A indicating that this gene itself can provide complete resistance to *H. avenae*.

T. boeoticum pau5088 showed resistance against a number of powdery mildew resistance isolates and *T. monococcum* pau14087 was completely susceptible. Inheritance studies in the RIL population *T. boeoticum* pau5088/*T. monococcum* pau14087 indicated the presence of two powdery mildew-resistant genes in *T. boeoticum* acc. pau5088. Analysis of powdery mildew infection and molecular marker data of the RIL population revealed that both powdery mildew-resistant genes are located on the long arm of chromosome 7A. Mapping was conducted using an integrated linkage map of 7A consisting of SSR, RFLP, STS, and DArT markers. These powdery mildew-resistant genes were tentatively designated as *PmTb7A.1* and *PmTb7A.2* (Chhuneja et al. 2012). The *PmTb7A.2* is closely linked to STS markers MAG2185 and MAG1759 derived from an RFLP probes which are linked to powdery mildew-resistant gene *Pm1*. This indicated that *PmTb7A.2* might be allelic to *Pm1*. The *PmTb7A.1,* flanked by a DArT marker *wPt4553* and an SSR marker *Xcfa2019* in a 4.3 cM interval, mapped proximal to *PmT7A.2*. *PmTb7A.1* is putatively a new powdery mildew-resistant gene. The powdery mildew-resistant genes from *T. boeoticum* have been transferred to cultivated wheat background through marker-assisted backcrossing, using *T. durum* as bridging species (Elkot et al. 2015). Besides disease-resistant genes, the RIL population also segregated for quality and domestication traits. The position of 13 QTL mapped so far in *T. boeoticum/T monococcum* RIL population is summarized in Fig. 12.4. This population is the evidence of the level of variability captured in the A genome of the wild species of wheat as from only two accessions 13 new genes/QTL could be identified and mapped.

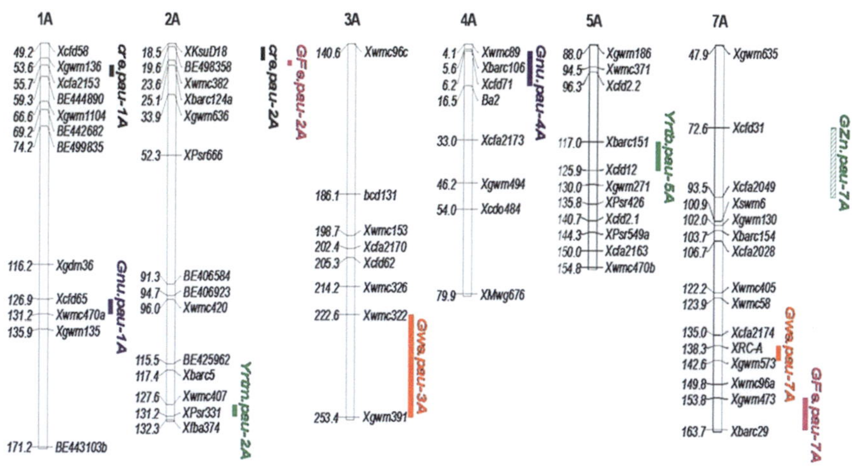

Fig. 12.4 Summary of the QTLs detected in *T. boeoticum/T. monococcum* RIL population for cereal cyst nematode resistance, *yellow* rust resistance, grain number/spikelet, grain weight, grain Fe and Zn concentration

12.8.2 Aegilops tauschii *the D Genome Donor of Wheat: A Mine of Biotic Stress Tolerance Genes*

Aegilops tauschii Coss., a diploid self-pollinating goatgrass species has contributed the D genome to common wheat. Hybridization of *Ae. tauschii* (DD) with tetraploid wheat, *T. turgidum* L. (AABB) about 7000 years ago, led to the development of hexaploid wheat *Triticum spelta* (L) Thell (AABBDD) from which common wheat evolved. The D genome of *Ae. tauschii* has much greater genetic diversity compared to the D genome of *T. aestivum*. The hybridization events that led to the evolution of bread wheat are thought to have involved only a few accessions of *Ae. tauschii.* In fact, *Ae. tauschii* represents more than 90 % of the total genetic variability present in the D genome (Lubbers et al. 1991). The morphological variation and ecological amplitude of *Ae. tauschii* exceeds that of any other diploid *Triticum* or *Aegilops* species (Zohary et al. 1969). Much of the genetic diversity in *Ae. tauschii* gene pool thus remained unutilized and may not be represented in the *T. aestivum* gene pool. Furthermore, the D genome of bread wheat has not undergone any major chromosomal restructuring in relation to the *Ae. tauschii* genome, ensuring high homology and ease of gene transfer. *Ae. tauschii* represents a rich reservoir of disease resistance (Dhaliwal et al. 1993; Villareal et al. 1994a, b; Cox 1998; Aseefa and Fehrmann 2004), productivity traits (Waines et al. 1987) and abiotic stress resistance (Trethowan and Mujeeb-Kazi 2008).

Based on spikelet morphology, *Ae. tauschii* has been divided into two subspecies, *tauschii* and *strangulata*. ssp. *tauschii* has a very wide geographic distribution extending westward to Turkey and eastward to Afghanistan and China, whereas ssp. strangulata has a narrow distribution occurring only in two disjoined regions, in southeastern Caspian Iran and Transcaucasia (Kihara et al. 1965; Yen et al. 1983; Jakaska 1995). Subspecies *strangulata* has been proposed as the D genome donor of wheat. *Strangulata* has been reported to have higher level of resistance than ssp. *tauschii* (Yildirim et al. 1995; Knaggs et al. 2000; Liu et al. 2010). Kihara and Tanaka (1958) and Kihara et al. (1965) assessed seedling responses to all three rusts among 167 accessions of *Ae. tauschii* from Pakistan, Afghanistan, and Iran. Resistant accessions were found mainly in the forms *meyeri* and *strangulata* collected from the Caspian Sea area of Iran while all accessions from Afghanistan and Pakistan were susceptible.

A commonly used route to transfer genes from *Ae. tauschii* is a two-step process of producing synthetic wheat via tetraploid × *Ae. tauschii* hybridization and colchicine doubling. These synthetics are then crossed to elite wheat lines and homologous recombination between *Ae. tauschii* and *T. aestivum* 'D' genome chromosomes can readily break undesirable linkages between target genes and alleles associated with linkage drag which might not be possible with introgressions from other species. CIMMYT recognized the potential of *Ae. tauschii* germplasm in broadening the gene pool of wheat and developed 1000 synthetics from 460 *Ae.*

tauschii accessions (Muzeeb-Kazi et al. 1987). These synthetics were later backcrossed to many CIMMYT and global elite breeding lines to introduce new variation for various morphological and agronomic traits (Villareal et al. 1994a, b), resistance to biotic stresses (Cox 1998; Ma et al. 1995) and abiotic stresses (Villareal et al. 2001). The synthetics were also shown to be diverse at the molecular level, and genetically distinct from cultivated wheats (Zhang et al. 2005).

Yellow rust-resistant gene *Yr28* is the only cataloged stripe rust resistance which has been derived from *Ae. tauschii.* It has been mapped on chromosome arm 4DS after its transfer to synthetic hexaploid wheat from *Ae. tauschii* (McIntosh et al. 2008, 2010; Singh et al. 2000a, b). Another gene temporarily designated as *YrAS2388* has been mapped on chromosome 4DS using an F_2 population between a resistant (*strangulata*) and susceptible *Ae. tauschii* accessions but this gene showed partial resistance when transferred in synthetic hexaploid wheat background (Huang et al. 2011).

Lr21 was transferred from *Ae. tauschii* accession TA1599 via a synthetic wheat (Rowland and Kerber 1974). This gene has been cloned from *Ae. tauschii* only (Huang et al. 2003). *Lr21* provides resistance to the current spectrum of *P. triticina* races in the Southern Great Plains. There are no confirmed reports of virulence to *Lr21. Lr1* is widespread in *Ae. tauschii* though there is no record of its transfer from *Ae. tauschii;* it may probably transferred to wheat at the time of the origin of wheat (Ling et al. 2004). Similarly, the *Lr34* haplotype was not detected in *Ae. tauschii*, and this gene probably arose during the few thousand years since the origin of common wheat (Gill et al. 2008).

An alternative method to transfer genes from *Ae. tauschii* is the direct hybridization between *Ae. tauschii* and *T. aestivum.* Combining gene transfer, genomic localization, and introgression (Olson et al. 2013) is an efficient method of expediting transfer of genes from *Ae. tauschii* into wheat-breeding germplasm. Adverse genetic interactions between the D genome of *Ae. tauschii* and the ABD genome of hexaploid wheat are uncommon (Gill and Raupp 1987). *Pm19* and *Pm*34 (Lutz et al. 1995a, b; Miranda et al. 2006) were transferred from *Ae. tauschii* into cultivated wheat. *Pm2* was physically mapped to chromosome 5DS by McIntosh and Baker (1970). *Pm35* is another gene derived from *Ae. tauschii* direct crosses with hexaploid wheat and is mapped on chromosome 5DL (Miranda et al. 2007).

Many leaf rust resistance genes have been identified in the *Ae. tauschii* including *Lr*22a (2DS), *Lr32* (3D), *Lr39*/*Lr41* (2DS), *Lr42* (1D) (Cox et al. 1994; Gill et al. 1991; Kerber 1987; Hiebert et al. 2007; Huang and Gill 2001; Huang et al. 2003; Rowland and Kerber 1974; Raupp et al. 2001) which were transferred to hexaploid wheat germplasm and later on mapped onto specific chromosomes (Table 12.5). Recently, new flanking markers have been identified in vicinity of *Lr*42 which will aid in MAS of this gene (Liu et al. 2013). *Lr21* and *Lr39* have been introgressed from *Ae. tauschii* ssp. *tauschii* and *Lr22a* and *Lr42* from ssp. *strangulata* (McIntosh et al. 2013). Two stem rust resistance genes such as *Sr33* and *Sr45* and powdery mildew resistance genes such as *Pm2* and *Pm19* have been transferred from *Ae. tauschii* (Hsam and Zeller 2002; Marais et al. 1998; McIntosh et al. 2013). Two germplasm lines were developed carrying genes *Pm34*, *Pm35* also from *Ae. tauschii* (Miranda et al. 2006, 2007) and mapped on chromosome 5D (Table 12.5).

At Punjab Agricultural University, Ludhiana, India, we are maintaining an active collection of 280 *Ae. tauschii* accessions. These accessions have been found to carry resistance genes for various biotic stresses including leaf rust, stripe rust, powdery mildew, and Karnal bunt. Stripe rust reactions of these accessions over last five years have shown a lot of variation in the rust reaction. Reaction of the stripe rust resistant accessions of *Ae. tauschii* varied from totally immune to traces of resistance (TR), moderately resistant, and moderately susceptible.

The stripe rust and leaf rust data of these accessions are summarized in Table 12.8. In 2013, 67 out of 203 accessions showed high level of stripe rust resistance and 52 were moderately resistant while in 2014 out of 281 accessions, 72 were highly resistant and 11 moderately resistant. Genome-wide Association studies based on SNP markers identified eight significant loci ($p < 0.01$) mapped on chromosome 7DL, 3DS, 2DS, 5DL for stripe rust resistance (our unpublished data). One of the accession pau14195 with multiple disease resistance was crossed with durum wheat cultivar PBW114 and resulting F_1 was crossed and backcrossed with leaf rust, stripe rust, and KB susceptible cultivar WH542 for transferring leaf rust and stripe rust resistance genes to hexaploid wheat background. One gene for leaf rust resistance and linked non-glaucousness was mapped on chromosome 2D in hexaploid background (Saluja personal communication).

Ae. tauschii has a very high level of KB resistance also and a subset of *Ae. tauschii* germplasm was screened for KB resistance. Over three years of screening under artificial inoculations, 20 accessions were identified to be resistant to KB and six accessions were moderately resistant (Chhuneja et al. 2008c). Almost 80 % of the KB-resistant accessions belonged to ssp. *tauschii* which is contrary to the rust resistance where almost 90 % of the resistant accessions belonged to ssp. *strangulata* (Chhuneja et al. 2010). For the transfer of KB resistance to cultivated wheat, an amphiploid was synthesized by crossing a KB-resistant *Ae. tauschii* acc. pau 3743 with KB susceptible *T. durum* cultivar WH890. The synthetic hexaploid was crossed with a KB susceptible hexaploid wheat cv. to transfer KB resistance to desirable agronomic background and homozygous introgression lines were developed (ILs). Introgressions of *Ae. tauschii* specific alleles were found on chromosomes 1D, 2D, 4D, and 6D (Fig. 12.5) which indicated that the KB resistance gene (s) may be located on these regions.

12.8.3 Ae. Speltoides *and Other S Genome Species*

Ae. speltoides has depicted very high levels of disease resistance for leaf rust and stripe rust from screening over many years in PAU, Ludhiana. Leaf and stripe rust data for 2012–2013 and 2013–2014 crop seasons recorded under field conditions is

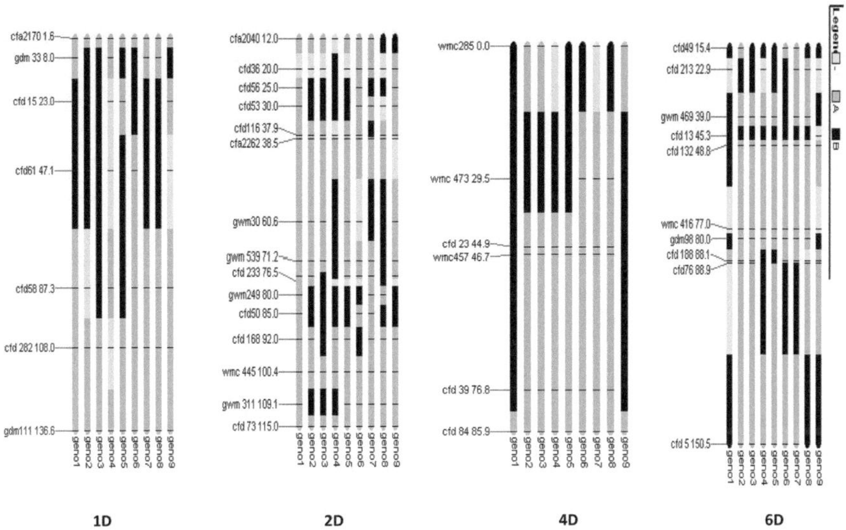

Fig. 12.5 Graphical genotyping of wheat–*Ae. tauschii* introgression lines using D genome-specific SSR markers. *Gray areas* represent wheat-specific alleles and *black areas* indicate introgression of *Ae. tauschii*-specific alleles on chromosomes 1D, 2D, 4D, and 6D. Map distances are according to Komugi composite wheat map

presented in Table 12.8. Leaf rust resistance genes *Lr28, Lr35, Lr36, Lr47, Lr51, Lr54,* and stem rust resistance genes *Sr32, Sr39, Sr47* have been transferred from different accessions of *Ae. speltoides* by different workers (Table 12.5; McIntosh et al. 2013). Leaf rust resistance gene *Lr28* located on chromosome 4AL was transferred from *Aegilops speltoides* into wheat by Riley et al. (1968b). It was subsequently backcrossed into different wheat backgrounds in India (Tomar and Menon 1998) and it provides effective resistance against all the Indian leaf rust pathotypes. The stem rust resistance gene *Sr39* was transferred to the hexaploid wheat cultivar Thatcher (Tc) from *Ae. speltoides*. The gene is also associated with adult plant leaf rust resistance gene *Lr35* which is highly effective against North American populations of *P. recondita* f. sp. *tritici*. Both the *Sr39* and the *Lr35* genes are located on the alien translocated segment on chromosome 2BS (Niu et al. 2011). *Sr47,* also from *Ae. speltoides*, has been mapped in durum wheat–*Ae. speltoides* chromosome translocation line T2BL-2SL&2SS which is different from *Sr32* and *Sr39* located in the same region of 2BS (Faris et al. 2008). Gene *Pm12* originating from *Ae. speltoides* is located on T6BS–6SS–6SL (Miller et al. 1988; Jia et al. 1996).

12.8.4 *Non-progenitor* Aegilops *Species for Transfer of Disease-Resistant Genes to Hexaploid Wheat*

The genus *Aegilops* consists of 22 species of which 10 are diploid, 10 are tetraploid, and 2 are hexaploid with basic chromosome number x = n = 7 and six different genomes as C, D, M, N, S, and U (van Slageren, 1994). *Aegilops* represents the largest part of the secondary gene pool of wheat and several species have been used in crop improvement programs. Non-progenitor *Aegilops* species with one of the genome as U has been found to be rich source of resistance genes (Mamluk and van Slageren 1994). *Ae. umbellulata* (UU), *Ae. geniculata* (UM), *Ae. triuncilais* (UC), *Ae. peregrina* (US), *Ae. kotschyi* (US), etc. are valuable sources of useful genes for wheat breeding. *Ae. caudata* (C), *Ae. comosa* (M), *Ae. uniaristrata* (NN) are also known to be rich sources of resistance to various pathogens and pests (Pasquini 1980; Gill et al. 1985; Manisterski et al. 1988; Anikster et al. 2005). Many genes conferring resistance to rust diseases, powdery mildew, cereal cyst nematode, and insect pests were transferred from *Aegilops* species into wheat (Jiang et al. 1994; Friebe et al. 1996; Dhaliwal et al. 2003; Marais et al. 2005; Kuraparthy et al. 2007a, b; Chhuneja et al. 2008a, b; Riar et al. 2012).

Some leaf rust and stripe rust resistance genes transferred from wild non-progenitor species to cultivated wheat have been shown in Table 12.6. Some of the genes for disease resistance transferred from distantly related species have been exploited commercially but others seem to be associated with reduced yield due to linkage drag (Young and Tanksley 1989). The most successful example of alien genome segment transfer is wheat–rye translocation. In this wheat chromosome, 1BL was replaced by rye chromosome 1RS. The 1RS arm in this translocation has many genes such as *Lr26* for leaf rust resistance, *Sr31* for stem rust resistance, *Yr9* for stripe rust resistance (Bartos and Bares 1971; Bartos et al. 1973, b), and *Pm8* for powdery mildew resistance. This translocation has been incorporated into 60 wheat varieties, including the prominent Veery lines, that occupied 50 % of all developing country wheat area, almost 40 million hectares. Translocation T3DS.3DL-3Ae#1L from *Agropyron elongatum* carrying *Lr24/Sr24* is the second most exploited alien introgression. *Lr9* from *Ae. umbellulata*, *Sr26, Lr19 and Sr25* from *Ag. elongatum* have also been exploited to some extent.

In the past several years, about 20 stem rust, 30 leaf rust, 10 stripe rust, and 15 powdery mildew resistance genes have been transferred from near and distant relatives of hexaploid wheat. Stem rust-resistant genes transferred from the tertiary gene pools of wheat include *Sr*24 from *Agropyron elongatum; Sr31*, *Sr1*, and *Sr50* (Anugrahwati et al. 2008) from *Secale cereale*; *Sr36* from *T. timopheevii* (Olson et al. 2010); *Sr3*8 from *Ae. ventricosa* (McIntosh et al. 1995a, b); *Sr26* from *A. elongatum*; and *Sr44* from *A. intermedium* (Liu et al. 2013). Stem rust resistance genes *Sr51, Sr52, and Sr53* transferred from the tertiary gene pool were identified in chromosome addition lines and Robertsonian translocations from *Ae. searsii*, *Dasypryum villosum*, and *Ae. geniculata*, respectively (Liu et al. 2011; Qi et al. 2011).

Mwale et al. (2014) reviewed all the cataloged and uncataloged genes of powdery mildew resistance from wild progenitor of wheat. Hsam et al. (1998) reported the presence of allele *Pm1d* in *T. spelta*. *Pm4a* and *Pm4b* on chromosome 2AL were first reported by The et al. (1979) on cultivars Khapli (*T. dicoccum*) and Armada (*T. carthlicum*), respectively, and subsequently reviewed by Huang and Roder (2004) and Alam et al. (2011). Alleles of *Pm5* were identified and mapped on wheat chromosome 7BL. These include gene *Pm5a* in the wheat cultivars Hope, and a recessive gene that originated from *T. dicoccum* L. *Pm5c* was derived from *T. sphaerococcum* var. *rotundatum* (Hasm et al., 2001). *T. diccocoides* is another important donor of powdery mildew resistance genes as *Pm26*, located on chromosome 2BS (Rong et al. 2000), *Pm42* (Hua et al. 2009a), and *Pm49*, located on chromosome 2BS were originated from *T. dicoccoides*. Other resistance genes transferred from *T. diccocoides* include *Pm16*, *Pm30*, *Pm31*, *Pm36*, *Pm41*, *Pm42*, *Pm49*, and *Pm50*, located on chromosomes 4A, 5B, 6A, 5B, 3B, 2B, and 2A, respectively (Piarulli et al. 2012; Mohler et al. 2013a). *Pm6* is originated from the 2G chromosome of *T. timopheevii* and was introgressed into chromosome 2BL of common wheat (Tao et al. 2000). *T. carthlicum* is another wild relative source of powdery mildew-resistant genes. They include genes *Pm4b* and *Pm33* (Zhu et al. 2005). *Secale cereale*, a distant relative of common wheat also contributed *Pm7*, *Pm8*, *Pm17*, and *Pm20* genes to common wheat (McIntosh et al. 2011). Two designated genes *Pm40* and *Pm43* were introgressed into common wheat from *Elytrigia intermedium* and *Th. intermedium*, respectively (Luo et al. 2009; He et al. 2009) while *Pm21* originated from *Haynaldia villosum* (Chen et al. 1995; Piarulli et al. 2012; Xiao et al. 2013).

Ernie Sears (1956) used *Ae. umbellulata* for the transfer of leaf rust resistance to wheat. He presented this remarkable work at the 1956 Brookhaven Symposium. This commenced with the addition to *T. aestivum* of a single chromosome of *Ae. umbellulata* which also carried a number of undesirable genetic information. Sears X-rayed addition line carrying *Ae. umbellulata* chromosome with leaf rust resistance. The irradiated plants were then used to pollinate normal wheat and resistance progeny was selected. Forty of these had one of at least seventeen different translocations between the *Aegilops* chromosome and the wheat chromosomes. There was one line with the resistance chromosome segment apparently incorporated in the form of an intercalary translocation. Further work, published in 1966, showed that the *Ae. umbellulata* segment was not in an intercalary position but that a long *Aegilops* segment had replaced the terminal part of the long arm of wheat 6B. This gene was later on designated as *Lr9*. The gene *Lr9* derived from *Ae. umbellulata* is a highly effective gene throughout the world except in North America (Shaner et al. 1972) and in Canada (Samborski and Dyck 1976). Another important leaf rust-resistant gene *Lr24* is tightly linked to stem rust-resistant gene *Sr24* on the long arm of the chromosome 3D. Both resistance genes were introduced from *Ag. elongatum* in a spontaneous translocation involving 3Ag from *Agropyron* and 3DL from wheat. The alien segment carrying *Lr24*/*Sr24* does not impose any deleterious effect on yield as several cultivars carrying *Lr24* have been released for cultivation in India.

Evaluation of the wild *Aegilops* species belonging to secondary and tertiary gene pool over many years led to the identification of potential sources of disease-resistant genes. The number of resistant accessions varied from year to year. One of the reasons could be change in the pathotype specificity which was observed in some other wheat lines also known genes.

12.9 Transfer of Rust Resistance Genes from Non-progenitor *Aegilops* Species

A leaf rust- and stripe rust-resistant disomic substitution line {DS5 M(5D)} with 5 M chromosome of *Ae. geniculata* substituted for 5D of wheat was developed through restricted backcrossing and selfing. The rust resistance of *Ae. geniculata* was transferred to wheat by induced homoeologous chromosome pairing between chromosomes 5 Mg of *Ae. geniculata* and 5D of wheat. The introgression lines were developed by crossing disomic substitution line DS 5 Mg(5D) with the Chinese Spring (CS) *Ph^I* stock (Chen et al. 1994) and crossing the F_1 with susceptible bread wheat cultivar WL711. Advanced backcross lines were characterized using molecular cytogenetic and molecular techniques, and translocation carrying these leaf and stripe rust-resistant genes was mapped on distal end of 5DS (Fig. 12.6). These co-segregating genes have been designated as *Lr57* and *Yr40* (Kuraparthy et al. 2007a; McIntosh et al. 2008). This translocation encompassing ~3.5 % of 5DS is the smallest alien introgression characterized so far.

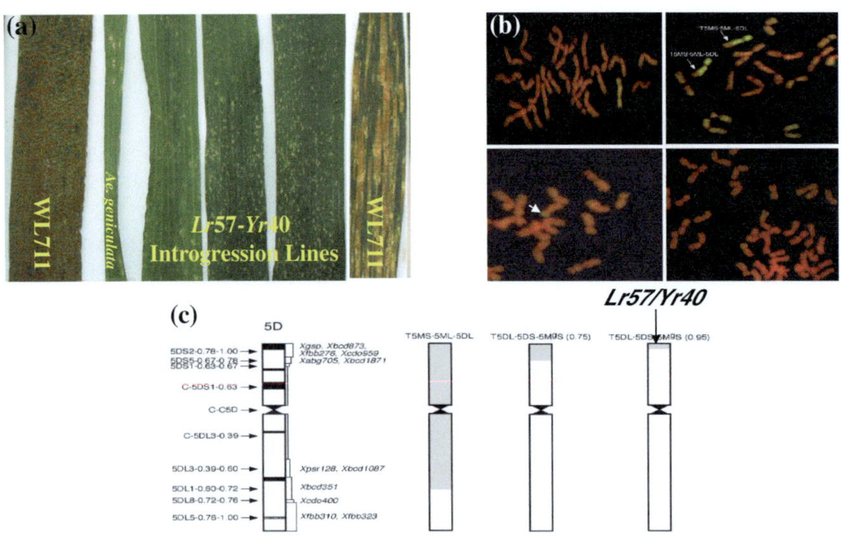

Fig. 12.6 Transfer and mapping of leaf rust and stripe rust resistance genes *Lr57-Yr40* introgressed from *Aegilops geniculata* (UUMM) through induced homoeologous pairing in *T. aestivum*. **a** Leaf and stripe rust reaction **b** Genomic in situ hybridization **c** mapping of *Lr57, Yr40* in introgression lines (Kuraparthy et al. 2007a, b)

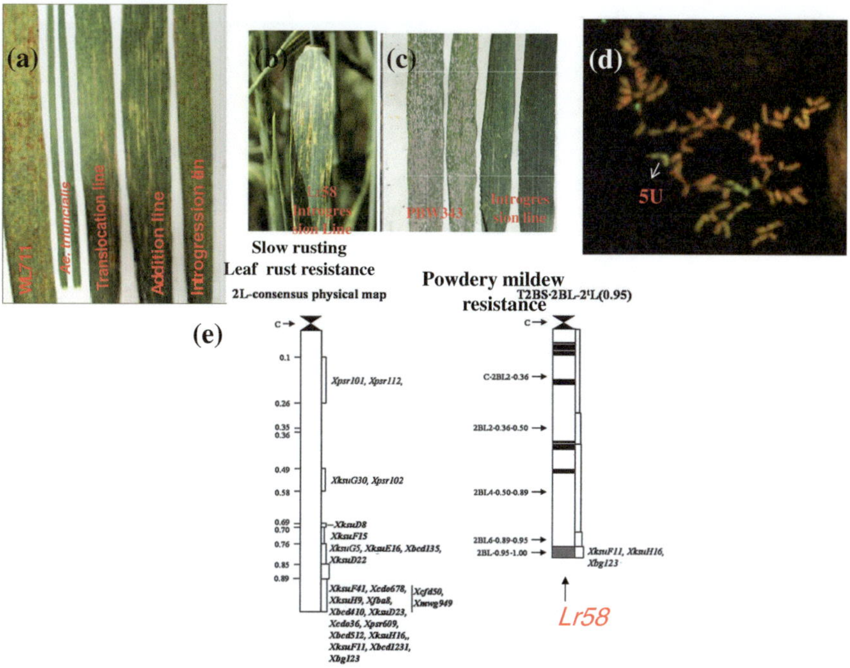

Fig. 12.7 **a** Leaf rust reaction of wheat–*Ae. triuncialis* translocation, addition, and introgression lines; **b** Slow rusting for leaf rust in wheat–*Ae. triuncialis* substitution line 5U-5A; **c** powdery mildew resistance in PBW343–*Ae. triuncilais* introgression lines; **d** Genomic in situ hybridization of wheat–*Ae. triuncialis* 5U-5A substitution line; **e** mapping of *Lr*58 introgressed from *Ae. triuncialis* through spontaneous translocation and characterized using molecular markers in wheat–*Ae. triuncialis* translocation line (Kuraparthy et al. 2007b)

PCR-based markers for these genes have also been developed. The introgression line with the smallest introgression carrying *Lr57-Yr40* is being used as a donor parent for mobilizing these genes to elite wheat backgrounds using MAS.

Similarly, a spontaneous translocation line developed from a cross of WL711 and *Ae. triuncialis* acc. 3549 was resistant to the most prevalent races of leaf rust in India. Genetic mapping in a segregating $F_{2:3}$ population showed that the rust resistance was monogenically inherited. Molecular analysis identified homeologous group 2 carrying the gene in question. The *Ae. triuncialis*-specific alleles of *XksuH16*, *XksuF11*, *Xbg123*, and one simple sequence repeat marker *Xcfd50* co-segregated with the rust resistance, suggesting that the wheat–*Ae. triuncialis* translocation occurred in the distal region of chromosome arm 2BL (Fig. 12.7 a,e) and was designated *Lr58* (Kuraparthy et al. 2007b).

From the same accession of *Ae. triuncialis* pau3549, a substitution line with resistance to leaf rust (slow rusting), powdery mildew, and cereal cyst nematode has also been developed. Alien chromosome in the WL711-*Ae. triuncialis* substitution line has been identified as 5U through C-banding . GISH studies also identified a satellite chromosome in the substitution line (Fig. 12.7 b,c,d). An alien addition line

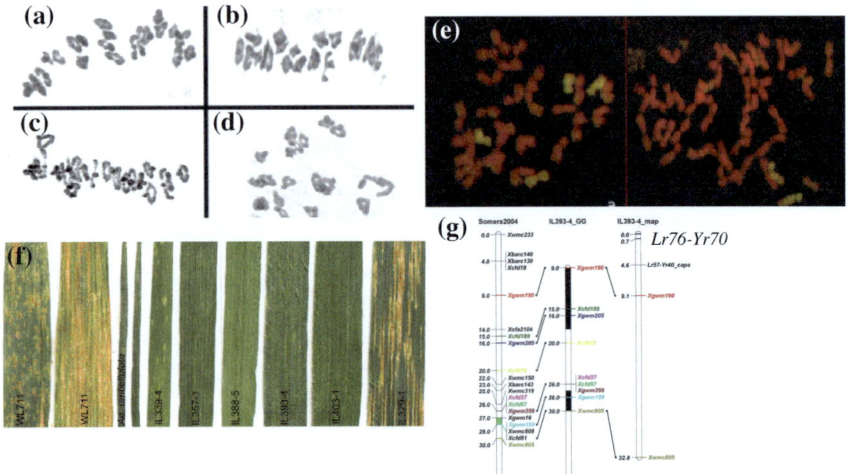

Fig. 12.8 **a–c** Meiotic analysis in wheat–*Ae. umbellulata* introgression lines; **d** Meiotic analysis in F_1 plant from the cross of IL393-4 with bread wheat cv. PBW343; **e** Genomic *in situ* hybridization of WL711–*Ae. umbellulata* addition lines; **f** Leaf rust and stripe rust reaction of WL711, *Ae. umbellulata,* and six different WL711–*Ae. umbellulata* introgression lines; **g** Molecular mapping of *Lr76-Yr70* transferred from *Ae. umbellulata* through induced homoeologous pairing in IL 393-4

from the same cross has resistance for leaf rust, powdery mildew, and Karnal bunt. Karnal bunt and powdery mildew resistance has been transferred from the alien chromosome to the wheat cultivar PBW343.

Leaf and stripe rust resistance genes have also been introgressed from diploid species *Ae. umbellulata* and *Ae. caudata* using *T. durum* as bridging species (Chhuneja et al. 2008b; Riar et al. 2012). Rust resistance of *Ae. umbellulata* was transferred to wheat by homoeologous pairing between *Ae. umbellulata* and wheat chromosomes. The resistant plants were backcrossed 2–3 times with a susceptible wheat cv. WL711 followed by selfing to develop homozygous introgression lines with a high level of resistance to leaf rust as well as stripe rust. One of these introgression lines was crossed with *T. aestivum* cv. PBW343 to generate a mapping population, and a BC-RIL population was developed which segregated for a single gene each for leaf rust and stripe rust resistance. Both the rust resistance genes, however, co-segregated and were mapped on short arm of wheat chromosome 5D (Fig. 12.8) and have been designated as *Lr76* and *Yr70* (Bansal et al. 2015). Similar strategy was used for transfer of leaf and stripe rust resistance from *Ae. caudata*. Two genes one each for leaf rust and stripe rust were again mapped on chromosome 5DS. The stripe rust resistance gene transferred from *Ae. caudata* was found to be an adult plant resistance gene (Riar et al. 2012; Kaur 2014).

Two leaf rust and one stripe rust resistance genes have also been transferred from a tetraploid non-progenitor species *Ae. peregrina* to wheat cultivar WL711 through induction of homoeologous pairing. *Ae. peregrina* accession pau3519 was crossed with Chinese Spring stock carrying inhibitor of Ph^{I} locus, F_1 was crossed with

WL711 and advance backcross introgression lines developed. Molecular mapping in two different introgression lines identified one leaf rust resistance gene on 2DL and a pair of linked leaf and stripe rust resistance genes on 5DS.

12.10 R Gene Clusters from Alien Germplasm

Most of the resistance genes introduced from wild progenitor and non-progenitor of wheat are major/race-specific genes. The short-lived nature of race-specific leaf rust resistance genes greatly compromises the efforts of scientists and breeders. Alternatively, a more durable form of resistance is attributed to slow leaf-rusting or durable resistance in the form of retarded disease progress in the field results from a longer latent period, smaller pustule size, and lower spore production (Ohm and Shaner 1976 ; Wilcoxson 1981 ; Das et al. 1992). These genes confer a slow rusting type of resistance (Caldwell 1968) despite a compatible host reaction and are effective across all races of the pathogen, and disease reaction is measured several times during the course of disease in a growing season as Area Under Disease progress Curve (AUDPC) (Jeger and Viljanen-Rollinson 2001). So far only five resistance gene blocks, Lr34/Yr18/Pm38/Sr57, Lr46/Yr29/Pm39/Sr58, Sr2/Yr30 Lr67/Yr46/Pm46/Sr55, and Lr68 known for partial but durable resistance; however, none is from wild wheat (Singh et al. 2000a , b ; Hiebert et al. 2010 ; Herrera-Foessel et al. 2011; Singh et al. 2011). Main reason of wild germplasm contributing major genes is their tedious process of resistance gene transfer into cultivated wheat involving the use of bridging cultivars, many crosses and backcrosses, selecting resistant plants throughout the process, and maintain recurrent parent background. This ends up transferring only major genes, as in most of the cases resistance genes are mapped after their transfer into cultivated backgrounds.

Genes from wild germplasm are inherited as blocks of multiple disease resistance due to the absence of recombination from alien chromosomal segments such as gene cluster *Lr26/Yr9/Sr31/Pm8* from rye chromosome IRS, ViRGA/ *Lr37/Yr17/Sr38* on chromosome 2B from *Ae. ventricosa* (McIntosh et al. 1995a, b ; Seah et al. 2001), Lr57/Yr40 on chromosome 5D from Ae. geniculata (Kurapathy et al. 2007a), *Lr76/Yr70* on chromosome 5D from *Ae. umbellulata* (Bansal et al. 2015), *Sr36/Pm6* from T. timopheevii on chromosome 2A, *Gb2/Pm17* from S. cereale, and two clusters of two genes each *Lr19/Sr25* and *Sr24/Lr24* from *A. elongatum* (Sears 1956; McIntosh et al. 1991; Delibes et al. 1993; Friebe et al. 1996). These introgressed segments of major genes clusters were shown to carry diverse and multiple genes that encode nucleotide-binding and leucine-rich repeat sequences, the most common class of plant disease resistance genes (Seah et al. 2001; Mago et al. 2005a, b). R genes evolve through a variety of molecular mechanisms: point mutations, unequal crossing over, gene conversion and recombination, illegitimate recombination, and insertion/deletions all contribute to variability (Kuang et al. 2004; Michelmore and Meyers 1998). Sequence variability in R genes/alleles was mostly found in the LRR region, which was shown to play a major role in pathogen recognition specificity (Shen et al. 2003a , b ; Yahiaoui et al. 2006). Tandem and segmental duplications have been reported as a source of structural plasticity of NBS-LRR genes in plant genomes. Clustering usually results from tandem duplications of paralogous sequences resulting in unequal crossing over. Analyses of the *Arabidopsis* genome indicate that numerous small-scale genomic duplications have copied or translocated one or several NBS-LRR genes from these clusters to distal and probably random locations in the genome. At some loci, tandem duplications have expanded gene families and the duplicated sequences have diverged through accumulated mutations, increasing the complexity of R gene sequences. Clusters of durable resistance genes differ from that of major genes as inactivation of *Lr34* mean inactivation of *Yr18, Pm38, and Sr57* also, while clusters of major genes *Lr26/Yr9/Sr31/Pm8* all are different loci inactivation of one cannot inactivate other (Mago et al. 2005a, b). These slow rusting genes boost the resistance of many major genes as German and Kolmer (1992) showed that *Lr34* enhance the effect of many major genes when present together. One way of prolonging the resistance of major genes is their combination with slow rusting genes. The combination of *Lr34* with *Lr12* and/or *Lr13* provided durable leaf rust resistance cultivars worldwide (Roelfs 1988). Singh and Huerta Espino (1995) showed an increased resistance of *Lr16* with the presence of slow rusting genes. Similarly, presence of slow rusting genes *Yr29* and *Yr30* increased the resistance of *Yr31* (Singh et al. 2003). Enhanced expression of major gene *Sr25* is reported in the presence of slow rusting gene *Sr2* in CIMMYT (Njau et al. 2010). The pleiotropic action of these genes on other diseases such as powdery mildew make them additionally valuable for breeding for broad spectrum resistance (Lillemo et al. 2008; Mago et al. 2011). Mostly, durable resistance genes are also found in clusters and unlike gene clusters from wild germplasm, these clusters are like single genes as described in Lr34/Yr18/Pm38 locus.

12.11 Next-Generation Sequencing Technologies for Monitoring Alien Introgressions

Identification of markers closely linked with disease resistance genes has progressed in the last decade through the development of high-throughput and cost-effective genotyping facilities. One of the first high-throughput platforms in wheat, diversity arrays technology (DArT) exploits independent chip hybridization of genome representation for diversity assessment of tested genomes and could test hundreds to thousands of genomic loci in parallel (Jaccoud et al. 2001). This approach can be more efficient using high-throughput next-generation sequencing (NGS) platforms for genome sequencing referred to as genotyping-by-sequencing (GBS) and can identify several hundred thousand genome tags (Poland et al. 2012). Another approach includes the use of advances in wheat genome sequencing and

NGS technologies to develop SNP chips for wheat. All these technologies individually or in combination can be used to fine map the genes of interest.

The International Wheat Genome Sequencing Consortium (IWGSC) was established in 2005 to sequence the wheat genome to enhance the knowledge of structure and function of wheat genome and create a platform for accelerating wheat improvement. With a genome sequence in hand, breeders can have access to complete, ordered gene catalog and an almost unlimited number of molecular markers that can be used for marker-assisted selection and precision-breeding approaches (Collard and Mackill 2008; Tester and Langridge 2010). Combined strategies are being deployed by the consortium to achieve a reference genome sequence of the hexaploid bread wheat genome cultivar Chinese Spring. These include physical mapping of Chinese Spring and *Aegilops tauschii* (the D genome progenitor of bread wheat), as well as survey sequencing and BAC-based (i.e., the minimum tiling path of the physical map) reference sequencing of Chinese Spring. The physical map of *Aegilops tauschii* was completed by Jia et al. (2013). The physical map of the largest wheat chromosome (3B, $\sim$1 Gb) was completed in 2008 (Paux et al. 2008). Physical mapping of the remaining chromosomes is underway. To facilitate anchoring, marker development, and to gain a first insight into the gene space composition, survey sequences were completed with the construction of the physical maps. The chromosome arm-based draft sequence of the bread wheat genome (IWGSC 2014) provided new insight into the structure, organization, and evolution of the large, complex genome of the world's most widely grown cereal crop. These arm-based sequences also became an immense resources for marker development. 7AL sequence was aligned against genic sequences of *Brachypodium* 7AL genic contig as well as those carrying NBS-LRR domains were identified. SSR and RGA-STS marker were developed from these contigs and mapped on 7AL of *T. boeoticum/T. monococcum* RIL population leading to fine mapping of two powdery mildew-resistant genes *PmTb7AL.1* and *PmTb7AL.2* (Chhuneja et al. 2015). Both the genes have been transferred to hexaploid wheat background using the linked RGA-STS marker (Elkot et al. 2015).

12.12 Flow Sorting of Chromosomes and Gene Identification

Dissecting and cloning individual chromosomes would largely facilitate genome analysis and gene cloning in wheat and other organisms with large and complex genomes. This approach reduces sample complexity and enables analysis at the subgenomic level. Flow cytometric chromosome sorting has been implemented successfully in many plant species, including cultivated cereals (such as bread and durum wheat), barley, rye, oats, rice, and maize (Dolezel et al. 2012). Molnár et al. (2011) reported flow sorting of individual chromosomes from *Ae. umbellulata*

(2n = 2x = 14, UU) and *Ae. comosa* (2n = 2x = 14, MM) and from their natural allotetraploid hybrids (*Ae. biuncialis* and *Ae. geniculata*). This study provided opportunity for the next-generation sequencing of individual *Aegilops* chromosomes for the development of sequence-based markers and their application in wheat breeding. Tiwari et al. (2014) flow-sorted short arm of chromosome 5 Mg of *Ae. geniculata* from a wheat line in which it was maintained as a telocentric chromosome. DNA of the sorted arm was amplified, sequenced, and used for SNP discovery against wheat homoeologous group-5 assemblies. A total of 2178 unique, 5 MgS-specific SNPs were discovered. Randomly selected samples of 59 5 MgS-specific SNPs were tested and of the selected SNPs, 97 % mapped to a chromosome 5 Mg addition to wheat, and 94 % to 5Mg introgressed from a different accession of *Ae. geniculata* substituting for chromosome 5D of wheat. The validated SNPs also identified chromosome segments of 5MgS origin in a set of T5D-5Mg translocation lines; eight SNPs (25 %) mapped to TA5601 and three to TA5602, the introgression lines carrying resistance to leaf rust (*Lr57*) and stripe rust (*Yr40*) identified by Kuraparthy et al. (2007a). The development of a large number of species/genome-specific SNP markers will facilitate the precise introgression and monitoring of alien introgressions in crop-breeding programs and enable fine mapping and cloning novel genes from the wild relatives of crop plants.

12.13 Cloning of Alien Disease Resistance Genes

A very large number of disease-resistant genes have been mapped and molecular markers suitable for marker-assisted selection (MAS) have been identified. It was considered very difficult to clone genes from wheat as it is an allohexaploid. Some success has been achieved and some rust-resistant genes (*Lr1, Lr10, Lr21, Lr34,* and *Yr36*) have been cloned and characterized. *Lr21* and *Yr36* are the only alien genes which have been cloned so far.

Map-based cloning of *Yr36* was done by crossing the susceptible durum wheat variety Langdon with the resistant isogenic recombinant substitution line RSL65. A population of 4500 F_2 plants was screened using *Yr36* flanking markers *Xucw71 and Xbarc136* and identified 121 lines with recombination events between these two markers. Based on genes from the rice collinear region, nine PCR markers were developed to construct a high-density map of *Yr36*. *Yr36* was mapped to a 0.14 cM interval delimited by markers *Xucw113* and *Xucw111*. BAC ends were used to rescreen the library and extend the contig by chromosome walking. BAC-end marker *Xucw127* was mapped proximal to *Yr36,* thereby completing the physical map. New markers were developed after sequencing of the BAC clones and *Yr36* resistance was mapped between *Xucw129* and *Xucw148* (0.02 cM). The gene includes a kinase and a putative START lipid-binding domain. Five independent mutations and transgenic complementations confirmed that both domains were necessary to confer resistance (Fu et al. 2009). Resequencing cloned genes can identify all mutations in single genes for population-based analyses of genetic changes in improved gene pools. This is useful to know about the genetic impacts of modern plant breeding on specific breeding target loci.

Another alien rust-resistant gene *Lr21* was cloned by diploid/polyploid shuttle mapping strategy (Huang et al. 2003). *Lr21* spanned 4318 bp and encoded a 1080-amino-acid protein containing a conserved nucleotide-binding site (NBS) domain, 13 imperfect leucine-rich repeats (LRRs), and a unique 151-amino-acid sequence missing from known NBS-LRR proteins at the N terminus. Fine-structure genetic analysis at the *Lr21* locus detected a non-crossover (recombination without exchange of flanking markers) within a 1415-bp region resulting from either a gene conversion tract of at least 191 bp or a double crossover. With the advancements in the wheat genome sequencing and flow sorting of the chromosomes, it will become comparatively easier to clone disease-resistant genes in wheat.

References

Aghaee-Sarbarzeh M, Singh H, Dhaliwal HS (2000) *Ph1* gene derived from *Ae. speltoides* induces homoeologous pairing in wide crosses of *Triticum aestivum*. J Heredity 91:417–421

Aghaee-Sarbarzeh M, Dhaliwal HS, Chhuneja P, Singh H (2001) Suppression of rust resistance genes from distantly related species in *Triticum durum-Aegilops* amphiploids. Wheat Inf Serv 92:12–16

Alam A, Xue F, Wang C, Wanquan J (2011) Powdery mildew resistance genes in wheat: identification and genetic analysis. J Mol Biol Res 1:21–38

Anikster Y, Manisterski J, Long DL, Leonard KJ (2005) Resistance to leaf rust, stripe rust and stem rust in *Aegilops* spp. in Israel. Plant Dis 89:303–308

Anker CC, Niks RE (2001) Prehaustorial resistance to wheat leaf rust in *Triticum monococcum* (s.s.). Euphytica 117:209–215

Anker CC, Buntjer JB, Niks RE (2001) Morphological and molecular characterisation confirm that *Triticum monococcum* s.s. is resistant to wheat leaf rust. Theor Appl Genet 103:1093–1098

Anugrahwati DR, Shepherd KW, Verlin DC, Zhang P, Mirzaghaderi G, Walker E, Francki MG, Dundas IS (2008) Isolation of wheat rye 1RS recombinants that break the linkage between the stem rust resistance gene *SrR* and secalin. Genome 51:341–349

Assefa S, Fehrmann H (2004) Evaluation of *Aegilops tauschii* Coss. for resistance to wheat stem rust and inheritance of resistance genes in hexaploid wheat. Genet Resour Crop Evol 51:663–669

Autrique E, Singh RP, Tanksley SD, Sorrells ME (1995) Molecular markers for four leaf rust resistance genes introgressed into wheat from wild relatives. Genome 38:75–83

Bai D, Knott DR, Zale JM (1998) The inheritance of leaf and stem rust resistance in *Triticum monococcum* L. Can J Plant Sci 78:223–226

Bansal M, Kaur S, Dhaliwal H, Chhuneja P, Bariana H, Bansal U (2015) Introgression of linked rust resistance genes *Lr76* and *Yr70* from *Aegilops umbellulata* to wheat chromosome 5DS. (under review)

Bartos P, Bares I (1971) Leaf and stem rust resistance of hexaploid wheat cultivars 'Salzmunder Bartweizen' and 'Weique'. Euphytica 20:435–440

Bartos P, Valkoun J, Kosner J, Slovencikova V (1973) Rust resistance of some European wheat cultivars derived from rye. In: Sears ER, Sears LM (eds) Proceedings of 4th international wheat genet symposium. University of Missouri, Columbia, USA, pp 145–46

Bennett FGA (1984) Resistance to powdery mildew in wheat: a review of its use in agriculture and breeding programmes. Plant Pathol 33:279–300

Blanco A, Gadaleta Cenci AA, Carluccio AV, Abdelbacki AMM, Simeone R (2008) Molecular mapping of the novel powdery mildew resistance gene *Pm*36 introgressed from *Triticum turgidum* var. *dicoccoides* in durum wheat. Theor Appl Genet 116:417–425
Bouhssini ME, Benlhabib O, Bentika A, Sharma HC, Lahloui S (1997) Sources of resistance in *Triticum* and *Aegilops* species to Hessian fly (Diptera: Cecidomyiidae) in Morocco. Arab J Plant Prot 15:126–128
Brown SM, Szewc-McFadden AK, Kresovich S (1996) Development and application of simple sequence repeat (SSR) loci for plant genome analysis. In: Jauhar PP (ed) Methods of genome analysis in plants. CRC Press Inc, Florida, pp 147–162
Brown-Guerdira GL, Singh S, Fritz AK (2003) Performance and mapping of leaf rust resistance to wheat from *Triticum timopheevii* subsp. *armeniacum*. Phytopathology 93:784–789
Caldwell RM (1968) Breeding for general and/or specific plant disease resistance. In: Finley KW, Shepherd KW (eds) 3rd International wheat genetics symposium. Canberra, Australia, pp 263–272
Cenci A, D'Ovidio R, Tanzarella OA, Ceoloni C, Porceddu E (1999) Identification of molecular markers linked to *Pm*13, an *Aegilops longissima* gene conferring resistance to powdery mildew in wheat. Theor Appl Genet 98:448–454
Ceoloni C, Del Signore G, Ercoli L, Donini P (1992) Locating the alien chromatin segment in wheat–*Aegilops longissima* mildew resistant transfers. Hereditas 116:239–245
Chen PD, Tsujimoto H, Gill BS (1994) Transfer of *PhI* genes promoting homoeologous pairing from *Triticum speltoides* to common wheat. Theor Appl Genet 88:97–101
Chen PD, Qi LL, Zhou B, Zhang SZ, Liu DJ (1995) Development and molecular cytogenetic analysis of wheat-*Haynaldia villosa* 6VS/6AL translocation lines specifying resistance to powdery mildew. Theor Appl Genet 91:1125–1129
Chen XM, Luo YH, Xia XC, Xia LQ, Chen X, Ren ZL, He ZH, Jia JZ (2005) Chromosomal location of powdery mildew resistance gene *Pm16* in wheat using SSR marker analysis. Plant Breed 124:225–228
Chen S, Rouse MN, Zhang W, Jin Y, Akhunov E, Wei Y, Dubcovsky J (2015) Fine mapping and characterization of *Sr21*, a temperature-sensitive diploid wheat resistance gene effective against the *Puccinia graminis* f. sp. *tritici* Ug99 race group. Theor Appl Genet 128:645–656
Cherukuri DP, Gupta PK, Charpe A, Koul S, Prabhu KV, Singh RB, Haq QMR (2005) Molecular mapping of *Aegilops speltoides* derived leaf rust resistance gene *Lr28* in wheat. Euphytica 143:19–26
Chhuneja P, Kaur S, Garg T, Ghai M, Kaur S, Prashar M, Bains NS, Goel RK, Keller B, Dhaliwal HS, Singh K (2008a) Mapping of adult plant stripe rust resistance genes in diploid a genome wheat species and their transfer to bread wheat. Theor Appl Genet 116:313–324
Chhuneja P, Kaur S, Goel RK, Aghaee-Sarbarzeh M, Prashar M, Dhaliwal HS (2008b) Transfer of leaf rust and stripe rust resistance from *Aegilops umbellulata* Zhuk. to bread wheat (*Triticum aestivum* L.). Genet Resour Crop Evol 55:849–859
Chhuneja P, Kaur S, Singh K, Dhaliwal HS (2008c) Evaluation of *Aegilops tauschii* (L.) germplasm for Karnal bunt resistance in a screen house with simulated environmental conditions. Plant Genet Resour Charact Util 6:79–84
Chhuneja P, Garg T, Kumar R, Kaur S, Sharma A, Bains NS, Ahuja M, Dhaliwal HS, Singh K (2010) Evaluation of *Aegilops tauschii* Coss. germplasm for agromorphological traits and genetic diversity using SSR loci. Indian J Genet 70:328–338
Chhuneja P, Kumar K, Stirnweis D, Hurni S, Keller B, Dhaliwal HS, Singh K (2012) Identification and mapping of two powdery mildew resistance genes in *Triticum boeoticum* L. Theor Appl Genet 124:1051–1058
Chhuneja P, Yadav B, Stirnweis D, Hurni S, Kaur S, Elkot AF, Keller B, Wicker T, Sehgal S, Gill BS, Singh K (2015) Fine mapping of powdery mildew resistance genes *PmTb7A.1* and *PmTb7A.2* in *Triticum boeoticum* (Boiss.) using the shotgun sequence assembly of chromosome 7AL. Theor Appl Genet. doi:10.1007/s00122-015-2570-5
Collard BCY, Mackill DJ (2008) Marker-assisted selection: an approach for precision plant breeding in the twenty-first century. Philos Trans R Soc Lond B Biol Sci 363:557–572

Cox TS (1998) Deepening the wheat gene pool. J Crop Prod 1:1–25
Cox TS, Hatchett JH (1994) Hessian fly resistance gene H26 transferred from *Triticum tauschii* to common wheat. Crop Sci 34:958–960
Cox TS, Raupp WJ, Gill BS (1994) Leaf rust-resistance genes *Lr41*, *Lr42* and *Lr43* transferred from *Triticum tauschii* to common wheat. Crop Sci 34:339–343
Dadkhodaie NA, Karaoglou H, Wellings CR, Park RF (2011) Mapping genes *Lr53* and *Yr35* on the short arm of chromosome 6B of common wheat with microsatellite markers and studies of their association with *Lr36*. Theor Appl Genet 122:479–487
Das MK, Rajaram S, Mundt CC, Kronstad WE (1992) Inheritance of slow-rusting resistance to leaf rust in wheat. Crop Sci 32:1452–1456
Dedryver F, Jubier MF, Thouvenin J, Goyeau H (1996) Molecular markers linked to the leaf rust resistance gene *Lr24* in different wheat cultivars. Genome 39:830–835
Delibes A, Romero D, Aguaded S, Duce A, Mena M, Lopez-Brana I, Andres MF, MartinSanchez JA, Garcia-Olmedo F (1993) Resistance to the cereal cyst nematode (*Heterodera avenae* Woll.) transferred from the wild grass *Aegilops ventricosa* to hexaploid wheat by a "stepping-stone" procedure. Theor Appl Genet 87:402–408
Dhaliwal HS, Singh H (1997) Breeding for resistance to bunts and smuts: Indian scenario. In: Proceedings bunts and smuts of wheat: an international symposium. North Carolina, North American Plant Protection Organization, Ottawa, pp 327–347
Dhaliwal HS, Singh H, Singh KS, Randhawa HS (1993) Evaluation and cataloguing of wheat germplasm for disease resistance and quality. In: Damania AB (ed) Biodiversity and wheat improvement. Wiley, London, pp 123–140
Dhaliwal HS, Singh Harjit, William M (2002) Transfer of rust resistance from *Aegilops ovata* into bread wheat (*Triticum aestivum* L.) and molecular characterization of resistant derivatives. Euphytica 126:153–159
Dhaliwal HS, Chhuneja P, Gill RK, Goel RK, Singh H (2003) Introgression of disease resistance genes from related species into cultivated wheats through interspecific hybridization. Crop Improv 29:1–18
Diaz-Salazar J, Orellana J (1995) *Aegilops searsii* species specific DNA and chromosome markers. Chromosome Res 3:99
Doležel J, Vrána J, Šafár J, Bartoš J, Kubaláková M, Šimková H (2012) Chromosomes in the flow to simplify genome analysis. Funct Integr Genomics 12:397–416
Duan X, Sheng B, Zhou Y, Xiang Q (1998) Monitoring of the virulence population of *Erysiphe graminis* f. sp. *tritici*. Acta Phytophylac Sin 25:31–36
Dubcovsky J, Luo MC, Dvorak J (1995) Differentiation between homoeologous chromosomes 1A of wheat and lAm of *Triticum monococcum* and its recognition by the wheat *Phl* locus. Proc Natl Acad Sci USA 92:6645–6649
Dvorak J (1977) Transfer of leaf rust resistance from *Aegilops speltoides* to *Triticum* aestivum. Can J Genet Cytol 19:133–141
Dvorak J, di Terlizzi P, Zhang HB, Resta P (1993) The evolution of polyploid wheats: Identification of the A genome donor species. Genome 36:21–31
Dvorak J, Luo M-C, Yang ZL, Zhang HB (1998) The structure of the *Aegilops tauschii* gene pool and the evolution of hexaploid wheat. Theor Appl Genet 97:657–670
Dyck PL, Sykes EE (1994) Genetics of leaf-rust resistance in three spelt wheats. Can J Plant Sci 74:231–233
Elkot AFA, Chhuneja P, Kaur S, Saluja M, Keller B, Singh K (2015) Marker assisted transfer of two powdery mildew resistance genes *PmTb7A.1* and *PmTb7A.2* from *Triticum boeoticum* (Boiss.) to *Triticum aestivum* (L.). PLoS One 10:e0128297. doi:10.1371/journal.pone.0128297
Faris JD, Xu SS, Cai X, Friesen TL, Jin Y (2008) Molecular and cytogenetic characterization of a durum wheat–*Aegilops speltoides* chromosome translocation conferring resistance to stem rust. Chromosome Res 16:1097–1105
Feldman M, Sears ER (1981) The wild gene resources of wheat. Sci Am 244:98–109

Feng J, Zuo LL, Zhang ZY, Lin RM, Cao YY, Xu SC (2011) Quantitative trait loci for temperature-sensitive resistance to *Puccinia striiformis* f. sp. *tritic*i in wheat cultivar Flinor. Euphytica 178:321–329

Feuillet C, Messmer M, Schachermayr G, Keller B (1995) Genetic and physical characterisation of the *Lr1* leaf rust resistance locus in wheat (*Triticum* aestivum L.). Mol Gen Genet 248:553–562

Friebe B, Mukai Y, Dhaliwal HS, Martin TJ, Gill BS (1991) Identification of alien chromatin specifying resistance to wheat streak mosaic virus and greenbug in wheat germplasm by C-banding and in situ hybridization. Theor Appl Genet 81:381–389

Friebe B, Heun M, Tuleen N, Zeller FJ, Gill BS (1994) Cytogenetically monitored transfer of powdery mildew resistance from rye into wheat. Crop Sci 34:621–625

Friebe B, Jiang J, Raupp WJ, McIntosh RA, Gill BS (1996) Characterization of wheat-alien translocations conferring resistance to diseases and pests: current status. Euphytica 91:59–87

Friebe B, Wilson DL, Raupp WJ, Gill BS, Brown-Guedira GL (2005) Notice of release of KS04WGRC45 leaf rust resistant hard winter wheat germplasm. Annu Wheat Newsl 51:188–189

Fu D, Uauy C, Distelfeld A, Blechl A, Epstein L, Chen X, Sela H, Fahima T, Dubcovsky J (2009) A Kinase-START gene confers temperature-dependent resistance to wheat stripe sust. Science 323:1357–1360

Fuentes-Davila G, Rajaram S, Singh G (1995) Inheritance of resistance to Karnal bunt (*Tilletia indica* Mitra) in bread wheat (*Triticum aestivum* L.). Plant Breed 114:250–252

Gall J, Pardue M (1969) Formation and detection of RNA-DNA hybrid molecules in cytological preparations. Proc Natl Acad Sci USA 63:378–383

Gerechter-Amitai AK, Wahl I, Vardi A, Zohary D (1971) Transfer of stem rust seedling resistance from wild diploid einkorn to tetraploid durum wheat by means of a triploid hybrid bridge. Eriplzyticu 20:281–285

Gerechter-Amitai ZK, Grama A (1974) Inheritance of resistance to stripe rust (*Puccinia striiformis*) in crosses between wild emmer (*Triticum dicoccoides*) and cultivated tetraploid and hexaploid wheats, I. *Triticum durum*. Euphytica 23:387–392

German SE, Kolmer JA (1992) Effect of *Lr34* in the enhancement of resistance to leaf rust of wheat. Theor Appl Genet 84:97–105

Gill BS, Raupp WJ (1987) Direct genetic transfers from *Aegilops squarrosa* L. to hexaploid wheat. Crop Sci 27:445–450

Gill BS, Sharma HC, Raupp WJ, Browder LE, Hatchett JH, Harvey TL, Moseman GJ, Waines JG (1985) Evaluation of *Aegilops* species for resistance to wheat powdery mildew, wheat leaf rust, Hessian fly and greenbug. Plant Dis 69:314–316

Gill BS, Raupp WI, Sharma HC, Browder LE, Hatchett JH, Harvey TL, Moseman JG, Waines JG (1986) Resistance in *Aegilops tauschii* to wheat leaf rust, wheat powdery mildew, greenbug, and Hessian fly. Plant Dis 70:550–556

Gill BS, Hatchett JH, Raupp WJ (1987) Chromosomal mapping of Hessian fly resistance gene *H13* in the D genome of wheat. J Heredity 78:97–100

Gill KS, Lubbers EL, Gill BS, Raupp WJ, Cox TS (1991) A genetic linkage map of *Triticum tauschii* (DD) and its relationship to the D genome of bread wheat (AABBDD). Genome 34:362–374

Gill BS, Huang L, Kuraparthy V, Raupp WJ, Wilson DL, Friebe B (2008) Alien genetic resources for wheat leaf rust resistance, cytogenetic transfer, and molecular analysis. Aus J Agric Res 59:197–208

Gold J, Hardner D, Towley-Smith F, Aung T and Procunier J (1999) Development of a molecular marker for rust resistance genes *Sr*39 and *Lr*35 in wheat breeding lines. Electron J Biotechnol 2. http://www.ejbiotechnology.info/content/vol2/issue1/full/1/

Gold J, Harder D, Townley-smith F, Aung T, Procunier JD (2002) Development of molecular marker for rust resistance genes *Sr*39 and *Lr*35 in wheat breeding lines. Electron J Biotechnol 2:35–40

Guo Q, Zhang ZJ, Xu YB, Li GH, Feng J, Zhou Y (2008) Quantitative trait loci for high temperature adult-plant and slow-rusting resistance to *Puccinia striiformis f. sp. tritici* in wheat cultivars. Phytopathology 98:803–809

Gupta SK, Charpe A, Prabhu KV, Haque QMR (2006) Identification and validation of molecular markers linked to the leaf rust resistance gene *Lr19* in wheat. Theor Appl Genet 113:1027–1036

Hao YF, Chen ZB, Wang YY, Bland D, Buck J, Brown-Guedira G, Johnson J (2011) Characterization of a major QTL for adult plant resistance to stripe rust in US soft red winter wheat. Theor Appl Genet 123:1301–1411

Harjit-Singh Tsujimoto H, Sakhuja PK, Singh T, Dhaliwal HS (2000) Transfer of resistance to wheat pathogens from *Aegilops triuncialis* to bread wheat. Wheat Inf Serv 91:5–10

Hartl L, Weiss H, Stephan U, Zeller FJ, Jahoor A (1995) Molecular identification of powdery mildew resistance genes in common wheat (*Triticum* aestivum L.). Theor Appl Genet 90:601–606

He R, Chang Z, Yang Z, Yuan Z, Zhan H, Zhang X, Liu J (2009) Inheritance and mapping of powdery mildew resistance gene *Pm43* introgressed from *Thinopyrum intermedium* into wheat. Theor Appl Genet 118:1173–1180

Helguera M, Khan IA, Dubcovsky J (2000) Development of PCR markers for wheat leaf rust resistance gene *Lr47*. Theor Appl Genet 100(7):1137–1143

Helguera M, Khan IA, Kolmer J, Lijavetzki D, Zhong-qi L, Dubcovsky J (2003) PCR assays for the *Lr37-Yr17-Sr38* cluster of rust resistance genes and their use to develop isogenic hard red spring wheat lines. Crop Sci 43(5):1839–1847

Helguera M, Vanzetti L, Soria M, Khan IA, Kolmer J, Dubcovsky J (2005) PCR markers for *Triticum speltoides* leaf rust resistance gene *Lr51* and their use to develop isogenic hard red spring wheat lines. Crop Sci 45:728–734

Herrera-Foessel SA, lagudah ES, Huerta-Espino J, Hayden M, Bariana HS, Singh D, Singh RP (2011) New slow rusting leaf rust and stripe rust resistance genes *Lr67* and *Yr46* are pleiotropic or closely linked. Theor Appl Genet 122:239–249

Herrera-Foessel SA, Singh RP, Huerta-Espino J, William HM, Garcia V, Djurle A, Yuen J (2008) Identification and characterization of leaf rust resistance gene *Lr14a* in durum wheat. Plant Dis 92:469–473

Herrera-Foessel SA, Singh RP, Lillemo M, Huerta-Espino J, Bhavani S, Singh S, Lan C, Calvo-Salazar V, Lagudah ES (2014) *Lr67/Yr46* confers adult plant resistance to stem rust and powdery mildew in wheat. Theor Appl Genet 127:781–789

Heslop-Harrison JS, Leitch AR, Schwarzacher T, Anamthawat- Jonsson K (1990) Detection and characterization of 1B/1R translocations in hexaploid wheat. Heredity 65:385–392

Heun M, Friebe B, Bushuk W (1990) Chromosomal location of the powdery mildew resistance gene of Amigo wheat. Phytopathology 80:1129–1133

Heun M, Scha fer-Pregl R, Klawan D, Castagna R, Accerbi M, Borghi B, Salamini F (1997) Site of einkorn wheat domestication identified by DNA fingerprinting. Science 278:1312–1314

Hiebert CW, Thomas JB, Somers DJ, McCallum BD, Fox SL (2007) Microsatellite mapping of adult plant leaf rust resistance gene in *Lr22* in wheat. Theor Appl Genet 115:877–884

Hiebert CW, Thomas JB, McCallum BD, Humphreys DG, DePauw RM, Hayden MJ, Mago R, Schnippenkoetter W, Spielmeyer W (2010) An introgression on wheat chromosome 4DL in RL6077 (Thatcher*6/PI 250413) confers adult plant resistance to stripe rust and leaf rust (*Lr67*). Theor Appl Genet 121:1083–1091

Hsam SLK, Zeller FJ (2002) Breeding for powdery mildew resistance in common wheat (*Triticum* aestivum L.). In: Belanger RR, Bushnell WR, Dik AJ, Carver TLW (eds) The powdery mildews, a comprehensive treatise. APS Press, St. Paul, pp 219–238

Hsam SLK, Huang XQ, Ernst F, Hartl L, Zeller FJ (1998) Chromosomal location of genes for resistance to powdery mildew in common wheat (*Triticum* aestivum L. em Thell.). 5. Alleles at the *Pm1* locus. Theor Appl Genet 96:1129–1134

Hsam SLK, Huang XQ, Zeller (2001) Chromosomal location of genes for resistance to powdery mildew in common wheat (*Triticum aestivum* L. em. Thell.) 6. Alleles at the *Pm5* locus. Theor Appl Genet 102:127–133

Hsam SLK, Lapochkina IF, Zeller FJ (2003) Chromosomal location of genes for resistance to powdery mildew in common wheat (*Triticum aestivum* L. em Thell.). 8. Gene *Pm32* in a wheat *Aegilops speltoides* translocation line. Euphytica 133:367–370

Hu WG, Wang YJ, Wang CY, Ji WQ (2007) Genetic analysis on the powdery mildew resistance of Shanxi wheat landraces. J Triticeae Crops 27:341–344

Hua W, Liu Z, Zhu J, Xie C, Yang T, Zhou Y, Duan X, Sun Q, Liu Z (2009) Identification and genetic mapping of *Pm42*, a new recessive wheat powdery mildew resistance gene derived from wild emmer (*Triticum turgidum* var. *dicoccoides*). Theor Appl Genet 119:223–230

Huang L (1998) Molecular markers linked to the leaf rust resistance genes *Lr39* and *Lr40* of wheat introgressed from *Aegilops tauschii*. Kansas State University, kansas, pp 178

Huang L, Gill BS (2001) An RGA-like marker detects all known *Lr*21 leaf rust resistance gene family members in *Aegilops tauschii* and wheat. Theor Appl Genet 103:1007–1013

Huang XQ, Roder MS (2004) Molecular mapping of powdery mildew resistance in wheat: a review. Euphytica 137:203–223

Huang L, Brooks SA, Li W, Fellers JP, Trick HN, Gill BS (2003) Map-Based cloning of leaf rust resistance gene *Lr21* from the large and polyploid genome of bread wheat. Genetics 164:655–664

Huang L, Zhang LQ, Liu BL, Yan ZH, Zhang B, Zhang YL, Liu DC (2011) Molecular tagging of a stripe rust resistance gene in *Aegilops tauschii*. Euphytica 179:313–318

Hussien T, Bowden RL, Gill BS, Cox TS, Marshall DS (1997) Performance of four new leaf rust resistance genes transferred to common wheat from *Aegilops tauschii* and *Triticum monococcum*. Plant Dis 81:582–586

IWGSC (2014) A chromosome-based draft sequence of the hexaploid bread wheat (*Triticum aestivum*) genome. Science 345:1251788-1–1251788-11

Jaccoud D, Peng K, Feinstein D, Kilian A (2001) Diversity arrays: a solid state technology for sequence information independent genotyping. Nucleic Acids Res 29(4):e25

Jakaska V (1995) Isoenzymes in the evaluation of germplasm diversity in wild diploid relatives of cultivated wheat. In: Damania AB (ed) Biodiversity and wheat improvement. ICARDA, Wiley-Sayce Publication, London, pp 247–257

Jarve K, Peusha HO, Tsymbalova J, Tamm S, Devos KM, Enno TM (2000) Chromosomal location of a *Triticum timopheevii*—derived powdery mildew resistance gene transferred to common wheat. Genome 43:377–381

Jeger MJ, Viljanen-Rollinson SLH (2001) The use of the area under the disease-progress curve (AUDPC) to assess quantitative disease resistance in crop cultivars. Theor Appl Genet 102:32–40

Jia J, Devos KM, Chao S, Miller TE, Reader SM, Gale MD (1996) RFLP-based maps of the homoeologous group-6 chromosomes of wheat and their application in the tagging of *Pm12*, a powdery mildew resistance gene transferred from *Aegilops speltoides* to wheat. Theor Appl Genet 92:559–565

Jia J, Zhao S, Kong X, Li Y, Zhao G et al (2013) *Aegilops tauschii* draft genome sequence reveals a gene repertoire for wheat adaptation. Nature 495:91–95

Jiang J, Gill BS (1994) Different species-specific chromosome translocations in *Triticum timopheevii* and *T. turgidum* support the diphyletic origin of polyploid wheats. Chr Res 2:59–64

Jiang J, Friebe B, Gill BS (1994) Recent advances in alien gene transfer in wheat. Euphytica 73:199–212

John H, Birnstiel M, Jones K (1969) RNA-DNA hybrids at the cytological level. Nat (Lond) 223:582–587

Johnson R (1988) Durable resistance to yellow (stripe) rust in wheat and its implications in plant breeding. In: Simmonds NW, Rajaram S (eds) Breeding strategies for resistance to the rusts of wheat. CIMMYT, DF, Mexico, pp 63–75

Johnson BL, Dhaliwal HS (1976) Reproductive isolation of *Triticum boeoticum* and *Triticum urartu* and the origin of the tetraploid wheats. Am J Bot 63:1088–1094

Jørgensen JH (1973) Gene *Pm6* for resistance to powdery mildew in wheat. Euphytica 22:43

Kaur P (2014) Molecular characterization of wheat-*Ae. caudata* introgression (s) carrying leaf and stripe rust resistance gene (s) using wheat genome sequence data. Punjab Agricultural University Thesis

Kerber ER (1987) Resistance to leaf rust in hexaploid wheat, *Lr32*, a third gene derived from *Triticum tauschii*. Crop Sci 27:204–206

Kerber ER, Aung T (1999) Leaf rust resistance gene *Lr34* associated with non suppression of stem rust resistance in the wheat cultivar Canthatch. Phytopathology 89:518–521

Kerber ER, Dyck PL (1969) Inheritance in hexaploid wheat of leaf rust resistance and other characters derived from *Aegilops squarrosa*. Can J Genet Cytol 11:639–647

Kerber ER, Dyck PL (1990) Transfer to hexaploid wheat of linked genes for adult-plant leaf rust and seedling stem rust resistance from an amphiploid of *Aegilops speltoides x Triticum monococcum*. Genome 33:530–537

Khan RR, Bariana HS, Dholakia BB, Naik SV, Lagu MD, Rathjen AJ, Bhavani S, Gupta VS (2005) Molecular mapping of stem and leaf rust resistance in wheat. Theor Appl Genet 111:846–850

Kihara H (1919) Ueber cytologische Studien bei einigen Getreidearten. 1. Spezies-Bastarde des Weizens und Weizenroggen-Bastard. Bot Mag 32:17–38

Kihara H (1944) Discovery of the DD-analyser, one of the ancestors of *Triticum vulgare* (Japanese). Agric Hortic 19:13–14

Kihara H, Tanaka M (1958) Morphological and physiological variation among Aegilops squarrosa strains collected in Pakistan, Afganistan and Iran. Preslia. 30:241–251

Kihara H, Yamashita K, Tanaka M (1965) Morphological, physiological, geographical and cytological studies in *Aegilops* and *Triticum* collected in Pakistan, Afghanistan and Iran. In: Yamashita K (ed) Cultivated plants and their relatives. Koei Printing, Japan, pp 1–118

Knaggs P, Ambrose MJ, Reader SM, Miller TE (2000) Morphological characterization and evaluation of the subdivision of *Aegilops tauschii* Coss. Wheat Inf Serv 91:15–19

Knight E, Greer E, Draeger T, Thole V, Reader S, Shaw P, Moore G (2010) Inducing chromosome pairing through premature condensation: Analysis of wheat interspecific hybrids. Funct Integr Genomics 10:603–608

Knott DR (1962) The inheritance of rust resistance:IX. The inheritance of resistance to races 15B and 56 of stem rust in the wheat variety Khapstein. Canm J Plant Sci 42:415–419

Kolmer J (2008) *Lr63*, *Lr64*. In: McIntosh RA, Dubcovsky J, Rogers WJ, Morris C, Appels R, Xia XC (eds) Catalogue of gene symbols for wheat: 2009 supplement, p 271 (Reference 10550, p 273). Ann Wheat Newsl 55:256–278

Kolmer JA, Anderson JE, Flor JM (2010) Chromosome location, linkage with simple sequence repeat markers, and leaf rust resistance conditioned by gene *Lr*63 in wheat. Crop Sci 50:2392–2395

Kuang H, Woo SS, Meyers BC (2004) Multiple genetic processes result in heterogeneous rates of evolution within the major cluster disease resistance genes in lettuce. Plant Cell 16:2870–2894

Kuraparthy V, Singh H, Singh S, Chhuneja P, Dhaliwal HS (2001) Microsatellite marker linked to a leaf rust resistance gene from *Triticum monococcum* L. transferred to bread wheat. J Plant Biochem Biotechnol 10:127–132

Kuraparthy V, Sood S, Chhuneja P, Dhaliwal HS, Kaur S, Bowden RL, Gill BS (2007) A cryptic wheat-*Aegilops triuncialis* translocation with leaf rust resistance gene *Lr58*. Crop Sci 47:1995–2003

Kuraparthy V, Sood S, See DR, Gill BS (2009) Development of a PCR assay and marker-assisted transfer of leaf rust and stripe rust resistance genes *Lr57* and *Yr40* into hard red winter wheats. Crop Sci 49:120–126

Kuraparthy V, Sood S, Guedira GB, Gill BS (2011) Development of a PCR assay and marker-assisted transfer of leaf rust resistance gene *Lr58* into adapted winter wheats. Euphytica 180:227–234

Kurapathy V, Chhuneja P, Dhaliwal HS, Kaur S, Bowden RL, Gill BS (2007) Characterization and mapping of cryptic alien introgression from *Aegilops geniculata* with novel leaf rust and stripe rust resistance genes *Lr57* and *Yr40* in wheat. Theor Appl Genet 114:1379–1389

Lan CX, Ni XW, Yan J, Zhang Y, Xia XC, Chen XM, He ZH (2010) Quantitative trait loci mapping of adult-plant resistance to powdery mildew in Chinese wheat cultivar Lumai 21. Mol Breed 25:615–622

Law CN, Wolfe MS (1966) Location for genetic factors for mildew resistance and ear emergence time on chromosome 7B of wheat. Can J Genet Cytol 8:462–470

Li G, Fang T, Zhang H (2009) Molecular identification of a new powdery mildew resistance gene *Pm41* on chromosome 3BL derived from wild emmer (*Triticum turgidum* var. *dicoccoides*). Theor Appl Genet 119:531–539

Lillemo M, Asalf B, Singh RP, Huerta-Espino J, Chen XM, He ZH, Bjornstad A (2008) The adult plant rust resistance loci *Lr34/Yr18* and *Lr46/Yr29* are important determinants of partial resistance to powdery mildew in bread wheat line Saar. Theor Appl Genet 116:1155–1166

Lillemo M, Singh RP, William M, Herrera-Foessel SA, Huerta-Espino J, Germa′n S, Campos P, Chaves M, Madriaga R, Xia X, Liang S, Liu D, Li Z, Lagudah E (2011) Multiple rust resistance and gene additivity in wheat: lessons from multi-location case studies in the cultivars Parula and Saar. In: Global rust initiative meeting, St. Paul, pp 111–120

Ling H-Q, Qiu J, Singh RP et al (2004) Identification and genetic characterization of an *Aegilops tauschii* ortholog of the wheat leaf rust disease resistance gene *Lr1*. Theor Appl Genet 109:1133–1138

Liu ZY, Sun QX, Ni ZF, Nevo E, Yang TM (2002) Molecular characterization of a novel powdery mildew resistance gene *Pm30* in wheat originating from wild emmer. Euphytica 123:21–29

Liu S, Yu L-X, Singh RP, Jin Y, Sorrells ME, Anderson JA (2010) Diagnostic and co-dominant PCR markers for wheat stem rust resistance genes *Sr*25 and *Sr*26. Theor Appl Genet 120:691–697

Liu W, Rouse M, Friebe B, Jin Y, Gill BS, Pumphrey MO (2011) Discovery and molecular mapping of a new gene conferring resistance to stem rust, *Sr53*, derived from *Aegilops geniculata* and characterization of spontaneous translocation stocks with reduced alien chromatin. Chromosome Res 19:669–682

Liu J, Chang Z, Zhang X, Yang Z, Li X, Jia J, Zhan H, Guo H, Wang J (2013) Putative *Thinopyrum intermedium*-derived stripe rust resistance gene *Yr50* maps on wheat chromosome arm 4BL. Theor Appl Genet 126:265–274

Lubbers EL, Gill KS, Cox TS, Gill BS (1991) Variation of molecular markers among geographically diverse accessions of *Triticum tauschii*. Genome 34:354–361

Luo PG, Luo HY, Chang ZJ, Zhang HY, Zhang M, Ren ZL (2009) Characterization and chromosomal location of *Pm40* in common wheat: a new gene for resistance to powdery mildew derived from *Elytrigia intermedium*. Theor Appl Genet 118:1058–1064

Lutz J, Hsam SLK, Limpert E, Zeller FJ (1995a) Chromosomal location of powdery mildew resistance genes in *Triticum aestivum* L. (common wheat). 2. Genes *Pm2* and *Pm19* from *Aegilops squarrosa* L. Heredity 74:152–156

Lutz J, Katzhammer M, Stephan U, Felsenstein FG, Oppitz K, Zeller FJ (1995b) Identification of powdery mildew resistance genes in common wheat (*Triticum aestivum* L. em. Thell.). V. Old German cultivars and cultivars released in the former GDR. Plant Breed 114(1):29–33

Ma H, Hughes GR (1993) Resistance to Septoria nodorum blotch in several *Triticum*. Euphytica 70:151–157

Ma H, Singh RP, Mujeeb-Kazi A (1995) Suppression/expression of resistance to stripe rust in synthetic hexaploid wheat (*Triticum turgidum/T. tauschii*). Euphytica 83:87–93

Mago R, Spielmeyer W, Lawrence GJ, Lagudah ES, Ellis JG, Pryor A (2002) Identification and mapping of molecular markers linked to rust resistance genes located on chromosome 1RS of rye using wheat-rye translocation lines. Theor Appl Genet 242(104):1317–1324

Mago R, Bariana HS, Dundas IS, Spielmeyer W, Lawrence GJ, Pryor AJ, Ellis JG (2005a) Development of PCR markers for the selection of wheat stem rust resistance genes *Sr24* and *Sr26* in diverse wheat germplasm. Theor Appl Genet 111:496–504

Mago R, Miah H, Lawrence GJ, Wellings CR, Spielmeyer W, Bariana HS, Mcintosh RA, Pryor AJ, Ellis JG (2005b) High-resolution mapping and mutation analysis separate the rust resistance genes *Sr31*, *Lr26* and *Yr9* on the short arm of rye chromosome 1. Theor Appl Genet 112:41–50

Mago R, Zhang P, Bariana HS, Verlin DC, Bansal UK, Ellis JG, Dundas IS (2009) Development of wheat lines carrying stem rust resistance gene *Sr39* with reduced *Aegilops speltoides* chromatin and simple PCR markers for marker-assisted selection. Theor Appl Genet 124:65–70

Mago R, Brown-Guedira G, Dreisigacker S, Breen J, Jin Y, Singh R, Appels R, S. Lagudah ES, Ellis J, Spielmeyer W (2011) An accurate DNA marker assay for stem rust resistance gene *Sr2* in wheat. Theor Appl Genet 122:735–744

Mago R, McIntosh R, Dodds P, Dvorak J, Lagudah E (2013a) The gene *Sr33*, an ortholog of Barley *Mla* genes, encodes resistance to wheat stem rust race Ug99. Science 341:786–788

Mago R, Verlin D, Zhang P, Bansal U, Bariana H S, Jin Y, Ellis J, Hoxha S and Dundas I (2013) Development of wheat–*Aegilops speltoides* recombinants and simple PCR based markers for *Sr*32 and a new stem rust resistance gene on the 2S#1 chromosome. Theor Appl Genet. doi:10.1007/s00122-013-2184-8

Mamluk OF, Van Slageren MW (1994). Sources of resistance to wheat diseases in *Aegilops* and *Amblyopyrum* spp. In: Proceedings of 9th congress of the Mediterranean Phytopathological Union. Kusadasi, Turkey, pp 269–270

Manisterski J, Segal A, Levy AA, Feldman M (1988) Evaluation of Israeli *Aegilops* and *Agropyron* species for resistance to wheat leaf rust. Plant Dis 72:941–944

Marais GF, Wessels WG, Horn M (1998) Association of a stem rust resistance gene (*Sr45*) and two Russian wheat aphid resistance genes (*Dn5* and *Dn7*) with mapped structural loci in common wheat. S Afr J Plant Soil 15:67–71

Marais GF, McCallum B, Snyman JE, Pretorius ZA, Marais AS (2005) Leaf rust and stripe rust resistance genes *Lr54* and *Yr37* transferred to wheat from *Aegilops kotschyi*. Plant Breed 124:538–541

Marais GF, McCallum B, Marais AS (2006) Leaf rust and stripe rust resistance genes derived from *Aegilops sharonensis*. Euphytica 149:373–380

Marais GF, Mccallum B, Marais AS (2008) Wheat leaf rust resistance gene *Lr59* derived from *Aegilops peregrina*. Plant Breed 127(4):340–345

Marais F, Marais A, Mccallum B, andPretorius Z (2009) Transfer of leaf rust and stripe rust resistance genes *Lr62* and *Yr42* from *Aegilops neglecta* Req. ex *Bertol*. to common wheat. Crop Sci 49(3):871–879

Marais GF, Bekker TA, Eksteen A, McCallum BD, Fetch TG, Marais AS (2009b) Attempts to remove gametocidal genes co-transferred to common wheat with rust resistance from *Aegilops speltoides*. Euphytica 171:71–85

Marais GF, Badenhorst PE, Eksteen A, Pretorius ZA (2010a) Reduction of *Aegilops sharonensis* chromatin associated with resistance genes *Lr56* and *Yr38* in wheat. Euphytica 171(1):15–22

Marais GF, Kotze L, Eksteen A (2010b) Allosyndetic recombinants of the *Aegilops peregrina*-derived *Lr59* translocation in common wheat. Plant Breed 129(4):356–361

May CE, Lagudah ES (1992) Inheritance in hexaploid wheat of *septoria tritici* blotch resistance and other characteristics derived from *Triticum tauschii*. Aust J Agric Res 43:433–442

McFadden ES, Sears ER (1946) The origin of *Triticum speltoides* and its free-threshing hexaploid relatives. J Hered 37:81–89

McGrann GR, Smith PH, Burt C, Mateos GR, Chama TN, MacCormack R, Wessels E, Agenbag G, Boyd LA (2014) Genomic and genetic analysis of the wheat race-specific yellow rust resistance gene *Yr*5. J Plant Sci Mol Breed 3:2

McIntosh RA, Baker EP (1970) Cytogenetic studies in wheat IV. Chromosome location and linkage studies involving the *Pm2* locus for powdery mildew resistance. Euphytica 19:71–77

McIntosh RA, Miller TE, Chapman V (1982) Cytogenetical studies in wheat. XII. *Lr28* for resistance to Puccinia recondita and *Sr34* for resistance to P. *graminis tritici*. Z. Pflanzenzuchtg 89:295–306

McIntosh RA, Dyck PL, The TT, Cusick J, Milne DL (1984) Cytogenetical studies in wheat XIII. *Sr35*—a third gene from *Triticum monococcum* for resistance to *Puccinia graminis tritici*. Z. Pflanzenzuecht. 92:1–14

McIntosh RA, Hart GE, Gale MD (1991) Catalogue of gene symbols: 1991 supplement. Wheat Newsl 37:200–216

McIntosh RA, Wellings CR, Park RF (1995a) Wheat rusts: an atlas of resistance genes. CSIRO, Melbourne

McIntosh RA, Wellings CR, Park RF (1995) Wheat rusts: an atlas of resistance genes. CSIRO publications, East Melbourne

McIntosh RA, Yamazaki Y, Dubcovsky J, Rogers J, Morris C, Somers DJ, Appels R, Devos KM (2008) Catalogue of gene symbols for wheat. http://www.shigen.nig.ac.jp/wheat/komugi/genes/download.jsp

McIntosh RA, Dubcovsky J, Rogers WJ, Morris C, Appels R, Xia XC (2009) Catalogue of gene symbols: 2009 supplement. In: KOMUGI-Integrated wheat science database. http://www.shigen.nig.ac.jp/wheat/komugi/genes/download.jsp

McIntosh RA, Yamazaki Y, Dubcovsky J, Roger J, Morris C, Somer DJ, Appels R, Devos KM (2010) Catalogue of gene symbols of wheat. In: KOMUGI-Integrated wheat science database. http://www.shigen.ac.jp/wheat/komugi/gene/macgene/2013/GeneCatalogueIntroduction.pdf

McIntosh RA, Yamazaki Y, Dubcovsky J, Rogers J, Morris C, Somers DJ, andDevos KM (2011) Catalogue of gene symbols for wheat: 2011 supplement. http://wwwshigen.nig.ac.jp/wheat/komugi/genes/download.jsp

McIntosh RA, Dubcovsky J, Rogers J, Morris C, Appels R, Xia X (2012) Catalogue of gene symbols for wheat: 2012 supplement. http://www.shigen.nig.ac.jp/wheat/komugi/genes/macgene/supplement2009.pdf, p 247

McIntosh RA, Yamazaki Y, Dubcovsky J, Rogers J, Morris C, Somers J, Appels R and Devos KM (2013) Catalogue of gene symbols for wheat. In: KOMUGI-integrated wheat science database at http://www.shigen.nig.ac.jp/wheat/komugi/genes/download.jsp

Mebrate SA, Oerke EC, Dehne HW, Pillen K (2008) Mapping of the leaf rust resistance gene *Lr*38 on wheat chromosome arm 6DL using SSR markers. Euphytica 162:457–466

Mehdi V, Joshi LM, Abrol YP (1973) Studies on chapattis quality: VI. Effect of wheat grains with bunts on the quality of chapattis'. Bull Grain Technol 11:195–197

Melichar JPE, Berry S, Newell C, MacCormack R, Boyd LA (2008) QTL identification and microphenotype characterisation of the developmentally regulated yellow rust resistance in UK wheat cultivar Guardian. Theor Appl Genet 117:391–399

Michelmore RW, Meyers BC (1998) Clusters of resistance genes in plants evolve by divergent selection and a birth-and-death process. Genome Res 8:1113–1130

Michelmore RW, Paran I, Kesseli RV (1991) Identification of markers linked to disease-resistance genes by bulked segregant analysis: a rapid method to detect markers in specific genomic regions by using segregating populations. Proc Natl Acad Sci USA 88:9828–9832

Miller TE, Reader SM, Singh D (1988) Spontaneous non-Robertsonian translocations between wheat chromosomes and an alien chromosome. In: Koebner R, Miller TE (eds) Proceedings of 7th international wheat genet symp. Institute of Plant Science Research, Cambridge, pp 387–390

Miranda LM, Murphy J, Marshall PD, Leath S (2006) *Pm34*: A new powdery mildew resistance gene transferred from *Aegilops tauschii* Coss. to common wheat (*Triticum aestivum* L.). Theor Appl Genet 113:1497–1504

Miranda LM, Murphy J, Marshall PD, Cowger C, Leath S (2007) Chromosomal location of *Pm35*, a novel *Aegilops tauschii* derived powdery mildew resistance gene introgressed into common wheat (*Triticum aestivum* L.). Theor Appl Genet 114:1451–1456

Mitra M (1931) A new bunt on wheat in India. Ann Appl Biol 18:178–179

Mohler V, Hsam SLK, Zeller FJ, Wenzel G (2001) An STS marker distinguishing the rye-derived powdery mildew resistance alleles at the *Pm8*/*Pm17* locus of common wheat. Plant Breed 120:448–450

Mohler V, Bauer C, Schweizer G, Kempf H, Hartl L (2013a) *Pm50*: a new powdery mildew resistance gene in common wheat derived from cultivated emmer. J Appl Genet 54(3):259–263

Molnar I, Kubalakova M, Simkova H, Cseh A, Molnar-Lang M, Dolezel J (2011) Chromosome isolation by flow sorting in *Aegilops umbellulata* and *Ae. comosa* and their allotetraploid hybrids *Ae. biuncialis* and *Ae. geniculata*. PLoS One 6:e27708. doi:10.1371/journal.pone.0027708

Monneveux P, Zaharieva M, Rekika D (2001) The utilisation of *Triticum* and *Aegilops* species for the improvement of durum wheat. Options Me´diterrane´nnes 40:71–81

Morgunov A, Montoya J, Rajaram S (1994) Genetic analysis of resistance to Karnal bunt (*Tilletia indica* (Mitra)) in bread wheat. Euphytica 74:41–46

Mujeeb-Kazi A, Roldan S, Suh DY, Sitch LA, Farooq S (1987) Production and cytogenetic analysis of hybrids between *Triticum aestivum* and some caespitose *Agropyron* species. Genome 29:537–553

Mukai Y, Friebe B, Hatchett JH, Yamamoto M, Gill BS (1993) Molecular cytogenetic analysis of radiation-induced wheat-rye terminal and intercalary chromosomal translocations and the detection of rye chromatin specifying resistance to Hessian fly. Chromosoma 102:88–95

Murata M, Nakata N, Yasumuro Y (1992) Origin and molecular structure of a midget chromosome in a common wheat carrying rye cytoplasm. Chromosoma 102:27–31

Mwale VM, Chilembwe EHC, Uluko HC (2014) Wheat powdery mildew (*Blumeria graminis* f. sp. *tritici*): Damage effects and genetic resistance developed in wheat (*Triticum aestivum*). Int Res J Plant Sci 5:1–16

Neu C, Stein N, Keller B (2002) Genetic mapping of the *Lr20–Pm1* resistance locus reveals suppressed recombination on chromosome arm 7AL in hexaploid wheat. Genome 45:737–744

Niu Z, Klindworth DL, Friesen TL, Chao S, Jin Y, Cai X, Xu SS (2011) Targeted introgression of a wheat stem rust resistance gene by DNA marker-assisted chromosome engineering. Genetics 187:1011–1021

Njau PN, Jin Y, Huerta-Espino J, Keller B, Singh RP (2010) Identification and evaluation of sources of resistance to stem rust race Ug99 in wheat. Plant Dis 94:413–419

Ohm HW, Shaner GE (1976) Three components of slow leaf-rusting at different growth stages in wheat. Phytopathology 66:1356–1360

Olson EL, Brown-Guedira G, Marshall D, Stack E, Bowden RL, Jin Y, Rouse M, Pumphrey MO (2010) Development of wheat lines having a small introgressed segment carrying stem rust resistance gene *Sr22*. Crop Sci 50:1823–1830

Olson EL, Rouse MN, Pumphrey MO, Bowden RL, Gill BS, Poland JA (2013) Introgression of stem rust resistance genes *Sr*TA10187 and *Sr*TA10171 from *Aegilops tauschii* to wheat. Theor Appl Genet 126(10):2477–2484

Pasquini M (1980) Disease resistance in wheat. II. Behaviour of *Aegilops* species with respect to *Puccinia recondita* f. sp. *tritici*, *Puccinia graminis* f. sp. *tritici* and *Erysiphe graminis* f. sp. *tritici*. Genet Agrar 34:133–148

Paull JG, Pallotta MA, Langridge P, The TT (1994) RFLP markers associated with *Sr22* and recombination between chromosome 7A of bread wheat and the diploid species *Triticum boeoticum*. Theor Appl Genet 89:1039–1045

Paux E, Sourdille P, Salse J et al (2008) A physical map of the 1-gigabase bread wheat chromosome 3B. Science 322:101–104

Peil A, Korzun V, Schubert V, Schumann E, Weber WE, Roder MS (1998) The application of wheat microsatellites to identify disomic *Triticum aestivum-Aegilops markgrafii* addition lines. Theor Appl Genet 96:138–146

Periyannan S, Moore J, Ayliffe M, bansal U, Wang X, Huang L (2013) The gene Sr33, an orthologue of Barley *Mla* genes, encodes resistance to stem rust race Ug99. Science 341:786–788

Periyannan S, Bansal U, Bariana H, Deal K, Luo MC, Dvorak J, Lagudah E (2014) Identification of a robust molecular marker for the detection of the stem rust resistance gene *Sr45* in common wheat. TAG 127:947–955

Perugini LD (2007) Genetic characterization of wheat germplasm with resistance to fusarium head. ProQuest, Ann Arbor

Piarulli L, Gadaletaa A, Manginia G, Signorilea MA, Pasquinib M, Blancoa A, Simeone R (2012) Molecular identification of a new powdery mildew resistance gene on chromosome 2BS from *Triticum turgidum* ssp. *dicoccum*. Plant Sci 196:101–106

Pietro JP, Caillaud CM, Chaubet B, Pierre JS, Trottet M (1998) Variation in resistance to the grain aphid, *Sitobion avenae* (Sternorhynca: Aphididae) among diploid wheat genotypes: multivariate analysis of agronomic data. Plant Breed 117:407–412

Pinkel D, Straume T, Gray JW (1986) Cytogenetic analysis using quantitative, high sensitivity, fluorescence hybridization. Proc Natl Acad Sci USA 83:2934–2938

Plamenov D, Belchev I, Kiryakova V, Spetsov P (2009) Fungal resistance of *Triticum durum—T. monococcum ssp. aegilopoides* amphiploid. J Plant Dis Protect 116:60–62

Poland JA, Rife TW (2012) Genotyping-by-sequencing for plant breeding and genetics. Plant Genome 5:92–102

Poland JA, Brown PJ, Sorrells ME, Jannink JL (2012) Development of high-density genetic maps for barley and wheat using a novel two-enzyme genotyping-by-sequencing approach. PLoS One 7:e32253. doi:10.1371/journal.pone.0032253

Prins R, Groenewald JZ, Marais GF, Snape JW, Koebner RMD (2001) AFLP and STS tagging of *Lr*19, a gene conferring resistance to leaf rust in wheat. Theor Appl Genet 103:618–624

Procunier JD, Townley-smith TF, Prashar S, Gray M, Kim WK, Czamecki E, Dyck PL (1995a) PCR-based RAPD/DGGE markers linked to leaf rust resistance genes *Lr29* and *Lr25* in wheat (*Triticum* aestivum L.). J Genet Breed 49(1):87–92

Purnhauser L, Gyulai G, Tar M, Csosz M, Mesterhazy A, Heszky L, Hrazdina G (2000) Use of molecular markers in wheat breeding for disease resistance. Use of agriculturally important genes in biotechnology. In: Proceedings of the NATO advanced research workshop. Szeged, Hungary, pp 52–57

Qi LL, Cao MS, Chen PD, Li WL, Liu DJ (1996) Identification, mapping, and application of polymorphic DNA associated with resistance gene *Pm*21 of wheat. Genome 39:191–197

Qi LL, Pumphrey MO, Friebe B, Zhang P, Qian C, Bowden RL, Rouse MN, Jin Y, Gill BS (2011) A novel Robertsonian translocation event leads to transfer of a stem rust resistance gene (*Sr52*) effective against race Ug99 from *Dasypyrum villosum* into bread wheat. Theor Appl Genet 123:159–167

Qiu YC, Sun XL, Zhou RH, Kong XY, Zhan SS, Jia JZ (2006) Identification of microsatellite markers linked to powdery mildew resistance gene *Pm2* in wheat. Cereal Res Commun 34(4):1267–1273

Ramburan VP, Pretorius ZA, Louw JH, Boyd LA, Smith PH, Boshoff WHP, Prins R (2004) A genetic analysis of adult plant resistance to stripe rust in the wheat cultivar Kariega. Theor Appl Genet 108:1426–1433

Raupp WJ, Singh S, Brown-Guerdira GL, Gill BS (2001) Cytogenetic and molecular mapping of the leaf rust resistance gene *Lr39* in wheat. Theor Appl Genet 102:347–352

Rayburn AL, Gill BS (1985) Use of biotin-labeled probes to map specific DNA sequences of wheat chromosomes. J Hered 76:78–81

Riar AK, Kaur S, Dhaliwal HS, Singh K, Chhuneja P (2012) Introgression of a leaf rust resistance gene from *Aegilops caudata* to bread wheat. J Genet 91:155–161

Riley R (1960) The diploidization of polyploid wheat. Heredity 15:407–429

Riley R, Unrau J, Chapman V (1958) Evidence on the origin of the B genome of wheat. J Hered 49:91–98

Riley R, Chapman V, Johnson R (1968a) Introduction of yellow rust resistance of *Aegilops comosa* into wheat by genetically induced homoeologous recombination. Nature 217:383–384

Riley R, Chapman V, Johnson R (1968b) The incorporation of alien disease resistance in wheat by genetic interference with the regulation of meiotic chromosome synapsis. Genet Res Camb 12:198–219

Robert O, Abelard C, Dedryver F (1999) Identification of molecular markers for the detection of the yellow rust resistance gene *Yr17* in wheat. Mol Breed 5:167–175

Röder MS, Korzun V, Wendehake K, Plaschke J, Tixier MH, Leroy P, anal MW (1998) A microsatellite map of wheat. Genetics 149:2007–2023

Roelfs AP, 1988. Resistance to leaf and stem rusts in wheat. In: Simmonds NW, Rajaram S (eds) Breeding strategies for resistance to the rusts of wheat. CIMMYT, Mexico 10:22

Rong JK, Millet E, Manisterski J, Feldman M (2000) A new powdery mildew resistance gene: introgression from wild emmer into common wheat and RFLP based mapping. Euphytica 115:121–126

Rouse MN, Jin Y (2011) Stem rust resistance in A-genome diploid relatives of wheat. Plant Dis 95:941–944

Rowland GG, Kerber ER (1974) Telocentric mapping in hexaploid wheat of genes for leaf rust resistance and other characters derived from *Aegilops squarrosa*. Can J Genet Cytol 16:137–144

Rush CM, Stein JM, Bowden RL, Riemenschneider R, Boratynski T, Royer MH (2005) Status of Karnal bunt of wheat in United States 1996–2004. Plant Dis 89:212–222

Saintenac C, Zhang W, Salcedo A, Rouse MN, Trick HN, Akhunov E, Dubcovsky J (2013) Identification of wheat gene *Sr35* that confers resistance to Ug99 stem rust race group. Science 341(6147):783–786

Sambroski DJ, Dyck PL (1976) Inheritance of virulence in *Puccinia recondita* in six backcross lines of wheat with single genes for resistance to leaf rust. Can J Bot 54:1666–1671

Santra DK, Chen XM, Santra M, Campbell KG, Kidwell K (2008) Identification and maping QTL for high-temperature adult-plant resistance to stripe rust in winter wheat (*Triticum aestivum* L.) cultivar 'Stephens'. Theor and Appl Genet 117:793–802

Sarkar P, Stebbins GL (1956) Morphological evidence concerning the origin of the B genome in wheat. Am J Bot 43:297–304

Saur L (1991) In search of sources of resistance to head blight caused by *Fusarium culmorum* in wheat and related species. Agronomie 11:535–541

Sax K (1922) Sterility in wheat hybrids. II. Chromosome behavior in partially sterile hybrids. Genetics 7:513–550

Schachermayr G, Siedler H, Gale MD, Winzeler H, Winzeler M, Keller B (1994) Identification and localization of molecular markers linked to the *Lr9* leaf rust resistance gene of wheat. Theor Appl Genet 88:110–115

Schmolk M, Mohler V, Hartl L, Zeller FJ, Sai L, Hsam K (2012) A new powdery mildew resistance allele at the *Pm4* wheat locus transferred from einkorn (*Triticum monococcum*). Mol Breed 29:449–456

Schwarzacher T, Leitch AR, Bennett MD, Heslop Harrison J S (1989) In situ localization of parental genomes in a wide hybrid. Ann Bot 64:315–324

Schwarzacher T, Anamthawat-Jonsson K, Harrison GE, Islam AKMR, Jia JZ, King IP, Leitch AR, Miller TE, Reader SM, Rogers WJ, Shi M, Heslop-Harrison JS (1992) Genome in situ hybridization to identify alien chromosomes and chromosome segments in wheat. Theor Appl Genet 84:778–786

Seah S, Bariana H, Jahier J, Sivasithamparam K, Lagudah ES (2001) The introgressed segment carrying rust resistance genes *Yr17*, *Lr37* and *Sr38* in wheat can be analysed by a cloned disease resistance gene-like sequence. Theor Appl Genet 102:600–605

Sears ER (1956) The transfer of leaf rust resistance from *Aegilops umbellulata* to wheat. Brookhaven Symp in Biol. No. 9, Genetics in Plant Breeding, pp 1–22

Sears ER (1972) Chromosome engineering in wheat. In: Stadler Symp, vol 4. University of Missouri, Columbia, pp 23–38

Sears ER (1977) Induced mutant with homoeologous pairing in common wheat. Can J Genet Cytol 19:585–593

Sears ER (1981) Transfer of alien genetic material to wheat. In: Evans LT, Peacock WJ (eds) Wheat science today and tomorrow, pp 75–89

Sears ER, Briggle LW (1969) Mapping the gene *Pm1* for resistance to *Erysiphe graminis* f. p. *tritici* on chromosome 7A of wheat. Crop Sci 9:96–97

Seyfarth R, Feuillet C, Schachermayr G, Winzeler M, Keller B (1999) Development of a molecular marker for the adult plant leaf rust resistance gene *Lr35* in wheat. Theor Appl Genet 99:554–560

Shaner GE, Roberts JJ, Finney RE (1972) A culture of *Puccinia recondita* virulent on the wheat cultivar Transfer. Plant Dis Rep 56:827–830

Sharma HC, Gill BS (1983) Current status of wide hybridization. Euphytica 32:17–31

Shen X, Zhou M, Lu W, Ohm H (2003a) Detection of Fusarium head blight resistance QTL in a wheat population using bulked segregant analysis. Theor Appl Genet 106:1041–1047

Shen XR, Ittu M, Ohm HW (2003b) Quantitative loci conditioning resistance to Fusarium head blight in wheat line F201R. Crop Sci 43:850–857

Shi A, Leath S, Murphy JP (1998) A major gene for powdery mildew resistance transferred to common wheat from wild einkorn wheat. Phytopathology 88:144–147

Simons K, Abate Z, Chao S, Zhang W, Rouse M, Jin Y, Elias E, Dubcovsky J (2011) Genetic mapping of stem rust resistance gene *Sr13* in tetraploid wheat (*Triticum turgidum* ssp. *durum* L.). Theor Appl Genet 122:649–658

Singh G, Rajaram S, Montoya J, Fuentes-Davila G (1995) Genetic analysis of Karnal bunt resistance in 14 Mexican bread wheat genotypes. Plant Breed 114:439–441

Singh H, Grewal TS, Dhaliwal HS, Pannu PPS, Bagga PS (1998) Sources of leaf rust and stripe rust resistance in wild relatives of wheat. Crop Improv 25:26–33

Singh H, Grewal TS, Pannu PPS, Dhaliwal HS (1999) Genetics of resistance to Karnal bunt (*Tilletia indica* Mitra) in bread wheat. Euphytica 105:125–131

Singh RP, Nelson JC, Sorrells ME (2000a) Mapping *Yr28* and other genes for resistance to stripe rust in wheat. Crop Sci 40:1148–1155

Singh H, Tsujimoto H, Sakhuja PK, Singh T, Dhaliwal HS (2000b) Transfer of resistance to wheat pathogens from *Aegilops triuncialis* into bread wheat. Wheat Inf Ser 91:5–10

Singh RP, William HM, Huerta-Espino J, Crosby M (2003) Identification and mapping of gene Yr31 for resistance to stripe rust in *Triticum aestivum* cultivar Pastor. In: Proceedings 10th international wheat genetics symposium, Instituto Sperimentale per la

Singh RP, William HM, Huerta-Espino J, Rosewarne G (2004a) Wheat rust in Asia: meeting the challenges with old and new technologies. In: New directions for a diverse planet. Proc 4th Int Crop Sci Cong, 26 Sept–1 Oct 2004, Brisbane, Australia

Singh S, Franks CD, Huang L, Brown-Guedira GL, Marshall DS, Gill BS, Fritz A (2004b) *Lr41*, *Lr39*, and a leaf rust resistance gene from *Aegilops cylindrica* may be allelic and are located on wheat chromosome 2DS. Theor Appl Genet 108:586–591

Singh K, Chhuneja P, Ghai M, Kaur S, Goel RK, Bains NS, Keller B, Dhaliwal HS (2007a) Molecular mapping of leaf and stripe resistance genes in *Triticum monococcum* and their transfer to hexaploid wheat. In: Buck H, Nisi JE, Solomon N (eds) Wheat production in stressed environments. Springer, Neatherlands, pp 779–786

Singh K, Ghai M, Garg M, Chhuneja P, Kaur P, Schnurbusch T, Keller B, Dhaliwal HS (2007b) An integrated molecular linkage map of diploid wheat based on a *Triticum boeoticum* X *T. monococcum* RIL population. Theor Appl Genet 115:301–312

Singh RP, Hodson DP, Huerta-Espino J, Jin Y, Njau P et al (2008) Will stem rust destroy the world's wheat crop? Adv Agron 98:271–309

Singh K, Chhuneja P, Singh I, Sharma SK, Garg T, Garg M, Keller B, Dhaliwal HS (2010) Molecular mapping of cereal cyst nematode resistance in *Triticum monococcum* L. and its transfer to the genetic background of cultivated wheat. Euphytica 176:213–222

Singh RP, Hodson DP, Huerta-Espino J, Jin Y, Bhavani S, Njau P, Herrera-Foessel S, Singh PK, Singh S, Govindan V (2011) The emergence of Ug99 races of the stem rust fungus is a threat to world wheat production. Annu Rev Phytopathol 49(1):465–481

Song QJ, Shi JR, Singh S, Fickus EW, Costa JM, Lewis J, Gill BS, Ward R, Cregan PB (2005) Development and mapping of microsatellite (SSR) markers in wheat. Theor Appl Genet 110:550–560

Spielmeyer W, Moullet O, Laroche A, Lagudah ES (2000) Highly recombinogenic region at seed storage protein loci on chromosome 1DS of *Aegilops tauschii*, the D genome donor of wheat. Genetics 155:361–367

Stepien L, Golka L, Chekowski J (2003) Leaf rust resistance genes of wheat: identification in cultivars and resistance sources. Theor Appl Genet 44:139–149

Sun GL, Fahima T, Korol AB, Turpeinen T, Grama A, Ronin YI, Nevo E (1997) Identification of molecular markers linked to the *Yr15* stripe rust resistance gene of wheat originated in wild emmer wheat, *Triticum dicoccoides*. Theor Appl Genet 95:622–628

Taketa S, Nakazaki T, Schwarzacher T, Heslop-Harrison JS (1997) Detection of a 4DL chromosome segment translocated to rye chromosome 5R in an advanced hexaploid triticale line Bronco 90. Euphytica 97:91–96

Tao W, Liu D, Liu J, Feng Y, Chen P (2000) Genetic mapping of the powdery mildew resistance gene *Pm6* in wheat by RFLP analysis. Theor Appl Genet 100:564–568

Tar M, Purnhauser L, Csõsz M (2008) Identification and localization of molecular markers linked to the *Lr52* leaf rust resistance gene of wheat. Cereal Res Commun 36:409–415

Tautz D, Renz M (1984) Simple sequences are ubiquitous repetitive components of eukaryotic genomes. Nucleic Acids Res 12:4127–4138

Tester M, Langridge P (2010) Breeding technologies to increase crop production in a changing world. Science 327:818–822

The TT, McIntosh RA, Bennett FGA (1979) Cytogenetical studies in wheat. Monosomic analysis, telocentric mapping and linkage relationship of gene *Sr21*, *Pm4* and *Mle*. Aust J Biol Sci 32:115–125

Thomas J, Nilmalgoda S, Hiebert C, McCallum B, Humphreys G (2010) Genetic markers and leaf rust resistance of the wheat gene *Lr32*. Crop Sci 50:2310–2317

Tiwari VK, Wang S, Sehgal S, Vrána J, Friebe B, Kubaláková M, Chhuneja P, Doležel J, Akhunov E, Kalia B, Sabir J, Gill BS (2014) SNP discovery for mapping alien introgressions in wheat. BMC Genom 15:273–283

Tomar SMS, Menon MK (1998) Introgression of alien genes for leaf rust (*Puccinia recondita*) resistance into bread wheat (*Triticum aestivum*) cultivars. Indian J Agric Sci 68:675–681

Trethowan RM, Mujeeb-Kazi A (2008) Novel germplasm resources for improving environmental stress tolerance of hexaploid wheat. Crop Sci 48:1255–1265

Tsilo TJ, Jin Y, Anderson JA (2008) Diagnostic microsatellite markers for the detection of stem rust resistance gene *Sr36* in diverse genetic backgrounds of wheat. Crop Sci 48:253–261

Vaccino P, Accerbi M, Corbellini M (1993) Cultivar identification in *Triticum aestivum* using highly polymorphic RFLP probes. Theor Appl Genet 86:833–836

Valkoun J, Kucerova D, Bartos P (1986) Transfer of leaf rust resistance from *Triticum monococcum* L. to hexaploid wheat. Z PXanzenzucht 96:271–278

Van Slageren MW (1994) Wild wheats: a monograh of *Aegilops* L. and *Amblyopyrum* (Jacub. And Spadi) Eig (Poaceae). Wageningen Agricultural University Papers. 94–7:512

Varshney RK, Graner A, Sorrells ME (2005) Genic microsatellite markers in plants: features and applications. Trends Biotechnol 23:48–55

Vikal Y, Chhuneja P, Singh R, Dhaliwal HS (2004) Tagging of an *Aegilops speltoides* derived leaf rust resistance genes *Lr28* with a microsatellite marker in wheat. J Plant Biochem Biotech 13:47–49

Villareal RL, Mujeeb-Kazi A, Deltoro E, Crossa J, Rajaram S (1994a) Agronomic variability in selected *Triticum turgidum* × *T. tauschii* synthetic hexaploid wheats. J Agron Crop Sci 173:307–317

Villareal RL, Mujeeb-Kazi A, Rajaram S, Deltoro E (1994b) Morphological variability in some synthetic hexaploid wheats derived from *Triticum turgidum* × *Triticum tauschii*. J Genet Breed 48:7–16

Villareal RL, Fuentes-Davila G, Mujeeb-Kazi A, Rajaram S (1995) Inheritance of resistance to *Tilletia indica* (Mitra) in synthetic hexaploids X *Triticum aestivum* crosses. Plant Breed 114:547–548

Villareal RL, Sayre K, Bañuelos O, Mujeeb-Kazi A (2001) Registration of four synthetic hexaploid wheat (*Triticum turgidum/Aegilops tauschii*) germplasm lines tolerant to water logging. Crop Sci 41:274

Waines JG, Payne PI (1987) Electrophoretic analysis of the high-molecular-weight glutenin subunits of *Triticum monococcum*, *T. urartu*, and the A genome of bread wheat *(T. aestivum)*. Theor Appl Genet 74:71–76

Wilcoxon RD (1981) Genetics of slow rusting in cereals. Phytopathology 71:989–993

Williams CE, Collier CC, Sardesai N, Ohm HW, Cambron SE (2003) Phenotypic assessment and mapped markers for H31, a new wheat gene conferring resistance to Hessian fly (Diptera: *Cecidomyiidae*). Theor Appl Genet 107:1516–1523

Wu S, Pumphrey M, Bai G (2009) Molecular mapping of stem rust-resistance gene *Sr40* in wheat. Crop Sci 49:1681–1686

Xiao M, Song F, Jiao J, Wang X, Xu H, Li H (2013) Identification of the gene *Pm47* on chromosome 7BS conferring resistance to powdery mildew in the Chinese wheat landrace Hongyanglazi. Theor Appl Genet 126:1397–1403

Xie CJ, Sun QX, Yang ZM (2003) Resistance of wild emmers from Israel to wheat rusts and powdery mildew at seedling stage. J Triticeae Crops 23:39–42

Xu SS, Jin Y, Klindworth DL, Wang R-C, Cai X (2009) Evaluation and characterization of seedling resistances to stem rust Ug99 races in wheat-alien species derivatives. Crop Sci 49:2167–2175

Yahiaoui N, Srichumpa P, Dudler R, Keller B (2004) Genome analysis at different ploidy level alows cloning of powdery mildew resistance gene *Pm3b* from hexaploid wheat. Plant J 37:528–538

Yahiaoui N, Brunner S, Keller B (2006) Rapid generation of new powdery mildew resistance genes after wheat domestication. Plant J 47:85–98

Yahiaoui N, Kaur N, Keller B (2009) Independent evolution of functional *Pm*3 resistance genes in wild tetraploid wheat and domesticated bread wheat. Plant J 57:846–856

Yamaleev AM, Dolotovskii IM, Noikonov VI (1989) Relationship between resistance of wheat root rots and genome composition. Doklady Vsesoyuznoi Ordena Lenina i Ordena Trudovogo Krasnogo Znameni Akademii Sel'skhozyaistvennykh Nauk Imeni V.I. Lenina 6:4–6

Yamamori M (1994) An N-band marker for gene *Lr18* for resistance to leaf rust in wheat 89 (5):643–646

Yao G, Zhang J, Yang L, Xu H, Jiang Y, Xiong L, Zhang C, Zhang Z, Ma Z, Sorrells ME (2007) Genetic mapping of two powdery mildew resistance genes in einkorn (*Triticum monococcum* L.) accessions. Theor Appl Genet 114:351–358

Yen C, Yang JL, Liu XD, Li LR (1983) The distribution of *Aegilops tauchii* Cosson in China with reference to the origin of the Chinese common wheat. In: Sakamoto S (ed) Proc VI Int Wheat Genet Symp, Kyoto. Maruzen, Kyoto, pp 55–58

Yildirim A, Jones SS, Murray TD, Cox TS, Line RF (1995) Resistance to stripe rust and eyespot diseases of wheat in *Triticum tauschii*. Plant Dis 79:1230–1236

Young ND, Tanksley SD (1989) RFLP analysis of the size of chromosomal segments retained around the Tm-2 locus of tomato during backcross breeding. Theor Appl Genet 77:353–359

Yu BS, Sun GR (1995) Preliminary study of several spring wheat varieties for resistance to Septoria diseases. Crop Genet Resour 1:27–29 (in Chinese)

Yu G, Zhang Q. Friesen T L, Rouse MN, Jin Y, Zhong S, Rasmussen JB, Lagudah ES, Xu SS (2011) Identification and mapping of *Sr*46 from *Aegilops tauschii* accession *CIae* 25 conferring resistance to race TTKSK (Ug99) of wheat stem rust pathogen. Theor Appl Genet 128:431–443

Zeller, FJ (1973) 1B/1R wheat-rye chromosome substitutions and translocations. In Sears ER, Sears LMS (eds) Proceedings of 4th int wheat genetics symposium, Columbia, Missouri, pp 209–221

Zeller FJ, Kong L, Hart L, Mohler V, Hsam SLK (2002) Chromosomal location of genes for resistance to powdery mildew in common wheat (*Triticum aestivum* L. em Thell.) gene *Pm29* in line Pova. Euphytica 123:187–194

Zhang P, Dreisigacker S, Meichinger AE, Reif JC, Mujeeb Kazi A, Van Ginkel M, Hoisington D, Warburton ML (2005) Quantifying novel sequence variation and selective advantage in synthetic hexaploid wheats and their backcross-derived lines using SSR markers. Mol Breed 15:1–10

Zhang Q, Klindworth DL Friesen TL, Chao S, Jin Y, Cai X, Xu SS (2012) Development and characterization of wheat lines with *Sr37* for stem rust resistance derived from wild *Timopheev*'s wheat. Meeting Abstract, p. 316

Zhu ZD, Zhou RH, Kong XY, Dong YC, Jia JZ (2005) Microsatellite markers linked to two genes conferring resistance to powdery mildew in common wheat introgressed from *Triticum carthlicum* acc. PS5. Genome 48:585–590

Zohary D, Harlan JR, Vardi A (1969) The wild diploid progenitors of wheat and their breeding value. Euphytica 18:58–65

Chapter 13
Genetic Improvement of Sugarcane Through Conventional and Molecular Approaches

Upendra Kumar, Priyanka and Sundip Kumar

Abstract In recent years, efforts to improve sugarcane have focused on the development of biotechnology tools for this crop. It has become clear that sugarcane lacks tools for the biotechnological route of improvement and that the initial efforts in sequencing ESTs had limited impact on breeding. Until recently, the models used by breeders in statistical genetic approaches have been developed for diploid organisms, which are not ideal for a polyploid genome such as that of sugarcane. Breeding programs deal with decreasing yield gains. The contribution of multiple alleles to complex traits such as yield is a basic question underlining the breeding efforts that could only be addressed by the development of specific tools for this grass. However, functional genomics has progressed, and gene expression profiling leads to the definition of gene networks. The sequencing of the sugarcane genome, which is underway, will greatly contribute to numerous aspects of research on grasses. We expect that both the transgenic and the marker-assisted routes for sugarcane improvement will contribute to increased sugar, stress tolerance, and higher yield and that the industry for years to come will be able to rely on sugarcane as the most productive energy crop.

Keywords Sugarcane · Nobilization · Chromosome · GISH · FISH

U. Kumar · Priyanka
Centre of Excellence for Mountain Biology, Uttarakhand Council for Biotechnology, Biotech Bhavan, Haldi 263146, India
e-mail: baliyan.upendra@gmail.com

Priyanka
e-mail: pripanwar211086@gmail.com

S. Kumar (✉)
Molecular Biology and Genetic Engineering, College of Basic Sciences and Humanities, G.B. Pant University of Agriculture and Technology, Pantnagar 263145, India
e-mail: malik.sundip@gmail.com

V.R. Rajpal et al. (eds.), *Molecular Breeding for Sustainable Crop Improvement*, Sustainable Development and Biodiversity 11,
DOI 10.1007/978-3-319-27090-6_13

13.1 Introduction

Sugarcane is an important tropical crop and has served as a source of sugar for hundreds of years. With an originally soft, watery culm sugarcane acquired a distinctive feature of partitioning carbon into sucrose in its stem. The striking ability of accumulating levels of sucrose that can reach around 0.7 M in mature internodes (Moore 1995) is an almost unique feature in cultivated plants. Sugarcane is generally used to produce sugar and has recently gained increased attention because of its use in the production of ethanol, which represents an important renewable biofuel source (Pandey et al. 2000). The main product of sugarcane is sugar. Several by-products are produced from crushing sugarcane at the sugar mill. These include refined sugar, raw sugar, bagasse, molasses, alcohol, dextran, crude wax and glucose. These by-products are used by various industries like Bagasse-based industries to produce pulp, paper, particle boards etc. Press mud-based industries mainly produce fertilizer and the wax and compost industries animal feed. Sugarcane bagasse is largely used for energy cogeneration at the mill or for the production of animal feed, thus increasing the overall efficiency of the crop system (Pandey et al. 2000; Sun et al. 2004; Sangnark and Noomhorm 2004; Paiva et al. 2004; Han and Wu 2004). It is expected that enzymatic and hydrolytic processes that allow the bagasse carbon units from cellulose and hemicellulose to be fermented will soon be scaled up for ethanol production, turning sugarcane into an efficient crop for energy production (Paiva et al. 2004; Han and Wu 2004). Sugarcane is cultivated in more than 50.12 lakh hectares in tropical and subtropical regions of the world, producing up to 3521.4 lakh tons of sugarcane and 245.5 lakh tones of sugar during the year 2014–2015. India is the second largest producer of sugar in the world after Brazil and shares 17 %of the world production (http://www.agricoop.nic.in).

Both naturally occurring polyploidization and human-mediated polyploidization have been central to sugarcane domestication and improvement. *Saccharum* and sorghum are thought to have diverged from a common ancestor between 5 and 9 million years ago (Al-Janabi et al. 1994; Wang et al. 2010), and intergeneric crosses between some genotypes can still be made (Dewet et al. 1976). *Saccharum* and sorghum share more extensive genome-wide colinearity, and fewer chromosomal rearrangements (Dufour et al. 1997a, b; Ming et al. 1998), than share with maize, wheat, and rice. Many regions of the sorghum genome correspond to four or more homologous regions of *S. officinarum*, showing that in the short period since their divergence from a common ancestor, *S. officinarum* has been through at least two whole-genome duplications (Ming et al. 1998). These recent genome duplications are superimposed on an additional duplication shared by most if not all cereals (Paterson et al. 2004). A basic chromosome number of $x = 10$ appears likely to be ancestral to the *Saccharinae*, being consistent with sorghum, and note that $x = 5$ sorghums are not ancestral but are recently derived from $x = 10$ types (Spangler et al. 1999).

A further unusual feature of *Saccharum* transmission genetics has introduced still additional complexity into its genetic composition. The interspecific crosses that were a hallmark in the evolution of modern cultivars were followed by backcrosses to *S. officinarum* clones to recover types adapted to increased sugar content (Price 1965). During this process, a high frequency of transmission of $2n$ chromosomes by the female (*S. officinarum*) parent was discovered (Bremer 1961), which facilitated the recovery of *S. officinarum* alleles for sugar production, while introgressing disease resistance, vigor, and adaptability from *S. spontaneum*. This "nobilization" process yielded interspecific polyaneuploid genotypes of a complexity exceeding that of most if not all other crops. The meiosis of modern sugarcane cultivars mainly involves bivalent pairing (Price 1963; Burner and Legendre 1994), and chromosome assortment results from general polysomy in some cases of preferential pairing (Grivet et al. 1996).

13.2 Centers of Origin and Diversity

The genus *Saccharum* probably originated before the continents assumed their current shapes and locations. The genus consists of 35–40 species and has two centers of diversity: the Old World (Asia and Africa) and the New World (North, Central, and South America). Asia has approximately 25 native species, North America six native species, four or five introduced species, and Central America three or four native and some introduced species (Webster and Shaw 1995). Africa has two native, and Australia has one naturalized species (Darke 1999; Bonnett et al. 2008).

The Brazilian *Saccharum* species have not been well characterized. Only regional floristic surveys have reported the presence of these species. One study described the native species *S. asperum*, *S. angustifolium*, *S. purpureum*, *S. biaristatum*, *S. glabrinodis*, *S. clandestinus*, and *S. villosum*, but the authors commented that these species were poorly defined so that it is possible that they all might be variations of a single species (Smith et al. 1982). In fact, from these species, only *S. asperum*, *S. angustifolium*, and *S. villosum* are currently accepted scientific names (The Plant List 2010). In another study, the native Brazilian species were identified as *S. villosum, S. asperum*, and *S. baldwinii* (Filgueiras and Lerina 2001).

The center of origin and diversity of *S. spontaneum* is the more temperate regions of subtropical India. However, because *S. spontaneum* can be grown in a wide range of habitats and altitudes (in both tropical and temperate regions), it is currently spread over latitudes ranging from 8°S to 40°N in three geographic zones: (a) east, in the South Pacific Islands, the Philippines, Taiwan, Japan, China, Vietnam, Thailand, Malaysia, and Myanmar; (b) central, in India, Nepal, Bangladesh, Sri Lanka, Pakistan, Afghanistan, Iran, and the Middle East; and (c) west, in Egypt, Kenya, Sudan, Uganda, Tanzania, and other Mediterranean countries. These zones roughly represent natural cytogeographical clusters because *S. spontaneum* tends to present a different number of chromosomes in each of these locations (Daniels and Roach 1987).

13.3 Taxonomy

Sugarcane belongs to the tribe *Andropogoneae* and to the subtribe *Saccharinae*. Sugarcane geneticists have adopted the term '*Saccharum*' complex, originally coined by Mukherjee (1957) to describe a subset of genera within *Saccharinae* closely related to *Saccharum* to have contributed to its genetic background. Genera within the *Saccharum* complex include *Erianthus, Miscanthus, Narenga, Saccharum, and Sclerostachya* (Amalraj and Balasundaram 2005). Using DNA sequences to assess the interrelationships of genera within the *Saccharum* complex, Hodkinson et al. (2002) concluded that *Saccharum* and *Miscanthus* are more closely allied to each other than they are to other genera.

Sugarcane geneticists have traditionally included six species in the genus *Saccharum,* namely *S. officinarum* ($x = 10$, $2n = 80$; sweet chewing cane found in native gardens in New Guinea and other South Pacific Islands), *S. robustum* ($x = 10$, $2n = 60, 80$; putative ancestor of *S. officinarum* found most commonly on river banks in the same region), *S. edule* ($2n = 60–80$, produces aborted tassels, a delicacy in the same region), *S. barberi* ($2n = 111–120$, semisweet Indian cane), *S. sinense* ($2n = 81–124$, semisweet Chinese cane), and *S. spontaneum* ($x = 8$, $2n = 40–128$, wild cane found throughout Asia).

13.4 Sugarcane Breeding

13.4.1 Classical Breeding

The cultivated varieties of sugarcane (complex hybrids) are developed by using *S. officinarum*, the noble sugarcanes; *S. barberi*, the Indian sugarcanes; *S. sinense*, the Chinese sugarcane; and two wild species, viz. *S. spontaneum* and *S. robustum* germplasm. The genes for sucrose accumulation in modern sugarcane varieties are derived from *S. officinarum*, *S. barberi*, and *S. sinense*. The wild species has contributed disease resistance, tolerance to environmental stress, and higher yield potential through higher biomass production. Prior to the twentieth century, the world sugarcane industry was dependent on the noble canes (*S. officinarum*) and the canes of India (*S. barberi*) and China (*S. sinense*). The varieties were limited in number and yield potential, were susceptible to disease and pests, and were not adaptable to unfavorable ecological conditions. It was the pioneering work of the Dutch breeders in Indonesia that provided the basis for breeding high-yielding cultivars. Since the resistance to stress environments and diseases was not available within the genetic variability of *S. officinarum*, *S. barberi*, and *S. sinense*, therefore, wild species of *Saccharum* were also brought into the breeding programs. In India, Imperial Sugarcane Breeding Station, Coimbatore, was established in the year 1912, and Dr. C.A. Barber who was the first in charge of this station crossed

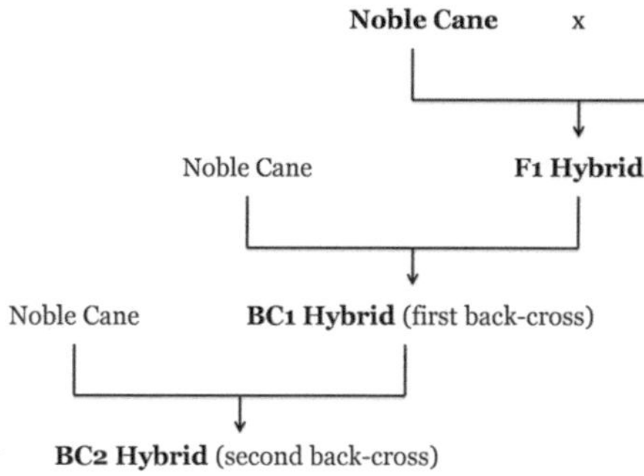

Fig. 13.1 Genetic base-broadening through "nobilization." The noble canes include the *S. officinarum* spp. or commercial hybrids with high sucrose content

S. officinarum with *S. spontaneum* (nobilization) and produced a commercial variety, Co. 205 in the first generation (Fig. 13.1). Later clones of *S. officinarum*, *S. spontaneum*, and *S. barberi* were hybridized, and their derivatives were back-crossed to *S. officinarum* (renobilization). Subsequently, the derivatives of these species were utilized objectively for the development of improved varieties of sugarcane. However, it was realized in the recent past that the genetic improvement in the newly developed varieties was not forthcoming as expected. It was probably because the narrow genetic base of the available breeding stocks is based on the limited number of clones of the above species. Therefore, emphasis was made to broaden the genetic base of the breeding stocks by the development of new interspecific hybrid (ISH) clones for utilization in breeding programs at Sugarcane Breeding Institute, Coimbatore, India. As a result, a total of 486 ISH clones have been produced. In the development of these ISH clones, a number of species, namely *S. officinarum* (33 clones), *S. spontaneum* (20 clones), *S. barberi, S. sinense* (9 clones), *S. robustum* (13 clones), and some indigenous (13 clones) and exotic (6 clones) ISH derivatives, were used as parents. The qualitative and quantitative characters of these ISH clones were studied, and some of these ISH clones (about 20) were included in the breeding population at National Hybridization Garden, Coimbatore, India, to be used as parents in the further breeding programs.

Intergeneric hybridization has also been tried as a means to broaden the genetic base, to obtain commercially useful characteristics, and to increase hybrid vigor. Although many attempts to cross between the intergeneric species may have been made in sugarcane research stations, limited publications are available. Two genera, namely *Erianthus* and *Miscanthus*, have received considerable attention of plant breeders. Among the *Erianthus* genus, *E. arundinaceus* has been of greatest interest because of its large stature, excellent ratoon yields, deep and extensive root system, tolerance to drought and floods, and resistance to diseases of importance in sugarcane. The genus *Miscanthus* has been attractive because of its superior over-wintering ability in temperate climates and as an energy cane (Tew and Cobill 2008). In addition, downy mildew (*Peronosclerospora sacchari*) resistance genes have been reported to be successfully transferred from *Miscanthus* to sugarcane (Chen and Lo 1989).

Despite all the promises, introgression breeding may hold; in general, it is difficult to estimate its impact or success in recent decades. It has also been noted that much effort has not led to commensurate commercial successes (Berding and Roach 1987; Stalker 1980). According to Wang et al. (2008), the process of introgression in sugarcane breeding is therefore traditionally a long-term and risky investment. The time and risk factors have clearly acted to reduce the level of resources devoted in most sugarcane breeding programs to introgression breeding despite general agreement among sugarcane breeders of its potential value. Much emphasis is laid on crosses that include *S. officinarum* hybrid parents with potentially high breeding values and appreciable agronomic characteristics.

13.4.2 Molecular Breeding

As such, modern sugarcane cultivars are highly heterozygous, with several different alleles at each locus. Such genomic redundancy may confer an evolutionary advantage (to buffer mutation load), or encourage the divergence of duplicated genes to adopt new functions. However, its large genome size, complicated genome organization and high level of diversity present special challenges for sugarcane genetic analysis, and generally slow rates of gain in crop improvement program through conventional breeding approach. However, biotechnology has arisen as a powerful tool to establish association between different traits and genes or molecular markers and has been used to facilitate genetic manipulation via marker-assisted selection (MAS). A number of candidate genes are now reported for various traits of sugarcane, and the work for the saturation of genetic maps of sugarcane is in progress in different laboratories. Marker techniques such as RAPD (Al-Janabi et al. 1993), RFLP (Da Silva et al. 1993; Lu et al. 1994a, b), AFLP (Lima et al. 2002), and SSR (Selvi et al. 2003) have already been used in sugarcane for genotyping or genetic mapping, and the efforts have been made to develop molecular markers using these techniques (Glaszmann 1994; Lu et al. 1994a, b; D'Hont et al. 1995; Alix et al. 1998; Jannoo et al. 1999b; Paulet et al. 2000).

13.5 Molecular Cytogenetics of Sugarcane

Sugarcanes are characterized by numerous (from 36 to more than 200) small- and variable-sized chromosomes (Fig. 13.2a). Classical cytogenetic studies have been essential in establishing a classification of the genus *Saccharum* and in understanding the nobilization process (reviewed by Sreenivasan et al. 1987). Modern sugarcane cultivars are derived from a few interspecific crosses performed a century ago between *S. officinarum* ($2n = 80$), the domesticated sugar-producing species,

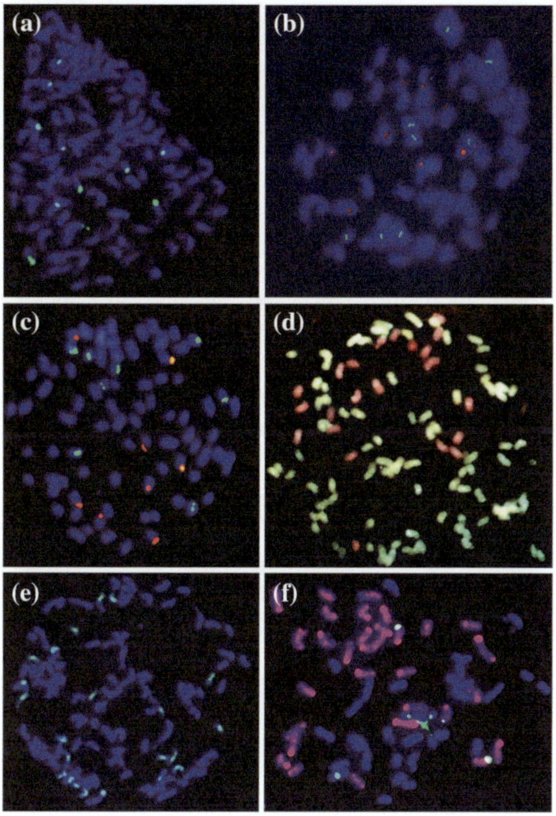

Fig. 13.2 Somatic chromosomes of sugarcane: (**a**) showing the functioning 2*n* gametes in an intergeneric hybrid of *S. officinarum* × *Erianthus*, the presence of 11 45S rDNA sites (8 *green* sites on the chromosomes of *S. officinarum* (due to functioning of 2*n* gametes) and rest of the three 45S rDNA sites on the chromosomes of *Erianthus* (due to functioning of *n* gametes), (**b**) $2n = 64$ chromosomes of *S. spontaneum* showing 45S rDNA sites (*red*) on 8 chromosomes and 5S rDNA sites on 8 chromosomes indicating $x = 8$, the basic chromosome number of *S. spontaneum*, (**c**) $2n = 80$ chromosomes of *S. officinarum* showing 45S rDNA sites (*red*) on 8 chromosomes and 5S rDNA sites on 8 chromosomes indicating $x = 10$, the basic chromosome number of *S. officinarum*, (**d**) $2n = 106$ chromosomes of a commercial hybrid (Co 419) after GISH using genomic probe of *S. spontaneum* labeled with rhodamine (*red*) and genomic probe of *S. officinarum* labeled with FITC (*green*) showing strong hybridization signals (*red*) on 22 chromosomes and on 84 chromosomes (*green*), (**e**) $2n = 108$ chromosomes of a newly developed interspecific hybrid (ISH 100) after GISH using genomic probe of *S. spontaneum* labeled with FITC (*green*) showing strong hybridization signals (*green*) on 22 chromosomes, (**f**) $2n = 62$ chromosomes of an intergeneric F_1 hybrid among *S. spontaneum* × *Erianthus* after GISH using genomic probe of *Erianthus* labeled with rhodamine (*red*) and 45S rDNA probe labeled with FITC (*green*) showing strong hybridization signals (*red*) on 30 chromosomes and 45S rDNA sites on 7 chromosomes indicating functioning of $n + n$ gametes of *S. spontaneum* and *Erianthus*

S. barberi ($2n$ = 81–124), a group of old Indian cultivars, and *S. spontaneum* ($2n$ = 36–128), the wild species. These interspecific crosses were followed by a few backcrosses to *S. officinarum* clones to recover types adapted to cultivation (Arceneaux 1965; Price 1965). During these crosses, breeders selected the results of a transmission of $2n$ chromosomes by the female parent, a phenomenon commonly observed in this type of combination (Bremer 1923, 1961, Fig. 13.2a), which facilitated the recovery of clones adapted to sugar production. These crosses were successful in introgressing disease resistance, vigor, and adaptability to the sugar-producing lines. This process yielded composite interspecific genomes, the complexity of which probably exceeds that of any other major crops. Modern cultivars are highly polyploid and aneuploid, with 100–130 chromosomes (Simmonds 1976).

A breakthrough in our understanding of sugarcane cytogenetics has been achieved over the last 20 years by using molecular cytogenetics in conjunction with diversity and genetic mapping studies. Molecular cytogenetics has been used to determine the origin of *S. barberi*, a group of canes involved in the origin of modern sugarcane cultivars. In addition, molecular cytogenetics revealed the size of the basic chromosome sets in *S. officinarum* and *S. spontaneum* and the genome structure of modern cultivars and related genera.

13.5.1 Determination of Basic Chromosome Numbers

The size of the basic chromosome set ($1x$) in sugarcane and related germplasm has been actively debated for a long time. Basic chromosome numbers of x = 5, 6, 8, 10, and 12 have been proposed for the *Saccharum* species (reviewed by Sreenivasan et al. 1987), and the possibility of several basic chromosome numbers in this genus has been suggested. The chromosome number of *S. officinarum* has been established as $2n$ = 80. Clones with the morphology of *S. officinarum* but with higher chromosome numbers are considered as typical or hybrids (reviewed by Sreenivasan et al. 1987) for *S. officinarum* and its wild progenitor, *S. robustum*, which exhibits 60 to 200 chromosomes with major cytotypes of $2n$ = 60 or 80; the most likely basic chromosome number is x = 10. This is consistent with the most common number in the *Andropogoneae* tribe (Bremer 1961), and the major cytotypes are more likely to represent euploid forms. For *S. spontaneum,* which displays a wide range of chromosome numbers from $2n$ = 36 to $2n$ = 128 with five major cytotypes: $2n$ = 64, 80, 96, 112, and 128 (Panje and Babu 1960), the series suggested a basic chromosome number of x = 8. However, because of the high polyploidy and the difficulty of differentiating the chromosomes based on their morphology, these hypotheses could not be tested with classical cytogenetics.

Fluorescence in situ hybridization (FISH) of ribosomal gene clusters was used to address the question of basic chromosome numbers in *S. officinarum* and *S. spontaneum*, the two species involved in the origin of modern cultivars either directly or through *S. barberi*. In plants, the 45S and 5S rRNA genes are arranged in long tandem arrays of repeat units containing the coding sequences and intergenic spacers. These two multigene families are organized in separated clusters, each one being located at one locus or several loci in the genome (Appels and Honeycutt 1986). The major 45S rDNA sites are usually associated with secondary constrictions and nucleolus organizer regions as opposed to non-expressed minor sites. The 45S rRNA loci consist of many copies (up to several thousands) of a DNA unit of several kbps. The 5S genes are smaller (several 100 bp) and less repeated. D'Hont et al. (1996, 1998) and Panwar et al. (2012) analyzed the number of 5S and 45S rDNA sites in different accessions of three of the *Saccharum* species, namely *S. officinarum*, *S. spontaneum*, and *S. robustum,* and on the basis of distribution of 45S and 5S rDNA sites, they have reported the basic chromosome numbers in these species 10, 8, and 10, respectively, and their ploidy level, assuming one site (45S rDNA/5S rDNA) per copy of genome which is also confirmed by Panwar et al. (2012) (Fig. 13.2b, c). Two distinct chromosome organizations coexist in modern cultivars. The genetic maps available so far suggest that the parental genomes are colinear and probably differ by only a small number of rearrangements (Grivet et al. 1996; Ming et al. 1998). One such case may reside on homology group (HG) VIII. The 45S rRNA genes were genetically mapped by Grivet and coworkers in cultivar R570 at an interstitial position on *S. spontaneum* cosegregation groups of HG VIII. This HG comprises two large *S. spontaneum* cosegregating groups together with two separate sets of smaller *S. officinarum* cosegregation groups that could not be merged. The structure of the chromosomes of this HG VIII may thus be different in *S. officinarum* and *S. spontaneum.*

Flow cytometry has been used to estimate the genome size of sugarcane (D'Hont and Glaszmann 2001). The size of the total genome is 7.7 pg (7440 Mbp) for *S. officinarum* ($2n = 8x = 80$), 6.2 pg (5990 Mbp) for *S. spontaneum* ($2n = 8x = 64$), and 11 pg (10,000 Mbp) for a typical model sugarcane cultivar (R570, $2n$ = 115). This is much larger than that in rice, 860 Mbp ($2n$ = 24); sorghum, 1600 Mbp ($2n$ = 20); or maize, 5500 Mbp ($2n$ = 20). However, taking into account the ploidy level, the size of the basic genome (1X), 930 Mbp (0.96 pg) for *S. officinarum* and 750 Mbp (0.78 pg) for *S. spontaneum* are close to sorghum, with 800 Mb (x = 10), as compared to 430 Mbp for rice (x = 12) and 2750 Mbp for maize (x = 10).

13.5.2 Genetic Constitution

Saccharum species present high ploidy levels. *S. officinarum* is octoploid ($2n$ = 80) having x = 10 chromosomes, which is the basic chromosome number of members of the *Andropogoneae* tribe (D'Hont et al. 1995; Cesnik and Miocque 2004; de Nobrega and Dornelas 2006; Panwar et al. 2012). *S. spontaneum* has x = 8 chromosomes (D'Hont et al. 1996; Panwar et al. 2012) but presents great variation in chromosome numbers with five main cytotypes: $2n$ = 62, 80, 96, 112, or 128 (Daniels and Roach 1987; Sreenivasan et al. 1987).

Modern sugarcane cultivars, which were derived from the hybridization between these two species, are considered allopolyploid hybrids (Daniels and Roach 1987), with most exhibiting a $2n + n$ constitution, representing two copies of the *S. officinarum* genome plus one copy of *the S. spontaneum* genome (Cesnik and Miocque 2004). The *S. officinarum* genome usually duplicates when it is hybridized with *S. spontaneum* and even with *Erianthus* (Kumar et al. unpublished). This phenomenon facilitated the work of the first breeders because nobilization consisted of increasing the ratio of the *S. officinarum* to that of the *S. spontaneum* genome (Bremer 1961). Genomic in situ hybridization (GISH) in interspecific and intergeneric hybrids has proved as a powerful tool to differentiate the chromosomes of different genomes and to identify the true interspecific/intergeneric hybrids (Fig. 13.2f). According to GISH studies, the genomes of modern hybrids are composed of 10–20 % of *S. spontaneum* chromosomes, 5–17 % of recombinant chromosomes containing part of *S. officinarum* and part of *S. spontaneum* chromosomes, and the remainder composed of *S. officinarum* chromosomes (Piperidis and D'Hont 2001; D'Hont 2005; Kumar et al. 2007, Fig. 13.2d, e). The hybrids are usually aneuploid, with a prevalence of bivalents, a significant proportion of univalents and rare multivalent associations during meiosis (Daniels and Roach 1987). Despite this genome complexity, evidence suggests a diploid-like mode of inheritance (Hogarth 1987).

13.6 Existing Genomic Resources

13.6.1 Genetic Diversity

Recent genomic data for evaluating genetic diversity within the genus suggest new relationships among accessions and may ultimately produce a definitive classification for the sugarcane species. The first molecular evidence came from restriction fragment patterns of nuclear ribosomal DNA that was used to separate accessions of *S. spontaneum*, which showed the widest within-species variation, from accessions of four other taxa often afforded species status: *S. robustum, S. officinarum, S. barberi, and S. sinense* (Glaszmann et al. 1990). RFLP analyses of mitochondrial genome showed an identical pattern among 18 *S. officinarum* clones and 15 of 17 *S. robustum* clones (D'Hont et al. 1993). RFLP patterns were similar among *S. officinarum, S. barberi, S. sinense, and S. edule,* all of which were distinctively different from *S. spontaneum.* Restriction patterns of the chloroplast genome suggested that, except for *S. spontaneum*, all the *Saccharum* species have the same chloroplast restriction sites (Sobral et al. 1994). RFLP analysis of nuclear genomic DNA confirmed observations about the cytoplasmic genomes that suggested distinctively greater diversity within *S. spontaneum* than among the four other species that were highly similar (Burnquist et al. 1992; Lu et al. 1994a; Nair et al. 1999). The most recent analysis, based on genomic in situ hybridization, is compatible with the hypothesis that *S. barberi* and *S. sinense* were derived from interspecific hybridization between *S. officinarum* and *S. spontaneum* (D'Hont et al. 2002). These authors conclude that genetic similarities among *S. barberi* and *S. sinense* accessions do not support the present classification of these being two distinct taxa.

Because of its polyploid nature, interspecific origin, and vegetative propagation, high levels of heterozygosity were detected among modern sugarcane cultivars using RFLP markers (Lu et al. 1994b; Jannoo et al. 1999a). The major part of this diversity was attributed to the 15–25 % chromosome complement that was inherited from *S. spontaneum* by random assortment of half of its chromosome, which has the greatest interspecific diversity (D'Hont et al. 1996). Similar patterns of molecular diversity were also detected using AFLP markers (Lima et al. 2002). On the other hand, modern sugarcane cultivars, derived from a small germplasm base contributed by only a few genotypes, show strong linkage disequilibrium.

13.6.2 Synteny with Other Members of the Grass Family

The conservation of gene repertoire and colinearity of gene order in the genomes of diverse grasses are well established (Freeling 2001). For sugarcane, the small diploid genome of sorghum has proven an especially facile model. Sorghum is the closest relative of sugarcane, and the two grasses differed from a common ancestor about five million years ago. Sorghum and sugarcane genomes share more extensive genome-wide colinearity and fewer chromosomal rearrangements (Dufour et al. 1997a, b; Guimaraes et al. 1997; Ming et al. 1998), than either share with any other known grass. Comparative mapping to establish colinearity between sugarcane and maize is complicated by segmental polyploidy of the maize genome and the resulting mapping of many sugarcane loci to two duplicated loci in maize (Grivet et al. 1996; Dufour et al. 1997a, b). Although it has not been through a genomic duplication event subsequent to its divergence from sugarcane, rice is much more distantly related and numerous chromosomal rearrangements are found when attempting to align their genomes.

Colinearity has been employed to evaluate the correspondence of QTLs affecting related traits in sugarcane and other grasses. Corresponding QTLs controlling plant height and flowering were found in sorghum and sugarcane (Ming et al. 2002). Several previously mapped maize and rice mutants and QTLs of the sugar metabolic pathway might be candidate genes for controlling sugar content in sugarcane (Ming et al. 2001). Sorghum, rice, and maize linkage maps and physical maps were used to identify potential markers for fine mapping and chromosome walking toward cloning the rust resistance gene in sugarcane (Asnaghi et al. 2000); sorghum RFLP markers played a key role in mapping this gene to a small interval. The close relationship between these grasses, a high degree of colinearity, and cross-hybridization of DNA probes are compelling reasons for using the more abundant information from the small genome sorghum to guide molecular mapping and positional cloning in sugarcane.

13.6.3 Mapping Quantitative Trait Loci for Economic Traits

Mapping quantitative trait loci (QTL) in autopolyploids is complicated by the potential for segregation of three or more alleles at a locus and by the lack of preferential pairing. As a consequence, different parental alleles of autopolyploids are not mutually exclusive alternatives. For the subset of polymorphic alleles that show simplex segregation ratios, the effect of an allele substitution can be estimated from the average phenotypic difference between the two possible genotypes (presence vs absence). Large-scale QTL mapping was conducted in two inter-specific populations (Ming et al. 2001, 2002) and in a segregating population from a selfed hybrid R570 (Hoarau et al. 2002). Most QTL alleles for sugar content showed phenotypic effects consistent with the parental phenotypes. However, the occasional transgressive QTLs revealed opportunities to purge unfavorable alleles from cultivars or to introgress valuable alleles from exotics (Ming et al. 2001). In many cases, QTLs controlling a given trait were mapped to corresponding genomic locations within the same genotype, across genotypes, and across species. This complex mapping of a given trait suggests that at least some QTLs on the same cluster might be different forms of the same gene or conserved homologous genes (Ming et al. 2001, 2002).

Multiplex segregation at QTL loci may be partly responsible for phenotypic buffering that is an important factor in the success of many autopolyploid crops. In several cases, two or more loci detected by the same DNA probe were each associated with variation in sugar content and plant height, and enabled to investigate the possibility of multiplex phenotypic buffering in sugarcane. "Stacking" of multiple doses of chromosomal segments containing favorable QTLs generally produced diminishing effects on phenotype, especially in cases where high-order duplications could be tested (Ming et al. 2001, 2002). This is similar to the results reported from stacking unlinked QTLs in the diploid tomato. The tomato results were attributed to epistasis (Eshed and Zamir 1996). Evaluating epistasis in sugarcane is complicated by the possibility of nonlinear interactions between alleles at homologous loci, in addition to nonlinear interactions between unrelated loci (Eshed and Zamir 1996). Detecting this type of phenotypic buffering has potential for cultivar improvement through marker-assisted selection in autopolyploid crops. Although diagnostic DNA markers are capable to pyramid multiple QTLs in a polyploid, incorporating just one copy of the multiple alleles may be sufficient to achieve most of the desired effects in the breeding population. Nonadditive gene action in multiple-dose QTLs also may have contributed to evolutionary opportunities. If single copy of a gene/QTL is physiologically sufficient, the additional copies are "extra" and thus free to collect mutations, often becoming nonfunctional, but perhaps occasionally resulting in a distinctive new function that improves fitness.

13.6.4 Map-Based Cloning

The first major gene of sugarcane mapped was the gene for resistance to brown rust (*Puccinia melanocephala* H & P Syd.) in "R570" (Daugrois et al. 1996). Mapping this gene with sugarcane c-DNA probe CDSR29 provided the first opportunity to evaluate the potential for map-based cloning in a complex polyploid plant. A bacterial artificial chromosome (BAC) library was constructed with 14 × basic genome or 1.3 × total genome coverage using genomic DNA from R570 (Tomkins et al. 1999). Meanwhile, a fine mapping project began to saturate the region surrounding the rust resistance gene. Using the synteny relationship between sugarcane and sorghum, maize, and rice, and selecting probes in the surrounding regions, this unlinked rust resistance gene was mapped to the end of a linkage group corresponding to sorghum linkage group D (Asnaghi et al. 2000). Bulk segregant analysis added eight markers surrounding the rust resistance gene with the two closest flanking markers placed 1.9 and 2.2 cM from the resistance gene (Asnaghi et al. 2004). Flanking markers were narrowed down to 0.1 and 0.3 cM on each side of the target gene, by chromosome walking using sugarcane, sorghum, and rice BAC resource. Beginning with an unlinked rust resistance gene with a tagged marker 10 cM away to produce a fine-mapped target gene flanked by sugarcane BACs, this work demonstrated the rapid advancement of sugarcane genomics.

13.6.5 Sugarcane ESTs

Sugarcane was not among early candidates for whole-genome sequencing due to the complexity and size of its genome, estimated at 10 Gb for a modern cultivar (D'Hont and Glaszmann 2001). However, mRNAs, the transcribed part of the genome, are much more tractable and constitute the bulk of DNA sequences currently available for sugarcane. An organism search of the NCBI nucleic acid databases on 15 May, 2015, using the term *Saccharum* yields 9717 nucleotide sequences, 285,216 expressed sequence tags (ESTs), and 83,138 genome survey sequences (GSSs). The ESTs can be subdivided by species into three major groups: two small groups of ESTs from *S. arundinaceum* and *S. officinarum* as well as an extremely large group of ESTs derived from *Saccharum* hybrid cultivars, the modern varieties of sugarcane. The bulk of the cultivar ESTs correspond to six cultivars from breeding programs in Brazil (SP80-3280, SP70-1143), India (CoS 767, Co 1148), Australia (Q117), and USA (CP72-2086). The mixed group contains ESTs from mixed tissue samples containing the Brazilian varieties CB47-89, RB855205, RB845298, RB805028, SP80-87432, PB5211 × P57150-4 or SP83-5077, SP80-185, SP87-396, SP80-3280, and SP803280 × SP81-5441. Sugarcane EST acquisition commenced in South Africa, with a small collection of ESTs being generated from both leaf roll and stem from the cultivar NCo376 (Carson and Botha 2002). The largest collection of ESTs was generated by

SUCEST, a consortium of Brazilian researchers who generated approximately 238,000 ESTs from 26 libraries constructed from diverse tissues from several Brazilian varieties (Vettore et al. 2001, 2003). EST collections were also generated in Australia from cultivar Q117 (Bower et al. 2005), in the USA from CP72-2086 research used their own nomenclature since they all entered the sequencing arena well before the large-scale public sequence clustering projects commenced. However, the variations in nomenclature and "species" designation impacted the ability of the entire sugarcane community to correctly utilize the vast resource. This was seen in various sugarcane EST cluster databases, including sugarcane UniGene, the sugarcane transcript assembly at JCVI, and the sugarcane transcript assembly at PlantGBD. This group made a decision in 2009 to engage with NCBI to harmonize sugarcane species and hybrid name in order to remove artificial discrepancies. Sugarcane EST data available at NCBI reflect this harmonization of nomenclature and therefore may differ in some cases from nomenclature used in the earlier publications. Most of the issues centered on the incorrect use of "*Saccharum officinarum*" as a species name for modern cultivars instead of "*Saccharum* hybrid cultivar." Data mining of the existing entries as well as publications associated with these cultivar names allowed for the reassignment of most entries into their correct cultivar groups. The species name "*Saccharum officinarum*" is now reserved for the ancestral species of sugarcane which has $2n = 80$ chromosomes, $x = 10$ (Sreenivasan et al. 1987), distinct from modern commercial varieties which are interspecific polyaneuploid hybrids with chromosome numbers varying between 100 and 130 (D'Hont et al. 1996). The correct assignment of species or cultivar name to the sequence entries will provide an extra layer of information that will be useful during sequence assembly and annotation. It may also assist in *in silico* SNP and microsatellite discovery.

Comparative analysis of 42,982 sugarcane aligned sequences (SASs) with the protein and DNA sequences from *Arabidopsis* and rice provided the first detailed estimates of the degree conservation/divergence between a monocot and eudicot (Vincentz et al. 2004). The 42,982 SASs represent possibly 33,620 unique genes (Vettore et al. 2003). Among them, 70.5 % have homologous sequences in Arabidopsis, 2 % in other eudicot, 14 % in monocot, and 13.5 % no matches. The 14 % monocot-specific cDNA sequences may represent novel genes on fast-evolving sequences that diverged from their eudicot counterparts. Another noticeable application of the EST resources is the identification of resistance gene analogs (RGAs) (Rossi et al. 2003). A total of 88 RGAs were identified based on their sequence homology to typical disease resistance genes. These sugarcane RGAs included representatives of the three major groups of resistance genes with a nucleotide-binding site (NBS), leucine-rich repeat (LRR), and a serine–methionine (S/T Kinase) domain. Fifty-five RGAs were used as RFLP probes for genetic mapping and identified 148 single-dose loci. Several RGA clusters were found, including one cluster of two loci mapped close to the sugarcane brown rust resistance gene. Detailed sequence analyses of these two RGAs with their rice and maize orthologs suggested a polyphyletic origin. These sugarcane RGAs are a useful resource for identifying and cloning disease resistance genes (Rossi et al. 2003).

References

Alix K, Baurens FC, Galszmann JC, D'Hont A (1998) Isolation and characterization of a satellite DNA family in the saccharum complex. Genome 41:854–864

Al-Janabi SM, Honeycutt RJ, McClelland M, Sobral BWS (1993) A genetic linkage map of Saccharum spontaneum L. 'SES 208'. Genetics 134:1249–1260

Al-Janabi SM, McClelland M, Petersen C, Sobral BWS (1994) Phylogenetic analysis of organellar DNA-sequences in the andropogoneae, saccharinae. Theor Appl Genet 88(8):933–944

Amalraj VA, Balasundaram N (2005) On the taxonomy of the members of 'Saccharum complex'. Genet Resour crop Evol 53:35–41

Apples R, Honeycutt R (1986) rDNA evolution over a billion years. In: Dutta SK (ed) DNA systematics. CRC Press, Boca Raton, New York, London, Tokyo, pp 81–125

Arceneaux G (1965) Cultivated sugarcanes of the world and their botanical derivation. Proc Int Soc Sugar Cane Technol 12:844–854

Asnaghi C, Paulet F, Kaye C, Grivet L, Deu M, Glaszmall JC, D'Hont A (2000) Application of synteny across Poaceae to determine the map location of a sugarcane rust resistance gene. Theor Appl Genet 101:962–969

Asnaghi C, Roques D, Ruffel S, Kaye C, Hoarau JY, Telismart H, Girard JC, Raboin LM, Risterucci AM, Grivet L, D'Hont A (2004) Targeted mapping of a sugarcane rust resistance gene (Bru1) using bulked segregant analysis and AFLP markers. Theor Appl Genet 108:759–764

Berding N, Roach BT (1987) Germplasm, collection and use. In: Heinz DJ (ed) Sugarcane improvement through breeding. Elsevier, Amsterdam, pp 143–210

Bonnett GD, Nowak E, Olivares-Villegas JJ, Berding N, Morgan T, Aitken KS (2008) Identifying the risks of transgene escape from sugarcane crops to related species, with particular reference to *Saccharum spontaneum* in Australia. Trop Plant Biol 1:58–71

Bower NI, Casu RE, Maclean DJ, Reverter A, Chapman SC, Manners JM (2005) Transcriptional response of sugarcane roots to methyl jasmonate. Plant Sci 168 (3):761–772

Bremer G (1923) A cytological investigation of some species and species-hybrids of the genus Saccharum. Genetica 5:273–326

Bremer G (1961) Problems in breeding and cytology sugarcane. Euphytica 10:59–78

Burner DM, Legendre BL (1994) Cytogenetic and fertility characteristics of elite sugarcane clones. Sugar Cane 1:6–10

Burnquist WL, Sorrellls ME, Tanksley S (1992) Characterization of genetic variability in Saccharum germplasm by means of restriction fragment length polymorphism (RFLP) analysis. Proc Int Soc Sugar Cane Technol 21:355–365

Carson DL, Botha FC (2002) Genes expressed in sugarcane maturing internodal tissue. Plant Cell Rep 20:1075–1081

Cesnik R, Miocque J (2004) Melhoramento da cana-de-açúcar. Embrapa, Brasília, p 307

Chen YH, Lo CC (1989) Disease resistance and sugar content in *Saccharum-Miscanthus* hybrids. Taiwan Sugar 36:9–12

D'Hont A (2005) Unraveling the genome structure of polyploids using FISH and GISH; examples of sugarcane and banana. Cytogenet Genome Res 109:27–33

D'Hont A, Glaszmann JC (2001) Sugarcane genome analysis with molecular markers, a first decade of research. Proc Int Soc Sugar Cane Technol 24:556–559

D'Hont A, Lu YH, Feldmann P, Glaszmann JC (1993) Cytoplasmic diversity in sugarcane revealed by heterologous probes. Sugar Cane 1:12–15

D'Hont A, Rao S, Feldmann P, Grivet L, Islam-Faridi N, Berding N, Glaszman JC (1995) Identification and characterisation of intergeneric hybrids, *Saccharum officinarum* × *Erianthus arundinaceus*, with molecular markers and in situ hybridisation. Theor Appl Genet 91:320–326

D'Hont A, Grivet L, Feldmann P, Rao S, Berding N, Glaszmann JC (1996) Characterization of the double structure of modern sugarcane cultivars (Saccharum spp.) by molecular cytogenetics. Mol Gen Genet 250:405–413

D'Hont A, Ison D, Alix K, Roux C, Glaszmann JC (1998) Determination of basic chromosome numbers in the genus Saccharum by physical mapping of ribosomal RLA genes. Genome 41:221–225
D'Hont A, Lu YH, Paulet F, Glaszmann JC (2002) Oligoclonal interspecific origin of 'North Indian' and 'Chinese' sugarcanes. Chromosomes Res 10:253–262
Da Silva J (1993) A methodology for genome mapping of autopolyploids and its application to sugarcane (Saccharum spp.). Ph.D. thesis, Cornell University, Ithaca, NY
Daniels J, Roach BT (1987) Taxonomy and evolution. In: Heinz DJ (ed) Sugarcane improvement through breeding, vol 1. Elsevier, Amsterdam, pp 7–84
Darke R (1999) The color encyclopedia of ornamental grasses: sedges, rushes, restios, cat-tails and selected bamboos. Timber Press, Portland
Daugrois JH, Grivet L, Roques D, Hoarau JY, Lombard H, Glaszmann JC, D'Hont A (1996) A putative major gene for rust resistance linked with a RFLP markers in sugarcane cultivar 'R570'. Theor Appl Genet 92:1059–1064
de Nobrega JCM, Dornelas MC (2006) Biotecnologia e Melhoramento da Cana-de-Açúcar. In: Segato SV, pinto AS, Jendiroba E, Nobrega JCM (eds) Atualizaçao em produçao de cana-de-açucar, vol 1. Piracicaba, pp 39–56
Dewet JMJ, Gupta SC, Harlan JR, Grassl CO (1976) Cytogenetics of introgression from Saccharum into sorghum. Crop Sci 16(4):568–572
Dufour P, Deu M, Grivet L, Dhont A, Paulet F, Bouet A, Lanaud C, Glaszmann JC, Hamon P (1997a) Construction of a composite sorghum genome map and comparison with sugarcane, a related complex polyploid. Theor Appl Genet 94(3–4):409–418
Dufour PDM, Grivet L, D'Hont A, Paulet F (1997b) Construction of a composite sorghum genome map and comparison with sugarcane, a related complex polyploid. Theor Appl Genet 94:409–418
Eshed Y, Zamir D (1996) Less-than-additive epistatic interactions of quantitative trait loci in tomato. Genetics 143:1807–1817
Filgueiras TS, Lerina R (2001) *Saccharum.* In: Longhi-Wagner HM (ed) Poacaeae, Flora Fanerogâmica do Estado de São Paulo, Fapesp-Hucitec, São Paulo. http://www.ibot.sp.gov.br/PESQUISA/florasp/florasp.htm. Cited 17 Mar 2008
Freeling M (2001) Grasses as a single genetic system. Plant Physiol 125:1191–1197
Glaszmann JC (1994) Molecular diversity and genome structure in modern sugarcane varieties. Euphytica 78:217–226
Glaszmann JC, Lu YH, Lanaud C (1990) Variation of nuclear ribosomal DNA in sugarcane. J Genet Breed 44:191–198
Grivet L, D'Hont A, Roques D, Feldmann P, Lanaud C, Glaszmann JC (1996) RFLP mapping in cultivated sugarcane (Saccharum spp.): genome organization in a highly polyploid and aneuploid interspecific hybrid. Genetics 142(3):987–1000
Guimaraes CT, Sills GR, Sobral BW (1997) Comparative mapping of Andropogoneae: Saccharum L. (sugarcane) and its relation to sorghum and maize. Proc Natl Acad Sci (USA) 94:14261–14266
Han G, Wu Q (2004) Comparative properties of sugarcane rind and wood strands for structural composite manufacturing. For Prod J 54(12):283–288
Hoarau J-Y, Grivet L, Offmann B, Raboin L-M, Diorflar J-P, Payet J, Hellamann M, D'Hont A, Glaszmann JC (2002) Genetic dissection of a modern sugarcane cultivar (Saccharum spp.). ll. Detection of QTLs for yield components. Theor Appl Genet 105:1027–1037
Hodkinson TR, Chase MW, Renvoize SA (2002) Characterization of a genetic resource collection for Miscanthus (Saccharinae, Andropogoneae, Poaceae) using AFLP and ISSR PCR. Ann Bot 89:627–636
Hogarth DM (1987) Genetics of sugarcane. In: Heinz DJ (ed) Sugarcane improvement through breeding. Elsevier, Amsterdam, pp 255–271
Jannoo N, Grivet L, Dookun A, D'hont A, Glaszmann JC (1999a) Linkage disequilibrium among modern sugarcane cultivars. Theor Appl Genet 99:1053–1060

Jannoo N, Grivet L, Seguin M, Paulet F, Doamingue R, Rao PS, Dookun A, D'Hont A, Glaszmann JC (1999b) Molecular investigation of the genetic base of sugarcane cultivars. Theor Appl Genet 99:171–184

Kumar S, Sharma B, Khan AK (2007) Characterization of interspecific hybrids of sugarcane for their use in sustainable breeding programme using genomic In-Situ hybridization. In: National seminar on varietal planning for improving productivity and sugar recovery in sugarcane at G. B. P. U A. & T Pantnagar, pp 142–148

Lima MLA, Garcia AAF, Oliveira KM, Matsuoka S, Arizona H, de Souza CL Jr, de Souza AP (2002) Analysis of genetic similarity detected by AFLP and coefficient of parentage among genotype of sugar cane (Saccharum spp.). Theor Appl Genet 104:30–38

Lu YH, D'Hont A, Paulet F, Grivet L, Arnaud M, Glaszmann JC (1994a) Molecular diversity and genome structure in modern sugarcane varieties. Euphytica 78:217–226

Lu YH, D'Hont A, Walker DIT, Rao PS, Feldmann P, Glaszmann JC (1994b) Relationships among ancestral species of sugarcane revealed with RFLP using single copy maize nuclear probes. Euphytica 78:7–8

Ming R, Liu SC, Lin YR, da Silva J, Wilson W, Braga D, van Deynze A, Wenslaff TF, Wu KK, Moore PH, Burnquist W, Sorrells ME, Irvine JE, Paterson AH (1998) Detailed alignment of *Saccharum* and sorghum chromosomes: comparative organization of closely related diploid and polyploid genomes. Genetics 150(4):1663–1682

Ming R, Liu SC, Moore PH, Irvine JE, Paterson AH (2001) QTL analysis in a complex autopolyploid: genetic control of sugar content in sugarcane. Genome Res 11:2075–2084

Ming R, DelMonte TA, Hernandez E, Moore PH, Irvine JE, Paterson AH (2002) Comparative analysis of QTLs affecting plant height and flowering among closely-related diploid and polyploid genomes. Genome 45:794–803

Moore PH (1995) Temporal and spatial regulation of sucrose accumulation in the sugarcane stem. Aust J Plant Physiol 22(4):661–679

Mukherjee SK (1957) Origin and distribution of *Saccharum*. Bot Gaz 119:55–61

Nair PNR, Sjogren U, Figdor D, Sundqvist G (1999) Persistent periapical radiolucencies of root-filled human teeth, failed endodontic treatments, and periapical scars. Oral Surg Oral Med Oral Pathol Oral Radiol Endod 87:617–627

Paiva JMF, Trindade WG, Frollini E, Pardini LC (2004) Carbon fiber reinforced carbon composites from renewable sources. Polymer-Plastics Technol Eng 43(4):1187–1211

Pandey A, Soccol CR, Nigam P, Soccol VT (2000) Biotechnological potential of agro-industrial residues—I: sugarcane bagasse. Bioresour Technol 74(1):69–80

Panje RR, Babu CN (1960) Studies in *Saccharum spontaneum*; distribution and geographical association of chromosome numbers. Cytologia 25:152–172

Panwar V, Sharma B, Kumar S (2012) Karyological studies and FISH landmarks on somatic chromosomes of *Saccharum officinarum* and *S. spontaneum.* VEGETOS 25(2):120–126

Paterson AH, Bowers JE, Chapman BA (2004) Ancient polyploidization predating divergence of the cereals, and its consequences for comparative genomics. Proc Natl Acad Sci USA 101:9903–9908

Paulet FC, Kaye C, Gay C, Rodier-Goud M, Galaszmann JC, D'Hont A (2000) Development of sugarcane (*Saccharum* spp.) microsatellite markers. References mSSCIR1 to mSSCIR78 in EMBL database (www.ebi.ac.uk)

Piperidis G, D'Hont A (2001) Chromosome composition analysis of various *Saccharum* interspecific hybrids by genomic in situ hybridization (GISH). Proc Int Soc Sugar Cane Technol 24:565–566

Price S (1963) Cytogeetics of modern sugarcanes. Econ Bot 17:97–105

Price S (1965) Interspecific hybridization in sugarcane breeding. Proc Int Soc Sugar Cane Technol 12:1021–1026

Rossi M, Araujo PG, Paulet F, Garsmeur O, Dias VM, Chen H (2003) Genomic distribution and characterization of EST-derived resistance gene analogs (RGAs) in sugarcane. Mol Genet Genomics 269:406–419

Sangnark A, Noomhorm A (2004) Effect of dietary fiber from sugarcane bagasse and sucrose ester on dough and bread properties. Lebensmittel-Wissenschaft und-Technologie 37(7):697–704

Selvi A, Nair NV, Balasundaram N, Mohapatra T (2003) Evaluation of maize microsatellite markers for genetic diversity analysis and fingerprinting in sugarcane. Genome 46:394–403

Simmonds NW (1976) Progress in sugarcane breeding. Sugarcane breeders' Newsl 38:75–78

Smith LB, Wasshausen DC, Klein RM (1982) Gramíneas. Gêneros: 85. Paspalum até Zea. Flora Ilustrada Catarinense. Herbário Barbosa Rodrigues, Itajaí, p 504

Sobral BWS, Braga DP, Lahood ES, Keim P (1994) Phylogenetic analysis of chloroplast restriction enzyme site mutations in the Saccharinae Griseb. Subtribe of the Andropogoneae Dumort. Tribe. Theor Appl Genet 87:843–853

Spangler R, Zaitchik B, Russo E, Kellogg E (1999) Andropogoneae evolution and generic limits in sorghum (Poaceae) using ndhF sequences. Syst Bot 24(2):267–281

Sreenivasan T, Ahloowalia B, Heinz D (1987) Cytogenetics. In: Heinz D (ed) Sugarcane improvement through breeding. Elsevier, New York, pp 211–254

Stalker HT (1980) Utilisation of wild species for crop improvement. Adv Agron 33:111–147

Sun JX, Sun XF, Zhao H, Sun RC (2004) Isolation and characterization of cellulose from sugarcane bagasse. Polym Degrad Stab 84(2):331–339

Tew TL, Cobill RM (2008) Genetic improvement of sugarcane (*Saccharum* spp.) as an energy crop. In: Vermerris W (ed) Genetic improvement of Bioenergy crops. Springer Science and Business Media, pp 249–272

The Plant List (2010) Version 1. http://www.theplantlist.org/. Cited 10 Jan 2011

Tomkins JP, Yu Y, Miller-Smith H, Frisch DA, Woo SS, Wing RA (1999) A bacterial artificial chromosome library for sugarcane. Theor Appl Genet 99:419–424

Vettore A, da Silva F, Kemper E, Arruda P (2001) The libraries that made SUCEST. Genet Mol Biol 24:1–7

Vettore AL, da Silva FR, Kemper EL, Souza GM, da Silva AM, Ferro MI, Henrique-Silva F, Giglioti EA, Lemos MV, Coutinho LL et al (2003) Analysis and functional annotation of an expressed sequence tag collection for tropical crop sugarcane. Genome Res 13(12):2725–2735

Vincentz M, Cara FAA, Okura VK, da Silva FR, Pedrosa GL, Hemerly AS, Capella AN, Marins M, Ferreira PC, França SC, Grivet L, Vettore AL, Kemper EL, Burnquist WL, Targon MLP et al (2004) Evaluation of monocot and eudicot divergence using the sugarcane transcriptome. Plant Physiol 134:951–959

Wang L-P, Jackson PA, Lu X, Fan Y-H, Foreman JW, Chen X-K, Deng H-H, Fu C, Ma L, Aitken KS (2008) Evaluation of Sugarcane × *Saccharum spontaneum* progeny for biomass composition and yield components. Crop Sci 48:951–961

Wang J, Roe B, Macmil S, Murray J, Tang H, Najar F, Wiley G, Bowers JE, Chen C, Rokhsar DS, Hudson ME, Moose SP, Paterson AH, Ming R (2010) Microcollinearity between diploid sorghum and autopolyploid sugarcane genomes. BMC Genomics 11:261

Webster R, Shaw RB (1995) Taxonomy of the native north American species of Saccharum (Poaceae: Andropogoneae). Sida 16(3):551–580

Chapter 14
Germline Transformation for Crop Improvement

Divya Mohanty, Atika Chandra and Rajesh Tandon

Abstract In present times, when genetically modified (GM) crops are creating a niche for themselves in the agricultural arena, germline transformation is likely to reduce the time and effort to produce transgenics. In the last three decades of transgenic research, phenomenal success has been achieved but has remained limited to species that lent themselves easily to genetic modification. More than 15 dicot and 11 monocot taxa have been tested for male germline transformation. On the other hand at least 23 three dicot and four monocot taxa have been tested for genetic modification through female germline. Amongst the male germ cells, cellular systems ranging from microspore, immature and mature pollen, pollinaria, pollen protoplasts, pollen tubes, exine detached pollen (EDP) and pollen derived embryos have been tested for transient or stable integration of foreign genes. A variety of methods and variants and combinations of methods such as agrolistics that combines *Agrobacterium* mediated transformation and biolistics, are available for the introduction of genes into the male germline, as it is accessible to treatments under a variety of conditions. Amongst the methods tested in male germline transformation, particle bombardment remains the most preferred method. The female germline, being largely inaccessible has lent itself to modification mostly via *Agrobacterium*-mediated methods. The success of *in planta* vacuum infiltration and floral dip exercise seems to be confined to Crucifers with ovule as the prime target. Applicability of germline transformation methods is being tested on a wider range of crop plants.

Keywords Germline · Transformation · Pollen · Ovule · Doubled haploids · *Agrobacterium* · Plant genotype · Target accessibility · Brassicaceae

---

D. Mohanty · R. Tandon (✉)
Department of Botany, University of Delhi, Delhi 110007, India
e-mail: tandon.raj@gmail.com

D. Mohanty
e-mail: divyamohanty13@gmail.com

A. Chandra
Maitreyi College, University of Delhi, Delhi 110021, India
e-mail: atikachandra@gmail.com

V.R. Rajpal et al. (eds.), *Molecular Breeding for Sustainable Crop Improvement*, Sustainable Development and Biodiversity 11,
DOI 10.1007/978-3-319-27090-6_14

14.1 Introduction

By the year 2050, the demand for food, feed, fibre and biomass is expected to increase by 70 % (FAO 2009). Unfortunately, many crops have stagnated in yield due to a variety of reasons including biotic and abiotic stresses (Grover and Pental 2003; Ray et al. 2012). Arable land remains limited and further expansions are severely curbed due to the deteriorating soil profiles, habitat modification and climate change. Conventional methods of crop breeding or the contemporary tools of biotechnology alone cannot meet the challenge of providing for billions in the immediate future. It is thus imperative to integrate the two and improve the tools of both the approaches to fulfil the needs.

Genetically modified (GM) crops such as soybean, cotton, maize and canola were the first to become commercially available in the year 1996 (Mannion and Morse 2013). United States of America is the major contributor of GM crops followed by Brazil and Argentina, both in acreage and production (Mannion and Morse 2013). World wide release and cultivation of transgenic crops has been limited due to various reasons, but the need for economic benefits and sustainable agriculture is breaking the barriers (Choudhary et al. 2014; Prado et al. 2014). GM technology has been the fastest adopted strategy, revolutionizing agriculture within the first sixteen years of its existence (James 2012). This is substantiated in the hundred-fold increase in the area under GM crop cultivation.

Recovering a 'useful' transgenic plant itself is a tenacious exercise requiring standardization at each step of the process and often aggravated by in vitro recalcitrance in some of the important crops. The challenge lies in optimizing multiple parameters that include identification of suitable genotype/s, selection of responsive tissue, screening compatible and most efficient method/s for stable delivery and integration of the desired gene that finally culminate in recovery of genetically stable transformants. Efficiency of transformations is further influenced by unpredictable transgene expression, silencing of genes and formation of chimeric plants (aberrant tissue formed by genetically distinct cells—untransformed and transformed). With advancement in transformation technologies, the last two decades have witnessed refinement of existing and the development of newer methods to recover transgenic crops. Amongst all the available transformation methods, the most preferred have been the *Agrobacterium*-mediated and microprojectile bombardment-mediated (biolistic) transformation methods (Barampuram and Zhang 2011). It has been realized that rapid production of GM crops would require simplification of transformation strategies and protocols and yet ensure higher efficiencies of recovering stable transformants.

Plant germline transformation, involving the introduction of desired genes into the male and/or female gametophytes, can prove to be a useful approach (Alwen et al. 1990; Roeckel et al. 1992). Essentially, the recovery of transgenics from transformed germline cells can follow either the pollination/pollen tube pathway or in vitro regeneration pathway. Theoretically, it is possible to bypass some intervening steps of in vitro regeneration and use either of the genetically transformed gametes for affecting fertilization and seed formation (*in planta* transformation). The seeds thus obtained would develop into transgenic plants in the usual way. Alternatively, protocols based on direct morphogenesis of shoots from the rescued transgenic embryos could alleviate the possible aberrations and variations that arise through indirect morphogenesis of callus. In former, pollination would ensure that the trait gets transmitted to the progeny while in the latter, regeneration from germline tissues can be accomplished [doubled haploids (DH)/ androgenesis/gynogenesis]. Both the strategies would be extremely useful in recovering transgenics especially from the recalcitrant crops and also in species where regeneration time is longer (Bolik and Koop 1991). The DH represent a unique genetic population where the haploid complement carried by pollen is doubled. Such plants have been advantageous in mapping and breeding of several useful agronomic traits. Contemporary interest in doubled haploids has surged so as to attain the dual purpose of introgressing genes and recovering homozygous lines for the transgene (Forster et al. 2007). Earlier, the researchers and breeders were captivated to reach the same end product of double haploids rather than improving the technique. Subsequently, with the improvement of technology and better understanding of the fundamental mechanisms, there is a renewed interest in the production of haploids and double haploids (Forster et al. 2007). At present, DH have been recovered from more than 200 plant species across the plant kingdom (Forster et al. 2007; Maluszynski et al. 2003). In Europe, nearly 50 % of modern-day barley cultivars are produced via DH technology (Forster et al. 2007). In comparison to the traditional plant breeding methods of backcrossing, DH along with marker-assisted selection save ample amount of time (Toojinda et al. 1998).

Techniques of transformation employed so far, such as floral dip (Clough and Bent 1998) and MAGELITR (Touraev et al. 1997), have proved successful in germline transformation, although obtaining high transformation efficiencies in many plant species is still a challenge. Also it appears that the methods to recover transgenics through germline transformation are yet to be attempted on many crops and there is need to refine the methods. In this review, we provide a conspectus and assessment of the progress made in male and female germline transformation techniques.

14.2 Male Germline Transformation

The male germline in plants essentially includes the sporogenous tissue and the post-meiotic products of microspore mother cell. Based on the developmental stage targeted for transformation, it may be broadly classified into two pathways: the sporophytic and the gametophytic (Resch and Touraev 2011; Brew-Appiah et al. 2013). When the immature pollen grains (microspores) are transformed and induced towards the embryogenesis (androgenesis), it is referred to as sporophytic path. Whereas when the microspores are transformed and matured in vitro or when the mature pollen grains are targeted and then either used for pollinations with the aim of producing GM seed (pollen tube pathway), it refers to the gametophytic path.

Pollen transformation has been proposed as a convenient method of gene transfer into higher plants (Eudes et al. 2014; Eapen 2011; Hess 1980, 1987; Ohta 1986; Resch and Touraev 2011). The key factors that influence the process of integrating and expressing genes via male germline include the microspore stage, several pre-treatments including osmotic treatment, the plant species/genotypes used, method of gene introduction, vector constructs and genes used. Uninucleate microspore (immature pollen) is the preferred stage for germline transformation, as it offers several advantages. Microspores and young pollen grains are devoid of exine, produced in large numbers and thus numerous targets are available for gene transfer in a single experiment raising the chances of recovering independent transformation events (Touraev et al. 2001). Additionally, all microspores exhibit physiological uniformity within an anther and can also be synchronously matured in vitro. This is especially useful in germline transformations via gametophytic pathway. The sporophytic pathway, wherein the microspores, can be induced to embark on the embryogenic route giving rise to haploid plants has an obvious advantage of producing homozygous doubled haploid plants in a single generation. Contrastingly, in conventional breeding programs the recombinant progeny obtained from the heterozygous individuals are selected and have to be backcrossed several times so that the desired trait/s gets fixed in the homozygous state. The androgenic strategy is time saving and has facilitated crop breeding programs extensively. The androgenic haploids can be later chemically induced to form homozygous diploid plants, thereby ensuring that the transgenic trait is stably inherited in the successive generations.

Experimental evidence has confirmed that mature pollen may also take up DNA and deliver it to the embryo sac, resulting in the integration of foreign DNA into the genome of the embryo and endosperm (Alwen et al. 1990). However, the presence of exine in mature pollen hinders gene delivery, particularly when *Agrobacterium*-mediated transformation or microinjection approaches are employed.

14.2.1 Methods of Gene Transfer into Pollen

As a standard practice, microspores are usually isolated at a stage before pollen mitosis is initiated. In some cases even the early binucleate stage has been found to be amenable for both haploid pollen culture and pollen transformation. At these stages, the cell fate is relatively undefined, and thus, genetic manipulation is easier. The gene of interest (GOI) should essentially be incorporated into the microspore genome at the single cell stage before first mitosis, so that it is transmitted to all the subsequent daughter cells, reducing the chance of obtaining chimeras. In case where the binucleate pollen grains are used, the success of transformation depends on which of the cells—the vegetative or the generative is transformed. In the case where successful production of transgenic plants via biolistic transformation of bicellular pollen had been reported (van der Leede-Plegt et al. 1995), transmission to the F2 generation was not found and may be due to damage of sperm nuclei inflicted by the microprojectiles (Touraev et al. 1997). As early as 1985, Sanford and co-workers attempted to validate the claim that simple imbibition of pollen with genomic DNA followed by normal pollination would result in production of transformed seed. However, no clear-cut transformation events could be recovered at that time. Consequently, the other methods were devised. Several direct and indirect methods of gene delivery have been developed for transformation of pollen/microspores such as *Agrobacterium*-mediated transformation, electroporation, microprojectile bombardment, microinjection, MAGELITR and sonication. The four key factors that may influence the process of integrating and expressing genes via male germline transformations include the developmental stage, plant species (and genotypes) used, method of gene introduction and the vector constructs used. Variable rates of transformation have been reported for a transient or stable integration of marker genes introduced into the microspore/pollen (Table 14.1). However, a reliable, efficient and general protocol applicable to a variety of plants is yet to be standardized.

14.2.1.1 *Agrobacterium*-Mediated Transformation (*Agroinfiltration*)

Agrobacterium sps. are well known as the 'natural genetic engineers'. Amongst these, *A. tumefaciens* (recently classified as *Rhizobium radiobacter*; Young et al. 2001) is most widely employed for transformations. The transformation method involving these bacterial species is termed 'Agroinfiltration' and is regulated by unique mechanism of gene transfer (Tinland 1996; Zupan and Zambryski 1997). Agroinfiltration is a comparatively simple method involving no specialized equipment but is dependent on the genotype of plant species as well as the host range of the bacterium. The method is one of the most efficient and extensively used procedures for transformations in a variety of plant tissues and species (Sharma et al. 2005). It has been used for transfer of single gene of interest or even large DNA sequences (Hamilton 1997). It primarily results in the integration of a single or low copy number of transferred DNA per transformed cell, giving way to stable expression. Further, it excludes the vector backbone sequences (the non-T-DNA portion of the Ti plasmid) from the introduced DNA into the plant genome. It also has high potential for independent integration of co-transformed DNA fragments.

Infection of rapeseed pollen (*Brassica napus*) with *Agrobacterium* had been tested (Pechan 1989; Huang 1992), even though no evidence of stable gene integration into the genome was presented. Vacuum infiltration method devised successfully for transformation of *Arabidopsis thaliana* (Bechtold et al. 1993; Bechtold and Bouchez 1995) has also been tested for pollen of *Petunia hybrida* resulting in 9 % transformation efficiency (Tjokrokusumo et al. 2000). The evidence for pollen transformation was observed by Ye et al. (1999), Tjokrokusumo et al. (2000), Xu et al. (2008) in *A. thaliana, P. hybrida, B. rapa* ssp. *chinensis,* respectively. However, Desfeux et al. (2000) could not reproduce the same in case of *Arabidopsis*. Nevertheless, some pollen grains did give positive expression and it cannot be completely ruled out, as the target of transformation during *in planta*

Table 14.1 List of plants species attempted for microspore/pollen transformation via different methods of gene introduction

Family	Plant species (genotype)	Explant/s	Expression/transformants	Transformation efficiency	Transformation method	Reference(s)
MONOCOTS						
Asparagaceae	*Asparagus officinalis*	Pollen	Transient	–	B	Kakuta et al. (2001)
Commelinaceae	*Tradescantia paludosa*	Pollen	Transient	–	B	Hamilton et al. (1992)
Iridaceae	*Freesia refracta*	Pollen	Transient	–	B	Tanaka et al. (1995)
Liliaceae	*Lilium longiflorum*	Pollen	Transient	17 per 1,00,000 viable pollen	B	van der Leede-Plegt et al. (1992)
		Mature Pollen	Transient	0.92 %	B	Nishihara et al. (1993)
		Pollen	Transient	0.72 ± 0.20 %	B	Tanaka et al. (1995)
		Pollen protoplast	Transient	70 %	E	Miyoshi et al. (1995)
		Pollen	Transient	–	W + A	Kim et al. (2007)
Liliaceae	*Tulipa gesneriana*	Pollen	Transient	–	B	Tanaka et al. (1995)

(continued)

Table 14.1 (continued)

Family	Plant species (genotype)	Explant/s	Expression/transformants	Transformation efficiency	Transformation method	Reference(s)
Poaceae	*Hordeum vulgare*	Microspores	Transient	5–17 %	E	Joersbo et al. (1990)
		Microspores	Stable	1 in 10^7 microspores	B	Jähne et al. (1994)
		Microspores	Transient, stable	–	B	Harwood et al. (1995)
		Microspores	Transient, stable	1 in 10^6 microspores (transient)	B	Yao and Kasha (1997), Yao et al. (1997)
				1 in 10^7 microspores (stable)		
		Microspores	Stable	1 in 10^6, 3 in 10^6	B	Carlson et al. (2001)
		Microspores	Stable	2.2 fertile transgenic plants per spike	A	Kumlehn et al. (2006)
		Microspores	Stable	1 in 2.4×10^5 to 1 in 4.8×10^5 microspores	B	Shim et al. (2009)
Poaceae	*Secale cereale*	Floral tillers	Transient	1 out of 1000 seeds	A	de la Pena et al. (1987)
Poaceae	*Sorghum bicolor*	Pollen	Transient, stable	–	S	Wang et al. (2007)
		Pollen	Transient	1–3.4 %	A	Elkonin et al. (2012)

(continued)

Table 14.1 (continued)

Family	Plant species (genotype)	Explant/s	Expression/transformants	Transformation efficiency	Transformation method	Reference(s)
Poaceae	*Triticale*	Microspores	Transient	5 % (Tat), 14 % (Tat2), 31 % (Pep1)	C	Chugh et al. (2009)
		Microspores	Stable	Regeneration ~10 %	C + A	Ziemienowicz et al. (2012)
		Microspores	Transient, stable	16 % (Rec1), 26 % (Tat2, Pep1)	C	Shim et al. (2013)
Poaceae	*Triticum aestivum*	Pollen	Transient	1 %	A	Hess (1987), Hess et al. (1991)
		Microspores and microspore-derived callus	Transient, stable	1.3 in 10^6 microspores	B	Folling and Olesen (2001)
		Microspores	Stable	–	A	Liu (2004)
		Microspores	Stable	–, 1.22 in 10^4 microspores	A, E	Brew-Appiah et al. (2013)
Poaceae	*Zea mays*	Pollen	Not observed	Nil	I	Sanford et al. (1985)
		Pollen	Transient	9.29 %	P	Ohta (1986)
		Microspores	Transient	–	E + PG	Fennell and Hauptmann (1992)
		Microspores, pollen	Transient	–	E	Jardinaud et al. (1995)
		Pollen	Transient	–	B	Kakuta et al. (2001)
		Pollen	Stable	–	S	Wang et al. (2001)

(continued)

Table 14.1 (continued)

Family	Plant species (genotype)	Explant/s	Expression/transformants	Transformation efficiency	Transformation method	Reference(s)
		Microspore	Transient	0.1–0.15 %	E	Obert et al. (2004)
		Pollen	Transient	39 %	B	Schreiber and Dresselhaus (2003)
DICOTS						
Brassicaceae	*Brassica juncea*	Pollen	Stable	–	S	Wang et al. (2008)
Brassicaceae	*Brassica napus*	Microspores	Stable	27–51 %	Mi	Neuhaus et al. (1987)
		Microspore, microspore-derived embryos	Stable	7.3 %	A	Pechan (1989)
		Microspore-derived embryos	Stable	0.005 %	A	Swanson and Erickson (1989)
		Microspores	Transient	–	E	Jardinaud et al. (1993)
		Microspores	Transient	2.75 %	Mi	Jones-Villeneuve et al. (1995)
		Microspores	Transient, stable	0.6 %	B	Shedden (1997)
		Microspores	Stable	0.2 %	B	Fukuoka et al. (1998)

(continued)

Table 14.1 (continued)

Family	Plant species (genotype)	Explant/s	Expression/transformants	Transformation efficiency	Transformation method	Reference(s)
		Microspores	Transient	0.17 %	B	Nehlin et al. (2000)
		Microspores	Transient, Stable	15.55 % (Bar), 11.11 % (*GUS*)	B + A	Abdollahi et al. (2009, 2011
Fabaceae	*Medicago sativa*	Pollen	Transient	–	B	Ramaiah and Skinner (1997)
Fabaceae	*Vigna unguiculata*	Pollen	Transient	0.36 %	A	Ilori and Pellegrineschi (2011)
Fagaceae	*Castanea dentata*	Pollen	Transient	4.1 %	B	Fernando et al. (2006)
Malvaceae	*Gossypium hirsutum*	Pollen	Stable	0.77 %	V + A	Li et al. (2004)
Paeoniaceae	*Paeonia lactiflora*	Immature Pollen	Transient	0.32 %	B	Nishihara et al. (1993)
Plantaginaceae	*Antirrhinum majus*	Microspores	Transient	0.01–0.2 % (*GFP*), 5–7 % (*GUS*)	B	Barinova et al. (2002)
Solanaceae	*Datura innoxia*	Pollen, Pollen-derived embryos	Not observed	Nil for pollen	A	Sangwan et al. (1993)
Solanaceae	*Lycopersicon esculentum*	Pollen	Not observed	Nil	I	Sanford et al. (1985)
		Pollen	Transient	–	B	Twell et al. (1989a)

(continued)

Table 14.1 (continued)

Family	Plant species (genotype)	Explant/s	Expression/transformants	Transformation efficiency	Transformation method	Reference(s)
Solanaceae	*Nicotiana glutinosa*	Pollen	Transient, Stable	3 %, 6.66×10^{-4} %	M	van der Leede-Plegt et al. (1992, 1995)
Solanaceae	*Nicotiana gossei*	Pollen	Transient, stable	–	E	Abdul-Baki et al. (1990)
Solanaceae	*Nicotiana rustica*	Immature pollen	Transient	0.36 %	M	Nishihara et al. (1993)
Solanaceae	*Nicotiana tabacum*	Pollen	Transient	0.1 %	B	Twell et al. (1989a)
		Microspores and pollen	Transient, stable	0.025 %, Nil, Nil	B, A, I	Stöger et al. (1992)
		Mature pollen	Transient	0.72 %	B	Nishihara et al. (1993)
		Pollen, pollen-derived embryos	Not observed	Nil for pollen	A	Sangwan et al. (1993)
		Pollen	Stable	5 in 10^4 pollen	B	Stöger et al. (1995)
		Pollen, exine-detached pollen (EDP)	Transient	Fivefold more in EDP than pollen	E	Shi et al. (1996)
		Microspores	Stable	0.017 %	Mg	Touraev et al. (1997)
			Transient		B	

(continued)

Table 14.1 (continued)

Family	Plant species (genotype)	Explant/s	Expression/transformants	Transformation efficiency	Transformation method	Reference(s)
		Pollen, exine-detached pollen (EDP)		More in EDP than pollen; 0.24 % (*GFP*) and 0.95 % (*GUS*) in EDP		Wang et al. (1998)
		Microspores	Stable	15 %	Mg	Aziz and Machray (2003)
		Pollen	Transient	0.1 %	B	Schreiber and Dresselhaus (2003)
		Microspores, pollen	Stable	2.79×10^{-2} %, 1.08×10^{-3} %	Mg, B	Aionesei et al. (2006)
		Microspores, pollen	Transient	>4 %, 10–20 %	B	Akhond and Machray (2009)
Solanaceae	*Petunia hybrida*	Pollen	Stable	9 %	V + A	Tjokrokusumo et al. (2000)

Abbreviations used for transformation method: *A* Agroinfiltration; *B* Bombardment; *C* CPP mediated; *E* Electroporation; *I* Imbibition; *Mi* Microinjection; *Mg* MAGELITR; *P* Pollen + exogenous DNA; *PG* PEG; *S* Sonication; *V* Vacuum infiltration; *W* Wounding transformations. Infiltration of *B. campestris* ssp. *chinensis*, *B. rapa* ssp. *chinensis*, and *A. lasiocarpa*, produced hemizygous transformants, which indicated that either the pollen or the ovule could be targets but much was not explored (Liu et al. 1998; Qing et al. 2000; Tague 2001).

Although a favoured method, agroinfiltration has been difficult in many monocots. Moreover, agroinfiltration of their microspores or pollen grains is problematic due to a thicker exine (Potrykus 1991; Sangwan et al. 1993; Dormann et al. 2001). The exine is known to adsorb bacteria and interferes with their elimination from the cultures at later stages. The alternatives to overcome this physical barrier are limited, providing one of the main reasons for preference of other methods such as microprojectile bombardment. Nevertheless, it has been observed that the flavonoid compounds extracted from the pollen and stigma of *P. hybrida* act as analogues of acetosyringone-like compounds (comparably less efficient) that are inducers of the virulence genes (in the *vir* region) of the Ti plasmid (Zerback et al. 1989).

Although earlier attempts of transforming microspores/pollen of monocots were largely unsuccessful (Dormann et al. 2001; Heberle-Bors 1995; Wu et al. 1998), modifications in the protocols have led to successful transformation of microspores/pollen in wheat (Liu 2004) and barley (Kumlehn et al. 2006). In wheat, transformation of the microspores by co-cultivation with *A. tumefaciens* strain AGL-1 produced transgenic plants including spontaneous doubled haploids (Liu 2004). Recently, conditions have been optimized for transformation of three spring wheat cultivars (Brew-Appiah et al. 2013). In *Hordeum vulgare* (L.), infection of androgenic pollen cultures with agrobacteria presented a novel approach for genetic transformation. Several parameters were identified and optimized including target cell survival, to establish an efficient and reproducible method of generating transgenic barley. Primary transgenics thus recovered, exhibited stable integration, expression and inheritance of the gene introduced and few were even found to be homozygous (Kumlehn et al. 2006).

Amongst the dicots, tobacco and Brassica plants are the favourable material but legumes have also been tested. Ilori and Pellegrineschi (2011) transformed twelve accessions of cowpea (*Vigna unguiculata*). In addition to the effect of genotype, role of flower pigmentation if any on pollen transformation was also analysed. After agroinfiltration, a high rate of flower abortion was encountered in cowpea. They concluded that transformation was genotype dependent but flower pigmentation independent. However, very low transformation frequencies observed were attributed to the *Agrobacterium* strain used, effect of temperature, failure of incorporation of the *Bar* gene, incorporation of *Bar* gene into a reading frame not recognized by the host plant genome and gene inactivation or silencing.

The renewed interest is paving the way to improve the efficiency of *Agrobacterium*-mediated transformation of male germline by inclusion of the pre-treatments. In particular, pre-treatments such as physical wounding by sonication (Trick and Finer 1997), silicon carbide fibres, sand, aluminium oxide (Kim et al. 2007) and microprojectile bombardment ('Agrolistics'; Abdollahi et al. 2009) have been tested with some degree of success (Tables 14.1 and 14.2) in different

Table 14.2 A comparison of male germline transformations using agroinfiltration

Plant species	Developmental stage	*Agrobacterium* strain	Plasmid vector (binary)	Selectable marker (agent/gene)	Transformation efficiency	References
Nicotiana tabacum	Pollen (microspores or mature pollen)	GV2260, LBA4404	pBIPA2b	*GUS*	Nil	Stöger et al. (1992)
Datura innoxia, Nicotiana tabacum	Isolated pollen and pollen-derived embryos	C58C1Rif	binary (pGS Gluc I, pGS TRN 943, pBI 121) and cointegrate (pGV 246) vectors	*GUS*	Nil for pollen	Sangwan et al. (1993)
Petunia hybrida	Freshly dehisced anthers	AGL0	pCGP1258	Basta resistance	55 %	Tjokrokusumo et al. (2000)
Gossypium hirsutum	Pollen	GV3101	pCAMBIA1301	*GUS*	0.77 %	Li et al. (2004)
Triticum aestivum	Microspores (mid-to-late-uninucleate stage)	AGL-1	RS 128'Xyl	Bialaphos	–	Liu (2004)
	Microspores	AGL-1	RS 128'Xyl	Bialaphos	–	Brew-Appiah et al. (2013)
Hordeum vulgare	Microspores before the first pollen mitosis (androgenetic pollen cultures)	LBA4404 GV3101	pSBI, pYF133 pUGAB7	*GUS*, *GFP*, *PAT* or *BAR* for Basta resistance	2.2 fertile transgenic plant per spike 31 %—LBA4404 69 %—GV3101	Kumlehn et al. (2006)
Lilium longiflorum	Pollen grains	LBA4404	pBI121	*GUS*	–	Kim et al. (2007)

(continued)

Table 14.2 (continued)

Plant species	Developmental stage	*Agrobacterium* strain	Plasmid vector (binary)	Selectable marker (agent/gene)	Transformation efficiency	References
Brassica napus	Microspores and microspore-derived proembryos	GV3850	pCV730	*NPTII* activity	7.3 %	Pechan (1989)
	Microspore-derived embryos	LBA4404	pBinl9	*NPTII* activity	0.005 %	Swanson and Erickson (1989)
	Microspores	LBA4404	pBGWFS7-64	*GUS*	–	Abdollahi et al. (2009, 2011)
Vigna unguiculata	Pollen	pGV2260	Ptjk 142	Bialaphos	0.36 %	Ilori and Pellegrineschi (2011)
Sorghum bicolor	Panicle	AGL0	p35SGIB	Bar, *GUS*	1–3.4 %	Elkonin et al. (2012)

systems thereby opening up future options for devising efficient, reproducible protocols in the future.

14.2.1.2 Electroporation

Electroporation relies on the use of electric pulse to create transient pore/s in the membrane to facilitate transmembrane movement of molecules. It is an effective and valuable tool for the direct delivery of naked DNA into a range of plant tissues including pollen. Uptake of dyes by electroporation of the germinating pollen grains had been confirmed earlier, by Mishra et al. (1987), suggesting that there is a possibility of introducing large molecular weight molecules with little loss of pollen viability. Matthews et al. (1990) demonstrated for the first time the potential of transforming pollen of tobacco by electroporation without any detrimental effects, strengthening the idea that transformation of mature pollen could be an effective method/technique of gene transfer in plants. By employing electroporation method, transient GUS activity in the intact microspores was reported in *B. napus* (Jardinaud et al. 1993).

Further optimization of the conditions for successful electroporation of tobacco pollen and stable integration of ß-glucuronidase (*GUS*) and chloramphenicol acetyltransferase (*CAT*) in the recovered transgenics was reported by Saunders and Matthews (1995). As the exine posed a barrier to transformation of pollen, a new (cell) system the exine-detached pollen (EDP) was developed in *Nicotiana tabacum*. Electroporation of the EDP resulted in a fivefold increase in expression of *GUS/uidI A* gene as compared to transformation of the pollen or germinating pollen grains (Shi et al. 1996).

Electroporation overcame the 'host range' limitation of *Agrobacterium*-mediated transformations. The monocot pollen was equally competent in uptake of exogenous DNA as those of dicots. A small molecule such as propidium iodide was shown to electropermeate into barley microspores (Joersbo et al. 1990). Combining electroporation with polyethylene glycol (PEG), successfully delivered free DNA into maize microspores (Fennell and Hauptmann 1992). In lily pollen, it was essential to use pollen protoplasts for the transfer of genes by electroporation (Miyoshi et al. 1995). Several parameters may influence the efficiency of electroporation and subsequent recovery of transformants. These include, concentration of plasmid DNA, promoter driving GOI (monocot-specific versus constitutive promoter), electroporation media, number and duration of pulse, voltage and frequency. Detailed studies have been done to study the effect of these parameters on microspore viability and induction of embryogenesis in maize (Obert et al. 2004). Positive GUS expression was detected only when the gene was placed under monocot-specific promoter. The ideal physical parameters were 200 mg/ml concentration of plasmid DNA, application of three pulses with frequency of 0.1 Hz and field strength of 400 V/cm for 20 ms. Brew-Appiah et al. (2013) studied the factors affecting microspore transformation by electroporation in seven different cultivars of wheat. They tested a range of voltages, 150–1000 V, and found a single pulse of ~375 V effective in producing maximum number of transformants.

14.2.1.3 *Microprojectile Bombardment*

Sanford et al. (1987) for the first time devised a method to introduce DNA coated on metal microjectiles, fired by helium gas under high pressure. The helium accelerates the gold or tungsten particles coated with DNA enabling them to cross any physical barrier by brute force. This method is largely genotype independent and is a promising tool for gene delivery into plants that are recalcitrant to *Agrobacterium* infections and those plant parts that were not amenable to other methods. The first successful application of particle bombardment method was reported in tobacco by Twell et al. (1989a) where transient expression of *GUS* marker gene was observed. Subsequently, particle bombardment-mediated entry of foreign DNA into the pollen grains was reported in *Nicotiana glutinosa* and *Lilium longiflorum* (van der Leede-Plegt et al. 1992). It further led to the development of transformants for *N. glutinosa* by bombarding the pollen with DNA-coated particles, pollinating and then selecting the resulting seeds with kanamycin for transformation events (van der Leede-Plegt et al. 1995).

The successful production of fertile and homozygous transformants by microprojectile bombardment was reported first by Stöger et al. (1995) in *N. tabacum*. Microspores at mid-binucleate stage were selected for transformation. However, survival of microspores and in vitro regeneration was poor and resulted in low transformation rates. Phenotypic, molecular and genetic evidence for the production of stable transformants was presented by directly transferring the gene into isolated microspores of rapeseed by bombarding, modifying the microspore culture conditions and adopting the firefly luciferase (*LUC*) gene as a non-destructive marker (Fukuoka et al. 1998).

Several monocot species have been successfully transformed using variants of the particle bombardment method. Transformation of pollen grains of lily (*L. longiftorum*), freesia (*Freesia refracta*) and tulip (*Tulipa gesneriana*) was achieved by the use of a pneumatic particle gun (Tanaka et al. 1995). In this study, successful expression of *gus* gene was found to be influenced by the developmental stage of pollen. In barley, selection of developmental stage and pre-treatments was crucial for success. Shim et al. (2009) tested the hypothesis that targeting the pollen at the S-phase of cell cycle (or just prior) coupled with arabinogalactans in the induction medium (pre-treatments) led to a higher frequency of homozygous gene insertion. Genes were successfully delivered into pollen of maize and asparagus with help of an improved particle gun and magnetic selection of the pollen after bombarding with magnetic particles (Kakuta et al. 2001).

Irrespective of the method employed, the developmental stage of pollen may greatly influence the uptake of foreign DNA. When mature pollen was bombarded with β-glucuronidase (*GUS*) gene cloned behind the pollen-specific *PA2* promoter of the chalcone isomerase gene of *P. hybrida*, expression was observed. While using immature pollen with same gene and technique, the gene expressed at a very low frequency. Co-culture of mature pollen with *A. tumefaciens* or the imbibition method did not express the *GUS* gene (Stöger et al. 1992). Till 1998, although the transient expression of introduced marker genes was observed following electroporation and PEG-mediated gene transfer, no stable transformation was confirmed (Fennell and Hauptmann 1992; Jardinaud et al. 1993). On the other hand, the particle bombardment technique had proven reliable for the introduction of genes into isolated pollen/microspores (Twell et al. 1989a; Nishihara et al. 1993) and stable transformation had been reported in *Nicotiana* (Stöger et al. 1995; Nishihara et al. 1995) and barley (Jähne et al. 1994). The first reproducible transformation of *Zea mays* pollen was achieved in 2003 by Schreiber and Dresselhaus.

14.2.1.4 Microinjection

Microinjection is a skill-dependent technique that has been employed successfully for the introduction of foreign DNA into animal cells. It does present a favourable option for delivering DNA directly into single cell systems like the microspores, bypassing the exine, intine and the plasma membrane. In practice, a fine injection needle is used to inject the exogenous DNA through the membrane and the cell is allowed to mature in vitro. The method is laborious, requiring training, patience and precision. If the accuracy is not met, the injection needle is likely to push the microspore to one side rather than penetrating it. Osmotic balance has also been found to be major factor affecting survival of treated microspores. Bolik and Koop (1991) reported bursting of pollen due to inadequate osmotic environment. A major limitation of this method is that only a small number of cells can be treated at any given time. Success of this method would largely depend on optimization of protocol/s for in vitro pollen maturation and further facilitate gametophytic pathway of transformation.

14.2.1.5 MAGELITR

Male germline transformation (MAGELITR), developed by Touraev et al. in 1997, was the first successful demonstration of recovering transgenic plants based on the gametophytic transformation pathway. They biolistically transferred foreign DNA into unicellular microspores of *N. tabacum* cv. Petite Havana SR1 that were allowed to develop into mature pollen in vitro and further used for in vivo pollination. MAGELITR is a fast, regeneration-independent and genotype-independent method, not prone to chimerism and somaclonal variation, which should be applicable to a wide range of species. It is dependent on the development of protocols that sustain in vitro maturation of pollen that would be genotype/species specific. The major drawback is the reported low transformation efficiency of 0.017 % (obtained 5 resistant seeds out of 30,000). On similar lines, Barinova et al. (2002) developed a novel transformation method for in vitro maturation of microspores followed by transient transformation of pollen at different stages in *Antirrhinum majus*; however, no transgenic seed was obtained.

An efficient male germline transformation protocol was devised for producing transgenic tobacco (*N. tabacum* L. cv. anthii) without selection, dispensing the need for a selectable marker (Aziz and Machray 2003). The high frequency (15 %) of stable transformation observed exceeded that reported by other methods in the previous studies (Negrutiu et al. 1986; Booy et al. 1989; Tjokrokusumo et al. 2000). In cases where higher transformation efficiencies were reported (Bechtold and Bouchez 1995), reproducibility and stable transformation across the successive generations were not achieved and DNA integration in the progeny was unstable (Hess et al. 1991; Langridge et al. 1992; Zeng et al. 1994). The method lays promise in its applicability to those plants for which the standardized protocols for in vitro maturation of pollen are available.

14.2.1.6 Sonication

The use of ultrasound to facilitate uptake of naked, exogenous DNA into a variety of plant tissues has been reported by several workers (Joersbo and Brunstedt 1990, 1992; Zhang et al. 1991). This method involves cavitation and has been used to overcome the barrier presented by the pollen exine of maize for uptake of exogenous DNA (Wang et al. 2001). They successfully recovered transgenic maize plants by disrupting the pollen wall by ultrasonication. The method was also successful in transformation of *Sorghum bicolor* (Wang et al. 2007) and *Brassica juncea* (Wang et al. 2008). Mild ultrasonication pre-treatment of pollen generated nearly 16 % efficiency in two inbred lines of *Sorghum* (Wang et al. 2007). This method, albeit simple, has not gained wide popularity, probably as it requires some degree of skill and employs specific instrumentation. It could be used in combination with the other methods to enhance the recovery of transgenics. Sonication has been found to assist and improve the agroinfiltration of the vegetative tissues of different plants known as SAAT (Sonication Assisted *Agrobacterium* Transformation; Trick and Finer 1997). This can also prove beneficial for the transformations of pollen grains.

14.2.1.7 Peptide Nanocarrier-Mediated Transfection/Transduction

The discovery that short peptide sequences of 9–30 amino acids could efficiently transport biologically active molecules across the cell membranes, led to a new method of gene delivery known as cell-penetrating peptides (CPPs) or protein transduction domains (PTDs) or peptide nanocarriers (e.g. Tat2, Pep1). Although this method was devised around 20 years ago, it has largely been used in animal systems to transport a variety of macromolecules that are generally excluded due to permeability barriers via physical interaction or chemical conjugation with CPPs/PTDs. They have recently been used for plant systems (Chugh and Eudes 2008a, b). The CPPs have been demonstrated to non-covalently deliver macromolecules like linear plasmid DNA or protein in the isolated microspores of *Triticale* cv. Alta. The delivery of ~272 kDa Gus enzyme in its active and tetramer form suggested that CPPs could carry macromolecules much larger than their own size (Chugh et al. 2009). This proved that CPPs could enter the microspores as efficiently as they translocate in other plant or mammalian cells. Ziemienowicz et al. (2012) successfully showed the delivery of an in vitro-prepared nanocomplex of single-stranded T-DNA molecule (independent of *Agrobacterium*), VirD2 (virulence protein D2), Rec A (recombination protein A) by a Tat2 nanocarrier into *Triticale* microspores. It was hypothesized that this approach would result in successful delivery and integration of low copy number of transgene into the genome, although no transgenic plants were generated. Recently, the efficiency of transport of double-stranded DNA by Tat2 and Pep1 CPP nanocarriers into *Triticale* microspores and the capacity of Rec A to protect the degradation of linear DNA was validated (Shim et al. 2013). A detailed mechanism of the delivery and stable inheritance is yet to be elucidated. These discoveries further confirm potential of CPPs in designing simple, cost-/time-effective strategies for genetic manipulation especially in those crops that are not amenable to *Agrobacterium*-mediated transformation.

14.3 Female Germline Transformation

The female gametophyte or the embryo sac is a unique genetic and cellular assemblage, with seven cells and eight nuclei (in the typical *Polygonum* type). Genetic modification of female germline cells offers several advantages over transformation of vegetative cells. Theoretically, ovules could be isolated and subjected to different methods of gene introduction and cultured in vitro till maturity. However, the method is extremely laborious and inefficient. Amenability of the female germline was first demonstrated through *in planta* transformation by successfully obtaining stable transformants after co-cultivating the germinating seeds of *Arabidopsis* (Feldmann and Marks 1987). This method of gene transfer was devised to bypass in vitro regeneration via organogenesis or embryogenesis, thereby preventing stress induced under in vitro conditions and the generation of somaclonal variants (Cullis 1990; Cullis and Cleary 1986; Cullis and Kolodynska 1975; Labra et al. 2004; Phillips et al. 1994). *In planta* transformation method overcame many limitations: First, it was relatively genotype independent than the other *Agrobacterium*-mediated protocols; second, as the young differentiating ovaries (with or without ovules) were targeted, many independently transformed events from a single transformation experiment could be recovered; third, it is a non-destructive method that was cost-effective (as compared to earlier protocols) allowing for rapid and easy transformation protocols; fourth, it could (theoretically) recover a large number of transgenic plants; and finally, it was time saving as numerous genes could be tested and transgenics analysed in a comparatively shorter duration. This method was particularly promising for recalcitrant plants. Using this basic strategy, several protocols/variants have been developed. It has also been used to generate a large pool of transgenics (T-DNA-tagged mutants) that have been instrumental in unravelling the function of many *Arabidopsis* genes.

However, the transformation efficiency is relatively low as compared to transformation of vegetative tissues, and the exact biological mechanisms involved behind this transformation are still obscure. The prime requirements affecting agroinfiltration of the generative organs depend on the ease with which the *Agrobacterium* interacts with the target cell and induct the virulence genes (Bechtold et al. 2000). Also, one of the major drawbacks of this *in planta* transformation is the inability of replicating the same in plants besides *A. thaliana* and beyond the family Brassicaceae. It is conjectured that in member of Brassicaceae some inducers of virulence are produced or allows the bacteria an easy access to the target.

There are also reports of recovering transgenic by pollen tube pathway (rice; Luo and Wu 1989) and ovarian injection in soybean, a recalcitrant legume crop (Hu and Wang 1999). Application of *Agrobacterium* onto previously isolated silks of maize followed by pollination with pollen of the same cultivar has also been reported to yield transgenic seeds (Chumakov et al. 2006).

14.3.1 The Methods for Altering (Gene Delivery into) the Female Germline

Agrobacterium-mediated transformation remains the most favoured amongst the different methods of gene introduction. The capability of *A. tumefaciens* to carry out genetic modification has several advantages (Alimohammadi and Bagherieh-Najjar 2009; Tzfira and Citovsky 2006; Bernhardt et al. 2012). The ability of this unique group of soil pathogens to infect different plant parts has been utilized in developing protocols that bypass the lengthy procedures involved in in vitro regeneration and selection. Successful agroinfections have been reported in a variety of plant parts/tissue systems like germinating seeds (Feldmann and Marks 1987); inflorescences (Clough and Bent 1998; Bechtold et al. 2000; Chang et al. 1994; Desfeux et al. 2000; Katavic et al. 1994; Trieu et al. 2000) and the complete plants of *Arabidopsis* in bloom (Bechtold et al. 1993). However, the reproducibility and efficiency of transformation varies across these tissues (Tables 14.3 and 14.4). Several modifications have been incorporated to reduce the time-, cost-, space- and achieve-enhanced transformation efficiency. This has led to formulation of a variety of protocols that are discussed below.

Table 14.3 List of plant species attempted for female germline transformation via different methods of gene introduction

Family	Plant species (genotype)	Common name	Transformation method	Transformation efficiency	Reference
MONOCOTS					
Poaceae	*Oryza sativa*	Rice	Pistil drip	20 %	Luo and Wu (1989)
Poaceae	*Sorghum bicolor*	Sorghum	Flower inoculation method	1.4–2.2 %	Elkonin et al. (2009)
Poaceae	*Triticum aestivum*	Common wheat	Flower inoculation method	Up to 2.6 %	Hess et al. (1991)
			Floral dip	6.8 %	Zale et al. (2009)
Poaceae	*Zea mays*	Maize	Flower inoculation method	6.8 %	Chumakov et al. (2006)
			Ovary-drip transformation	6.47 %	Yang et al. (2009)
			Flower inoculation method	32.7 %	Mamontova et al. (2010)
DICOTS					
Amaranthus	*Amaranthus*	Amaranth	Drop-by-drop	0.2, 0.6, 1.8 % with 3 constructs	Munusamy et al. (2013)
Brassicaceae	*Arabidopsis griffithiana*	Dwarf rocket	Floral dip, Vacuum infiltration	Nil	Tague (2001)
Brassicaceae	*Arabidopsis lasiocarpa*	Rock cress	Floral dip	0.03–0.5 %	Tague (2001)
			Vacuum infiltration	0.02–0.7 %	Tague (2001)
Brassicaceae	*Arabidopsis lyrata ssp. petraea*	Northern rock cress	Floral dip, vacuum infiltration	Nil	Tague (2001)
Brassicaceae	*Arabidopsis thaliana*	Thale cress	Vacuum infiltration	1 %	Bechtold et al. (1993)
			Vacuum infiltration, floral dip	0.5-3 %	Clough and Bent (1998)
			Vacuum infiltration	6 % ovules and 1 % pollen	Ye et al. (1999)
			Floral dip	0.48 %	Desfeux et al. (2000)

(continued)

Table 14.3 (continued)

Family	Plant species (genotype)	Common name	Transformation method	Transformation efficiency	Reference
			Vacuum infiltration (V), floral dip (D), floral spray (repeated, S)	0.07–13.6 %	Bechtold et al. (2000)
			Floral dip	0.86–2.09 %	Chung et al. (2000)
			Floral spray	0.95–3.86 %	
			Vacuum infiltration	0.76–1.76 %	
			Vacuum infiltration, floral dip	1 %	Bechtold et al. (2003)
			Floral dip	Checked DNA polymorphism	Labra et al. (2004)
			Floral dip	0.57 ± 0.18	Martinez-Trujillo et al. (2004)
			Drop-by-drop method	1.03–2.57	
			Floral dip	~3.5 %	Harrison et al. (2006)
			Floral dip	>100 per transformation of ~16 plants	Logemann et al. (2006)
			Vacuum infiltration	1.37–1.55 %	Wiktorek-Smagur et al. (2009)
			Floral dip	1.73–2.01 %	
			Floral dip	0.1–0.7 %	Davis et al. (2009)
			Floral inoculation	0.3–1 %	Narusaka et al. 2010
			Floral dip	0.76–2.28 %	Ghedira et al. (2013a)
			Floral dip	1.64 %	Mireault et al. (2014)
Brassicaceae	*Brassica carinata*	Abyssinian mustard	Floral dip	1.49 %	Verma et al. (2008)

(continued)

Table 14.3 (continued)

Family	Plant species (genotype)	Common name	Transformation method	Transformation efficiency	Reference
Brassicaceae	*Brassica juncea*	Indian mustard or brown mustard	Vacuum infiltration	0.8 % (pollen)	Chhikara et al. (2012)
Brassicaceae	*Brassica napus*	Rapeseed/canola	Vacuum infiltration	0.18 %	Wang et al. (2003)
			Floral dip	1.86 %	Verma et al. (2008)
			Floral dip	2–3 %	Li et al. (2010)
Brassicaceae	*Brassica rapa* ssp. *chinensis*	Pak choi	Vacuum infiltration	–	Liu et al. (1998)
			Vacuum infiltration	0.01 %	Qing et al. (2000)
			Vacuum infiltration	0.01–0.03 %	Xu et al. (2008)
Brassicaceae	*Brassica rapa* ssp. pekinensis	Chinese cabbage	Floral dip	0.1 %	Gao et al. (2012)
Brassicaceae	*Camelina sativa*	False flax/gold-of-pleasure	Vacuum infiltration	1.3 %	Lu and Kang (2008)
			Floral dip	0.08–0.8 %	Liu et al. (2012)
Brassicaceae	*Capsella bursa-pastoris*	Shepherd's purse	Floral dip, vacuum infiltration	Nil	Tague (2001)
			Floral dip	0.01–0.48 %	Bartholmes et al. (2008)
Brassicaceae	*Raphanus sativus*	Radish	Floral dip	0.1–1.4 %	Curtis and Nam (2001)
Brassicaceae	*Thellungiella halophila*	Salt cress	Floral dip	0.1–0.3 %	Inan et al. (2004)
Cactaceae	*Notocactus scopa* cv. *Soonjung* and	Silver ball cactus	Vacuum infiltration, pin-prickling, pin-prickling combined with vacuum infiltration (pre-top-cutting-off), pin-prickling combined with vacuum infiltration (post-top-cutting-off)	67–100 %	Seol et al. (2008)
Cactaceae	*Hylocereus trigonus*	Night blooming cactus	Vacuum infiltration, pin-prickling, pin-prickling combined with a syringe injection and directly injected	Not available	Seol et al. (2008)
Fabaceae	*Glycine gracilis*	Soybean	Pollen tube pathway, ovarian injection	1.3 %	Hu and Wang (1999)

(continued)

Table 14.3 (continued)

Family	Plant species (genotype)	Common name	Transformation method	Transformation efficiency	Reference
Fabaceae	*Glycine max*	Soybean	Pistil drip	Nil	Shou et al. (2002)
			Ovary-drip transformation	3.2 %	Liu et al. (2009a)
			Ovary-drip transformation at 4 levels, removal of: stigma, 1/2–2/3 style, whole style, whole style, 1/5–2/5 ovary	1-11.3 %	Liu et al. (2009b)
Fabaceae	*Medicago truncatula*	Barrel medic	Vacuum infiltration of flower	4.7–76 %	Trieu et al. (2000)
Fabaceae	*Melilotus alba* (wt, *opf* mutant)	White sweetclover, white melilot	Floral dip	Nil	Hirsch et al. (2010)
Malvaceae	*Gossypium hirsutum*	Upland cotton	Vacuum infiltration with pollen followed by pollination	0.8 %	Li et al. (2004)
			Pistil drip	0.01–0.93 %	Chen et al. (2010)
Solanaceae	*Petunia hybrida*	Petunia	Drop-by-drop method	7.5 %	Tjokrokusumo et al. (2000)
			Vacuum infiltration with pollen followed by pollination	9 %	Tjokrokusumo et al. (2000)
Solanaceae	*Solanum lycopersicum*	Tomato	In vitro fruit injection, in vivo fruit injection, floral dip	Floral dip—0, 12, 23 %	Yasmeen et al. (2009)

Table 14.4 A comparison of female germline transformations using agroinfiltration

Plant	Agrobacterium strain	Binary vector	Surfactant (vol./vol.) S—Silwet L-77 T—Tween 20	Sucrose concentration (wt/vol.)	% transformation	Author
MONOCOTS						
Sorghum bicolor	GV3101	pTd33	–	–	1.4–2.2 %	Elkonin et al. (2009)
Triticum aestivum	C58C1	pGV3850::1103neo	–	–	Up to 2.6 %	Hess et al. (1991)
	C58C1, or AGL1	pDs(Hyg)35S, or pBECKSred	0.4 % S	5 %	6.8 %	Zale et al. (2009)
Zea mays	GV3101	pTd33	–	–	6.8 %	Chumakov et al. (2006)
	AGLO	E35S-*licBM2*	–	–	32.7 %	Mamontova et al. (2010)
DICOTS						
Amaranthus	AGL1	p5b5, p5d9, p5f7 in pDRB6b	0.01 % T	5 %	0.6, 0.2, 1.8 %	Munusamy et al. (2013)
Arabidopsis lasiocarpa	LBA4404	pBIN-mgfp5-ER	0.1–0.2 % S	0–10 %	0.6–0.7 %	Tague (2001)
Arabidopsis thaliana	MP5-1	pGKB5			1 %	Bechtold et al. (1993)
	GV3101(pMP90)	pBIN-mgfp5-ER	0.05 % S	5 %	0.5–3 %	Clough and Bent (1998)
	ABI	pMON15737 or 15726	0.02 % S	5 %	6 % ovules and 1 % pollen	Ye et al. (1999)
	GV3101(pMP90)	pBIN-mgfp5-ER	0.03 % S	5 %	0.48 %	Desfeux et al. (2000)
	C58C1(pMP90)	SLJ6585 (p6585), pGKB5 (MP5-1), pJD121	0.05 % S	5 %	0.07–13.6 %	Bechtold et al. (2000)
	GV3101	pBI121C	0.02 % S	5 %	0.76-3.86 % (by all 3 methods)	Chung et al. (2000)
	C58C1 (pGV2260), C58C1 (pMP90)	pDFJ48-*GUS*, pJD121	0.05 % S	5 %	1 %	Bechtold et al. (2003)

(continued)

Table 14.4 (continued)

Plant	Agrobacterium strain	Binary vector	Surfactant (vol./vol.) S—Silwet L-77 T—Tween 20	Sucrose concentration (wt/vol.)	% transformation	Author
	GV3101 (pMP90)	pBIN-mgfp5-ER	0.05 S	5 %	–	Labra et al. (2004)
	GPV2260	p46UTR-BIN	0.05 % S	5 %	2–3 %	Martinez-Trujillo et al. (2004
	GV3101	pBINPLUS, pSKI015, pBIG-HYG	0.05 % S	5 %	~3.5 %	Harrison et al. (2006)
	GV3101 (pMP90RK)	pKWS05	0.03 % S	5 %	>100 per transformation of ~16 plants	Logemann et al. (2006)
	GV3101	pCAMBIA 2301, pCAMBIA 1305.1	0.01 % S;	5 %	1.37–2.01 %	Wiktorek-Smagur et al. (2009)
	ABI, GV3101	CCR2:LUC	0.02 % S	2.5 %	0.1–0.7 %	Davis et al. (2009)
	–	pBI101, pGWB1	0.02 % S	5 %	0.3–1 %	Narusaka et al. (2010)
	C58C1Rif[r](pMP90), LBA4404 compared with 8 more strains	pTJK136	0.05 % S	10 %	0.76–2.28 %	Ghedira et al. (2013a, b)
	C58C1	pB7FWG2	0.01 %, 0.05 % XIAMETER OFX-0309	5 % glucose	1.64 %	Mireault et al. (2014)
Brassica carinata	GV3101	pBinAR-P_{LEA1}: BcZF1	0.05 % S	5 %	1.49 %	Verma et al. (2008)
Brassica juncea	EHA105	pCAMBIA2301	0.1 % T	3 %	0.8 %	Chhikara et al. (2012)
Brassica napus	C58CIRif[R]	pNOV264	0.01 % S	5 %	0.18 %	Wang et al. (2003)
	GV3101	pBinAR-P_{LEA1}: BcZF1	0.05 % S	5 %	1.86 %	Verma et al. (2008)

(continued)

Table 14.4 (continued)

Plant	Agrobacterium strain	Binary vector	Surfactant (vol./vol.) S—Silwet L-77 T—Tween 20	Sucrose concentration (wt/vol.)	% transformation	Author
	LBA4404	pCAMBIA2200, pCAMBIA1300	0.1 % % S	3 %	2–3 %	Li et al. (2010)
Brassica rapa ssp. *chinensis*				5 %	–	Liu et al. (1998)
	C58(pMP90)	pDHB-NIa1	0.05 % S		0.01 %	Qing et al. (2000)
	C58C1 (pMP90)	pBBBast-*gus*-intron	–	–	0.01–0.03 %	Xu et al. (2008)
Brassica rapa ssp. pekinensis	GV3101	pCAMBIA1391-*gus*A	0.05 % S	5 %	0.1 %	Gao et al. (2012)
Camelina sativa	GV3101 (pMP90)	pGDP-FAH12	0.05 % S	5 %	1.3 %	Lu and Kang (2008)
	GV3101, EHA105, At503	TG_Cs 14, TG_Cs 19	0.025 % S	5 %	0.08–0.8 %	Liu et al. (2012)
Capsella bursa-pastoris	GV3101/pMP90, LBA4404	mGFP5-ER, pGPTV-bar or pFGC5941	0.01–0.05 % S	3–10 %	0.01–0.48 %	Bartholmes et al. (2008)
Raphanus sativus	AGL1	pCAMBIA3301	0.05 % S/0.1 % Pluronic F-68/0.05 % T	5 %	1.4 %/0.1 %/ 0.3 %	Curtis and Nam (2001)
Thellungiella halophila	–	pSK1015, pE1829	0.05 % S	5 %	0.1–0.3 %	Inan et al. (2004)
Notocactus scopa cv. *Soonjung*	LBA4404	pBI-121	–	5 %	67–100 %	Seol et al. (2008)
Hylocereus trigonus	LBA4404	pBI-121	–	5 %	Not available	
Medicago truncatula	ASE1,EHA105, GV3101	pSLJ525 or pSKI006, pBI121-bar or PKYLX7-Gus or pBINmgfp-ER-bar or pGA482-bar, pSKI015	0.02 % S	–	4.7–76 %	Trieu et al. (2000)
Melilotus alba (wild type)	EHA105, LBA4404	pGKB5	–	10 mM glucose	Nil	Hirsch et al. (2010)

(continued)

Table 14.4 (continued)

Plant	Agrobacterium strain	Binary vector	Surfactant (vol./vol.) S—Silwet L-77 T—Tween 20	Sucrose concentration (wt/vol.)	% transformation	Author
Melilotus alba (*opf* mutant)	LBA4404	pCAMBIA1301	0.01 % S	10 mM Glucose	Nil	Hirsch et al. (2010)
Gossypium hirsutum	GV3101	pCAMBIA 1301	–	–	0.8 %	Li et al. (2004)
	EHA105	pKF111	0.05 % S	10 %	0.01–0.93 %	Chen et al. (2010)
Petunia hybrida	AGLO	pCGP1258	–	20 %	7.5, 9 %	Tjokrokusumo et al. (2000)
Solanum lycopersicum	EHA 105, EHA 101	pROKIIAP1GUSint or pROKIILFYGUSint, p35SGUSint	–	30 %	Floral dip—0, 12, 23 %	Yasmeen et al. (2009)

14.3.1.1 In Planta Vacuum Infiltration

Vacuum infiltration had been an important tool for plant physiologists to study plant–pathogen interaction. By applying vacuum, the air spaces between the cells in the plant tissue are decreased and allow better penetration of pathogenic bacteria into the inter cellular spaces. The vacuum generates a negative atmospheric pressure causing the infiltration medium to relocate into the plant tissue. Bechtold et al. (1993) devised the method of *Agrobacterium* vacuum infiltration of adult *Arabidopsis* plants. This involved six main steps—growing *Arabidopsis* to flowering stage, uprooting of plants, application of *Agrobacterium* to whole plants via vacuum infiltration in a sucrose/hormone growth medium, replanting, collection of seeds, and selection on antibiotic or herbicide media or other selectable markers. As this method did not require any sophisticated instrument, it was adopted rapidly and became very popular (Bent 2000; Hansen and Wright 1999; Bechtold et al. 1993). Further improvisations to the basic method included infiltrating only the floral buds/shoots that removed the cumbersome step of uprooting the whole plants and the subsequent replanting steps.

Vacuum infiltration has been successful in crucifers like *B. rapa* ssp. *chinensis* (Qing et al. 2000; Xu et al. 2008), *Arabidopsis lasiocarpa* (Tague 2001), *B. napus* (Wang et al. 2003), *B. juncea* (Chhikara et al. 2012), *Camelina sativa* (Lu and Kang 2008). A few species from the other families also exhibited positive results, including *P. hybrida* (Solanaceae; Tjokrokusumo et al. 2000) and *Medicago truncatula* (Fabaceae; Trieu et al. 2000). Several factors including pressure, duration of infection, duration and intensity of vacuum applied, infiltration medium used, pre-treatment of parent plant (used for infiltration) are crucial for vacuum infiltration. Two variations on this theme have been equally successful (in *Petunia*) —infiltration of the pollen before pollination and subsequent recovery of transformed seed; or application of *Agrobacterium* suspension onto the stigma followed by pollination with untransformed pollen. Two sequential *Agrobacterium* vacuum infiltrations of the young buds at interval of a week have been successful in transforming only the anther and pollen of *B. juncea,* with an efficiency of 0.8 % and not the female germline (Chhikara et al. 2012).

The pressure due to vacuum adversely effects the survival of the cells. The most appropriate inoculation time for highest efficiency in case of *Arabidopsis* is 4 min. (Wiktorek-Smagur et al. 2009). In *A. lasiocarpa*, the infiltration medium was modified, and both vacuum infiltration and floral dip were performed. In comparison with floral dip vacuum infiltration gave higher rate of transformation without any affecting the seed set (Tague 2001). In *B. napus*, transgenic plants could be obtained by using both the plants that started flowering naturally or those in which flowering was induced by low temperature. The most effective condition was infiltration at 25–27 in. Hg vacuum applied twice for duration of more than 5 min either continuously or at an interval of 1 week. Lower vacuum pressure (<25 in. Hg) or infiltration of germinating seedlings did not yield any transgenic (Wang et al. 2003). In *P. hybrid*, two methods were employed that yielded almost similar efficiency. Pollen was transformed by vacuum infiltration and applied on the stigma or *Agrobacterium* suspension was directly applied on the stigma before pollination and then pollination performed with untreated pollen. In former, pollen was the target, while target for latter was not clear (Tjokrokusumo et al. 2000). For *B. rapa*, the original method was modified with prolongation of vacuum step (Liu et al. 1998; Qing et al. 2000). In *C. sativa,* an attempt for floral dip failed to produce any transgenic plant. While 85 kPa vacuum for 5 min. was successful for recovering the transgenic plants (Lu and Kang 2008). Use of vacuum did not have any adverse effect on the survival of the treated plant and subsequent seed set.

14.3.1.2 *Floral Dip*

The original *in planta* transformation method was labour-intensive even though it avoided the intricacies of in vitro cultures. In addition, reproducibility and extrapolation of this method to the other plant species were limited. These drawbacks could be attributed to several factors, but the vacuum step was considered a major hurdle as it affected survival of cells. To omit this step, the protocol was simplified to develop the 'floral dip' method. This modified method involved submerging the developing inflorescence only into the infiltration medium containing *Agrobacterium* in a buffered solution and a surfactant. Importantly, the method eliminated the need for uprooting and replanting of plants during agroinfiltration (Clough and Bent 1998). This protocol was simple, fast and proved to be efficacious. Several parameters affected the efficiency of transformation such as the components of the infiltration medium, inoculation frequency, inoculation time, co-cultivation in different plant species (Wiktorek-Smagur et al. 2009). Analyses of the transformants demonstrated that insertion is random and there was need to overcome the uncontrolled gene expression (Bernhardt et al. 2012).

In monocots also, which are usually difficult systems for regeneration, floral dip proved useful in producing stable, low copy number transgenics of wheat over three to six generations. The first transgenic was produced after infiltrating six unemasculated mid-to-late-uninucleate-stage spikes with clipped florets with agrobacteria for two minutes. The ideal stage for dipping identified was mid-to-late-uninucleate microspore stage, when the spike is yet to emerge from the sheath. Dipped the inflorescence at other stages resulted in decreased seed set (Zale et al. 2009).

Furthermore, with the aim of improving the protocol many manipulations have been attempted. For *Arabidopsis* Logemann et al. (2006) proposed obliterating the need for large volumes of bacterial cultures in liquid medium to save time and space. They demonstrated that similar transformation efficiency could be achieved by using the bacteria grown on culture plates even after one week of storage at 4 °C. It also reduced the risk of large-scale contamination to some extent. Manipulation of the *Agrobacterium* culture by pelleting and resuspending it into a buffered medium (mostly sucrose) is a prerequisite before dipping the plants. Davis et al. (2009) and Li et al. (2010) have devised a simple method for transforming *A. thaliana* and *B. napus*, respectively, by floral dip method. They developed a bacterial growth medium supplemented with surfactant which supported floral dip, thereby circumventing the need for exchange into a buffer medium for resuspending the bacterial cells before inoculation. Davis et al. (2009) also showed that by using two *Agrobacterium* cultures harbouring two different vectors, it was possible to generate a double-transformation event at once by dipping into a mixture of the two cultures. This saves cost and time by facilitating insertion of two different transgenes into a single plant in one generation.

14.3.1.3 Floral Spray

This technique was primarily devised to omit the vacuum step and avoid handling of large volumes of *Agrobacterium* culture (Chung et al. 2000). The bolting plants were sprayed once or up to three times with the bacterial inoculum per transformation until the suspension began to drip off and then covered with a plastic dome to maintain the humidity. The primary transformants were then allowed to grow and set seeds. Spraying the bacterial culture in the form of aerosols yielded transformation rate comparable to floral dip.

14.3.1.4 Drop-by-Drop Method

Modification of floral dip was needed to overcome the adverse effects of the detergent on silique development. In this modification, the inoculum was directly and selectively applied ‘drop-by-drop’ onto the new closed floral buds (without opening the flowers). This reduced the exposure to infiltration medium and damage to the developing siliques (Martinez-Trujillo et al. 2004). A comparative study revealed that a twofold higher efficiency could be achieved in drop-by-drop method as compared to the conventional floral dip method. *Amaranthus* has been successfully transformed using drop-by-drop method (Munusamy et al. 2013). Inoculum density and number of applications influence the transformation efficiency of this method.

14.3.1.5 Floral or Flower Inoculating Method

A hypothesis put forth by a Zhou et al. (1979) engendered a new technique of transformation via the pollen tube pathway. In 1983, this method was successfully used to transfer DNA into cotton (Zhou et al. 1983). Transformation via the pollen tube pathway involves applying the bacterial inoculum to the inflorescence or the stigma directly. This could be before, after or during the pollination event. The germinated pollen grains along with *Agrobacterium* are transferred into close vicinity of its target. In some cases, even the pollen may get transformed, which would still lead to the formation of a heterozygous transformant after fertilization. There have been successful reports of transforming some monocot species, namely

Sorghum bicolor (Elkonin et al. 2009), *Triticum aestivum* (Hess et al. 1991), *Z. mays* (Mamontova et al. 2010), by applying *Agrobacterium* culture to the pistil (on the stigmatic surface) suggesting transformation via the pollen tube pathway. On the same lines, a floral inoculating method was developed by Narusaka et al. (2010) in *A. thaliana*. The inoculation of bacterial cell suspension on independent flower buds allowed many transformation events to be accomplished simultaneously.

14.3.1.6 Ovary/Pistil Drip Transformation

Ovary-drip transformation is biosafe transformation system which comprises self-pollination followed by making a wound site on the ovary wall through excision of whole of the style, onto which the exogenous DNA solution supplemented with surfactant is directly dripped. This method, initially developed for soybean cultivars, has been established for maize also. In case of soybean, it has been found that for improving the efficiency dripping should be synchronized with syngamy (fusion of gametes) and the division of the zygote. The earlier attempt for transforming soybean through the pollen tube pathway was irreproducible (Shou et al. 2002). Reproducible and improved transformation efficiency has been established in soybean through ovary-drip transformation (Liu et al. 2009a, b). It is a simple and direct method of transformation with the help of a linear gene cassette and does not involve *Agrobacterium* or *vir*-gene machinery to integrate the DNA, producing low copy number, vector- and marker-free stable transgenic. At present, this is the only strategy available for effecting transformation in legume (Liu et al. 2009a, b).

In case of maize, a window is incised into the central portion of the ear with the help of a sterile scalpel, styles are completely removed, and then, exogenous DNA with sucrose and surfactant is directly applied to each ovary drop-by-drop with help of a micropipette (Yang et al. 2009). Pistil drip transformation has also been applied for transforming *Gossypium hirsutum* (Chen et al. 2010). In this case, agroinfiltration was performed by inoculating *Agrobacterium* at different times on the first and second day of flowering on pistils either with intact stigma or after decapitating it. They found that the excision of stigma resulted in a shorter travel distance of *Agrobacterium* and easier accessibility to the ovarian locule which eventually yielded enhanced transformation efficiency.

For both soybean and maize, 0.05 % surfactant and 5 % sucrose promote the exogenous DNA to enter the embryo sac (Liu et al. 2009a, b; Yang et al. 2009). In addition to reducing the surface tension, surfactant protects the linear gene cassette from degradation. This showed that an enhanced efficiency could be obtained if the passage to the embryo sac in the ovules is shortened due to less DNA degradation. The time of application of extraneous DNA to the wound site is very important for success in ovary-drip transformation and for soybean it is 6–8 h after self-pollination. The other factors, which influence the efficiency, include the physiological traits of the host soybean plants like the pod-bearing rate. Excessive wounding of the flowers results into abortion, reducing the pod-bearing rate. But inadequate infiltration may not lead to target, which reiterates the importance of pathway length that gives highest frequency and pod-bearing rates. In the study on soybean, the complete removal of style without causing wounding of ovary gives maximum frequency. Besides transformation of barrel medic by the floral dip method (Trieu et al. 2000), this is the only one strategy for successful transformation of a legume (Liu et al. 2009a, b).

14.3.2 Factors Influencing Female Germline Transformation Efficiencies

Success of different transformation protocols has been attributed to several parameters either singly or in combination thereof. These include the physiological state of the plant, the infiltration medium, number of inoculations performed, duration of vacuum treatment, and inoculation density (concentration of *Agrobacterium*) (Grabowska and Filipecki 2004; Bechtold et al. 1993). Clough and Bent (1998) have identified sucrose and surfactant/vacuum to be the main parameters influencing transformation efficiency of *in planta* methods, although species-/genotype-specific optimization has to be carried out for all the systems tested. Inclusion of surfactant in the infiltration medium could seamlessly replace the use of vacuum. Altering the composition of media, growth regulators and even changes in pH or bacterial density did not have significant effect on the transformation efficiencies.

14.3.2.1 Plant Genotype, Physiological State and Developmental Stage

One of the most important variables in germline transformation is the developmental stage of the plant at the time of inoculation with *Agrobacterium*. Plants that reach either at the primary or secondary bolt development are suitable for supporting high and reproducible transformation efficiency (Curtis and Nam 2001). Clough and Bent (1998) reported that *Arabidopsis* plants in which primary bolts were clipped and secondary bolts were 2–10 cm, having well-developed inflorescence with maximum unopened floral bud clusters, represented the ideal stage. Seeds harvested from the plants that had tertiary bolt development at time of transformation were not transformed (Curtis and Nam 2001). Partly open-flower buds gave the best results during alfalfa transformation (Trieu et al. 2000). With *B. rapa* ssp. *chinensis*, 50- to 60-cm-long floral shoots with few open flowers were the best (Liu et al. 1998; Qing et al. 2000). In *A. lasiocarpa* and *Raphanus sativus*, the floral shoots with closed buds were considered the ideal stage (Tague 2001; Curtis and Nam 2001).

Different plants or the ecotypes of the same species may exhibit variation in transformation efficiencies even after being transformed by the same *Agrobacterium* strain. A study with different *Arabidopsis* ecotypes showed resistance of some ecotypes to agroinfection (Nam et al. 1997). This observation was supported by Ghedira et al. (2013a) who provided the evidence that physiology and the ecotype of the dipped plant influences transformation efficiency of *A. thaliana* along with the bacterial strain.

14.3.2.2 Sucrose

Sucrose is a crucial component of the inoculation medium. Absence of sucrose in medium invariably results in the failure of transformation. Usually, 5 % sucrose is added to the inoculation medium. However, for transforming *Capsella bursa-pastoris* infiltration medium having 10 % sucrose was found to be the best (Bartholmes et al. 2008). In case of *Arabidopsis*, 10 % sucrose gave highest efficiency, but it could not be reproduced consistently (Clough and Bent 1998). Low-grade sucrose or 5 % glucose can also be used as substitutes, but mannitol has deleterious effects on the growth of plant (Clough and Bent 1998).

14.3.2.3 Surfactant

In floral dip method, a non-ionic, organosilicone-based surfactant-like Silwet L-77 (0.05–0.1 %) was vital for successful transformation. In case of *in planta* vacuum transformations, surfactant was not required. Concentrations of Silwet L-77 beyond 0.1 % were detrimental causing necrosis of plant tissue, destruction of flower buds, inhibiting fruit development (Clough and Bent 1998) and also resulting in abortive flowers and low seed set (Tague 2001). In *A. thaliana*, *B. rapa* and *M. truncatula,* the concentration of Silwet L-77 was optimized at 0.02 %, but for *A. lasiocarpa*, it was five times greater (Clough and Bent 1998; Bechtold et al. 2000; Chung et al. 2000; Qing et al. 2000; Trieu et al. 2000; Ye et al. 1999; Tague 2001). Bartholmes et al. (2008) reported that Silwet L-77 at concentrations between 0.02 and 0.05 % was best for *C. bursa-pastoris.*

Detergent surfactants like Tween 20 or Pluronic F-68 have also been used. However, Silwet L-77 is preferred as compared to others as it reduces surface tension more effectively, keeps bacterial suspension viable for longer duration on plant and has low phytotoxicity, supporting intensive penetration of bacteria to relatively inaccessible plant tissues. Use of Silwet L-77 also replaces the vacuum step in floral dip experiment. In radish, Pluronic F-68 at 0.1 % (w/v) on primary bolted plants gave a superior efficiency than 0.01 and 0.1 % Silwet L-77 (Curtis and Nam 2001). The reason is attributed to the fact that Pluronic F-68 can increase the permeability of plasma membrane (Lowe et al. 1993) and may facilitate the movement of agrobacteria to sites responsible for seed development. Recently, XIAMETER OFX-0309, a non-ionic, non-cytotoxic and silicon-based surfactant, has been reported to attain enhanced transformation frequency of ~1.6 % in *A. thaliana*, at a concentration of 0.05 % in floral dip experiments (Mireault et al. 2014).

14.3.2.4 Agrobacterium Strain

Host-strain specificity is well established in case of *Agrobacterium* strains. The host genotype and bacterial strain should be compatible for obtaining high transformation efficiency. Such an interaction has been recorded for *Arabidopsis* transformation via *Agrobacterium* (Clough and Bent 1998; Bent 2000). Use of acetosyringone, a phenolic compound, has been tested to achieve higher transformations even for vegetative tissues. Phenolic compounds induce the activity of virulence genes of *Agrobacterium* (Hirooka et al. 1987), thereby enhancing transformation efficiencies. Correct choice of *Agrobacterium* strain resulted in successful transformation of *C. bursa-pastoris* (Bartholmes et al. 2008), on which the earlier attempts had failed (Tague 2001).

14.3.2.5 Number of Applications of Agrobacterium

When plants with indeterminate inflorescences are infiltrated, repeated applications of *Agrobacterium* have been found to enhance the transformation efficiency. As the flower buds in *Arabidopsis* belonging to Brassicaceae are borne acropetally on the inflorescence, new unopened buds differentiate successively providing opportunity for reapplication. Excessive reapplication of *Agrobacterium* at less than 4-day intervals causes necrosis and may even lead to death (Clough and Bent 1998). The number of applications required for higher efficiency varies in different species. In floral dip experiments of barrel medic, two inoculations at a seven-day interval increased the efficiency on one the hand, but on the other hand decreased viability of plants (Trieu et al. 2000). Whereas in *A. lasiocarpa* (Tague 2001) only one application yielded efficiency than in *A. thaliana*. In floral spray experiments of *Arabidopsis*, spraying for thrice gave higher efficiency than once (Chung et al. 2000). Liu et al. (2012) also used two floral dip steps with an interval of one week for successful transformation of *C. sativa* without the use of vacuum.

14.3.2.6 Humidity Chamber

Maintaining high humidity during the first 12–24 h after *Agrobacterium* inoculation has been found to be beneficial in most cases (Clough and Bent 1998). A plastic dome or wrap/sheet is put around the inflorescence or plant. It is assumed that this prolongs availability of surface water through which *Agrobacterium* can swim eventually helps in reaching the target cells. In addition, it also maintains host tissues in good condition.

14.3.2.7 Accessibility of Germline Cells

It is a widely known fact that floral dip is a success with very few plant species beyond the family Brassicaceae. This is largely because of the developmental pattern of the floral organs. Within the family also, there is enormous variation in the pattern of flower development which accounts for the differences in transformation frequency. For example, Pak choi (*B. rapa* ssp. *chinensis*) and *A. thaliana* vary in the development pattern of pistil. In Pak choi, the access to gynoecium is difficult than in *Arabidopsis,* thus reducing the transformation efficiency. In addition, the density and viability of bacterial population around the target cell, as well as the presence of virulence-inducing factors, affect the rate of transformations. Continued transformation was observed even weeks after the first transformation event by increasing the amount of time of bacterial exposure to the plant tissue (Xu et al. 2008). Ghedira et al. (2013b) reported that the target accessibility to *Agrobacterium* is a limiting factor for successful transformation as compared to integration of T-DNA. If the target is not accessible to the agrobacteria, any attempt to transform may become futile even if all the other parameters are fulfilled. It appears that a combination of different methods supported by crucial levels of factors may hold the key to successful germline transformations.

14.3.3 Establishing the Target of Female Germline Transformation

In various protocols based on *Agrobacterium*-mediated transformations, access to the embryo sac requires the bacteria to traverse through a tortuous maze of tissues systems and their exudates. In spite of these tissue barriers, the possibility of ovules being a target was considered in *Arabidopsis*. As high transformation efficiency was achieved by *in planta* transformation, it was opined that the target to be transformed must have been easily accessible and present on the exterior of the plant part. Thus, it was proposed that the *Agrobacterium* could be transforming the pollen grains either on the stigmatic surface or within the anthers at some developmental stage (Bechtold et al. 2000; Desfeux et al. 2000; Ye et al. 1999). Conclusive evidence came from the analysis of the transformants. The progeny recovered after infiltration were hemizygous for the transgene indicating that transformation occurred at a late stage of male and female germline development (Feldmann 1992; Bechtold et al. 1993).

Experimental evidence for confirming the target of transformation was provided by Ye et al. (1999). In genetic crosses, trans-seeds were obtained only when the infiltrated plants were used as pollen recipient and not as pollen donor. This suggested that *Agrobacterium* transformation occurred through the ovule pathway (Desfeux et al. 2000; Ye et al. 1999). Further, the expression of the scorable marker gene *GUS* was found to be more in several ovules especially of unopened agroinfiltrated flowers than to the pollen, in which it was almost negligible. During microscopic analyses, two conditions were revealed even within the same ovarian locule: one, in which the entire ovule/ovules turned blue indicating *GUS* expression throughout ovule; second, in which ovules were stained only at the site of embryo, and more specifically at the micropylar end. The transgenic seeds thus recovered also tested positive for *GUS* (including seed coats and parts of the interior of seed tissues). In general, the final seed transformation efficiencies determined after selection correlated well with those of ovule transformation. This demonstrated clearly that the various developmental stages from ovule primordia to the mature gametophyte served as the prime and effective targets. Thus, the most probable cellular target of vacuum infiltration or floral dip transformation is likely to be the female reproductive tissue, particularly the ovule and the genetic target is female chromosome set (Bechtold et al. 2000).

Further investigations by Bechtold et al. (2003) were done to establish the optimum developmental stage required to achieve higher transformation efficiencies. *GUS* analyses of the transgenic progeny obtained after infiltrations revealed two distinct categories of transformants: first, where the *GUS* positive tissues could be either the embryo or the endosperm; the second, where both tested positive (co-transformation). This observation led to the conclusion that transformation could be an early- or a late-event.

Early transformation events, where both the endosperm and embryo scored positive for expression, indicated that transformation occurred before the gametophyte differentiation. This observation also suggested that in addition to the egg cell even the endosperm was a prospective target, as was previously established (Bechtold et al. 2003). In certain seeds, where both the embryo and endosperm were transformed, it was difficult to establish whether they were the result of single or two independent transformation events. Late transformation events were cases where either the embryo or the endosperm tested positive for GUS expression. In such cases, the transformation occurred after differentiation of the gametophyte. Conclusive proof for the above was established by performing Southern blot analyses on transformants in *C. bursa-pastoris* (Bartholmes et al. 2008). Here, the mechanism of infection is similar to that of *A. thaliana*. Southern blot analysis revealed unique band patterns in the different transformants indicating infection after the formation of ovules. Moreover, the embryo and the endosperm were analysed for the expression of a selectable marker and more independent transformations were observed in which either one was transformed as compared to co-transformation. The transfer of T-DNA occurs during gametophyte development and on repeating the transformation, the second T-DNA gets randomly integrated in the genome which is not influenced by previous transformation event (Bechtold et al. 2000).

Besides the relatively more responsive crucifers, *in planta* method has also been tested in the forage legume *M. truncatula* (Trieu et al. 2000). A vernalization treatment (4 °C for 2 weeks) was included in the procedure to induce early flowering. Most of the transformants were homozygous and sibling transformants, even though independent transformants were also produced. Sibling transformants exhibited identical hybridization patterns through Southern Blot, as they were derivatives of the same T-DNA integration event. Infiltration of plants in bloom as well as those of seedlings exhibited remarkable transformation frequencies. However, the exact cellular target and the mechanism remained elusive (Trieu et al. 2000; Bent 2000). It has been proposed that the meristematic cells of the axillary buds which later give rise to the developing bolts get transformed, leading to production of sibling transformants. Investigation of *R. sativus* transformation also suggests infection of meristem cells (Curtis and Nam 2001). Thus, transformation in *Medicago* probably occurred at developmental stages earlier than those observed in *Arabidopsis*, ensuring female germline modification.

Higher transformation efficiencies achieved in different members of the Brassicaceae as compared to species from the other families distinctly points to the presence of special attributes in Brassicaceae. One common point would be the pattern of inflorescence and floral development. An indeterminate inflorescence as in *Arabidopsis*, Pak choi, radish, cabbage and *Capsella* provides flowers at different developmental stages, thus providing an advantage for successful transformation. Thus, inoculation of the entire inflorescence with *Agrobacterium* increases the possibility that at least some flowers might be at the right developmental stage.

Bowman (1994) studied the floral development of *Arabidopsis* by scanning electron microscopy and proposed the probable route of *Agrobacterium* entry/passage into the interior of gynoecium. In *Arabidopsis*, the floral meristem differentiates organs in a concentric pattern giving rise to the four whorls. This developmental pattern positions the gynoecium in the centre, surrounded by androecium, petals and sepals in the second, third and fourth whorls, respectively. As further growth continues, the gynoecium elongates and develops like a cylinder but remains open at the apical region. The stigmatic cap seals the locular cavity (Bowman 1994) only three days prior to anthesis. This pattern of development is different from other plants like soybean, where the locule closes more than 10 days prior to anthesis (Johns and Palmer 1982). Comparison of the transformation frequencies in the two species clearly provides the insight for success of floral dip in case of *Arabidoposis* and the other members of Brassicaceae. In *Arabidopsis*, ovule primordium arises one day earlier and megasporocyte formation occurs one day after the closure of gynoecium (Bowman 1994). Flowers at younger developmental stages ($\sim$6–11 days from anthesis) exhibit higher rates of transformations, which could be due to the fact that an open gynoecium (at the time of infiltration) allows the bacteria an easy access to the target tissue. Such an entry is prevented by the stigmatic cap in the mature flowers. Similar findings have been reported in radish, where transgenic seeds were obtained only when flowers were dipped at the early stage of bolting (Curtis and Nam 2001). In Pak choi, the ideal stage for transformation is around 10 days before anthesis (period prior to closure of ovarian locule) (Xu et al. 2008).

14.4 Nature and Selection of Transformants

The recovery of 'useful' transgenics is the final measure of success for any method of genetic transformation. Transgenics containing single copy inserts of gene of interest (GOI) and exhibiting optimum levels of transgene expression are the desired end products. These can be deployed directly for agronomic ends or taken forward for building the germplasm base in breeding programs.

As discussed in earlier sections, every method/protocol employed for germline transformations has its own merit/s and demerit/s. A spectrum of variables needs to be optimized for successful gene delivery, integration and expression. Amongst these, an important parameter is the nature of exogenous DNA. Published information spanning over two decades has shown a gradual progress in the nature of DNA used for transforming pollen, microspores germline from genomic preparations to purified plasmid vectors. There has also been a significant improvement in vector/gene constructs that includes testing promoters—constitutive to tissue specific.

Agrobacterium is the most favoured agent of gene delivery as it is naturally tailored to deliver single copy of transgene that usually exhibits optimum expression (Tinland 1996; Zupan and Zambryski 1997). Success has been reported for transformation of both male and female germlines (Eapen 2011; Chumakov and Moiseeva 2012). The major limitation of this method remains the preferential infection of a given host by the bacterial strain (strain dependent host specificity) (Grabowska and Filipecki 2004). Due to this limitation, studies based on testing efficiencies of transformation with several types of constructs are tedious. The other methods like microinjection (major limitation is that only a small number of cells can be treated at any given time) and electroporation are good for analyses of transient expression rather than recovery of transgenics from male and female germline.

Most of the analyses for studies in dicots that assess role of promoters have focussed on using particle bombardment as the method of gene delivery and tobacco as a model system. Although this method gives rise to integration of transgene at multiple sites in the genome and these could be tandem inserts, it allows for screening a variety of constructs simultaneously (Travella et al. 2005). Multiple inserts can lead to gene silencing and have been reported although studies indicate that such incidence is similar in both the *Agrobacterium*-mediated transformations and particle bombardment (Birch 1997). The insertion of truncated, duplicated and/or rearranged transgenes could also be due to the fact that the DNA delivered is naked and not protected from the nucleases. Nevertheless, this method has been used for screening a variety of constructs simultaneously (Travella et al. 2005). It has been considered to be genotype independent and a very promising tool for plants as well as plant parts that are recalcitrant to other methods of gene delivery.

Vector constructs also influence detection, selection and recovery of transformants. As compared to vegetative tissues, developmental stage-specific promoters are required for expression in male and female germline. Traditionally most transgenes were put under control of the constitutive promoter *CaMV35S* from cauliflower mosaic virus. However, transient expression of the *GUS* gene driven by *CaMV35S* promoter was not detected histochemically in pollen grains of lily and *N. glutinosa* (van der Leede-Plegt et al. 1992) and detected rarely in those of tobacco (Guerrero et al. 1990, Stöger et al. 1991; Twell et al. 1989a), petunia (Mascarenhas and Hamilton 1992) and *Tradescantia* (Hamilton et al. 1992). A comparison of the activity of different promoters in pollen has been performed using the particle bombardment system (Twell et al. 1989a; Hamilton et al. 1992; van der Leede-Plegt et al. 1992; Stöger et al. 1992; Nishihara et al. 1993). Twell et al. (1989a, b) were the first to show that anther-specific *LAT52* promoter from tomato could drive expression of the *GUS* gene in tobacco and tomato pollen using particle bombardment. This promoter has since then been successfully deployed to drive the expression of foreign genes in the pollen of various species, including tobacco (*N. glutinosa, N. tabacum, N. rustica and Paeonia* (McCormick et al. 1991; Nishihara et al. 1993; van der Leede-Plegt et al. 1992). Several other anther-specific promoters *LAT56* and *LAT59* (from tomato) and a *PA2* promoter (from *Petunia)* have also been tested successfully in tobacco pollen (van Tunen et al. 1990; McCormick et al. 1991; Twell et al. 1991; Stöger et al. 1992). Chimeric genes containing a pollen-specific promoter from tomato (*Lycopersicon esculentum*) *LAT52* or the *CaMV35S* promoter were transiently expressed following their introduction into tobacco (*N. tabacum*) pollen using high-velocity microprojectiles (Twell et al. 1989a).

Van der Leede-Plegt et al. (1992) studied the differential use of various promoters, namely *CaMV 35S*, *LAT52*, *chiA*, *PA2* and *TR2'*, in pollen grains of a dicot (*N. glutinosa*) and a monocot (*L. longiflorum*) plant species. Gene constructs in which the β-glucuronidase (*GUS*) gene was placed under the control of these promoters was introduced in pollen using a particle delivery system. No activity of the cauliflower mosaic virus (*CaMV35S*) promoter was detected in pollen of both *N. glutinosa* and *L. longiflorum*. The promoter of the tomato flower-specific *LAT52* gene was highly active in *N. glutinosa* pollen but remained silent in *L. longiflorum* pollen. Same was true for the pollen-specific chalcone–flavanone isomerase (*chiAPA2*) promoter originally isolated from petunia. This showed the differential activity of *LAT52* and *chiAPA2* in dicots and monocots, respectively. Interestingly, the *TR2'* mannopine synthase promoter of *Agrobacterium tumefaciens* was active in pollen from Solanaceous species (*N. glutinosa*) and also in pollen from the monocot *L. longiflorum* suggesting that it is active in vegetative and sporogenous tissues of both dicot and monocot plant species.

Nishihara et al. (1993) re-evaluated constructs tested earlier by Twell et al. (1989b) and van der Leede-Plegt et al. (1992) by particle bombardment of pollen grains of lily (*L. longiflorum*), two species of tobacco (*N. tabacum* and *N. rustica*) and peony (*Paeonia lactiflora*) by a pneumatic particle gun device (Iida et al. 1990). They reported that *35S CaMV* induced expression of *GUS* gene in all the species studied, albeit significantly lower levels as compared to expression driven by *LAT52*. The pollen-specific *Zm13* promoter from maize was tested by microprojectile bombardment of pollen of *Tradescantia* and tobacco (Hamilton et al. 1992). Miyoshi et al. (1995) reported 800-fold higher activity of *Zm13* in lily pollen as compared to *LAT52* promoter from tomato. Essentially, these studies confirmed the differential activity of promoters in pollen tested from dicots and monocots (van der Leede-Plegt et al. 1992). Thus, vector constructs need to be designed such that transient as well as stable expression is achieved depending on the plant genotype to be targeted and whether sporophytic or gametophytic pathway is to be employed.

In studies aimed at genetic modification of female germline, focus has been on evaluating bacterial strains and other parameters for infiltration. The binary plasmids tested, range from the C58ClRif(3850:1003) cointegrate vector (Feldman and Marks 1987) to pBIN/pBI121(Tague 2001) and the more recent pCAMBIA and pBECKSred vectors. Detection of gene expression also interferes with calculations of transformation efficiency and even comparison of constructs. The most favoured marker gene is the *GUS* gene (*uidA* from *Escherichia coli* K-12). Such a discrepancy between the results/observations of GUS expression in bombarded pollen of lily has been attributed to the presence (Nishihara et al. 1993) or absence (Twell et al. 1991; van der Leede-Plegt et al. 1992) of 20 % methanol in the X-Gluc solution used for the assay (Jefferson et al. 1987; Kosugi et al. 1990). Discrepancies such as 'background stain' have also been reported in vegetative tissues as well as pollen (Hu et al. 1990; Shi et al. 1995). The modified green fluorescent protein (*mgfp*) has also been tested in *Arabidopsis* and *Medicago* (refs from Chumakov and Moiseeva 2012). Anthocyanin pigmentation has been used as marker in wheat floral bud inoculation (Zale et al. 2009) where the transcription factors *Lc* and *C1* have been included in the binary vector. This is a novel and non-destructive visual, scorable marker. Different selection regimes determine the recovery of transgenic plants after genetic modification. In most cases of recovering transformants from male germline, Basta selection has been used.

The selection of seedlings is a crucial, but time-consuming process after the floral dip technique and requires 7-to 10-day selection period. This duration is enough to allow fungal pathogens to flourish due to the humidity provided and presence of sucrose in the residual infiltration medium. Fungal infections jeopardize the recovery of transformants; thus, it is imperative to include antibiotics. A method that segregates the kanamycin-, phosphinothricin- and hygromycin-resistant seedlings from the susceptible ones in only 3.25 days was put forth (Harrison et al. (2006). It also minimizes the risk of seedling loss by pathogen infection. Selection of kanamycin and phosphinothricin is on the basis of presence of chlorophyll in the expanded cotyledon as the former inhibits plastid protein synthesis (Gray et al. 1984), while latter inhibits glutamine synthase activity (Bayer et al. 1972; Tachibana et al. 1986), respectively. For hygromycin B, it is based on hypocotyls length as it suppresses cytosolic protein synthesis (Cabanas et al. 1978).

As per the conventional selection method, first the plants are allowed to dry and seeds are harvested. The seeds are surface sterilized by liquid- or vapour-phase methods and suspended in plates with half-strength Murashige and Skoog's medium (1/2X MS; Murashige and Skoog 1962) supplemented with 0.8 % agar to prepare plates. The medium also contains antibiotics kanamycin or hygromycin B or the herbicide phopshinothricin, for selection, medium cold treated for 2 days, grown for 7–10 days in a controlled environment at 24 °C under 23 h light 50–100 Einsteins m^{-2} s^{-1}. Resistant seedlings are transplanted into pots and grown to maturity (Clough and Bent 1998).

In the modified method, the seeds produced after transformations are surface sterilized and pipetted onto 1 % agar plates containing MS medium and kanamycin monosulphate (50 μg ml^{-1}) or phosphinothricin (50 μM) or hygromycin B (15 μg ml^{-1}). Seeds are stratified for 2 days in dark at 4 °C and then incubated for 4–6 h at 22 °C in continuous white light (80–200 μmol m^{-2} s^{1}) to stimulate germination in growth chamber. Then, plates are wrapped in foil and incubated for 2 days at 22 °C followed by removal of foil and incubation of seedlings for 24–48 h at 22 °C in continuous white light (80–200 μmol m^{-2} s^{1}). There after resistant/transformed seedlings are identified from non-resistant/non-transformed seedlings as kanamycin-resistant seedlings have long hypocotyls and green cotyledons, while non-resistant seedlings have long hypocotyls but pale cotyledons. Similarly, phosphinothricin-resistant seedlings have long hypocotyls and green cotyledons, whereas non-resistant seedlings have long hypocotyls and pale cotyledons. In contrast, hygromycin B-resistant seedlings have long hypocotyls and green cotyledons, whereas non-resistant seedlings have short hypocotyls but also have green cotyledons (Harrison et al. 2006).

Davis et al. (2009) tested an alternative method of transgenic selection on chromatography sand evading the need for surface sterilization of seeds. Another easy and quicker selection protocol has been put forth by Li et al. (2010) that involved only soaking of seeds with antibiotic for 24–36 h, followed by sowing in soil and selecting after germination of the seedlings.

14.5 Conclusion and Future Perspective

Germline transformation is a promising technique with fundamental and applied facets. However, this strategy has not gained widespread use especially with respect to crop improvement. Rapid advances in the DNA and RNA methodologies are churning out extensive expression profiles and regulatory networks are being constructed. Transcriptome analyses of microspores (at different stages of development), pollen grain, germinating pollen and the sperm nuclei are now providing valuable information on gene expression and the key regulators of germline differentiation in the model plant *Arabidopsis* (Grennan 2007). An interesting fact that emerges from the above analysis is that at most stages of development, DNA methylation is employed to ensure that male genome is transmitted without errors and the mechanisms to silence any extra genetic material are enhanced (Twell 2011). Developing germline transformation for any plant species would therefore require a very intense selection of a developmental stage, design of vectors and genes of interest, method of gene introduction and investigative procedures for analyses of gene expression. With the advent of new technologies, it opens up new avenues for crop improvement.

Currently, the entire research on germline transformations has been successful mostly on a very few monocots, members of the Brassicaceae family, more specifically *Arabidopsis* and *Brassica* and a Solanaceous member, tobacco. In future, it would be important to find the applicability of methods discussed to a wider range of crop plants and carry out in-depth experiments. The general mechanisms discovered in *Arabidopsis* may not hold true for other members as has been shown in case of Pak choi (Xu et al. 2008) and *Medicago* (Trieu et al. 2000). Transformation efficiencies may vary amongst the germline transformed, the plant species, the methods employed, and vector constructs tested.

Many findings are yet to be resolved like the questions as to why transgene expression levels decrease with plant age or over generations. The movement of *Agrobacteri*um in plants via vascular bundle or plasmodesmata remains to be explored, as in case of Pak choi (Xu et al. 2008). Even the ovary-drip method beckons further investigation to assess the stability and inheritance of the transgene as well as to determine the cause of non-Mendelian inheritance (Liu et al. 2009a, b). The *arabidopsis crabs-claw* (*crc*) mutants have a more accessible ovarian locule with an open gynoecium and thus served as an improved target for transformation by nearly sixfold (Desfeux et al. 2000). The *open-flower* (*opf*) mutant of *M. alba* has been studied for its potential of floral dip transformation. It is also found to be refractory just like its wild type (Hirsch et al. 2010) and trial to transform it has met with no success. This indicates that some additional factors beyond the plant genotype and accessibility of target might be affecting the transformation process. These aspects are yet to be deciphered.

References

Abdollahi MR, Corral-Martínez P, Mousavi A, Salmanian AH, Moieni A, Seguí-Simarro JM (2009) An efficient method for transformation of pre-androgenic, isolated *Brassica napus* microspores involving microprojectile bombardment and *Agrobacterium*-mediated transformation. Acta Physiol Plant 31:1313–1317

Abdollahi MR, Moieni A, Mousavi A, Salmanian AH (2011) High frequency production of rapeseed transgenic plants via combination of microprojectile bombardment and secondary embryogenesis of microspore-derived embryos. Mol Biol Rep 38:711–719

Abdul-Baki AA, Sanders JA, Matthews BF, Pittarelli GW (1990) DNA uptake during electroporation of germinating pollen grains. Plant Sci 70:181–190

Aionesei T, Hosp J, Voronin V, Heberle-Bors E, Touraev A (2006) Methotrexate is a new selectable marker for tobacco immature pollen transformation. Plant Cell Rep 25:410–416

Akhond MAY, Machray GC (2009) Development of an efficient transient gene expression assay based on tobacco (*Nicotiana tabacum* var. Xanthi) male gametophytes. Plant Tissue Cult Biotechnol 19:9–23

Alimohammadi M, Bagherieh-Najjar MB (2009) *Agrobacterium*-mediated transformation of plants: basic principles and influencing factors. Afr J Biotechnol 8:5142–5148

Alwen A, Eller N, Kastler M, Benito Moreno RM, Heberle-Bors E (1990) Potential of in vitro pollen maturation for gene transfer. Physiol Plant 79:194–196

Aziz N, Machray GC (2003) Efficient male germline transformation for transgenic tobacco production without selection. Plant Mol Biol 51:203–211

Barampuram S, Zhang ZJ (2011) Recent advances in plant transformation. Methods Mol Biol 701:1–35

Barinova I, Zhexembekova M, Barsova E, Lukyanov S, Heberle-Bors E, Touraev A (2002) *Antirrhinum majus* microspore maturation and transient transformation in vivo. J Exp Bot 53:1119–1129

Bartholmes C, Nutt P, Theissen G (2008) Germline transformation of Shepherd's purse (*Capsella bursa-pastoris*) by the 'floral dip' method as a tool for evolutionary and developmental biology. Gene 409:11–19

Bayer E, Gugel KH, Hägele K, Hagenmaier H, Jessipow S, König WA, Zähner H (1972) Stoffwechselprodukte von mikroorganismen. 98. Mitteilung. Phosphinothricin und phospinothricyl-alanylalanin. Helv Chim Acta 55:224–239

Bechtold N, Bouchez D (1995) In planta *Agrobacterium*-mediated transformation of *Arabidopsis thaliana* plants by vacuum infiltration. In: Potrykus I, Spagenberg G (eds) Gene transfer to plants, vol 1. Springer, Berlin, pp 19–23

Bechtold N, Ellis J, Pelletier G (1993) In planta *Agrobacterium*-mediated gene transfer by infiltration of adult *Arabidopsis thaliana* plants. C R Acad Sci Paris, Life Sci 316:1194–1199

Bechtold N, Jaudeau B, Jolivet S, Maba B, Vezon D, Voisin R, Pelletier G (2000) The maternal chromosome set is the target of the T-DNA in the in planta transformation of *Arabidopsis thaliana*. Genetics 155:1875–1887

Bechtold N, Jolivet S, Voisin R, Pelletier G (2003) The endosperm and the embryo of *Arabidopsis thaliana* are independently transformed through infiltration by *Agrobacterium tumefaciens*. Transgenic Res 12:509–517

Bent AF (2000) *Arabidopsis* in planta transformation. Uses, mechanisms, and prospects for transformation of other species. Plant Physiol 124:1540–1547

Bernhardt K, Vigelius SK, Wiese J, Linka N, Weber APM (2012) *Agrobacterium*-mediated *Arabidopsis thaliana* transformation: an overview of T-DNA binary vectors, floral dip and screening for homozygous lines. Endocytobiosis Cell Res 22:19–28

Birch RG (1997) Plant transformation: problems and strategies for practical application. Annu Rev Plant Physiol Plant Mol Biol 48:297–326

Bolik M, Koop HU (1991) Identification of embryogenic microspores of barley (*Hordeum vulgare* L.) by individual selection and culture and their potential for transformation by microinjection. Protoplasma 162:61–68

Booy G, Krens FA, Huizing HJ (1989) Attempted pollen mediated transformation of maize. J Plant Physiol 135:319–324

Bowman J (1994) *Arabidopsis*: an atlas of morphology and development. Springer, New York

Brew-Appiah RAT, Ankrah N, Liu W, Konzak CF, von Wettstein D, Rustgi S (2013) Generation of doubled haploid transgenic wheat lines by microspore transformation. PLoS ONE 8:e80155

Cabanas MJVD, Modolell J (1978) Dual interference of hygromycin B with ribosomal translocation and with aminoacyl tRNA recognition. Eur J Biochem 87:21–27

Carlson AR, Letarte J, Chen J, Kasha KJ (2001) Visual screening of microspore-derived transgenic barley (*Hordeum vulgare* L.) with green-fluorescent protein. Plant Cell Rep 20:331–337

Chang SS, Park SK, Kim BC, Kang BJ, Kim DU, Nam HG (1994) Stable genetic transformation of *Arabidopsis thaliana* by *Agrobacterium* inoculation in planta. Plant J 5:551–558

Chen TZ, Wu SJ, Zhao J, Guo WZ, Zhang TZ (2010) Pistil drip following pollination: a simple in planta *Agrobacterium*-mediated transformation in cotton. Biotechnol Lett 32:547–555

Chhikara S, Chaudhary D, Yadav M, Sainger M, Jaiwal P (2012) A non-tissue culture approach for developing transgenic *Brassica juncea* L. plants with *Agrobacterium tumefaciens*. In Vitro Cell Dev Biol Plant 48:7–14

Choudhary B, Gheysen G, Buysse J, van der Meer P, Burssens S (2014) Regulatory options for genetically modified crops in India. Plant Biotech J 12:135–146

Chugh A, Eudes F (2008a) Cellular uptake of cell-penetrating peptides pVEC and transportan in plants. J Pept Sci 14:477–481

Chugh A, Eudes F (2008b) Study of uptake of cell penetrating peptides and their cargoes in permeabilized wheat immature embryos. FEBS J 275:2403–2414
Chugh A, Amundsen EJ, Eudes F (2009) Translocation of cell-penetrating peptides and delivery of their cargoes in triticale microspores. Plant Cell Rep 28:801–810
Chumakov MI, Moiseeva EM (2012) Technologies of *Agrobacterium* plant transformation in planta. Appl Biochem Microbiol 48:657–666
Chumakov MI, Rozhok NA, Velikov VA, Tyrnov VS, Volokhina IV (2006) *Agrobacterium*-mediated in planta transformation of maize via pistil filaments. Genetika 42:1083–1088
Chung MH, Chen MK, Pan SM (2000) Floral spray transformation can efficiently generate *Arabidopsis*. Transgenic Res 9:471–486
Clough SJ, Bent AF (1998) Floral dip: a simplified method for *Agrobacterium*-mediated transformation of *Arabidopsis thaliana*. Plant J 16:735–743
Cullis CA (1990) DNA rearrangements in response to environmental stress. Adv Genet 28:73–97
Cullis CA, Cleary W (1986) DNA variation in flax tissue culture. Can J Genet Cytol 28:247–251
Cullis CA, Kolodynska K (1975) Variation in the isozymes of flax (*Linum usitatissimum*) genotrophs. Biochem Genet 13:687–697
Curtis IS, Nam HG (2001) Transgenic radish (*Raphanus sativus* L. *longipinnatus* Bailey) by floral dip method-plant development and surfactant are important in optimizing transformation efficiency. Transgenic Res 10:363–371
Davis AM, Hall A, Millar AJ, Darrah C, Davis SJ (2009) Protocol: streamlined sub-protocols for floral dip transformation and selection of transformants in *Arabidopsis thaliana*. Plant Meth 5:1–7
de la Pena A, Lörz H, Schell J (1987) Transgenic rye plants obtained by injecting DNA into young floral tillers. Nature 325:274–276
Desfeux C, Clough SJ, Bent AF (2000) Female reproductive tissues are the primary target of *Agrobacterium*-mediated transformation by the *Arabidopsis* floral dip method. Plant Physiol 123:895–904
Dormann M, Wang HM, Oelck M (2001) Transformed embryogenic microspores for the generation of fertile homozygous plants. US Patent, 6,316,694 B1
Eapen S (2011) Pollen grains as a target for introduction of foreign genes into plants: an assessment. Physiol Mol Biol Plants 17:1–8
Elkonin LA, Ravin NV, Leshko EV, Volokhina IV, Chumakov MI, Skryabin KG (2009) In planta agrobacterial transformation of sorghum plants. Biotekhnologiya 1:23–30
Elkonin LA, Nosova ON, Italianskaya JV (2012) *Agrobacterium*-mediated genetic transformation of *Sorghum* using tissue culture-based and pollen-mediated approaches. J Stress Physiol Biochem 8:S28
Eudes F, Shim Y, Jiang F (2014) Engineering the haploid genome of microspores. Biocatal Agric Biotechnol 3:20–23
FAO (2009) How to feed the world in 2050. Discussion paper, High-Level Expert Forum. The Food and Agricultural Organization, Rome, Italy
Feldmann K (1992) T-DNA insertion mutagenesis in *Arabidopsis*: seed infection transformation. In: Koncz C, Chua NH, Schell J (eds) Methods in *Arabidopsis* research. World Scientific, Singapore, pp 274–289
Feldmann KA, Marks MD (1987) *Agrobacterium*-mediated transformation of germinating seeds of *Arabidopsis thaliana*: a non-tissue culture approach. Mol Gen Genet 208:1–9
Fennell A, Hauptmann R (1992) Electroporation and PEG delivery of DNA into maize microspores. Plant Cell Rep 11:567–570
Fernando DD, Richards JL, Kikkert JR (2006) In vitro germination and transient *GFP* expression of American chestnut (*Castanea dentata*) pollen. Plant Cell Rep 25:450–456
Folling L, Olesen A (2001) Transformation of wheat (*Triticum aestivum* L.) microspore-derived callus and microspores by particle bombardment. Plant Cell Rep 20:629–636
Forster BP, Heberle-Bors E, Kasha KJ, Touraev A (2007) The resurgence of haploids in higher plants. Trends Plant Sci 12:368–375

Fukuoka H, Ogawa T, Matsuoka M, Ohkawa Y, Yano H (1998) Direct gene delivery into isolated microspores of rapeseed (*Brassica napus* L.) and the production of fertile transgenic plants. Plant Cell Rep 17:323–328

Gao Y, Ren X, Ruan S, Wang H (2012) Genetic transformation of chinese cabbage (*Brassica rapa pekinensis*) with floral dip method. Agric Biotechnol 1(22–24):27

Ghedira R, De Buck S, Nolf J, Depicker A (2013a) The efficiency of *Arabidopsis thaliana* floral dip transformation is determined not only by the Agrobacterium strain used but also by the physiology and the ecotype of the dipped plant. Mol Plant Microbe Interact 26:823–832

Ghedira R, De Buck S, Van Ex F, Angenon G, Depicker A (2013b) T-DNA transfer and T-DNA integration efficiencies upon Arabidopsis thaliana root explant cocultivation and floral dip transformation. Planta 238:1025–1037

Grabowska A, Filipecki M (2004) Infiltration with *Agrobacterium*—the method for stable transformation avoiding tissue culture. Acta Phys Plant 26:451–458

Gray JC, Phillips AL, Smith AG (1984) Protein synthesis by chloroplasts. In: Elis RJ (ed) Chloroplast biogenesis. Cambridge University Press, Cambridge, pp 137–163

Grennan AK (2007) An analysis of the *Arabidopsis* pollen transcriptome. Plant Physiol 145:3–4

Grover A, Pental D (2003) Breeding objectives and requirements for producing transgenics for major field crops of India. Curr Sci 84:310–320

Guerrero FD, Crossland L, Smutzer GS, Hamilton DA, Mascarenhas JP (1990) Promoter sequences from a maize pollen-specific gene direct tissue-specific transcription in tobacco. Mol Gen Genet 224:161–168

Hamilton CM (1997) A binary-BAC system for plant transformation with high-molecular-weight DNA. Gene 200:107–116

Hamilton DA, Roy M, Rueda J, Sindhu RK, Sanford J, Mascarenhas JP (1992) Dissection of a pollen-specific promoter from maize by transient transformation assays. Plant Mol Biol 18:211–218

Hansen G, Wright MS (1999) Recent advances in the transformation of plants. Trends Plant Sci 4:226–231

Harrison SJ, Mott EK, Parsley K, Aspinall S, Gray JC, Cottage A (2006) A rapid and robust method of identifying transformed *Arabidopsis thaliana* seedlings following floral dip transformation. Plant Meth 2:19

Harwood WA, Bean SJ, Chen DF, Mullineaux PM, Snape JW (1995) Transformation studies in *Hordeum vulgare* using a highly regenerable microspore system. Euphytica 85:113–118

Heberle-Bors E (1995) Transformation via microspore embryogenesis. Rice Biotechnol Q 23:17–18

Hess D (1980) Investigations on the intra- and interspecific transfer of anthocyanin genes using pollen as vectors. J Plant Physiol 98:321–337

Hess D (1987) Pollen-based techniques in genetic manipulation. Int Rev Cytol 107:169–190

Hess D, Dressler K, Nimmrichter R (1991) Transformation experiments by pipetting *Agrobacterium* into the spikelets of wheat (*Triticum aestivum* L.). Plant Sci 72:233–244

Hirooka T, Rogowsky PM, Kado CI (1987) Characterization of the *virE* locus of *Agrobacterium tumefaciens* plasmid pTiC58. J Bacteriol 169:1529–1536

Hirsch AM, Lee A, Deng W, Tucker SC (2010) An open-flower mutant of *Melilotus alba*: potential for floral dip transformation of a papilionoid legume with a short life cycle? Am J Bot 97:395–404

Hu CY, Wang L (1999) In planta soybean transformation technologies developed in China: procedure, confirmation and field performance. In Vitro Cell Dev Biol Plant 35:417–420

Hu C, Chee PP, Chesney RH, Zhou JH, Miller PD, O'Brien WT (1990) Intrinsic GUS-like activities in seed plants. Plant Cell Rep 9:1–5

Huang B (1992) Genetic manipulation of microspores and microspore-derived embryos. In Vitro Cell Dev Biol 28:53–58

Iida A, Seki M, Kamada M, Yamada Y, Morikawa H (1990) Gene delivery into cultured plant cells by DNA-coated gold particles accelerated by a pneumatic particle gun. Theor Appl Genet 80:813–816

Ilori CO, Pellegrineschi A (2011) Transgene expression in cowpea (*Vigna unguiculata* (L.) Walp.) through *Agrobacterium* transformation of pollen in flower buds. Afr J Biotechnol 10:11821–11828

Inan G, Zhang Q, Li P, Wang Z, Cao Z, Zhang H, Zhang C, Quist TM, Goodwin SM, Zhu J, Shi H, Damsz B, Charbaji T, Gong Q, Ma S, Fredricksen M, Galbraith DW, Jenks MA, Rhodes D, Hasegawa PM, Bohnert HJ, Joly RJ, Bressan RA, Zhu JK (2004) Salt cress. A halophyte and cryophyte Arabidopsis relative model system and its applicability to molecular genetic analyses of growth and development of extremophiles. Plant Physiol 135:1718–1737

Jähne A, Becker D, Brettschneider R, Lörz H (1994) Regeneration of transgenic, microspore-derived, fertile barley. Theor Appl Genet 89:525–533

James C (2012) Global status of commercialized biotech/GM crops: 2012. ISAAA Brief No. 44. Ithaca (New York): ISAAA

Jardinaud M, Souvré A, Alibert G (1993) Transient *GUS* gene expression in *Brassica napus* electroporated microspores. Plant Sci 93:177–184

Jardinaud MF, Souvré A, Beckert M, Alibert G (1995) Optimisation of DNA transfer and transient ß-glucuronidase expression in electroporated maize (*Zea mays* L.) microspores. Plant Cell Rep 15:55–58

Jefferson RA, Kavanagh TA, Bevan MW (1987) *GUS* fusions: β-glucuronidase as a sensitive and versatile gene fusion marker in higher plants. EMBO J 6:3901–3907

Joersbo M, Brunstedt J (1990) Direct gene transfer to plant protoplast by mild sonication. Plant Cell Rep 9:207–210

Joersbo M, Brunstedt J (1992) Sonication: a new method for gene transfer to plants. Physiol Plant 85:230–234

Joersbo M, Jørgensen RB, Olesen P (1990) Transient electropermeabilization of barley (*Hordeum vulgare* L.) microspores to propidium iodide. Plant Cell Tiss Org Cult 23:125–129

Johns CW, Palmer RG (1982) Floral development of a flower-structure mutant in soybeans, *Glycine max* (L.) Merr. (Leguminosae). Am J Bot 69:829–842

Jones-Villeneuve E, Huang B, Prudhomme I, Bird S, Kemble R, Hattori J, Miki B (1995) Assessment of microinjection for introducing DNA into uninuclear microspores of rapeseed. Plant Cell Tiss Org Cult 40:97–100

Kakuta H, Maeda T, Akiyama M, Hashimoto R, Horikawa Y (2001) Transient gene expression in maize and *Asparagus* pollen using magnetic particles by particle gun. International Society for Horticultural Science

Katavic V, Haughn GW, Reed D, Martin M, Kunst L (1994) In planta transformation of *Arabidopsis thaliana*. Mol Gen Genet 245:363–370

Kim SS, Shin DI, Park HS (2007) Transient B-glucuronidase expression in lily (*Lilium longiflorum* L.) pollen via wounding-assisted *Agrobacterium*-mediated transformation. Biotechnol Lett 29:965–969

Kosugi S, Ohash Y, Nakajima K, Arai Y (1990) An improved assay for β-glucuronidase in transformed cells: methanol almost completely suppresses a putative endogenous B-glucuronidase activity. Plant Sci 70:133–140

Kumlehn J, Serazetdinova L, Hensel G, Becker D, Loerz H (2006) Genetic transformation of barley (*Hordeum vulgare* L.) via infection of androgenetic pollen cultures with *Agrobacterium tumefaciens*. Plant Biotech J 4:251–261

Kung S, Wu R (1992) Transgenic plants: present status and social and economic impacts, vol 2. Academic Press Inc., San Diego

Labra M, Vannini C, Grassi F, Bracale M, Balsemin M, Basso B, Sala F (2004) Genomic stability in *Arabidopsis thaliana* transgenic plants obtained by floral dip. Theor Appl Genet 109:1512–1518

Langridge P, Brettschneider R, Lazzeri P, Lorz H (1992) Transformation of cereals via *Agrobacterium* and the pollen pathway: a critical assessment. Plant J 2:631–638

Li X, Wang XD, Zhao X, Dutt Y (2004) Improvement of cotton fiber quality by transforming the *acsA* and *acsB* genes into *Gossypium hirsutum* L. by means of vacuum infiltration. Plant Cell Rep 22:691–697

Li J, Tan X, Zhu F, Guo J (2010) A rapid and simple method for *Brassica napus* floral dip transformation and selection of transgenic plantlets. Int J Biol 2:127–131

Liu W (2004) Transformation of microspores for generating doubled haploid transgenic wheat (*Triticum aestivum* L.). A dissertation submitted in partial fulfillment of the requirements for the degree of Doctor of Philosophy in Crop Science, Department of Crop and Soil Sciences, Washington State University

Liu F, Cao MQ, Yao L, Li Y, Robaglia C, Tourneur C (1998) In planta transformation of pak choi (*Brassica campestris* L. ssp. *chinensis*) by infiltration of adult plants with *Agrobacterium*. Acta Hortic 467:187–192

Liu J, Su Q, An L, Yang A (2009a) Transfer of a minimal linear marker-free and vector-free smGFP cassette into soybean via ovary-drip transformation. Biotechnol Lett 31:295–303

Liu M, Yang J, Cheng Y, An L (2009b) Optimization of soybean (Glycine max (L.) Merrill) in planta ovary transformation using a linear minimal gus gene cassette. J Zhejiang Univ Sci B 10:870–876

Liu X, Brost J, Hutcheon C, Guilfoil R, Wilson AK, Leung S, Shewmaker CK, Rooke S, Nguyen T, Kiser J, Rocher JD (2012) Transformation of the oilseed crop *Camelina sativa* by *Agrobacterium*-mediated floral dip and simple large-scale screening of transformants. In Vitro Cell Dev Biol-Plant 48:462–468

Logemann E, Birkenbihl RP, Ulker B, Somssich IE (2006) An improved method for preparing *Agrobacterium* cells that simplifies the *Arabidopsis* transformation protocol. Plant Meth 2:16

Lowe KC, Davey MR, Power JB, Mulligan BJ (1993) Surfactant supplements in plant culture systems. Agro-food Ind Hi-Tech 4:9–13

Lu C, Kang J (2008) Generation of transgenic plants of a potential oilseed crop *Camelina sativa* by *Agrobacterium*-mediated transformation. Plant Cell Rep 27:273–278

Luo Z, Wu R (1989) A simple method for the transformation of rice via the pollen-tube pathway. Plant Mol Biol Rep 7:69–77

Maluszynski M, Szarejko I, Kasha KJ (2003) Published protocols for other crop plant species. In: Maluszynski M, Kasha KJ, Forster BP, Szarejko I (eds) Doubled haploid production in crop plants: a manual. Kluwer Academic Publishers, Dordrecht, pp 309–336

Mamontova EM, Velikov VA, Volokhina IV, Chumakov MI (2010) *Agrobacterium* mediated in planta transformation of maize germ cells. Genetika 46:568–572

Mannion AM, Morse S (2013) GM crops 1996–2012: a review of agronomic, environmental and socio-economic impacts. University of Surrey, Centre for Environmental Strategy Working Paper 04/13, pp 1–40

Martinez-Trujillo M, Limones-Briones V, Cabrera-Ponce JL, Herrera-Estrella L (2004) Improving transformation efficiency of *Arabidopsis thaliana* by modifying the floral dip method. Plant Mol Biol Rep 22:63–70

Mascarenhas JP, Hamilton DA (1992) Artifacts in the localization of *GUS* activity in anthers of petunia transformed with a CaMV 35s-GUS construct. Plant J 2:405–408

Matthews BF, Abdul-Baki AA, Saunders JA (1990) Expression of a foreign gene in electroporated pollen grains of tobacco. Sex Plant Reprod 3:147–151

McCormick S, Yamaguchi J, Twell D (1991) Deletion analysis of pollen-expressed promoters. In Vitro Cell Dev Biol 27:15–20

Mireault C, Paris L, Germain H (2014) Enhancement of the *Arabidopsis* floral dip method with XIAMETER OFX-0309 as alternative to Silwet L-77 surfactant. Botany 92:523–525

Mishra KP, Joshua DC, Bhatia CR (1987) In vitro electroporation of tobacco pollen. Plant Sci 52:135–139

Miyoshi H, Usami T, Tanaka I (1995) High level of *GUS* gene expression driven by pollen-specific promoters in electroporated lily pollen protoplasts. Sex Plant Reprod 8:205–209

Munusamy U, Abdullah SNA, Aziz MA, Khazaai H (2013) Female reproductive system of *Amaranthus* as the target for *Agrobacterium*-mediated transformation. Adv Biosci Biotechnol 4:188–192

Murashige T, Skoog F (1962) A revised medium for rapid growth and bioassays with tobacco tissue culture. Physiol Plant 15:473–497

Nam J, Matthysse AG, Gelvin SB (1997) Differences in susceptibility of *Arabidopsis* ecotypes to crown gall disease may result from a deficiency in T-DNA integration. Plant Cell 9:317–333

Narusaka M, Shiraishi T, Iwabuchi M, Narusaka Y (2010) The floral inoculating protocol: a simplified *Arabidopsis thaliana* transformation method modified from floral dipping. Plant Biotechnol 27:349–351

Negrutiu I, Heberle-Bors E, Potrykus I (1986) Attempts to transform for kanamycin-resistance in mature pollen of tobacco. In: Mulcahy DL, Bergamini-Mulcahy G, Ottaviano E (eds) Biotechnology and ecology of pollen. Springer, New York, pp 65–70

Nehlin L, Möllers C, Bergman P, Glimelius K (2000) Transient *ß-gus* and *gfp* gene expression and viability analysis of microprojectile bombarded microspores of *Brassica napus* L. J Plant Physiol 156:175–183

Neuhaus G, Spangenberg G, Scheid OM, Schweiger HG (1987) Transgenic rapeseed plants obtained by the microinjection of DNA into microspore-derived embryoids. Theor Appl Genet 75:30–36

Nishihara M, Ito M, Tanaka I, Kyo M, Ono K, Irifune K, Morikawa K (1993) Expression of the β-Glucuronidase Gene in Pollen of lily (*Lilium longiflorum*), Tobacco (*Nicotiana tabacum*), *Nicotiana rustica*, and Peony (*Paeonia lactiflora*) by particle bombardment. Plant Physiol 102:357–361

Nishihara M, Seki M, Kyo M, Irifune K, Morikawa H (1995) Transgenic haploid plants of *Nicotiana rustica* produced by bombardment-mediated transformation of pollen. Transgenic Res 4:341–348

Obert B, Pónya Z, Pret'ová A, Barnabás B (2004) Optimization of electroporation conditions for maize microspores. Maydica 49:15–19

Ohta Y (1986) High-efficiency transformation of maize by a mixture of pollen and exogenous DNA. PNAS (USA) 83:715–719

Pechan P (1989) Successful cocultivation of *Brassica napus* microspores and proembryos with *Agrobacterium*. Plant Cell Rep 8:387–390

Phillips RL, Kaeppler SM, Olhoft P (1994) Genetic instability of plant tissue cultures: breakdown of normal controls. Proc Natl Acad Sci USA 91:5222–5226

Potrykus I (1991) Gene transfer to plants-assessment of published approaches and results. Annu Rev Plant Physiol Plant Mol Biol 42:205–225

Prado JR, Segers G, Voelker T, Carson D, Dobert R, Phillips J, Cook K, Cornejo C, Monken J, Grapes L, Reynolds T, Martino-Catt S (2014) Genetically engineered crops: from idea to product. Annu Rev Plant Biol 65:769–790

Qing CM, Fan L, Lei Y, Bouchez D, Tourneur C, Yan L, Robaglia C (2000) Transformation of Pak choi (*Brassica rapa* L. ssp *chinensis*) by *Agrobacterium* infiltration. Mol Breed 6:67–72

Ramaiah SM, Skinner DZ (1997) Particle bombardment: a simple and efficient method of alfalfa (*Medicago sativa* L.) pollen transformation. Current Sci 73:674–682

Ray DK, Ramankutty N, Mueller ND, West PC, Foley JA (2012) Recent patterns of crop yield growth and stagnation. Nat Commun 3:1293. doi:10.1038/ncomms2296

Resch T, Touraev A (2011) Pollen transformation technologies. In: Stewart CN Jr, Touraev A, Citovsky V, Tzfira T (eds) Plant transformation technologies. Wiley-Blackwell, UK, pp 83–92

Roeckel P, Molony MM, Drevert JR (1992) Plant transformation using sexual route. Int Rev Cytol 160:425–446

Sanford JC, Skubik KA, Reisch BI (1985) Attempted pollen-mediated plant transformation employing genomic donor DNA. Theor Appl Genet 69:571–574

Sanford JC, Klein TM, Wolf ED, Allen N (1987) Delivery of substances into cells and tissues using a particle bombardment process. J Part Sci Tech 6:559–563

Sangwan RS, Ducrocq C, Sangwan-Norreel B (1993) *Agrobacterium*-mediated transformation of pollen embryos in *Datura innoxia* and *Nicotiana tabacum*—production of transgenic haploid and fertile homozygous dihaploid plants. Plant Sci 95:99–115

Saunders JA, Matthews BF (1995) Pollen electrotransformation in tobacco. In: Nickoloff JA (ed) Methods in molecular biology, vol 55. Plant Cell Electroporation and Electrofusion Humana Press Inc., Totowa

Schreiber DN, Dresselhaus T (2003) In vitro pollen germination and transient transformation of *Zea mays* and other plant species. Plant Mol Biol Rep 21:31–41

Seol E, Jung Y, Lee J, Cho C, Kim T, Rhee Y, Lee S (2008) In planta transformation of *Notocactus scopa* cv. Soonjung by *Agrobacterium tumefaciens*. Plant Cell Rep 27:1197–1206

Sharma K, Bhatnagar-Mathur P, Thorpe TA (2005) Genetic transformation technology: status and probems. In Vitro Cell Dev Biol Plant 41:102–112

Shedden J (1997) Transient expression of ß-glucuronidase in embryogenic *Brassica napus* microspores following particle bombardment. A thesis submitted for the degree of Master of Science, Department of Biology, Queen's University, Kingston, Ontario, Canada

Shi HZ, Yang HY, Zhou C (1995) Intrinsic *GUS* activity in various tissues and during pollen development of *Nicotiana tabacum*. Acta Botanica Sinica 37:134–139

Shi HZ, Wang J, Yang HY, Zhou C, Chen ZH (1996) Exine detached pollen of *Nicotiana tabacum* as an electroporation target for gene transfer. Acta Botan Sinica 38:626–630

Shim YS, Pauls KP, Kasha KJ (2009) Transformation of isolated barley (*Hordeum vulgare* L.) microspores: II. Timing of pretreatment and temperatures relative to results of bombardment. Genome 52:175–190

Shim YS, Eudes F, Kovalchuk I (2013) dsDNA and protein co-delivery in triticale microspores. In Vitro Cell Dev Biol Plant 49:156–165

Shou H, Palmer RG, Wang K (2002) Irreproducibility of the soybean pollen-tube pathway transformation procedure. Plant Mol Biol Rep 20:325–334

Stöger E, Moreno RMB, Ylstra B, Vicente O, Heberle-Bors E (1992) Comparison of different techniques for gene transfer into mature and immature tobacco pollen. Transgenic Res 1:71–78

Stöger E, Fink C, Pfosser M, Heberle-Bors E (1995) Plant transformation by particle bombardment of embryogenic pollen. Plant Cell Rep 14:273–278

Swanson EB, Erickson LR (1989) Haploid transformation in *Brassica napus* using an octopine-producing strain of *Agrobacterium tumefaciens*. Theor Appl Genet 78:831–835

Tachibana KWT, Sekizuwa Y, Takematsu T (1986) Action mechanism of bialaphos. 2. Accumulation of ammonia in plants treated with bialaphos. J Pest Sci 11:33–37

Tague BW (2001) Germ-line transformation of *Arabidopsis lasiocarpa*. Transgenic Res 10:259–267

Tanaka T, Nishihara M, Seki M, Sakamoto A, Tanaka K, Irifune K, Morikawa H (1995) Successful expression in pollen of various plant species of in vitro synthesized mRNA introduced by particle bombardment. Plant Mol Biol 28:337–341

Tinland B (1996) The integration of T-DNA into plant genomes. Trends Plant Sci 1:178–184

Tjokrokusumo D, Heinrich T, Wylie S, Potter R, McComb J (2000) Vacuum infiltration of *Petunia hybrida* pollen with *Agrobacterium tumefaciens* to achieve plant transformation. Plant Cell Rep 19:792–797

Toojinda TE, Baird E, Booth A, Broers L, Hayes P, Powell W, Thomas W, Vivar H, Young G (1998) Introgression of quantitative trait loci (QTLs) determining stripe rust resistance in barley: an example of marker-assisted line development. Theor Appl Genet 96:123–131

Touraev A, Stöger E, Voronin V, Heberle-Bors E (1997) Plant male germline transformation. Plant J 12:949–956

Touraev A, Prosser M, Heberle-Bors E (2001) The microspore: a haploid multipurpose cell. Adv Bot Res 35:53–109

Travella S, Ross SM, Harden J, Everett C, Snape JW, Harwood WA (2005) A comparison of transgenic barley lines produced by particle bombardment and Agrobacterium-mediated techniques. Plant Cell Rep 23:780–789

Trick HN, Finer JJ (1997) SAAT: sonication-assisted *Agrobacterium*-mediated transformation. Transgenic Res 6:329–336

Trieu AT, Burleigh SH, Kardailsky IV, Maldonado-Mendoza IE, Versaw WK, Blaylock LA, Shin H, Chiou TJ, Katagi H, Dewbre GR, Weigel D, Harrison MJ (2000) Transformation of *Medicago truncatula* via infiltration of seedlings or flowering plants with Agrobacterium. Plant J 22:531–541

Twell D (2011) Male gametogenesis and germline specification in flowering plants. Sex Plant Reprod 24:149–160

Twell D, Klein TM, Fromm ME, McCormick S (1989a) Transient expression of chimeric genes delivered into pollen by microprojectile bombardment. Plant Physiol 91:1270–1274

Twell D, Wing R, Yamaguchi J, McCormick S (1989b) Isolation and expression of an anther-specific gene from tomato. Mol Gen Genet 217:240–245

Twell D, Yamaguchi J, Wing RA, Ushiba J, McCormick S (1991) Promoter analysis of genes that are coordinately expressed during pollen development reveals pollen-specific enhancer sequences and shared regulatory elements. Genes Dev 5:496–507

Tzfira T, Citovsky V (2006) *Agrobacterium*-mediated genetic transformation of plants: biology and biotechnology. Curr Opin Biotechnol 17:147–154

van der Leede-Plegt LM, van de Ven BCE, Bino RJ, van der Salm TPM, van Tunen AJ (1992) Introduction and differential use of various promoters in pollen grains of *Nicotiana glutinosa* and *Lilium longiflorum*. Plant Cell Rep 11:20–24

van der Leede-Plegt LM, van de Ven BCE, Schilder M, Franken J, van Tunen AJ (1995) Development of a pollen-mediated transformation method for *Nicotiana glutinosa*. Transgenic Res 4:77–86

van Tunen AJ, Mur LA, Brouns GS, Rienstra JD, Koes RE, Mol JNM (1990) Pollen-and anther-specific chi promoters from petunia: tandem promoter regulation of the *chiA* gene. Plant Cell 2:393–401

Verma SS, Chinnusamy V, Bansal KC (2008) A simplified floral dip method for transformation of *Brassica napus* and *B. carinata*. J Plant Biochem Biotechnol 17:197–200

Wang J, Shi HZ, Zhou C, Yang HY, Zhang XL, Zhang RD (1998) β-glucuronidase gene and green fluorescent protein gene expression in de-exined pollen of *Nicotiana tabacum* by particle bombardment. Sex Plant Reprod 11:159–162

Wang J, Sun Y, Cui G, Hu J (2001) Transgenic maize plants obtained by pollen-mediated transformation. Acta Bot Sinica 43

Wang WC, Menon G, Hansen G (2003) Development of a novel *Agrobacterium*-mediated transformation method to recover transgenic *Brassica napus* plants. Plant Cell Rep 22:274–281

Wang W, Wang J, Yang C, Li Y, Liu L, Xu J (2007) Pollen-mediated transformation of *Sorghum bicolor* plants. Biotechnol Appl Biochem 48:79–83

Wang J, Li Y, Liang C (2008) Recovery of transgenic plants by pollen-mediated transformation in *Brassica juncea*. Transgenic Res 17:417–424

Wiktorek-Smagur A, Hnatuszko-Konka K, Kononowicz AK (2009) Flower bud dipping or vacuum infiltration-two methods of *Arabidopsis thaliana* transformation. Russian J Plant Physl 56:560–568

Wu H, McCormac AC, Elliott MC, Chen DF (1998) *Agrobacterium*-mediated stable transformation of cell suspension cultures of barley (*Hordeum vulgare*). Plant Cell Tiss Org 54:161–171

Xu H, Wang X, Zhao H, Liu F (2008) An intensive understanding of vacuum infiltration transformation of pak choi (*Brassica rapa* ssp. *chinensis*). Plant Cell Rep 27:1369–1376

Yang A, Su Q, An L (2009) Ovary-drip transformation: a simple method for directly generating vector- and marker-free transgenic maize (*Zea mays* L.) with a linear GFP cassette transformation. Planta 229:793–801

Yao QA, Kasha KJ (1997) Potential of biolistic transformation of barley microspores based on viability and transient ß-glucuronidase activity. Genome 40:639–643

Yao QA, Simion E, William M, Krochko J, Kasha KJ (1997) Biolistic transformation of haploid isolated microspores of barley (*Hordeum vulgare* L.). Genome 40:570–581

Yasmeen A, Mirza B, Inayatullah S, Safdar N, Jamil M, Ali S, Choudhry MF (2009) In planta transformation of tomato. Plant Mol Biol Rep 27:20–28

Ye GN, Stone D, Pang SZ, Creely W, Gonzalez K, Hinchee M (1999) *Arabidopsis* ovule is the target for *Agrobacterium* in planta vacuum infiltration transformation. Plant J 19:249–257

Young JM, Kuykendall LD, Martínez-Romero E, Kerr A, Sawada H (2001) A revision of Rhizobium Frank 1889, with an emended description of the genus, and the inclusion of all species of Agrobacterium Conn 1942 and *Allorhizobium undicola* de Lajudie et al. 1998 as new combinations: *Rhizobium radiobacter, R. rhizogenes, R. rubi, R. undicola* and *R. vitis*. Int J Syst Evol Microbiol 51:89–103

Zale JM, Agarwal S, Loar S, Steber CM (2009) Evidence for stable transformation of wheat by floral dip in *Agrobacterium tumefaciens*. Plant Cell Rep 28:903–913

Zeng JZ, Wang DJ, Wu YQ, Zhang J, Zhou WJ, Zhu XP, Xu NZ (1994) Transgenic wheat plants obtained with pollen-tube pathway method. Sci China B 37:319–325

Zerback R, Dressler K, Hess D (1989) Flavonoid compounds from pollen and stigma of *Petunia hybrida*: inducers of the *vir* region of the *Agrobacterium tumefaciens* Ti plasmid. Plant Sci 62:83–91

Zhang L, Cheng XN, Zhao N, Li C, Yuan J, Jia S (1991) Efficient transformation of tobacco by ultrasonication. Nat Biotechnol 9:996–997

Zhou GY, Gong ZZ, Wang ZF (1979) The molecular basis of remote hybridization: an evidence of the hypothesis that DNA segments of distantly related plants may be hybridized. Acta Genetica Sinica 6:405–413

Zhou GY, Weng J, Zeng YS, Huang JG, Qian SY, Liu GL (1983) Introduction of exogenous DNA into cotton embryos. In: Wu R, Grossman L, Moldave K (eds) Methods in enzymology, vol 101. Academic Press, New York, pp 433–481

Ziemienowicz A, Shim YS, Matsuoka A, Eudes F, Kovalchuk I (2012) A novel method of transgene delivery into triticale plants using the *Agrobacterium* transferred DNA-derived nano-complex. Plant Physiol 158:1503–1513

Zupan J, Zambryski PC (1997) The *Agrobacterium* DNA transfer complex. Crit Rev Plant Sci 16:279–295

Chapter 15
Advances in Molecular Breeding of Pearl Millet

Deepmala Sehgal

Abstract Pearl millet [*Pennisetum glaucum* (L.) R. Br.] ($2n = 2 \times = 14$) is the sixth most important global cereal crop (after rice, wheat, maize, barley, and sorghum) which is grown in the hottest and driest regions of sub-Saharan Africa and the Indian subcontinent. It produces grains with high nutritive value even under hot, dry conditions, and on infertile soils of low water-holding capacity, where other cereal crops fail. This makes pearl millet a highly desirable crop for farmers in such harsh environments. Pearl millet became the focus of genome research almost at the same time as other major crops but then lagged behind as major crops dominated the genomics era. However, in the last decade, several efforts were initiated to rekindle the genomic research of this orphan crop resulting into generation of vast amounts of genomic information. Particularly, the recent whole-genome sequencing efforts taken for pearl millet by an international pearl millet genome sequencing consortium are remarkable. This chapter reviews the advances made in generating the genetic and genomics resources in pearl millet and their integration into molecular breeding. A successful example of marker-assisted selection (MAS) culminating in a product release is cited.

Keywords Pearl millet · SSRs · ESTs · TILLING · Whole-genome sequencing

15.1 Introduction

Pearl millet [*Pennisetum glaucum* (L.) R. Br.] is the sixth most important global cereal crop and the main source of food for 500 million of the poorest people living predominantly in parts of Asia and Africa. It is grown primarily by subsistence farmers in areas with very limited rainfall (300–500 mm), environments with high

D. Sehgal (✉)
Wheat Geneticist and Molecular Breeder, CIMMYT, CIMMYT Headquarters,
Km. 45, Carretera, Mex-Veracruz, El Batan, 56130 Texcoco, CP, Mexico
e-mail: d.sehgal@cgiar.org

V.R. Rajpal et al. (eds.), *Molecular Breeding for Sustainable Crop Improvement*, Sustainable Development and Biodiversity 11,
DOI 10.1007/978-3-319-27090-6_15

mean temperatures and frequent droughts and even on soils with poor fertility. Pearl millet grain has relatively high nutritional value compared to wheat, rice, and maize in terms of both protein content and amino acid composition (Sawaya et al. 1984; Ejeta et al. 1987). It also has superior levels of grain Fe and Zn (Velu et al. 2007; Govindaraj et al. 2013). The energy density of pearl millet is also relatively high, arising from its higher oil content relative to maize, wheat, or sorghum (Hill and Hanna 1990). These properties have made it a central component of the food security of the rural poor in dry areas.

Pearl millet became the focus of genome research almost at the same time (the early 1990s) as the other major crops. This genomics research led to the development of first linkage map in 1993 (Liu et al. 1994) by restriction fragment length polymorphism (RFLP) markers. However, as major crops dominated the genomics era, pearl millet lagged behind and remained an orphan crop for many years in terms of genetic and genomic resources development. The limited amount of genomic sequence information in pearl millet has limited progress in gene discovery and characterization, global transcript profiling, probe design for the development of gene arrays, and generation of molecular markers and their application in crop improvement programs. However, in the last decade, substantial investments have been made in large-scale genetic and genomic resources development in this crop, including development of simple sequence repeats (SSRs), diversity array technology (DArT), single-nucleotide polymorphism (SNPs) markers, expressed sequence tags (ESTs) or transcript reads, bacterial artificial chromosome (BAC) libraries, linkage maps, comparative maps, and genetic stocks such as core reference sets, mapping populations, near-isogenic lines, and an association mapping panel. These resources have not only accelerated gene discovery in this crop but also provided impetus to initiate molecular breeding.

This chapter will first review the progress made in the development of first and second generation of DNA markers, linkage maps, and genetic and genomic resources, and then provide briefings on the recent efforts put together in using the next-generation sequencing (NGS) technologies for genome-wide SNP marker development to accelerate fine mapping and molecular breeding.

15.2 Progress in Marker Development

15.2.1 First- and Second-Generation Markers and Their Use in Diversity Studies

A detailed description of different marker types available for plant genome research is beyond the scope of this review. Various reviews have been published which can be referred to by the readers (Sehgal et al. 2008; Kesawat and Das 2009; Jonah et al. 2011). Briefly, DNA markers can be classified into the following: (i) the first-generation molecular markers, including RFLPs, random amplified polymorphic

DNA (RAPD), and their modifications; (ii) the second-generation molecular markers, including SSRs, amplified fragment length polymorphism (AFLP), and their modified forms; and (iii) the third-generation molecular markers, SNPs from ESTs or from transcriptome/whole-genome sequencing.

Like in any other crop, initial genetic studies in pearl millet were conducted with isozymes-based markers (Tostain et al. 1987, Tostain and Marchais 1989; Tostain 1994) but soon replaced with RFLP markers (Clegg et al. 1984; Gepts and Clegg 1989). Chloroplast and nuclear DNA (rDNA and *Adh* sequences) probes were used to assess diversity in cultivated and wild accessions (Gepts and Clegg 1989), where profiles generated using nuclear DNA probes correlated with geographic distribution of genotypes, and chloroplast probes remained highly monomorphic. RFLP markers remained popular for quite some time for diversity assessments because of their codominant and highly reproducible nature (Bhattacharjee et al. 2002). These markers, however, were found difficult to adopt for large-scale studies that are hallmark of any worthwhile genetic study on genetic improvement. Therefore, soon PCR-based markers such as RAPDs and inter-simple sequence repeats (ISSRs) were adopted and became popular as they proved highly polymorphic and more informative in assessing diversity of cultivars and landraces, cytoplasmic male sterile and restorer lines (Chowdari et al. 1998; Kale and Munjal 2005; Yadav et al. 2007a, b), and for the identification of genetically diverse lines for hybridization programs (Govindaraj et al. 2009). At the same time, AFLP markers were also employed for the assessment of genetic diversity in landraces from India and West Africa (Busso et al. 2000; Vom Brocke et al. 2003). AFLP markers are dominant markers such as RAPDs, but the large number of loci amplified and their high reproducibility made them attractive over RAPDs.

In 2000, initiatives were taken for the development of sequence-based markers in pearl millet. It started with the development of BAC libraries in line Tift 23DB (Allouis et al. 2001). The library contained a total of 159,100 clones with an average insert size of 90 kb and corresponded to 14,200 Mb of genomic DNA (5.8 haploid genome equivalents). The BAC clones were used to develop the second generation of markers called microsatellites or SSRs (Allouis et al. 2001). More SSR markers were further designed from 40 BAC pools using 3′ end-anchored SSR primers which proved to be highly polymorphic with polymorphism information content (PIC) values up to 0.84 (Qi et al. 2001). With this advancement, more SSR markers were subsequently designed (Yadav et al. 2007) and various genetic diversity studies were initiated worldwide using SSR markers, and various gene bank and landrace collections were characterized (Budak et al. 2003; Kapila et al. 2008; Oumar et al. 2008; Stich et al. 2010). Many of these above-mentioned investigations also provided new and important insights into the origin of the cultivated species, phylogenetic relationships, and indication of domestication centers of cultivated pearl millet and/or possible location of domestication genes. For example, Kapila et al. (2008) revealed linkage group (LG) 6 to be least diverse and with least number of SSR markers. Such highly conserved nature of LG 6 was attributed to the presence of important genes involved in domestication. Oumar et al. (2008) analyzed 84 wild accessions and 355 cultivated accessions originating from the whole pearl millet distribution area in Africa and Asia and suggested a monophyletic origin of cultivated pearl millet in West Africa. This study also indicated eastern Mali and western Niger as the possible regions of domestication of pearl millet.

15.2.2 Development of EST-SSRs and Third-Generation Markers

15.2.2.1 Development of Expressed Sequence Tags (ESTs)

A comprehensive stress-response transcriptome resource for pearl millet was first developed by a group at International Centre for Genetic Engineering and Biotechnology (ICGEB), India. A drought-tolerant line 863B was used to generate subtractive cDNA libraries derived from leaves and root tissues of plants grown under controlled conditions and imposed to various drought, salinity, and cold stresses. The libraries were sequenced using traditional Sanger sequencing approach. A total of 2494 ESTs in response to drought, salinity, and cold stresses were made publicly available in NCBI database in 2003 (Mishra et al. 2007). They were assembled into a collection of 1850 unique sequences with 224 contigs and 1626 singleton sequences. Sequence comparison of these ESTs using BLASTX algorithm revealed many genes with stress-related functions. Based on their functions, they were divided into various categories, for example, transcription factors, kinases and phosphatases, secondary messengers, chaperons, proteinases, and those involved in scavenging and/or prevention of reactive oxygen species. The information on GenBank accession number and dbEST number for all the ESTs can be accessed from CD724312 to CD726805 in NCBI. These EST sequences are a rich source of stress-related genes and represent a major part of the stress-response transcriptome that will provide the foundation for further studies into understanding *Pennisetum*'s adaptability to harsh environmental conditions. The number of ESTs in NCBI database is rising for pearl millet. Hitherto, more than 6000 ESTs are available at NCBI's dbEST. More recently, high-temperature responsive ESTs have been added to this list (Padaria et al. 2012 unpublished).

15.2.2.2 Development of EST-SSRs

The exploitation of EST databases to develop microsatellite markers was first attempted in rice (Miyao et al. 1996) and has subsequently been reported from many plant species (Kantety et al. 2002; Varshney et al. 2002; Holton et al. 2002). This discovery provided the opportunity to develop markers (EST-SSRs) in a simple way just by electronic searches (data mining) of EST databases and primer designing using a freely available tool SSR identification tool (SSRIT) integrated in GRAMENE database. EST-SSRs constitute a novel source of markers that are physically associated with coding regions of the genome. In contrast to genomic SSRs, EST-SSRs enable assaying of the variation in transcribed sequences and genes where function is known. In pearl millet, EST-SSRs were developed by Senthilvel et al. (2004, 2008), Mariac et al. (2006), and Yadav et al. (2007). While Mariac et al. (2006) and Yadav et al. (2007) developed 16 and 19 EST-SSRs, respectively, Senthilvel et al. (2008) reported 58 new EST-SSRs which were also used for mapping (see later section on 15.3).

15.2.2.3 Development of Single-Strand Conformational Polymorphism (SSCP)-SNP Marker System

The ESTs were also utilized to develop the first SNP-based marker system called SSCP-SNP (Bertin et al. 2005) in pearl millet with the two objectives: (a) to develop a codominant and robust system that is moderately throughput, amenable to multiplexing such as SSRs and transferrable to breeding laboratories and (b) to design new markers for pearl millet with maximum comparative utility. Hence, ESTs showing good homology with rice single-copy genes were selected, and alignment of these selected ESTs with rice genomic sequences was done to obtain information on intron–exon boundaries in the millet ESTs. Primers were then designed in such a way so that they would amplify across the introns. A preliminary analysis showed that the millet SSCP-SNP primers amplified with a success rate of about 50 % in other cereals (rice, wheat, barley, and finger millet).

15.3 Development of Genetic Linkage Maps and Their Saturation

The first molecular genetic linkage map of pearl millet was created in 1992 (Liu et al. 1992, 1994a, b). It was based on RFLP markers using homologous (from the same crop) pearl millet probes. A few heterologous probes from rice, wheat, and barley along with several isozymes and known function probes were also included in this base map resulting in a map of 181 loci covering approximately 303 cM. This initial map was transferred to several additional crosses (Busso et al. 1995; Liu et al. 1996) in studies of sex-specific recombination rates in cultivated × cultivated and cultivated × wild crosses, and a pearl millet reference mapping population was developed based on the cross 81B × ICMP 451 (Hash and Witcombe 1994). This map has been used for high saturation marker genotyping using dominant AFLP markers, additional homologous probes from pearl millet, and heterologous probes from other grasses. The latter group of markers has improved our understanding of the complex relationships between the pearl millet genome and those of other cultivated gramineous species such as rice and foxtail millet (Devos and Gale 2000; Devos et al. 2000). It was revealed that compared to most other grasses, the pearl millet genome has undergone a large number of structural rearrangements (Devos and Gale 2000).

Nevertheless, regions of colinearity between the pearl millet and rice (grass model genome) could be clearly identified (Devos et al. 2000). These regions formed a framework for exploitation of the rice genomic sequence as a source of new markers and candidate genes underlying traits in pearl millet. This work extended the total pearl millet genetic linkage map length to approximately 600 cM.

The first SSR-based integrated map was reported by Qi et al. (2004). A new set of 44 genomic SSRs from a (CA)n-enriched small-insert genomic library was generated. Thirty-five of these SSRs from enriched library were integrated with 28 SSRs developed from BAC clones (Qi et al. 2001) and 353 RFLP markers to build a consensus map with mapping data from four pearl millet mapping populations; LGD 1-B-10 × ICMP 85410, 81B × ICMP 451, ICMB 841 × 863B, and PT 732B × P1449-2. The LGD 1-B-10 × ICMP 85410 cross comprised 133 F_2 progeny and was the original mapping population used by Liu et al. (1994). The parents of this cross differed by a translocation (Liu et al. 1994). The second mapping population 81B × ICMP 451 was a cross between two inbred lines and consisted of 157 F_2 progeny. Two further crosses, ICMB 841 × 863B (Yadav et al. 2004) consisting of 149 F_2 progeny, and PT 732B × P1449-2 consisting of 131 F_2 progeny, showed segregation for drought tolerance and downy mildew resistance, respectively. The consensus map contained a total of 418 markers in which 85 % of the markers were clustered and occupied less than one-third of the total map length. Also, the map contained big gaps of about 30 cM in the distal regions of some chromosomes.

Subsequently, EST-SSRs were developed (described previously in subsection development of EST-SSRs) and were added to the map using the F_2 mapping population of the cross ICMB 841-P3 and 863B-P2 (Senthilvel et al. 2008). Out of 58 EST-SSRs, 17 could be mapped on five linkage groups. Since the consensus map by Qi et al. (2004) contained RFLP and SSR markers mostly in the centromeric regions, EST-SSRs were expected to map in non-centromeric regions, thereby filling large gaps on distal regions of chromosomes. Some of the large gaps observed on LGs 1, 2, 6, and 7 on the map of this cross were filled by the EST-SSRs. For example, the 63 cM gap between *Xpsm*52 and *Xpsm*196.1 on LG 1 was covered by EST-SSRs *Xicmp*3085, *Xicmp*3088, *Xctm*112, and *Xctm*27. Further, new EST-SSRs mapped to distal ends of LGs 3 and 5 and increased their map lengths by 8 and 39 cM, respectively. To further increase the coverage of linkage groups and to increase the number of easy to use PCR-based markers on the map, another map was generated which contained 66 sequence-related amplified polymorphisms (SRAPs), 63 RAPDs, 27 ISSRs, 31 pearl millet, 6 sorghum, and 3 maize SSRs (Pedraza-Garcia et al. 2010). The resulting map consisted of nine linkage groups that spanned about 1796 cM.

Supriya et al. (2011) developed a map integrating EST-SSRs with diversity array technology (DArT) markers with an objective to increase the number of high-throughput markers which can be utilized for whole-genome breeding strategies. DArT allows simultaneous scoring of hundreds of restriction site-based polymorphisms between genotypes and does not require DNA sequence information or site-specific oligonucleotides. A genomic representation from 95 diverse pearl millet genotypes was used to develop a DArT array with circa 7000 clones following *Pst*I/*Ban*II complexity reduction. This array was used to genotype a set of 24 diverse pearl millet inbreds, and 574 polymorphic DArT markers were identified. A mapping population of 140 F_7 recombinant inbred lines (RILs) from cross H 77/833-2 × PRLT 2/89-33 was used to generate a map integrating DArT and EST-SSR marker data. This map contained 321 loci (258 DArTs and 63 SSRs) and spanned 1148 cM.

To facilitate search for association between candidate genes and QTLs underlying important agronomic traits, the map of H 77/833-2 × PRLT 2/89-33 cross was saturated with functionally important genes (Sehgal et al. 2012). Such maps are called molecular-function maps and have been generated in many species (Rostoks et al. 2005; Kota et al. 2008). These have facilitated the identification of candidate genes for various biotic stress resistance and abiotic stress tolerance QTLs in many crops (Nguyen et al. 2004; Tondelli et al. 2006; Diab et al. 2008). The molecular-function map in pearl millet was initially developed to saturate LG2 where a major QTL for drought tolerance (DT) resided. Subsequently the gene-based markers were extended to all LGs. To saturate LG2 with gene-based markers, published information of synteny between pearl millet LG2 and rice chromosomes 2S, 3L, 6S, and 10L (Devos et al. 2000) was utilized. The genomic sequences of 100 selected genes within the rice BAC contigs from each of the four syntenic rice chromosomes were retrieved, and primer pairs were designed for genes which showed significant homologies with pearl millet ESTs. To saturate other LGs with genes, another in silico approach was used wherein 200 pearl millet ESTs that were homologous to drought and other abiotic stress genes reported in other model or non-model crops were identified using the BLAST2GO program (http://www.blast2go.org/). For this set of 200 ESTs, primers were designed to amplify part of the 3′ untranslated region (3′UTR). These approaches led to mapping of 75 genes as SNP and CISP (conserved intron spanning primers) markers across seven LGs. Most importantly, DT-QTL region on LG2 was saturated with 18 genes which became candidates for future study (see Sect. 15.5).

Recently, a consensus map containing the largest set of mapped SSRs reported to date in pearl millet was developed by Rajaram et al. (2013). Briefly, 116 EST-SSR markers, 53 genomic SSRs, and 2 STS markers were used to construct linkage maps of four F_7 recombinant inbred populations (RIPs) based on crosses ICMB 841-P3 × 863B-P2, H 77/833-2 × PRLT 2/89-33, 81B-P6 × ICMP 451-P8, and PT 732B-P2 × P1449-2-P1. Eighty-nine EST-SSR marker loci from this consensus map had significant BLAST hits (top hits with e-value ≤ 1E-10) with the genome sequences of rice, foxtail millet, sorghum, maize, and Brachypodium with 35, 88, 58, 48, and 38 loci, respectively.

The most comprehensive map with the third-generation markers, i.e., SNPs has been generated using the most advanced NGS technology called genotyping-by-sequencing (GBS). The current NGS technologies are capable of analyzing anywhere from hundreds of thousands to tens of millions of DNA molecules in parallel, thus allowing the rapid identification of a large numbers of genetic markers, mainly SNPs (Imelfort et al. 2009). GBS is an advanced and highly cost-effective NGS approach for genotyping which produces up to a million

SNPs per genotype at a cost as low as \$20–\$40. Moumouni et al. (2015) constructed a GBS map in pearl millet using a F_2 population of 93 progenies from a wild × cultivated pearl millet cross. A total of 3321 SNPs were generated and 2809 high-quality SNPs exhibited a minor allele frequency ≥0.3. These formed a total of 314 non-redundant haplotypes for which a single representative SNP marker was used for map construction. This resulted in a genetic map with 314 SNP markers spanning a total distance of 640 cM. The SNPs were evenly distributed over seven linkage groups with an average density of 0.51 SNP/cM.

15.4 QTL Mapping in Pearl Millet

15.4.1 QTLs for Resistance to Biotic Stresses

15.4.1.1 QTLs for Downy Mildew Resistance

Downy mildew (DM) caused by the obligate biotrophic pathogen *Scelerospora graminicola* can cause devastating yield losses in pearl millet and is a major constraint to productivity (Singh et al. 1993). The inheritance of resistance to downy mildew is a quantitative character. QTLs for resistance to DM have been mapped by using phenotypic data from field as well as glasshouse screens established in UK and India (Jones et al. 1995, 2002). This was done to test whether the glasshouse screens established in India and UK are effective indicators of resistance in the field. Jones et al. (1995) screened pathogen populations of *S. graminicola* from India, Nigeria, Niger, and Senegal on F_4 mapping population of a cross LGD-1-B-10 × ICMP 85410. RFLP markers (Liu et al. 1994) were used to construct the map using F_2 plants. Independent inheritance of resistance to pathogen populations from India, Senegal, and populations from Niger to Nigeria was shown. These results demonstrate the existence of differing virulences in the pathogen populations from within Africa and between Africa and India. A major QTL for resistance against pathogen population from India was detected on LG1, against the pathogen populations from Nigeria to Niger on LG4, and against pathogen population from Senegal on LG2. There was no QTL that was effective against all four pathogen populations, suggesting that pathotype-specific resistance is a major mechanism of downy mildew resistance in this cross (Jones et al. 1995). The glasshouse experiments were done in India and UK using F_4 mapping population of another cross 7042(S)-1 × and P 7-3 (Jones et al. 2002). Again RFLP markers were employed for map construction. Two consistent QTLs were detected on LG1 and 2 in both field and glasshouse screens (UK and India) with LG1 QTL showing higher percentage variation (up to 60 %) than LG2 QTL (up to 16 %). The LG1 QTL detected in both these studies (Jones et al. 1995, 2002) was same. QTLs for DM resistance have also been identified on these LGs in other studies (Breese et al. 2002; Gulia 2004; Gulia et al. 2007)

15.4.1.2 QTLs for Rust and Pyricularia Leaf Spot Resistance

Rust caused by *Puccinia substriata* var. *indica* and pyricularia leaf spot caused by *Pyricularia grisea* are the two most-destructive diseases of pearl millet in the USA. Attempts have been made to identify closely linked markers/QTLs for disease resistances. Three populations were developed for mapping disease resistance, two for rust and one for pyricularia leaf spot. The first rust-resistant population segregated for the *Rr1* gene as well as rust resistance from the line ICMP 83506 developed at ICRISAT, India. ICMP 83506 was crossed to Tift 85DB, and a single F_1 individual was selfed to produce an F_2 population of 54 individuals segregating for two sources of rust resistance. This population was screened with two different races, 93-3 (avirulent to *Rr1*) and 92-1 (avirulent to ICMP 83506). The second population was made by pollinating the susceptible parent Tift 23DB with the resistant parent Tift 89D2. A single F_1 individual was selfed to produce an F_2 population of 62 individuals which was screened with the rust race 92-1 (avirulent to Tift 89D2). The third population segregating for pyricularia resistance was made by crossing the resistant parent *P. glaucum* ssp. *monodii* Ps34 (*rp1rp1* Tr*Tr*, pyricularia resistant) onto the pyricularia-susceptible genetic stock, red trichomeless (*Rp1Rp1 trtr*) in a Tift 23 background (Hanna and Burton 1992). A red, trichomed F_1 plant was selfed to produce an F_2 population of 62 individuals segregating for pyricularia disease resistance. Three segregating populations were screened for RAPDs using random decamer primers and for RFLPs using a core set of probes detecting single-copy markers on the pearl millet map (Liu et al. 1994).

The rust resistance gene *Rr1* from the pearl millet subspecies *P. glaucum* ssp. *monodii* was linked 8.5 cM from the RAPD OP-G8350. The linkage of two RFLP markers, *Xpsm108* (15.5 cM) and *Xpsm174* (17.7 cM), placed the *Rr1* gene on LG3 of the pearl millet map. Rust resistance genes from both Tift 89D2 and ICMP 83506 were placed on LG4 by determining genetic linkage to the RFLP marker *Xpsm716* (4.9 and 0.0 cM, respectively). Only one RAPD marker (OP-D11700, 5.6 cM) was linked to pyricularia leaf spot resistance. The use of these markers linked to rust- and pyricularia-resistant loci is currently limited to marker-assisted selection (MAS). Map-based cloning of these resistance genes was not feasible before due to the absence of a saturated genetic map for pearl millet and the presence of large gaps between markers. Now that the highly saturated maps are available (see above section on 15.3, cloning these resistance loci for designing markers for MAS is possible.

15.4.2 QTLs for Resistance to Abiotic Stresses

Post-flowering drought stress is one of the most important environmental factors reducing the grain yield (GY) and yield stability of pearl millet and increasing the incidence of crop failure in dryland production environments (Mahalakshmi et al. 1987). Terminal drought stress (flowering through grain filling) is more damaging to pearl millet productivity than stress at the vegetative stages. This is because pearl millet's asynchronous tillering behavior and rapid growth rate allow it to recover rapidly from intermittent drought stress during vegetative stages of plant development, but provide no advantages under terminal drought stress (Bidinger et al. 1987). Therefore, breeding for terminal drought tolerance has been a major goal. Significant progress has been made in mapping a number of QTLs for components of grain and stover yield, as well as yield maintenance, under terminal drought stress conditions in pearl millet (Yadav et al. 2002, 2003, 2004; Bidinger et al. 2007). Two sets of mapping population progeny, one from a cross between two elite inbred pollinators (H 77/833-2 and PRLT 2/89-33) and the other from a cross between two elite inbred seed parents (ICMB 841 and 863B), were used to map terminal drought tolerance of GY and their component traits (Yadav et al. 2002, 2003, 2004; Hash et al. 2003; Bidinger et al. 2007). Most importantly, both crosses identified a major QTL for terminal drought tolerance (DT) on LG2 explaining 23 % of the variation. The QTLs for biomass yield and harvest index (HI) also colocated with this major QTL on LG2 which suggested that increased DT conferred by this QTL on GY and its components might have been achieved by the effect of this QTL on both increased dry matter production and increased partitioning of dry matter to the grain (Yadav et al. 2002).

Kholova et al. (2010a, b; 2011) carried out important physiological and biochemical studies to dissect the physiological processes underlying this QTL. Various physiological (ABA concentration and transpiration efficiency) and biochemical (activities of ROS scavenging enzymes and photosynthetic pigment content) traits were analyzed in drought-sensitive and drought-tolerant parental lines and QTL NILs. The authors concluded that ROS machinery and pigment content do not play a key role in terminal drought tolerance in pearl millet (Kholova et al. 2011). They further discovered that DT-QTL on LG2 is correlated with a lower transpiration rate and a higher ABA levels under well-watered conditions (constitutive water-saving mechanism) and concluded that these traits contribute to water saving in the soil profile when water is non-limiting. This "extra" water, available for the later stage of the crop, becomes critical to guarantee water supply to the plants at the time of grain filling and therefore for GY under terminal drought. The effect of this QTL was also tested under saline and alkaline conditions which revealed positive effects of this QTL on GY and yield components (Sharma et al. 2011, 2014).

This QTL was identified as a major QTL for MAS in pearl millet. However, the size of this QTL was too large (circa 30 cM) to be taken up for MAS. Recently, efforts were made to delimit the QTL interval, and success has been achieved in identifying markers from candidate genes that could be used for MAS (see Sect. 15.5).

15.4.3 QTLs for Domestication Traits

QTLs involved in the domestication syndrome in pearl millet were dissected by studying the morphological differences between cultivated pearl millet (*Pennisetum glaucum* ssp. *glaucum*) and its wild ancestor (*Pennisetum glaucum* ssp. *monodii*) and mapping them using a F_2 population generated from wild × cultivated cross by means of RFLP markers (Poncet et al. 2000). Many morphological differences could be attributed to the effect of a small number of loci with relatively large effects. These loci were detected on four LGs (2, 5, 6, and 7). The loss of shedding ability, due to the absence of a functional abscission layer, was reported to be controlled by a single locus on LG6 (*al6*). Genetic control of the other spikelet traits involved factors with large effects was located close to *al6* and to an esterase gene, *Esterase-E*. QTLs with large effects on plant and spike morphology traits such as plant height, number of spikes, and weight of the spike were also mapped on LGs 6 and 7.

Poncet et al. (2002) also conducted a comparative mapping of QTLs involved in domestication of adaptative syndrome traits of pearl millet in two F_2 populations derived from domesticated (*Pennisetum glaucum* ssp. *glaucum*) × wild (*Pennisetum glaucum* ssp. *monodii*) crosses. In both populations, two regions of the genome were identified on LGs 6 and 7 that controlled most of the key morphological differences. Thus, these LGs play a central role both in the developmental control of spikelet structure and in the domestication process of this crop. The correspondence of the mapped QTLs on LGs 6 and 7 with those identified in other cereals was also determined using comparative maps of rice, maize, sorghum, and pearl millet (Whitkus et al. 1992; Ahn and Tanksley 1993; Ahn et al. 1993; Wilson et al. 1999; Devos et al. 2000). The region of LG6 involved in the shattering and spikelet structure in pearl millet corresponded to the ESTI-2-Mal I interval of rice chromosome 1 (Causse et al. 1994). This rice interval harbored a QTL for shattering (Xiong et al. 1999; Cai and Morishima 2000) and corresponded to regions of maize chromosomes 3 and 8 affecting seed dispersal ability. Similarly, the QTL on LG7 associated with spikelet architecture was reported to be syntenic to a QTL affecting seed dispersal located on rice chromosome 9 and sorghum LG C.

15.4.4 QTLs for Stover Yield and Quality

Two pearl millet mapping populations have been used for mapping pearl millet stover quality-related traits: ICMB 841 × 863B (also used previously for mapping drought-tolerant traits) and PT 732B × P 1449-2 (Hash et al. 2003). Important QTLs were obtained on LGs 3, 5, 6, and 7. The best putative QTL was detected for gas production from the leaf blade fraction of the stover, which maps to the top of LG7 of 863B, and accounts for ca. 20 % of the observed variation for this trait.

A marker-assisted backcross program was also initiated to transfer this leaf blade QTL from LG7 of 863B to ICMB 841 (Hash et al. 2003).

Nepolean et al. (2006) dissected QTLs for stover ruminant nutritional quality traits using one of the previously used populations (863B × ICMB 841). Ground stover samples were subjected to near-infrared reflectance spectroscopic (NIRS) analysis, and many traits were evaluated including gas volume (mL) produced after 24 h of in vitro digestion of 200 mg dry matter (GAS24), in vitro organic matter digestibility (IVOMD), nitrogen content (NDM), metabolizable energy content (ME), and sugar content (SUGSDM) on dry matter basis. Most important finding of this investigation was that a major pleiotropic QTL on LG2 was detected controlling GAS24, IVOMD, ME, and SUGSDM. This region was reported to be associated with drought tolerance previously by Yadav et al. (2002, 2004). These results suggested that transferring this stover quality QTL from 863B to 841B will improve both stover quality and terminal drought tolerance.

15.4.5 QTLs for Sink-Size Traits

In pearl millet, poor sink capacity with low HI (15–20 %), which in turn leads to low GY, has been a major issue (Yagya and Bainiwal 2001). To identify the stable regions in genome controlling sink-size traits, QTL analysis was conducted recently using a cross of two inbred lines having large differences in sink-size traits (panicle size and grain size). The mapping population, consisting of 188 F_2 individuals and their $F_{2:3}$ progenies, was produced from a cross between two diverse inbred lines: (81B × 4025-3-2-B)-11-5-2-2-B-2 used as the female parent and HHVBC II D2 HS-302-3-1-6-8-2-6-2-B used as male parent. These lines differed primarily for grain size (5 g 1000 grain female parent versus 13 g 1000 grain male parent) and panicle diameter (16 mm female parent versus 38 mm male parent). SSCP-SNP markers were used and a linkage map with 44 markers was used to map the QTLs. Genomic regions associated with panicle length, panicle diameter, and grain size were comapped on LG6, indicating the existence of a gene or gene cluster. The QTLs for panicle length on LG2 and LG6 (LOD > 3in both F2 and F2:3 data sets), for panicle diameter on LG2 and LG3 (LOD > 14 in the F2:3 data set), and for grain size on LG3 and LG6 (LOD > 3 in both F2 and F2:3 data sets), were identified as promising candidates for validation prior to possible application in marker-assisted breeding (Vengadessan et al. 2013).

15.5 From QTLs to Genes

The ultimate goal of all QTL studies is to identify candidate gene(s) to design functional markers for MAS. In pearl millet, although many QTLs have been reported for many traits, candidate gene identification has not met with much success mainly due to lack of sufficient genome sequence data. However, two important recent studies where a little success has been achieved using traditional fine mapping or association mapping are briefed here.

The recessive *d2* dwarfing gene has been deployed widely in commercial germplasm grown in India, the USA, and Australia. The *d2* dwarfing gene was previously located to a 23.2-cM interval on pearl millet LG4 (Azhaguvel et al. 2003) and Padi (2002) delimited the *d2* gene to a 2.8 cM between markers PSMP344 and the cosegregating markers B224C4P2 and RGR1963. Parvathaneni et al. (2013) identified the likely candidate for *d2* using a combination of approaches: genetic mapping in two F_2 populations with some 1500 progeny, haplotype analysis of three tall and three dwarf inbred lines to delineate the *d2* region, comparative information, and identification of annotated genes in sorghum in this delineated region. F_2 populations used were derived from crosses Tift 23DB (the d2 dwarf inbred) × ICMP 451 (the tall inbred) and PT 732B (d2d2) × P1449-2 (D2D2). Twenty-two $F_{2:3}$ plants of cross PT 732B × P1449-2 were grown and analyzed with markers for the region of interest on LG4, and a heterozygous F_3 plant was selfed to produce a 552 progeny population. This population was phenotyped for *d2* and mapped with RFLP markers to identify the important recombination events. Of the 19 recombination events that could be allocated, 18 events occurred between PSMP344 and *d2*, and 1 occurred between B224C4P2/RGR1963 and *d2*, indicating a tight linkage of *d2* with B224C4P2/RGR1963. Pearl millet BAC library (Allouis et al. 2001) was screened with RGR1963 marker and eleven positive clones were identified, of which BAC 293B22 was selected for sequencing. The sequence of BAC 293B22 was analyzed for gene prediction using the standard in silico approaches. A total of 915 F_2 individuals from the cross Tift 23DB × ICMP 451 were genotyped with markers to identify recombinants in the *d2* region. Plants carrying a recombination event in the *d2* region were selfed to produce F3 seed. 13–25 F3 plants were phenotyped for plant height, and their allelic composition was determined. This analysis delimited the gene to a 1.6-cM interval. A comparative mapping approach was then utilized to identify the candidate gene (for detailed information, readers are requested to refer to Parvathaneni et al. 2013). Briefly, haplotype analysis of three tall and three dwarf inbred lines delineated the *d2* region by two genetic markers that, in sorghum, defined a region of 410 kb with 40 annotated genes. One of the sorghum genes annotated within this region was *ABCB1*, which encodes a P-glycoprotein involved in auxin transport. This gene had previously been shown to underlie the economically important *dw3* dwarf mutation in sorghum. The cosegregation of *ABCB1* with the *d2* phenotype, its differential expression in the tall inbred ICMP 451 and the dwarf inbred Tift 23DB, and the similar phenotype of stacked lower internodes in the sorghum *dw3* (ortholog of *d2* in sorghum) and pearl millet *d2* mutants suggested that ABCB1 is a likely candidate for *d2*.

Similarly, the major drought tolerance QTL (DT-QTL) on LG2 has been identified as a major QTL for MAS in pearl millet. Recently, efforts were taken to fine

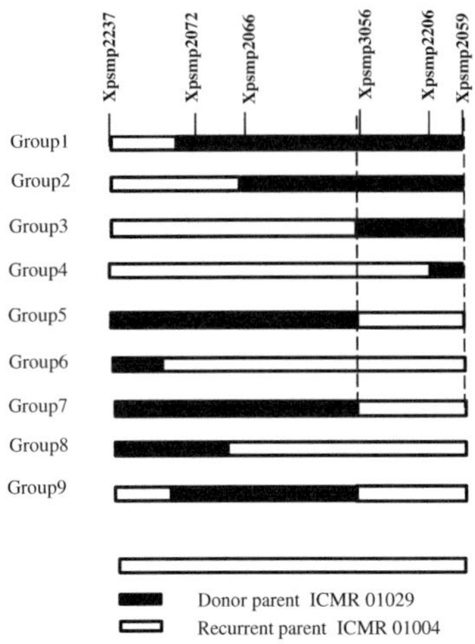

	No. of recombinants	GY	PHI	HI	TGW	DTGY	DTPHI	DTHI	DTTGW
Group1	10	1570.90* (1435.80)*	51.98* (51.80)*	0.29* (0.27)*	7.13* (7.42)*	0.45* (0.40)*	0.73* (0.75)*	0.65* (0.66)*	0.75* (0.73)*
Group2	12	1506.67* (1396.80)*	51.58* (49.31)*	0.28* (0.25)*	7.43* (7.61)*	0.46* (0.38)*	0.74* (0.71)*	0.64* (0.62)*	0.78* (0.73)*
Group3	8	1516.86* (1465.00)*	51.74* (52.19)*	0.27* (0.27)*	7.07* (6.93)*	0.46* (0.44)*	0.73* (0.76)*	0.63* (0.68)*	0.75* (0.67)
Group4	2	1409.00 (1317.00)	53.66* (51.51)*	0.22 (0.24)	6.02 (7.39)*	0.40 (0.38)	0.69 (0.69)	0.60 (0.61)	0.73 (0.76)*
Group5	5	1222.60 (1157.75)	42.06 (44.16)	0.22 (0.22)	5.92 (6.04)	0.36 (0.33)	0.62 (0.66)	0.53 (0.57)	0.69 (0.67)
Group6	5	1243.00 (1090.00)	43.14 (41.87)	0.23 (0.21)	6.10 (6.01)	0.36 (0.30)	0.61 (0.62)	0.53 (0.52)	0.71 (0.65)
Group7	9	1288.56 (1192.29)	45.29 (44.24)	0.24 (0.22)	6.25 (6.43)	0.36 (0.31)	0.64 (0.62)	0.55 (0.54)	0.71 (0.66)
Group8	1	1219.00 (1179.00)	41.81 (42.30)	0.23 (0.21)	5.59 (6.16)	0.33 (0.33)	0.59 (0.61)	0.53 (0.51)	0.72 (0.71)
Group9	3	1349.00 (1235.23)	48.36 (46.34)	0.26 (0.24)	6.23 (6.42)	0.39 (0.35)	0.69 (0.66)	0.61 (0.55)	0.72 (0.67)
	ICMR 01004	1218.58 (1113.97)	45.12 (42.36)	0.25 (0.21)	5.94 (5.94)	0.36 (0.32)	0.65 (0.63)	0.58 (0.53)	0.73 (0.69)

Fig. 15.1 Recombination break points in DT-QTL region on LG2 identified by six SSR markers. *GY* grain yield, *PHI* panicle harvest index, *TGW* thousand grain weight, DTGY drought tolerance of grain yield, *DTPHI* drought tolerance of panicle harvest index, *DTHI* drought tolerance of harvest index, *DTTGW* drought tolerance of thousand grain weight (Sehgal et al. unpublished)

map this QTL using the traditional fine-mapping population and a complementary association mapping approach. A high-resolution cross segregating specifically for the DT-QTL interval on LG 2 was developed at ICRISAT by crossing a DT-QTL NIL (ICMR 01029) with another NIL of the H 77/833-2 parent introgressed with DM QTLs. F_1 plants of each of the seven plant-by-plant crosses made between ICMR 1029 and ICMR 1004 were selfed to produce F_2 seeds. Seeds from two F_2 families were taken forward for the development of a fine-mapping population of 2500 individuals segregating for the DT-QTL interval on LG2. A schematic diagram of development of this population is outlined in Yadav et al. (2011). A smaller subset of the 55 most informative recombinants, representing all probable parental combinations, was identified from within this fine-mapping population by genotyping it with six SSR markers bracketing the entire DT-QTL region (Fig. 15.1). Each marker was used to divide the population into three genotypic classes (homozygous ICMR 1029, heterozygous ICMR 1029/ICMR 1004, and homozygous ICMR 1004). The phenotypic means of these genotypic groups were then compared with the phenotypic mean of ICMR 01004 (drought susceptible parent) using t-test. Using this strategy, the QTL was fine mapped to a 10-cM interval between markers *Xpsmp*3056 and *Xpsmp2059* and showed 24.3, 15.2, 9.0, 18.9, 24.9, 11.9, 9.5, and 3.6 % increased GY, panicle harvest index (PHI), HI, thousand grain weight (TGW) and drought tolerance (DT) of GY, PHI, HI, and TGW, respectively, compared to the recurrent parent ICMR 01004 (Fig. 15.1).

To further fine map the interval, new gene-based markers were designed to saturate the DT-QTL region (Sehgal et al. 2012) and a complementary approach of candidate gene-based association mapping was utilized to identify the genes for

MAS (Sehgal et al. 2015). To accomplish this, a pearl millet germplasm association panel (PMiGAP), comprising 250 inbred lines, was utilized. PMiGAP has been recently assembled from a large set of 1000 diverse breeding lines and accessions of landraces, elite cultivars, and mapping population parents, representing wide geographical range in Africa and Asia. Assemblage of PMiGAP is a further progress in terms of genetic and genomic resources development in pearl millet (Sehgal et al. 2015). It is anticipated that PMiGAP will provide the pearl millet community with a high-resolution platform for fine mapping of QTLs and (or) for allele mining of favorable genes of agronomic importance. The 250 PMiGAP entries have been assigned to four precocity groups (61, 63, 63, and 63 entries in early, medium early, medium, and late maturity groups, respectively) to reduce the confounding effects of flowering time.

The genes mapping in DT-QTL interval (Sehgal et al. 2012) were sequenced in all 250 PMiGAP lines and were studied for association with several traits including GY and yield components [panicle harvest index (PHI), grain harvest index (GHI), thousand grain weight (TGW), panicle diameter (PD), panicle yield (PY), etc.) and morpho-physiological traits (leaf senescence, leaf rolling) under both well-watered and drought conditions (Sehgal et al. 2015). Additionally, expression patterns of the genes were analyzed to shed a light on physiological mechanisms involved. Interestingly, many genes showed association with traits, and it became arduous to pinpoint a single candidate gene based on the results. However, two SNPs in two genes were identified which were suggested by the authors as promising candidates for MAS. A SNP in putative acetyl CoA carboxylase gene showed constitutive association with GY, GHI, and PY under all treatments, and an InDEl in putative chlorophyll a-/b-binding protein gene was significantly associated with both stay-green and GY traits under drought stress (Sehgal et al. 2015).

15.6 Development of TILLING Populations

Targeting induced local lesions in genomes (TILLING) is a method in molecular biology that allows directed identification of mutations in a specific gene. It is a non-transgenic reverse genetics approach that is applicable to all animal and plant species. This approach requires prior DNA sequence information and takes advantage of a mismatch endonuclease to locate and detect induced mutations. Ultimately, it can provide an allelic series of silent, missense, nonsense, and splice site mutations to examine the effect of various mutations in a gene. TILLING has proven to be a practical, efficient, and an effective approach for functional genomic studies in numerous plant and animal species. For a detailed knowledge on the protocols for TILLING, readers are requested to read Barkley and Wang (2008).

Since the inception of TILLING, this method has been widely used for the study of functional genomics in plants, especially for the model plants *Arabidopsis thaliana* (Greene et al. 2003) and *Lotus japonicas* (Perry et al. 2003; Horst et al.

2007). Today, TILLING populations are available for many crops including pea, maize, wheat, rice, and soybean (Barkley and Wang 2008). In pearl millet, initiatives to develop a TILLING population were taken in 2005 at ICRISAT for mining allelic variants in drought-responsive candidate genes (http://www.icrisat.org/bt-gene-discovery.htm). A total of 31,000 seeds of inbred line "P1449-2-P1" were mutagenized in three different batches using 5, 7.5, 9, and 10 mM ethylmethane sulfonate (generation M1). The M1 seeds were selfed to produce a M2 generation. To date, DNA from 9938 M2 TILLING lines is available. In order to increase the throughput during allele mining, eightfold pooling of normalized genomic DNA from 9938 lines has been accomplished in 12 pooled plates. The pools of DNA from mutant population are available to the researchers worldwide for allelic mining of candidate genes. Research efforts have been made to mine allelic variants for drought-responsive candidate genes (*DREB2A*, *EDR2*) with success (Jalaja 2011). For example, a mutant line 84081 has been identified, where changes in nucleotide sequences and the corresponding amino acids are clearly evident. This *DREB* mutant line should be tested for drought tolerance or susceptibility.

15.7 Success Story of Product Release Through MAS

The discovery of QTLs for downy mildew resistance led not only to identification of genomic region conferring resistance to the disease but also to the identification of donors of naturally occurring host-plant resistance genes, and a well-chosen set of lines to incorporate genes into cultivars grown by resource-poor farmers. A variety called *HHB 67* was released from CCS Haryana Agricultural University in 1989 (Kapoor et al. 1989). *HHB 67* possessed many traits that farmers appreciated, including early maturity that allowed the plants to escape drought stress at the end of the season. It became the most popular public sector pearl millet hybrid in India occupying over half of the pearl millet area in Haryana (over 300,000 ha during the rainy season of 2001). However, it soon became vulnerable to an epidemic of downy mildew. To improve the disease resistance of the parental lines of *HHB 67*, marker-assisted backcrossing (MABC) was used (Sharma 2001) to pyramid resistance genes. MABC is a rapid and more effective strategy than conventional breeding. In conventional breeding, once a single effective resistance gene is included it is often impossible to detect the presence of a second without expensive and time-consuming progeny testing every backcross generation. An improved version of *HHB 67* was developed which showed GY gains of 15 % along with significant improvements in downy mildew resistance without adversely affecting the early maturity of *HHB 67*. This was a remarkable achievement as yield gains from conventional yield-focused hybrid breeding were typically on the order of 1–2 % per year. This improved version of *HHB 67* was called *HHB 67 Improved,* and it was validated by ICRISAT, All India Coordinated Pearl Millet Improvement Project (AICPMIP) and CCS Haryana Agricultural University (CCS HAU). *HHB 67 Improved* was released for commercial cultivation in 2005 as the first public-bred product of marker-assisted breeding in India (Hash et al. 2006; Khairwal and Hash 2007). *HHB 67 Improved* is much more resistant to downy mildew disease and yields 2.0 tons ha^{-1}, while the original *HHB 67* yielded only 1.79 tons ha^{-1}.

15.8 Whole-Genome Sequencing Efforts in Pearl Millet

In the last decade, advances in DNA sequencing technologies have enabled the generation of a wealth of sequence information including whole-genome sequences. NGS platforms such as Roche 454GS FLX Titanium (http://www.454.com/) or Illumina Solexa Genome Analyzer (http://www.illumina.com/) can carry out high-capacity sequencing at reduced costs and increased rates compared to conventional Sanger sequencing (Varshney et al. 2009). Through NGS technologies, sequencing and resequencing of even large genomes have become feasible. Accordingly, reference or draft genome sequences for a number of species, including the model species *Arabidopsis thaliana* and *Brachypodium distachyon*, along with important crop species such as rice, sorghum, soybean, and maize, have been published (Morrell et al. 2012).

For whole-genome sequencing of pearl millet, ICRISAT established an informal consortium of research organizations with expertise in genome sequencing, crop genomics, and pearl millet research. The consortium members include the International Crops Research Institute for the Semi-Arid Tropics (ICRISAT, http://www.icrisat.org), India; L'institut de recherché pour le développement (IRD, http://www.ird.fr), France; Indian Council of Agricultural Research (http://www.icar.org.in/), India; University of Georgia (http://www.uga.edu), USA; Cornell University (http://www.cornell.edu/), USA; University of Florida (http://agronomy.ifas.ufl.edu), USA; L'Institut Sénégalais de Recherches Agricoles (ISRA; http://www.isra.sn/), Pioneer Overseas Corporation, India (http://www.pioneer.com/web/site/india), Fort Valley State University, USA (http://www.fvsu.edu), University of Vienna, Austria (http://www.univie.ac.at), Oklahoma State University, USA (http://biochemistry.okstate.edu), and Consiglio Nazionale delle Ricerche, Italy (http://www.cnr.it/sitocnr/home.html). Tift 23D2B1 was chosen for developing the draft genome sequence as it is an important ancestral genotype of many seed parents of pearl millet hybrids which are currently in use for forage and grain all over. A high-quality draft genome assembly has been developed using a hybrid sequence assembly approach [whole-genome shotgun sequence (WGS) data and BAC sequencing data together with a restriction site associated DNA (RAD) sequence tag-based genetic map]. In parallel, resequencing of 993 pearl millet germplasm lines (including 606 B and R-lines and 387 PMiGAP lines) by using whole-genome resequencing (WGRS) and/or RAD sequencing has also been accomplished.

Briefly, a total of 1.49 TB raw data has been generated from 9 insert libraries as well as from BAC pools using whole-genome shotgun sequencing approach. A genome assembly of 1.79 GB with 25,241 scaffolds (N50 = 884.94 kb) and >92 % scaffolds was anchored onto 7 linkage groups. A total of 38,579 genes have been identified. A comprehensive repeat annotation of the assembly showed that 73.6 % of the genome is comprised of transposable elements. A hapmap of pearl millet has also been developed based on resequencing of 993 germplasm lines. Further analysis of the data has shown up to ~29.5 million SNPs and 3.84 million InDEls in pearl millet.

15.9 Conclusions

Pearl millet genomics research boomed in the last decade leading to the development of highly saturated genetic maps, well-tailored segregating biparental and fine-mapping populations, association mapping panel, core sets, and whole-genome sequence. These genetics and genomics resources will open up new avenues for molecular breeding and application of genomic selection (predictive breeding) in pearl millet taking it to a new status as a crop. Transcriptomics applied to pearl millet has revealed an insight into mechanisms of drought/salt/cold stress tolerance developmental processes. The future will certainly see much more impact of transcriptomics in pearl millet breeding including identification of genes controlling complex traits. The saturation of genetic maps with gene-based markers will also facilitate identification of candidate genes for various biotic and abiotic stress-related traits. TILLING populations generated in pearl millet will make validation of candidate genes easy, and implementation of reverse genetics approaches will be a routine in future. Once the candidate genes are identified and validated, the complete genome sequence of the genes will facilitate primer designing from different parts of the genes (exons, introns, UTRs, etc.) to initiate allele mining projects. Some of the other more immediate opportunities lend by whole-genome sequence may be listed as follows: (a) innumerable number of molecular markers in genomic regions of choice to facilitate large-scale cloning of new genes, analysis of quantitative trait loci, and association studies of traits, (b) a plethora of approaches for understanding the function of each and every gene, (c) understanding temporal and tissue-specific gene expression in response to developmental and environmental cues, and (d) designing of a genome-wide perfect marker system based on SNPs in entire gene space of the species.

References

Ahn S, Tanksley S (1993) Comparative linkage maps of the rice and maize genomes. Proc Natl Acad Sci USA 90:7980–7984

Ahn S, Anderson J, Sorrells M et al (1993) Homoeologous relationships of rice, wheat and maize chromosomes. Mol Gen Genet 241:483–490

Allouis S, Qi X, Lindup S et al (2001) Construction of a BAC library of pearl millet, *Pennisetum glaucum*. Theor Appl Genet 102:1200–1205

Azhaguvel P, Hash C, Rangasamy P et al (2003) Mapping the d1 and d2 dwarfing genes and the purple foliage color locus P in pearl millet. J Hered 94:155–159

Barkley N, Wang M (2008) Application of TILLING and EcoTILLING as reverse genetic approaches to elucidate the function of genes in plants and animals. Curr Genomics 9:212

Bertin I, Zhu J, Gale M (2005) SSCP-SNP in pearl millet—A new marker system for comparative genetics. Theor Appl Genet 110:1467–1472

Bhattacharjee R, Bramel P, Hash C et al (2002) Assessment of genetic diversity within and between pearl millet landraces. Theor Appl Genet 105:666–673

Bidinger F, Mahalakshmi V, Rao GDP (1987) Assessment of drought resistance in pearl millet (*Pennisetum americanum* (L.) Leeke) II. Estimation of genotype response to stress. Crop Pasture Sci 38:49–59

Bidinger F, Nepolean T, Hash C et al (2007) Quantitative trait loci for grain yield in pearl millet under variable postflowering moisture conditions. Crop Sci 47:969–980

Breese, WA, Hash CT, Devos KM et al (2002) Pearl millet genomics- an overview with respect to breeding for resistance to downy mildew. Leslie JF (ed) Sorghum and millets pathology 2000. Iowa State Press, Ames, pp 243–246

Budak H, Pedraza F, Cregan P et al (2003) Development and utilization of SSRs to estimate the degree of genetic relationships in a collection of pearl millet germplasm. Crop Sci 43:2284–2290

Busso C, Liu C, Hash C et al (1995) Analysis of recombination rate in female and male gametogenesis in pearl millet (*Pennisetum glaucum*) using RFLP markers. Theor Appl Genet 90:242–246

Busso CS, Devos KM, Ross G et al (2000) Genetic diversity within and among landraces of pearl millet (*Pennisetum glaucum*) under farmer management in West Africa. Genet Res Crop Evol 47:561–568

Cai H-W, Morishima H (2000) Genomic regions affecting seed shattering and seed dormancy in rice. Theor Appl Genet 100:840–846

Causse MA, Fulton TM, Cho YG et al (1994) Saturated molecular map of the rice genome based on an interspecific backcross population. Genetics 138:1251

Chowdari K, Davierwala A, Gupta V et al (1998) Genotype identification and assessment of genetic relationships in pearl millet [*Pennisetum glaucum* (L.) R. Br] using microsatellites and RAPDs. Theor Appl Genet 97:154–162

Clegg MT, Rawson JR, Thomas K (1984) Chloroplast DNA variation in pearl millet and related species. Genetics 106:449–461

Devos KM, Gale MD (2000) Genome relationships: the grass model in current research. The Plant Cell Online 12:637–646

Devos K, Pittaway T, Reynolds A et al (2000) Comparative mapping reveals a complex relationship between the pearl millet genome and those of foxtail millet and rice. Theor Appl Genet 100:190–198

Diab AA, Kantety R, Ozturk N et al (2008) Drought-inducible genes and differentially expressed sequence tags associated with components of drought tolerance in durum wheat. Sci Res Essay 3:009–026

Ejeta G, Hassen M, Mertz E (1987) In vitro digestibility and amino acid composition of pearl millet (*Pennisetum typhoides*) and other cereals. Proc Natl Acad Sci USA 84:6016–6019

Gepts P, Clegg M (1989) Genetic diversity in pearl millet (*Pennisetum glaucum* [L.] R. Br.) at the DNA sequence level. J Hered 80:203–208

Govindaraj M, Selvi B, Prabhu DA et al (2009) Genetic diversity analysis of pearl millet (*Pennisetum glauccum* [L.] R. Br.) accessions using molecular markers. Afr J Biotechnol 8

Govindaraj M, Rai K, Shanmugasundaram P et al (2013) Combining ability and heterosis for grain iron and zinc densities in pearl millet. Crop Sci 53:507–517

Greene EA, Codomo CA, Taylor NE et al (2003) Spectrum of chemically induced mutations from a large-scale reverse-genetic screen in Arabidopsis. Genetics 164:731–740

Gulia SK (2004) QTL mapping for improvement of downy mildew [Sclerospora graminicola (Sacc.) J. Schroet.] resistance (DMR) in pearl millet [*Pennisetum glaucum* (L.) R. Br.] hybrid parental line ICMB 89111. In: Chaudhary Charan Singh Haryana Agricultural University, Hisar

Gulia SK, Hash CT, Thakur RP et al (2007) Mapping new QTLs for downy mildew [*Sclerospora graminicola* (Sacc.) J. Schroet.] resistance in pearl millet (*Pennisetum glaucum* (L.) R. Br.)," in Crop Production in Stress Environments. In: Singh, SP, Tomar, VS, Behl, RK, Upadhyaya, SD, Bhale, MS, and Khare, D (eds) Genetic and Management Options. Agrobios Publishers, Jodhpur (India), pp 373–386

Hanna W, Burton G (1992) Genetics of red and purple plant color in pearl millet. J Hered (USA) 5:386–388

Hash C, Witcombe J (1994) Pearl millet mapping populations at ICRISAT. In: Witcombe JR, Duncan RR (eds) Use of molecular markers in sorghum and pearl millet breeding for developing countries. Overseas Development Administration, London, pp 69–75

Hash C, Raj AB, Lindup S et al (2003) Opportunities for marker-assisted selection (MAS) to improve the feed quality of crop residues in pearl millet and sorghum. Field Crops Res 84:79–88

Hash CT, Sharma A, Kolesnikova-Allen MA et al (2006) Teamwork delivers biotechnology products to Indian small-holder crop-livestock producers: pearl millet hybrid "HHB 67 Improved" enters seed delivery pipeline. J SAT Agric Res 2:1–3

Hill G, Hanna W (1990) Nutritive characteristics of pearl millet grain in beef cattle diets. J Animal Sci 68:2061–2066

Holton TA, Christopher JT, Mcclure L et al (2002) Identification and mapping of polymorphic SSR markers from expressed gene sequences of barley and wheat. Mol Breed 9:63–71

Horst I, Welham T, Kelly S et al (2007) TILLING mutants of *Lotus japonicus* reveal that nitrogen assimilation and fixation can occur in the absence of nodule-enhanced sucrose synthase. Plant Physiol 144:806–820

Imelfort M, Duran C, Batley J et al (2009) Discovering genetic polymorphisms in next-generation sequencing data. Plant Biotechnol J 7:312–317

Jalaja N (2011) Development of transgenics for fungal resistance and discovery of chemically induced mutations in Pearl Millet (*Pennisetum glaucum* L.) population by TILLING. PhD thesis, Osmania University, India

Jonah P, Bello L, Lucky O et al (2011) Review: the importance of molecular markers in plant breeding programmes. Global J Sci Front Res 11

Jones E, Liu C, Gale M et al (1995) Mapping quantitative trait loci for downy mildew resistance in pearl millet. Theor Appl Genet 91:448–456

Jones E, Breese W, Liu C et al (2002) Mapping quantitative trait loci for resistance to downy mildew in pearl millet. Crop Sci 42:1316–1323

Kale A, Munjal S (2005) Molecular analysis of mitochondrial DNA of lines representing a specific CMS-fertility-restorer system of pearl millet [*Pennisetum glaucum* (L.) R. Sr.] by RAPD markers. Indian J of Genet Plant Breed 65:1–4

Kantety RV, La Rota M, Matthews DE et al (2002) Data mining for simple sequence repeats in expressed sequence tags from barley, maize, rice, sorghum and wheat. Plant Mol Biol 48:501–510

Kapila RK, Yadav RS, Plaha P et al (2008) Analysis of genetic diversity in pearl millet inbreds using microsatellite markers. Plant Breed 127:33–37

Kapoor R, Kakkar P, Khairwal I et al (1989) Bajra hybrid HHB 67: a major break through. Haryana Farming 18:17–21

Kesawat M, Das Kumar B (2009) Molecular markers: it's application in crop improvement. J Crop Sci Biotechnol 12:169–181

Khairwal I, Hash C (2007) HHB 67-improved–The first product of marker-assisted crop breeding in India. Asia-Pacific Consortium on Agricultural Biotechnology (APCoAB) e-News

Kholová J, Hash C, Kumar PL et al (2010a) Terminal drought-tolerant pearl millet [*Pennisetum glaucum* (L.) R. Br.] have high leaf ABA and limit transpiration at high vapour pressure deficit. J Exp Bot 61:1431–1440

Kholová J, Hash CT, Kakkera A et al (2010b) Constitutive water-conserving mechanisms are correlated with the terminal drought tolerance of pearl millet [*Pennisetum glaucum* (L.) R. Br.]. J Exp Bot 61:369–377

Kholová J, Hash CT, Kočová M et al (2011) Does a terminal drought tolerance QTL contribute to differences in ROS scavenging enzymes and photosynthetic pigments in pearl millet exposed to drought? Env Exp Bot 71:99–106

Kota R, Varshney R, Prasad M et al (2008) EST-derived single nucleotide polymorphism markers for assembling genetic and physical maps of the barley genome. Funct Integr Genomics 8:223–233

Liu CJ, Witcombe JR, Pittaway TS et al (1992) Complexes d'especes, flux de genes et ressources genetiques des plantes. In: Mounolou JC, Sarr A (eds) Colloque international Restriction fragment length polymorphism in pearl millet, *Pennisetum glaucum*, 8–10 janvier. Publications du Bureau des Ressources Genetiques, Paris, pp 233–241

Liu CJ, Witcombe JR, Hash CT et al (1994a) Use of molecular markers in sorghum and pearl millet breeding for developing countries. In: Witcombe JR, Duncan RR (eds) Overseas Development Administration, London, pp 57–64

Liu C, Witcombe J, Pittaway T et al (1994b) An RFLP-based genetic map of pearl millet (*Pennisetum glaucum*). Theor Appl Genet 89:481–487

Liu C, Devos K, Witcombe J et al (1996) The effect of genome and sex on recombination rates in *Pennisetum* species. Theor Appl Genet 93:902–908

Mahalakshmi V, Bidinger F, Raju D (1987) Effect of timing of water deficit on pearl millet (*Pennisetum americanum*). Field Crops Res 15:327–339

Mariac C, Luong V, Kapran I et al (2006) Diversity of wild and cultivated pearl millet accessions (*Pennisetum glaucum* [L.] R. Br.) in Niger assessed by microsatellite markers. Theor Appl Genet 114:49–58

Mishra RN, Reddy PS, Nair S et al (2007) Isolation and characterization of expressed sequence tags (ESTs) from subtracted cDNA libraries of *Pennisetum glaucum* seedlings. Plant Mol Biol 64:713–732

Miyao A, Zhong HS, Monna L et al (1996) Characterization and genetic mapping of simple sequence repeats in the rice genome. DNA Res 3:233–238

Morrell PL, Buckler ES, Ross-Ibarra J (2012) Crop genomics: advances and applications. Nat Rev Genet 13:85–96

Moumouni K, Kountche B, Jean M et al (2015) Construction of a genetic map for pearl millet, *Pennisetum glaucum* (L.) R. Br., using a genotyping-by-sequencing (GBS) approach. Mol Breed 35:1–10

Nepolean T, Blummel M, Raj AB et al (2006) QTLs controlling yield and stover quality traits in pearl millet. J SAT Agric Res 2:4

Nguyen T, Klueva N, Chamareck V et al (2004) Saturation mapping of QTL regions and identification of putative candidate genes for drought tolerance in rice. Mol Genet Genomics 272:35–46

Oumar I, Mariac C, Pham J-L et al (2008) Phylogeny and origin of pearl millet (*Pennisetum glaucum* [L.] R. Br) as revealed by microsatellite loci. Theor Appl Genet 117:489–497

Padi FK (2002) Genetic analyses of adaptive traits in pearl millet (*Pennisetum glaucum* (L.) R. Br.). In: University of East Anglia

Parvathaneni RK, Jakkula V, Padi FK et al (2013) Fine-mapping and identification of a candidate gene underlying the d2 dwarfing phenotype in pearl millet, *Cenchrus americanus* (L.) Morrone. G3: Genes| Genomes| Genetics 3:563–572

Pedraza-Garcia F, Specht J, Dweikat I (2010) A new PCR-based linkage map in pearl millet. Crop Sci 50:1754–1760

Perry JA, Wang TL, Welham TJ et al (2003) A TILLING reverse genetics tool and a web-accessible collection of mutants of the legume *Lotus japonicus*. Plant Physiol 131:866–871

Poncet V, Lamy F, Devos K et al (2000) Genetic control of domestication traits in pearl millet (*Pennisetum glaucum* L., Poaceae). Theor Appl Genet 100:147–159

Poncet V, Martel E, Allouis S et al (2002) Comparative analysis of QTLs affecting domestication traits between two domesticated × wild pearl millet (*Pennisetum glaucum* L., Poaceae) crosses. Theor Appl Genet 104:965–975

Qi X, Lindup S, Pittaway T et al (2001) Development of simple sequence repeat markers from bacterial artificial chromosomes without subcloning. Biotechniques 31(355):358–362

Qi X, Pittaway T, Lindup S et al (2004) An integrated genetic map and a new set of simple sequence repeat markers for pearl millet, *Pennisetum glaucum.* Theor Appl Genet 109:1485–1493

Rajaram V, Nepolean T, Senthilvel S et al (2013) Pearl millet [*Pennisetum glaucum* (L.) R. Br.] consensus linkage map constructed using four RIL mapping populations and newly developed EST-SSRs. BMC Genom 14:159

Rostoks N, Mudie S, Cardle L et al (2005) Genome-wide SNP discovery and linkage analysis in barley based on genes responsive to abiotic stress. Mol Genet Genomics 274:515–527

Sawaya W, Khalil J, Safi W (1984) Nutritional quality of pearl millet flour and bread. Plant Food Hum Nutr 34:117–125

Sehgal D, Bhat V, Raina SN (2008) Advent of DNA markers to decipher genome sequence polymorphism. In: Kirti PB (ed) Handbook of new technologies for genetic improvement of grain legumes. CRC Press, New York, pp 477–495

Sehgal D, Rajaram V, Armstead IP et al (2012) Integration of gene-based markers in a pearl millet genetic map for identification of candidate genes underlying drought tolerance quantitative trait loci. BMC Plant Biol 12:9

Sehgal D, Skot L, Singh R et al (2015) Exploring potential of pearl millet germplasm association panel for association mapping of drought tolerance traits. PLoS One. doi:10.1371/journal.pone.0122165

Senthilvel S, Mahalakshmi V, Sathish Kumar P et al (2004) New SSR markers for pearl millet from data mining of Expressed Sequence Tags. In: Proceedings of 4th international crop science congress, 26th Sept to 1st Oct, p 232

Senthilvel S, Jayashree B, Mahalakshmi V et al (2008) Development and mapping of simple sequence repeat markers for pearl millet from data mining of expressed sequence tags. BMC Plant Biol 8:119

Sharma A (2001) Marker-assisted improvement of pearl millet (*Pennisetum glaucum*) downy mildew resistance in elite hybrid parental line H 77/833-2. In: Chaudhary Charan Singh Haryana Agricultural University, Hisar, India

Sharma PC, Sehgal D, Singh D et al (2011) A major terminal drought tolerance QTL of pearl millet is also associated with reduced salt uptake and enhanced growth under salt stress. Mol Breed 27:207–222

Sharma PC, Singh D, Sehgal D et al (2014) Further evidence that a terminal drought tolerance QTL of pearl millet is associated with reduced salt uptake. Env Exp Bot 102:48–57

Singh SD, King S, Werder J (1993) Downy mildew disease of pearl millet. Information Bulletin no. 37. International Crops Research Institute for the Semi-Arid Tropics

Stich B, Haussmann BI, Pasam R et al (2010) Patterns of molecular and phenotypic diversity in pearl mille [*Pennisetum glaucum* (L.) R. Br.] from West and Central Africa and their relation to geographical and environmental parameters. BMC Plant Biol 10:216

Supriya A, Senthilvel S, Nepolean T et al (2011) Development of a molecular linkage map of pearl millet integrating DArT and SSR markers. Theor Appl Genet 123:239–250

Tondelli A, Francia E, Barabaschi D et al (2006) Mapping regulatory genes as candidates for cold and drought stress tolerance in barley. Theor Appl Genet 112:445–454

Tostain S (1994) Isozymic classification of pearl millet (*Pennisetum glaucum*, Poaceae) land races from Niger (West Africa). Plant Syst Evol 193:81–93

Tostain S, Marchais L (1989) Enzyme diversity in pearl millet (*Pennisetum glaucum*). Theor Appl Genet 77:634–640

Tostain S, Riandey M-F, Marchais L (1987) Enzyme diversity in pearl millet (*Pennisetum glaucum*). Theor Appl Genet 74:188–193

Varshney RK, Thiel T, Stein N et al (2002) In silico analysis on frequency and distribution of microsatellites in ESTs of some cereal species. Cell Mol Biol Lett 7:537–546

Varshney RK, Nayak SN, May GD et al (2009) Next-generation sequencing technologies and their implications for crop genetics and breeding. Trends Biotechnol 27:522–530

Velu G, Rai K, Muralidharan V et al (2007) Prospects of breeding biofortified pearl millet with high grain iron and zinc content. Plant Breed 126:182–185

Vengadessan V, Rai K, Kannan Bapu J et al (2013) Construction of genetic linkage map and QTL analysis of sink-size traits in pearl millet (*Pennisetum glaucum*). Int Sch Res Not 2013

Vom Brocke K, Christinck A, Weltzien R et al (2003) Farmers' seed systems and management practices determine pearl millet genetic diversity patterns in semiarid regions of India. Crop Sci 43:1680–1689

Whitkus R, Doebley J, Lee M (1992) Comparative genome mapping of Sorghum and maize. Genetics 132:1119–1130

Wilson WA, Harrington SE, Woodman WL et al (1999) Inferences on the genome structure of progenitor maize through comparative analysis of rice, maize and the domesticated panicoids. Genetics 153:453–473

Xiong L, Liu K, Dai X et al (1999) Identification of genetic factors controlling domestication-related traits of rice using an F_2 population of a cross between *Oryza sativa* and *O. rufipogon*. Theor Appl Genet 98:243–251

Yadav RS, Hash C, Bidinger F et al (2002) Quantitative trait loci associated with traits determining grain and stover yield in pearl millet under terminal drought-stress conditions. Theor Appl Genet 104:67–83

Yadav R, Bidinger F, Hash C et al (2003) Mapping and characterisation of QTL × E interactions for traits determining grain and stover yield in pearl millet. Theor Appl Genet 106:512–520

Yadav RS, Hash C, Bidinger F et al (2004) Genomic regions associated with grain yield and aspects of post-flowering drought tolerance in pearl millet across stress environments and tester background. Euphytica 136:265–277

Yadav O, Mitchell S, Zamora A et al (2007a) Development of new simple sequence repeat markers for pearl millet. J SAT Agric Res 3:34

Yadav S, Jain S, Jain V et al (2007b) Genetic analysis of CMS, restorer, hybrid and open-pollinated genotypes of Indian pearl millet [*Pennisetum glaucum* (L.) R. Br.] using ISSR markers. Indian J Biotechnol 6:340–348

Yadav RS, Sehgal D, Vadez V (2011) Using genetic mapping and genomics approaches in understanding and improving drought tolerance in pearl millet. J Exp Bot 62:397–408

Yagya D, Bainiwal C (2001) Genetic analysis of crosses among pearl millet populations. Int Sorghum Millets Newsl 42:68–70

Chapter 16
Molecular Marker-Based Selection Tools in Spring Bread Wheat Improvement: CIMMYT Experience and Prospects

Susanne Dreisigacker, Sivakumar Sukumaran, Carlos Guzmán, Xinyao He, Caixa Lan, David Bonnett and Jose Crossa

Abstract Wheat is a staple food for the major part of the world's population. For wheat and other crops, it is generally agreed that in order to meet future challenges in food production, multifaceted breeding approaches are needed, including the use of current available genomics resources. Since more than three decades, molecular markers have acted as a versatile genomics tool for fast and unambiguous genetic analysis of plant species of both diploid and polyploid origin. Together with decreasing marker assay costs and interconnected genotyping service facilities, the opportunity to apply marker-assisted selection (MAS) strategies is becoming accessible to more and more breeding programs. We describe the use of molecular markers in wheat breeding with emphasis on the status of MAS in the CIMMYT global wheat program and will share our experience on recently developed prediction methods using genome-wide markers to archive genetic gain for more complex traits.

S. Dreisigacker (✉) · S. Sukumaran · C. Guzmán · X. He · C. Lan · J. Crossa
CIMMYT Int., Apdo. Postal 6-641, 06600 Mexico, DF, Mexico
e-mail: S.Dreisigacker@cgiar.org

S. Sukumaran
e-mail: S.Sukumaran@cgiar.org

C. Guzmán
e-mail: C.Guzman@cgiar.org

X. He
e-mail: X.He@cgiar.org

C. Lan
e-mail: C.Lan@cgiar.org

J. Crossa
e-mail: J.CROSSA@cgiar.org

D. Bonnett
Bayer Crop Science, Fargo, ND 58104, USA
e-mail: david.bonnett@bayer.com

V.R. Rajpal et al. (eds.), *Molecular Breeding for Sustainable Crop Improvement*, Sustainable Development and Biodiversity 11,
DOI 10.1007/978-3-319-27090-6_16

Keywords Genomic selection · Marker-assisted selection · Wheat · Wheat abiotic stresses · Wheat diseases · Wheat quality

16.1 Introduction

Wheat is a staple food for the major part of the world's population. Significant growth in productivity and the production of wheat has been achieved in the past through conventional plant breeding. However, the wheat breeding community still faces formidable challenges in advancing yield potential and yield stability, especially in light of future climate change scenarios (Reynolds et al. 2009). It is therefore generally agreed that in order to meet future challenges in food production, multi-disciplinary, multifaceted breeding approaches are needed, including the use of current available genomics resources. This chapter describes the use of molecular markers in spring bread wheat breeding with emphasis on the status of marker-assisted selection (MAS) at the International Maize and Wheat Improvement Centre (CIMMYT). The number of markers known to be associated with quantitative trait loci (QTL) or genes for major economic traits in wheat has been growing during the last decade, and marker discovery will be further accelerated with the availability of a high-quality reference sequence of the wheat genome (Choulet et al. 2014). Together with decreasing marker assay costs and interconnected genotyping service facilities, the opportunity to apply MAS strategies is becoming accessible to more and more breeding programs. We have not attempted a comprehensive review of the literature related to the future potential of genomics resources in wheat improvement nor on the detailed biology of each described trait. In the context of wheat production challenges, this chapter seeks to provide insights into the current use of molecular markers as a progressing selection tool in the hands of wheat breeders. We will briefly describe how to optimize MAS strategies and how MAS is currently used at CIMMYT for major trait categories such as biotic stresses and quality traits and will share our experience on recently developed prediction methods using genome-wide markers to archive genetic gain for more complex traits.

16.2 Marker-Assisted Selection (MAS) Strategies

MAS allows for the selection of QTL or genes that control traits of interest and can supplement conventional breeding to increase genetic gain. The application of efficient MAS strategies in a breeding program can substantially cut down population sizes, allow selection for a maximum number of target loci, and thus reduce the time and cost needed to recover a desirable genotype. However, for a MAS strategy to be appropriate, several factors related to the trait and breeding approach need to be considered.

Depending on the trait selected for, empirical comparisons of MAS and phenotypic selection for increasing genetic gain revealed different results. In some studies, MAS has reported to archive higher selection gains than phenotypic selection (Abalo et al. 2009; Kuchel et al. 2007a; Miedaner et al. 2009). Other studies considered the two methods as equally effective (Moreau et al. 2004). In a third group of studies, phenotypic selection proved to be more efficient than MAS (Davis et al. 2006; Wilde et al. 2007). Most of the studies concluded that using MAS is most appropriate when the target trait (1) shows low heritability, (2) is difficult and cost-prohibitive to measure, or (3) requires desired pyramiding of a number of genes. Every breeding program has its own set of breeding objectives and its own way to measure a trait; therefore, the choice of traits for MAS and to be combined with phenotypic selection is individual for each breeding program and might vary between programs. The CIMMYT wheat breeding program targets breeding objectives of global relevance. Traits that have been targeted for MAS in CIMMYT spring bread wheat include mainly biotic stresses, quality traits, traits related to grain yield and plant development.

In every breeding program, modern varieties are combinations of alleles that have been assembled over multiple cycles of crossing and selection. A cross made with the aim of producing a variety will have parents with many alleles in common controlling these characters, and simple crosses or top crosses will be made. If parents have a lower coancestry and differ for a greater number of alleles, genetic variation of the progenies will increase, but it will be difficult to produce a line suitable for release as a variety from a simple biparental cross (Longin and Reif 2014). In these latter cases, or where one parent contributes only a small number of desirable attributes and the other contribute many more, one or more backcrosses may be necessary to recover a commercially viable line. Therefore, in addition to the target trait, a careful planning of the integration of marker and phenotypic selection depending on the breeding approach is crucial to maximize overall genetic gains.

16.2.1 Marker-Assisted Allele-Enrichment Strategies in Early Generations

In the commonly used breeding methods for self-pollinating crops, selecting desirable plants begins in early generations for traits of higher heritability. For traits of low heritability, selection is often postponed until the lines become more homozygous in later generations (F_5 or F_6). Selection of superior plants involves visual assessment for agronomic traits or resistance to stresses, as well as laboratory tests for quality and other traits. To improve early generation selection, markers should decrease the number of plants retained due to their early generation performance, and at the same time, they should ensure a high probability of retaining superior lines for selection in later generations (Eathington et al. 2007). If markers are cheaper, more heritable, easier, or more accurate to select for than the target trait, they should be used in earlier stages of a breeding program to increase the frequency of the alleles favorable for the trait under selection.

The principle of F_2-enrichment introduced by Bonnett et al. (2005) is applied at CIMMYT after parental material is first characterized with markers for known genes to identify those parents with favorable alleles, which are then selectively combined in crosses. The concept of F_2-enrichment is illustrated in Fig. 16.1. In a F_2-population from a cross, at every polymorphic locus, 3/4 of F_2-individuals will carry at least one copy of the preferred A allele. Both AA and Aa individuals will produce the preferred AA homozygous (fixed) progeny and should be retained in the population. Individuals with the aa genotype cannot produce the AA progeny and should be culled from the population. Culling aa and retaining both AA and Aa increase the frequency of the A allele from 1/2 to 2/3 and thus enrich the frequency of the A allele in the population. If no further selection was applied and the population was progressed to homozygosity by inbreeding or production of double haploids (DHs) from selected F_2's, the frequency of AA genotypes in the final population would be 2/3 and the frequency of aa only 1/3. Alternatively, to retaining all carriers of the desirable allele (AA, Aa), only the AA homozygous individuals could be retained in the population. The locus would not further segregate in the progeny; however, as the frequency of the AA individuals is only 1/4, one half less of F_2-individuals would be overall retained.

The advantage of F_2-enrichment becomes even more apparent with greater numbers of polymorphic loci (B, C, D, etc.), the breeding target being to pyramid or combine several genes into one single genotype. The difference in the frequencies needed to recover only homozygotes (fixation of the target allele) versus carriers (enrichment of the target allele) of all desirable alleles becomes larger. For example,

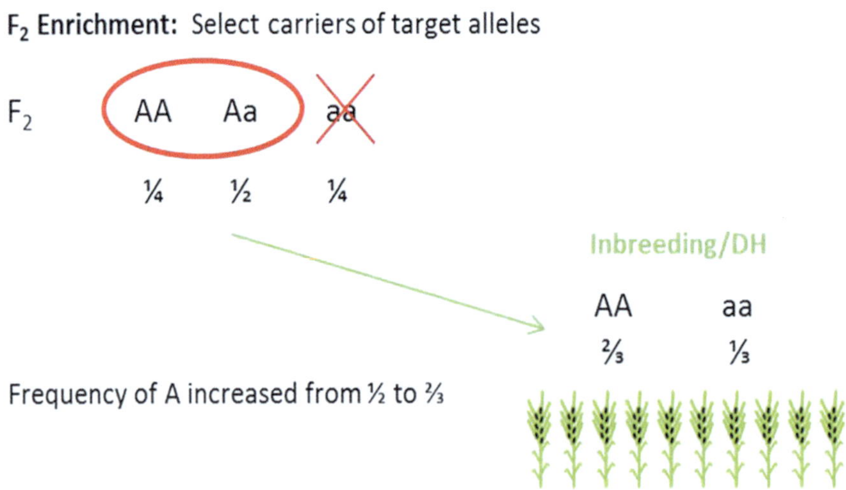

Fig. 16.1 Schematic representation of the F_2-enrichment strategy

in a population segregating at two loci, A and B, the frequency of the preferred AABB homozygote is much smaller at just 1/16 than the frequency of A-B-carriers (AABB, AABb, AaBB, AaBb) at 9/16. With n polymorphic loci, the frequency of homozygotes in F_2 is $(1/4)^n$ and the frequency of carriers is $(3/4)^n$. F_2-enrichment will increase the frequency of desirable homozygotes to $(2/3)^n$ in inbred or DH lines produced from the selected F_2's that carry at least one copy of the target allele at all loci. At CIMMYT, A-B-carriers are usually advanced via bulk breeding. In cases when three loci are combined, several bulks are sometimes advanced, e.g., a bulk that includes the A-B-carriers of all three loci but also a bulk that includes the A-B-carriers of two of the three loci.

Each selected F_2 will need to produce several progenies to make up the required number of lines in subsequent generations. Each selected F_2 should contribute equal numbers of progeny to the subsequent population in order to avoid changes in allele frequencies due to genetic drift. Table 16.1 shows the minimum number of progenies needed to recover one genotype homozygote for the target locus when using F_2-enrichment in a biparental cross in the F_2-generation and in later generation populations derived from the selected F_2's. For comparison, it also shows the population sizes needed to recover homozygotes in different generations when enrichment has not been applied.

In certain cases, backcross (BC_1F_1) or topcross (TCF_1) populations are made to combine genes of interest. If markers are going to be used in BC_1F_1 or TCF_1-populations, the desired alleles or allele combinations are of lower frequency. For example, desirable alleles coming from the non-recurrent or donor line will have a frequency of 1/4 in BC_1F_1 or TCF_1-populations, and half of the population will lack the allele. Selection among BC_1F_1 or TCF_1 populations will increase the frequency of target alleles from donors from 1/4 to 1/2 and ensure all selected individuals carry one copy of all target allele. If followed by F_2-enrichment, the frequency of donor alleles is increased from 1/4 to 2/3. Table 16.2 shows the frequencies of carriers and homozygotes for target alleles at single loci with a range of common initial allelic frequencies in different generations. This table can be used to calculate frequencies of carriers or homozygotes that can be selected in a desired generation.

In populations with differing frequencies of target alleles at different polymorphic loci, the frequency of an individual with a particular genotype across all loci can be calculated by multiplying the individual frequencies at each locus. For example, in a biparental population in which F_2-enrichment has been applied for target alleles at 6 loci, the frequency of a genotype homozygous at all loci in the F_4-generation is $0.583^6 = 0.060$. In a similar backcross population in which target alleles at 4 loci coming from the recurrent parent and 2 from the donor with enrichment applied in the BC_1F_1 for donor alleles and in F_2 for donor and recurrent parent alleles, the frequency of an individual in a DH population developed following F_2-enrichment would be 0.67^2 (donor alleles) $\times$ 0.857^4 (recurrent parent alleles) in both BC_1F_1 (increasing frequency at each locus from 1/4 to 1/2) and subsequent enrichment in F_2 increasing the frequency of these donor alleles from 1/2 to 2/3. Enrichment of the recurrent parent alleles in BC_1F_2 increases their frequency from 3/4 to 7/8. In spite of the relatively high frequency of homozygotes

Table 16.1 Population sizes required for fixation (fix) versus enrichment (enrich) of target alleles in biparental F_2-populations and to obtain at least one target homozygous genotype in later generation enriched (enrich) and non-enriched (rand) populations for different numbers of segregating loci (Bonnett et al. 2005)

	Pop. required for fix versus enrich (p = 0.05)		Population size required to obtain a target homozygote at all loci in rand and enrich populations (p = 0.05)									
Gen:	F_2		$F_{2:3}$		$F_{3:4}$		$F_{4:5}$		$F_{5:6}$		DH	
Loci:	Fix	Enrich	Rand	Enrich	Rand	Enrich	Rand	Enrich	Rand	Enrich	Rand	Enrich
1	11	3	7	5	6	4	5	3	5	3	5	3
2	47	4	20	11	15	8	13	6	12	6	11	6
3	191	6	56	23	35	14	28	11	25	10	23	9
4	766	8	151	47	81	25	61	18	53	16	47	14
5	3067	11	403	95	186	43	131	30	111	26	95	22
6	12,270	16	1076	191	426	75	281	49	231	40	191	33
7	49,081	21	2872	382	975	129	601	79	478	63	382	50
8	196,327	29	7659	766	2231	222	1284	128	988	98	766	76
9	785,312	39	20427	1533	5100	382	2741	205	2040	152	1533	114
10	3,141,252	52	54473	3067	11660	656	5848	329	4213	236	3067	172

Table 16.2 Frequencies of homozygotes (homo) and carriers of a target allele (A) for different allele frequencies and levels of inbreeding (Bonnett et al. 2005)

Allelic frequency	1/4 (e.g., non-recurrent parent allele in BC_1)		1/2 (e.g., biparental cross)		3/4 (e.g., recurrent parent allele in BC_1)		2/3 (e.g., following F_2-enrichment of biparental cross)		7/8 (e.g., following F_2-enrichment of recurrent parent allele in BC_1)	
Gen	Homo (AA)	Carrier (A−)	Homo (AA)	Carrier (A−)	Homo (AA)	Carrier (A−)	Homo (AA)	Carrier (A−)	Homo (AA)	Carrier (A−)
F_2	0.125	0.375	0.25	0.75	0.625	0.875	0.333	1	0.714	1
F_3	0.188	0.313	0.375	0.625	0.688	0.813	0.5	0.833	0.786	0.929
F_4	0.219	0.281	0.438	0.563	0.719	0.781	0.583	0.75	0.821	0.893
F_5	0.234	0.266	0.469	0.531	0.734	0.766	0.625	0.708	0.839	0.875
F_6	0.242	0.258	0.484	0.516	0.742	0.758	0.646	0.688	0.848	0.866
F_7	0.246	0.254	0.492	0.508	0.746	0.754	0.656	0.677	0.853	0.862
F_8	0.248	0.252	0.496	0.504	0.748	0.752	0.661	0.672	0.855	0.859
F_9	0.249	0.251	0.498	0.502	0.749	0.751	0.664	0.669	0.856	0.858
F_{10}	0.25	0.25	0.499	0.501	0.75	0.75	0.665	0.668	0.857	0.858
DH	0.25	0.25	0.5	0.5	0.75	0.75	0.667	0.667	0.857	0.857

for the recurrent parent alleles in a backcross, enrichment still requires smaller population sizes than selection of homozygotes. For more information on application of allele enrichment, refer to the publication of Bonnett et al. (2005). On high priority materials, at CIMMYT, we apply marker assays additionally at the F_4- or F_5-generation to ensure a high frequency of advanced progeny containing the alleles of interest.

In reality, markers or efficient phenotypic screens will rarely be available for allele's at all important loci segregating in a cross and it will not be possible to enrich frequencies of these alleles in early generations. Early generation selection strategies must therefore be designed to retain important allelic variation until later stages of the breeding process to select for more complex traits like yield that require homogeneous lines, large seed quantities, and expensive phenotypic screens to achieve acceptable heritability. Estimating the number of important polymorphic loci or deciding on a certain number of inbred lines to be retained for phenotypic selection such as in Table 16.1 can be implemented to optimize overall required population sizes. Required population sizes to recover an individual with a target genotype are inversely related to the frequency of those individuals. A formula for calculating population size for any frequency and desired level of confidence of recovery was given by Hanson (1959):

$$N = \frac{\log_n(x)}{\log_n(1 - G)},$$

where N is the population size, x is the specified probability of failure, and G is the genotypic frequency. A useful rule of thumb is to multiply the inverse of the frequency by 3 to achieve a commonly desired 95 % probability of recovery. For example, with a frequency of 1/16, the population size needed for 95 % probability of recovering the target genotype is 16 × 3 = 48. In other words, population size = (1/frequency of target genotype) × 3. This formula applies regardless of whether the target genotype is homozygous or heterozygous. Often, the number of important loci contributing variation to important traits in a cross will not be known, and partial enrichment is applied by estimating the number of important polymorphic loci or deciding on a certain number of inbred lines to retain for phenotypic selection. For example, measures can be translated to a partial enrichment strategy where, e.g., six loci are taken into account for enrichment plus additional four important polymorphic loci for which markers are not available.

16.2.2 Marker-Assisted Backcrossing (MABC)

As outlined above, if parents have a low coancestry, genetic variation of the progeny will increase, but it will be difficult to produce a line suitable for release as a variety from a simple biparental cross. Various generations of backcrossing are therefore used to transfer a desired trait from a rather unadapted donor plant into an elite genotype (recurrent parent) until most of the genes stemming from the donor are eliminated (Becker 1993). For example, wild relatives in wheat are crossed with current breeding lines to transfer a desired trait.

Markers can be used in the context of MABC to either control the target gene (foreground selection) or accelerate the reconstruction of the recurrent parent genotype (background selection). According to Tanksley et al. (1989), in traditional backcross breeding, the reconstruction of the recurrent parent genotype requires more than six generations, while this may be reduced to only three generations in MABC. Similarly, Hospital et al. (1992) and Ribaut and Hoisington (1998) concluded that employing molecular markers with known map position can speed up the recovery of the recurrent parent genome by about two to three generations. These findings are confirmed by the results of Frisch et al. (1999), who showed in a computer simulation that MAS can reconstruct a level of recurrent parent genome in BC_3 which would only be reached in BC_7 without the use of markers. Prigge et al. (2008) compared simulated and experimental data of a MABC program in rice and revealed good agreement.

The effectiveness of MABC depends on the availability of closely linked markers/flanking markers for the target loci, the size of the population, the number of backcrosses, and the position and number of markers for background selection. A straight forward way to accomplish MABC is the two-stage selection strategy. In BC_1F_1 populations, heterozygote genotypes at the target loci are first identified reducing the population size for further screening (foreground selection). For the background selection step, individuals with the fewest number of background markers from the donor parent are then selected. The upper limit of the number of background markers is defined by the number and length of the chromosomes. In rice and sugar beet, 50–60 background markers resulted in efficient selection response (Frisch and Melchinger 2005; Prigge et al. 2008). Markers should be evenly distributed to reflect all proportions of the genome. In subsequent backcross generations, selection is carried out to the same scheme, but only those markers are analyzed which have not been fixed for the recurrent parent in the preceding generation.

In BC_1F_1 populations, MABC would be more efficient for larger populations. Larger population sizes in earlier generations are also of advantage for more quantitative traits. However, larger population increase the number of marker data points required and hence the cost. In comparison with BC_1F_1 populations, the number of markers that needs to be analyzed in later backcross generations is lower. In a two-stage selection strategy, increasing the population size with the number of backcross generations reduces the number of marker loci and cost with comparable percentages of recovery of the recurrent parent genome (Frisch et al. 1999). Prigge et al. (2009) additionally showed that the approach of increasing population sizes in advanced backcross generations can be refined by sequentially increasing marker densities.

Two additional selection steps can follow the two-stage selection strategy. As a third step, after preselecting the individuals with the target loci, individuals can be analyzed for the two markers flanking the target locus. Individuals with one or two flanking markers fixed for the recurrent parent's allele are retained and then analyzed for the remaining markers. In some publication, this three-stage selection strategy is also called ‘recombinant selection’. As a fourth step, individuals with the maximum number of markers fixed on the chromosome of the target locus can be selected before analyzing all other remaining markers. These two steps provide an option which significantly reduces the number of data points required in comparison with the two-stage selection strategy.

16.2.3 Other Factors to Consider When Applying MAS

16.2.3.1 Imperfect Linkage Between Markers and Target Alleles

Due to the increased marker availability, recent genetic maps are dense; however, markers are mostly not perfectly linked with the target allele, which reflects the accuracy of MAS. For example, if the genetic distance between the marker and the target allele is 5 cM, on average five recombinants occur in a set of 100 progenies. In such cases, flanking markers can be very useful to decrease the probability of recombinants between target alleles and markers. If two flanking markers with a genetic distance of 5 cM to the target allele are applied, on average only 1 recombinant occurs in a set of 100 progenies. If imperfect markers are used in F_2-enrichment, the change in allele frequency will be slightly less than if markers were perfect. In spite of a slight reduction in efficiency, the use of imperfect markers still increases allele frequencies and is very worthwhile.

16.2.3.2 Dominant Versus Codominant Markers

Markers can be dominant or codominant, the latter being able to distinguish heterozygote and homozygote carriers of the target allele. Due to improved marker technologies, most of the more recent developed single-nucleotide polymorphism (SNP) markers for relevant genes in wheat are codominant. The advantages of codominant markers in F_2-enrichment are that they allow a more direct assessment of the frequencies of target alleles that they remove the need for progeny testing of selected later generation individuals (e.g., F_5 or F_6) to recover homozygotes. When dominant markers are used and progeny testing is not done, some selected individuals will be heterozygous for some of the target alleles. However, because the frequency of heterozygous individuals is halved with each generation of inbreeding, only relatively small numbers of selected F_6-individuals would be heterozygous at any of the target loci. In MABC, the advantages of codominant markers are more evident. For the background selection step in MABC, loci homozygote for the recurrent parent can be identified.

16.2.3.3 Linkage Between Two Alleles

With greater numbers of markers available for selection, it is inevitable at some point that a cross will involve target alleles that are linked. If the alleles are linked in coupling, they will behave more like a single gene and required population sizes will be smaller than if they were unlinked. If the alleles are linked in repulsion and a crossover between the loci is necessary to bring the target alleles together on the same chromosome, required population sizes will be considerably larger. For example, the wheat stem rust gene *Sr2* and the fusarium head blight gene *Fhb1* are linked in repulsion on chromosome 3BS (Anderson et al. 2007).

16.2.3.4 Polymorphism and Genetic Backgrounds

Ideally, a marker should be highly polymorphic in breeding materials and discriminate between different genotypes. In some cases, the target polymorphism of a marker is only specific in certain donors (e.g., for the stem rust resistance genes *Sr13* or *Sr47*, http://maswheat.ucdavis.edu/) and therefore not diagnostic in all genetic backgrounds. These markers cannot be used for the screening of unknown sets of germplasm of a breeding program. They can be used to follow a target allele in segregating populations including the parental line known to carry the target allele and the marker showing polymorphism between the carrier and non-carrier of the allele.

In QTL mapping experiments, parents that represent the extreme ends of a trait phenotype are chosen. The effect of the QTL might therefore be less significant when used for introgression into an elite breeding line. In other cases, the effect of a locus may differ in different genetic backgrounds due to the interaction with other loci (epistasis) (Holland 2001).

16.2.3.5 Environmental Effects and Genome Structure

While the effect of a QTL appears to be consistent across environments, the magnitude of the effect may vary. The extent of the QTL × environment interaction is often unknown because the mapping studies have been limited to only a few years or locations (Wang et al. 2007).

Markers can be identified and developed using populations where parents do not represent adapted germplasm, such as diploid or tetraploid wheat species. In such cases, the known polymorphism can be of little practical value although it is transferred to wheat through interspecific hybridization. Many key traits for wheat improvement present in alien segments have been transferred to wheat. The alien segments, however, are often large and can carry undesired characters in addition to the favorable trait, such as the high grain protein content gene *Gpc-B1* transferred from *Triticum turgidum* ssp. *dicoccoides*, which is negatively correlated with grain yield (Uauy et al. 2006). Recombination within these alien segments is very low, and advanced approaches that reduce the large linkage blocks are needed.

16.2.3.6 Logistic and Cost of MAS

Typically, breeding programs grow hundreds of populations and many thousands of individual plants. Given the extent and the complexity of selection required in breeding programs, one can easily appreciate the usefulness of new tools that may assist breeders in plant selection. The scale of the breeding programs, however, also underlines the challenges of incorporating MAS. A close relationship between breeders and molecular biologists supports the level of integration of MAS. For example, it is vital that the robustness and reliability of the markers available for genes or alleles of interest are evaluated before considering their routine application. Lack of confidence in published information is cited as one of the reasons that limit the use of markers in practical plant breeding (Kuchel et al. 2003). Leaf tissue collected in the field has also to be brought to the laboratory in time to provide the marker data to the breeders prior to selection or harvest.

Despite the recent shift to SNP-based platforms, e.g., KASP in wheat (http://www.cerealsdb.uk.net/) at least at CIMMYT, the cost of marker assays remains the rate-limiting factor for the adoption of MAS. Taking advantage of that present specialized genotype service, providers can evade the requirements of large capital investments for the acquisition of equipment and the regular labor expenses, and drastic reductions in assay costs are, however, difficult to achieve unless very large numbers of marker assays are deployed. For service providers and genotyping platforms, the cost per marker assay is associated with the sample volume. With its current sample volume, CIMMYT is reaching a minimum of 0.2 USD cost per SNP assay, while one-tenth of the cost would be desired.

16.3 Current Use of MAS for Wheat Biotic Stresses

16.3.1 Rust Resistance

Wheat leaf (brown) rust, stripe (yellow) rust, and stem (black) rust caused by *Puccinia triticina* (*Pt*), *P. striiformis* f. sp. *tritici* (*Pst*), and *P. graminis* f. sp. *tritici* (*Pgt*), respectively, are the most widely distributed wheat (*Triticum aestivum* L.) diseases in the world. They can cause yield loss up to 40 % of the wheat production area in countries such as Mexico, India, Pakistan, Bangladesh, and China (Dubin and Brennan 2009), and more than 20 stripe rust epidemics have been documented worldwide (Wellings 2011). Stem rust has historically been a big threat to global wheat production (Saari and Prescott 1985). A new *Pgt* race, TTKSK (commonly referred to as Ug99), detected in Uganda in 1998, had virulence to most of the widely deployed specific resistance genes and was seen as threat to global food security (Li et al. 2014). There are ways to manage these diseases, e.g., fungizide treatments, however, development and using resistant cultivars is the most efficient method to control them.

Two types of rust resistance genes are often defined in wheat. Race-specific resistance genes usually confer protection throughout the growth cycle, and therefore, resistance conferred by them is also called all-stage resistance (Chen et al. 2013). These resistance genes cause various degrees of hypersensitive reactions in the host if the pathogen possesses corresponding avirulence genes (Flor 1942). In contrast, race non-specific minor genes that confer adult plant resistance (APR) are usually present together with other similar genes and therefore associated with quantitative inheritance (Johnson and Law 1973). Most cultivars with multiple genes for APR are susceptible at the seedling stage but later display resistance to a number of races (Bjarko and Line 1988). For the last decade, identification of new sources of rust resistance has gained high priority in the wheat scientific community, e.g., via the Borlaug Global Rust Initiative (http://globalrust.org/). As a result, several new sources of resistance have been identified in global wheat germplasm, including CIMMYT by QTL mapping and association mapping approaches.

Tables 16.3, 16.4, and 16.5 provide an overview of available molecular markers for seedling resistance genes, APR genes with pleiotropic effect, and APR genes, respectively, present in CIMMYT wheat germplasm. Single gene resistance can usually be selected phenotypically in the greenhouse. For quantitative disease resistance, MAS can be very useful for pyramiding individual small-effect QTL or small and large-effect QTL in elite wheat lines.

Due to the vast availability of molecular markers demonstrated in Tables 16.3, 16.4, and 16.5, the screening of parental material, key breeding germplasm, and MAS for improved rust resistance represents a main part of molecular breeding work at CIMMYT. During the last years, CIMMYT has started to routinely evaluate its spring bread wheat international screening nurseries and other set of elite lines with some of the associated markers. The race non-specific slow rusting APR genes, such as *Lr34/Yr18/Sr57/Pm38*, *Lr46/Yr29/Sr58/Pm39*, *Lr67/Yr46/Sr55/Pm46*, and *Sr2/Yr30,* have been the major foundation of durable resistance in CIMMYT's germplasm. Early generation MAS is applied on a project bases to combine these genes with each other or with additional seedling resistance genes. Table 16.6 shows the results of a F_2-enrichment strategy for UG99 stem rust resistance. The total 305 lines in the 31st Semi-Arid Wheat Yield Screening Nursery (SAWYSN) are listed according to the environment they have been selected in and according to the selection method, conventional (Conv) selection or MAS, that has been used. The lines within each selection environment and scheme were subsequently classified according their disease rating. The main target for MAS was to combine two to three markers linked to the genes *Sr25*, *Sr1A1R*, *Sr24*, *Sr26*, and *Sr42/SrCad* in diverse CIMMYT wheat backgrounds. The lowest number of the SAWYSN, in total 40 lines, was derived from the MAS program in Mexico. However, within these 40 lines, the percentage of lines with R and R-MR ratings was highest (90 %) and about

Table 16.3 Available molecular markers for seedling resistance genes to rusts in the CIMMYT wheat breeding program

Gene	Marker name	Source	Reference
Lr16	*gwm210, wmc661*	Francolin#1	Lan et al. (2014)
Lr19/Sr25	*Psy1-E, wmc221*	Agatha, Misr#1	Zhang and Dubkovsky (2008), Dreisigacker, pers. comm.
Lr21	*D14*		Talbert et al. (1994)
Lr42	*cfd15, wmc432*	Quaiu#3	Basnet et al. (2014a)
Lr47	*CIMwMAS0055*		Helguera et al. (2000), Dreisigacker, pers. comm.
Lr51	*S30-13L/AGA7-759R*		Helguera et al. (2005)
Yr17/Sr38/Lr37	*CIMwMAS004*	Milan	Helguera et al. (2003), Dreisigacker, pers. comm.
Yr24/26	*We173*	Chuanmai 42	Wang et al. (2008)
Yr41	*gwm410, gwm374*	Chuannong 19	Luo et al. (2008)
Yr43	*wgp110, wgp103, barc139*	ID0377S	Cheng and Chen (2010)
Yr44	*pWB5/N1R1, wgp100, gwm501*	Zak	Cheng and Chen (2010)
Yr50	*gwm540, barc1096, wmc47, wmc310*	CH223	Liu et al. (2013a, b)
Yr60	*wmc776, wmc313, wmc219*	Lal Bahadur	Herrera-Foessel, pers, comm.
YrF	*gwm374, wmc474*	Francolin#1	Lan et al. (2014)
YrSuj/Yr67	*cfa2040, wmc526*	Sujata, C591	Lan, pers. comm.
Sr1BL.1RS	*SCM9*	Seri 82	Weng et al. (2007)
Sr13	*barc104, dupw167, CD926040, BE471213*	Kofa, Kronos	Simons et al. (2011)
Sr22	*wmc633, cfa2123*	Sr22 Tb, Steinwedel	Olson et al. (2010)
Sr23	*gwm210*	AC Domain	McCartney et al. (2005)
Sr26	*Sr26#43, BE518379*	WA1	Mago et al. (2005)
Sr33	*barc152, cfd15, BE405778, BE499711*	RL5288	Sambasivam et al. (2008), Periyannan et al. (2013)
Sr35	*cfa2170, cfa2076, wmc169, wmc559*	G2919	Zhang et al. (2010a, b)
Sr42/SrCad	*FSD_RSA*	Norin 40	Ghazvini et al. (2012)
Sr-6DS	*gpw5182, cfd49*	Niini, Coni, Blouk	Lopez-Vera et al. (2014a, b)
SrND643	*gwm350, wmc776, wmc219*	ND643	Basnet et al. (2014b)
SrHuw234	*wmc332*	Huwa	Lopez-Vera et al. (2014a, b)
SrYanac	*barc200*	Yaye	Lopez-Vera et al. (2014a, b)

Table 16.4 Available molecular markers for pleiotropic adult plant resistance (PAPR) genes to rusts in the CIMMYT wheat breeding program

Gene	Marker name	Source	Reference
Lr34/Yr18/Pm38/Sr57	*wMAS000003, wMAS000004*	Parula	Lagudah et al. 2009, http://www.cerealsdb.uk.net/
Lr46/Yr29/Pm39/Sr58	*csLv46, csLV46G22*	Pavon 76, Kenya Kongoni	Lagudah, pers. comm.
Lr67/Yr46/Pm46/Sr55	*csSNP856*	RL6077	Forrest et al. (2014)
Sr2/Yr30	*wMAS000005*	Pavon76	Mago et al. (2011), http://www.cerealsdb.uk.net/
Sr2/Yr30	*gwm533*	Pavon76	Spielmayr et al. (2003)

Table 16.5 Available molecular markers for adult plant resistance (APR) genes to rusts in the CIMMYT wheat breeding program

Gene	Marker name	Source	Reference
Lr68	*cs7BLNLRR, CIMwMAS0056*	Parula	Herrera-Foessel et al. (2012), Dreisigacker, pers. comm.
Yr54	*gwm301*	Quaiu#3	Basnet et al. (2014a)
Yr36	*wMAS000017*	Glupro	Uauy et al. 2005, http://www.cerealsdb.uk.net/
Yr39	*wgp36, wgp45, gwm18, gwm11*	Alpowa	Lin and Chen (2007)
Yr52	*barc182, wgp5258*	PI 183527	Ren et al. (2012)
Yr59	*wgp5175, bac32, bac182*	PI178759, PI 660061	Chen, pers. comm.
Sr56	*Sun209, Sun320*	Arina	Bansal et al. (2014)

Table 16.6 Disease rating for stem rust in the 31st Semi-Arid Wheat Yield Screening Nursery

Selection environment		Total	Disease rating for stem rust						
			R	R-MR	MR	MR-MS	MS	MSS	S
Mexico—Conv	No	159	25	52	30	27	22	3	0
	%		15.7	48.4	67.3	84.3	98.1	100.0	
Mexico—MAS	No	40	23	13	4	0	0	0	0
	%		57.5	90.0	100.0				
Kenya—Conv	No	106	44	46	10	5	1	0	0
	%		41.5	84.9	94.3	99.1	100.0		
ALL	No	305	92	111	44	32	23	3	0
	%		30.2	66.6	81.0	91.5	99.0	100.0	

Lines were classified according the selection environment and using conventional selection or MAS twofold larger than the percentage of lines with the same ratings derived from conventional selections in Mexico (48.4 %), clearly showing the effectiveness of MAS. A high percentage of lines with R and R-MR ratings were also derived from conventional selection in Kenya due to the large selection pressure that is obtained at that site underlining the importance of disease hot spot regions for selection.

Some genes currently not present in the CIMMYT elite germplasm are additionally introgressed. For example, *Yr36*, *Lr67/Yr46/Sr55*, *Sr33,* and *Sr35* will be introduced into advanced breeding lines using MABC. Resistance genes introduced to or derived from bread wheat *Lr19/Sr25* and *Lr47* were also introgressed in different durum genetic backgrounds via MABC. Furthermore, *Lr19/Sr25* and *Sr22*, both genes tightly linked on the long arm of chromosome 7A were combined in durum wheat backgrounds via the F_2-enrichment strategy. MAS is therefore enhancing and strengthening the resistance diversity in CIMMYT wheat germplasm with multiple gene pyramiding and will provide a better foundation for future breeding to reach the goal of developing durably resistant high-yielding wheat varieties.

16.3.2 Fusarium Head Blight (FHB)

Fusarium head blight (FHB) is a globally important wheat disease, with major epidemic regions being North America, Europe, East Asia, and the South Cone of South America. *Fusarium graminearum* (teleom. *Gibberella zeae*) is the most important causal agent worldwide. Besides yield reduction, FHB produces a set of mycotoxins, particularly deoxynivalenol (DON), which is harmful to both human and livestock. In most developed countries, legally enforceable DON limits in wheat grain and food products have been set, reflecting concerns for food safety.

Host resistance is the most important component in the disease management system, although other measures such as fungicide and cultural practices should also be considered to achieve a satisfactory control (Gilbert and Haber 2013). There are three major difficulties for breeding FHB resistance varieties: (1) the multigenic control of host resistance and a lack of functional markers; (2) limited resistance sources in adapted elite germplasm; and (3) multiple resistance components. Numerous host resistance mechanisms have been proposed, each having its own evaluation methods. The most famous resistance components are Type I for initial infection and Type II for disease spread in spike tissues (Liu et al. 2009).

FHB is a quantitatively inherited disease, making the application of MAS in this disease more difficult than in qualitatively inherited traits. Until now, more than hundred published studies have been performed to identify FHB resistance QTLs, which have been mapped to all the 21 wheat chromosomes, with various phenotypic effects (Buerstmayr et al. 2009; Liu et al. 2009). Until now, five QTLs have been fine-mapped and designated (Table 16.7), but none has been cloned and mostly only flanking SSR markers are available.

Table 16.7 Information on the five nominated FHB resistance genes

Gene	Source	Chromosome	Resistance type	Flanking markers	References
Fhb1	Sumai 3	3BS	II	*gwm533* and *gwm493*	Cuthbert et al. (2006)
Fhb2	Sumai 3	6BS	II	*gwm133* and *gwm644*	Cuthbert et al. (2007)
Fhb3	*Leymus racemosus*	T7AL·7Lr#1S	II	*BE586744, BE404728, BE586111*	Qi et al. (2008)
Fhb4	Wangshuibai	4B	I	*barc20* and *wmc349*	Xue et al. (2010)
Fhb5	Wangshuibai	5A	I	*barc56* and b*arc100*	Xue et al. (2011)

Fhb1 identified in Sumai 3 is the most well-studied FHB resistance gene. After its fine mapping, a codominant STS marker, *umn10*, was developed based on a polymorphic site near the candidate gene region. Recently, SNP markers, *Xsnp3BS-8 and Xsnp3BS-11*, were developed to facilitate high-throughput genotyping (Bernardo et al. 2012). It is noteworthy that this gene is usually in repulsive phase with the stem rust resistance gene, *Sr2*, compromising its application in breeding practices. This situation is being changed with the availability of lines in which the two genes are in coupling phase (Thapa et al. 2013). Similar lines with the two genes in coupling are recently available at CIMMYT. The 3BS chromosome region was transferred in the background of a set of adapted spring wheat lines, some of which are currently being introgressed via MABC into elite breeding germplasm.

In addition to these FHB resistance QTLs, dwarfing genes such as *Rht-B1b*, *Rht-D1b,* and *Rht8* have also been proved to be associated with FHB resistance, based on either pleiotropy, tight linkage, or disease escape (Buerstmayr et al. 2009). Although all the three aforementioned dwarfing genes reduce Type I resistance, *Rht-B1b* and *Rht8* confer less FHB susceptibility compared with *Rht-D1b*, and there was evidence showing that *Rht-B1b* is able to confer Type II resistance. Therefore, it is recommended to use *Rht-B1b* and *Rht8* in breeding practices to combine reduced plant stature with acceptable FHB resistance (Gilbert and Haber 2013).

Despite the extensive efforts on identification of resistance QTLs, limited progress has been obtained regarding the utilization of those QTLs in MAS, primarily due to the lack of functional markers. Nevertheless, there were examples where MAS was employed to develop FHB-resistant cultivars, such as the newly registered Canadian wheat cultivar 'Cardale' (Gilbert and Haber 2013). The utilization of MAS in FHB-resistant breeding at CIMMYT dates back to 2008 when a collaborative project with USDA-ARS Small Grains Genotyping Center, Fargo, was initiated (Duveiller et al. 2008). And nowadays, the 'haplotyping' system at CIMMYT comprises 17 markers for 10 validated QTLs on seven chromosomes (Table 16.8), which has facilitated the genotypic characterization of numerous CIMMYT elite lines, including the 13th and 14th FHB screening nursery (He et al.

Table 16.8 FHB markers used in CIMMYT's haplotyping system

Source	Chromosome	Resistance type	Flanking markers	Locus
Sumai 3	3BS	II	*umn10, Xsnp3BS-11, Xsnp3BS-8*	*Fhb1*
	5AS	I	*barc186* and *barc180*	*Fhb5*
	6BS	II	*gwm133* and *wmc179*	*Fhb2*
Frontana	3A	I	*dupw227*	
	5AS	I	*barc197* and *gwm129*	*Fhb5*
Wuhan 1	2DL	II	*wmc144* and *wmc245*	
	4BS	II	*wmc238* and *gwn149*	*Fhb4*
CJ 9306	2DL	II	*gwm157* and *gwm539*	
T. dicoccoides	3A	II	*gwm2*	
	7A	II	*barc121* and *wmc488*	

This table was modified from He et al. (2013b)

2013a, b). However, this system is being upgraded to incorporate markers in closer linkage with several of the QTLs, and it is predictable that the SSR and STS markers will be replaced by high-throughput SNP markers in near future.

16.3.3 Septoria Tritici Blotch (STB)

Septoria tritici blotch (STB) is a foliar blight disease that reduces yields up to 60 % under conducive environmental conditions, with Europe, North America, South America, Australia, and Central West Asia and North Africa (CWANA) region being the major epidemic regions (Raman and Milgate 2012). This disease is caused by the ascomycete fungal agent *Zymoseptoria tritici* (anamorph: *Septoria tritici*).

Host resistance to STB is reported to be both qualitative and quantitative. Although gene-for-gene interactions exist between a certain resistance genes and the corresponding pathogen isolates, the resistance conferred by each gene is weak and cannot provide sufficient protection to wheat as those in rusts and powdery mildew (Goodwin 2012). Like in other diseases, breakdown of STB resistance genes has been observed, e.g., resistance of the wheat cultivar 'Gene' was defeated only five years after its release, implying its resistance nature of 'race-specific' (Cowger et al. 2000). Thus, it is recommended to pyramid both qualitative and quantitative resistance genes in breeding materials to achieve durable resistance (Raman and Milgate 2012).

The first STB resistance gene, *Stb1*, was discovered in 1966 and designated in 1985, followed by *Stb2* and *Stb3* in 1985, and *Stb4* in 1994 (Goodwin 2012). But it is *Stb5* that was firstly mapped on a genetic map (Arraiano et al. 2001). Soon after this landmark work, 12 more resistance genes were reported in 2000s as reviewed by Goodwin (2012). In the last few years, *Stb16*, *Stb17,* and *Stb18* were identified and

Table 16.9 STB markers used in CIMMYT's haplotyping system

Gene	Flanking markers	Chromosome	Source
Stb1	*barc74*	5BL	Bulgaria 88
Stb2	*wmc230*	1BS	Veranopolis
Stb3	*gdm132*	7AS	Israel 493
Stb4	*gwm111*	7DS	Tadinia
Stb5	*gwm44*	7DS	CS/Synthetic 6x
Stb6	*gwm369*	3AS	Flame
Stb7	*wmc313*	4AL	Estanzuela Federal, Kavkaz-K4500 L.6.A.4
Stb8	*gwm577, gwm146*	7BL	Synthetic W7984, Opata m85
Stb9	*wmc317*	2B	Courtot
Stb10	*wms848*	1D	Kavkaz-K4500 L.6.A.4
Stb11	*barc008*	1BS	TE 9111
Stb12	*wmc219*	4AL	Estanzuela Federal, Kavkaz-K4500 L.6.A.4
Stb13	*wmc396*	7B	Salamouni
Stb14	*wmc500, wmc623*	3B	Salamouni
Stb16	*wmc494*	6B	SH M3, Opata M85
Stb17	*hbg247*	5A	Kavkaz-K4500 L.6.A.4
Stb18	*gpw3087, gpw5176*	6DS	Balance

mapped (Ghaffary et al. 2011, 2012), and the chromosomal localizations of *Stb2* and *Stb3* were adjusted from 3BS to 1BS and from 6DS to 7AS, respectively (Liu et al. 2013a, b; Goodwin et al. 2015). In addition to major resistance genes, several quantitative loci with minor effects (Simon et al. 2012; Kelm et al. 2012; Risser et al. 2011; Kosellek et al. 2013) have been identified. Add CIMMYT, recently two new resistant QTL were mapped on chromosomes 5AL and 3BS in two populations evaluated under field conditions in Mexico (Dreisigacker et al. 2015).

All mapped *Stb* genes but *Stb15* have at least one linked SSR marker, which enables the application of MAS (Ghaffary et al. 2011, 2012; Goodwin 2012). At CIMMYT, the haplotyping work on STB started from late 2000s on elite breeding materials and parents of mapping populations, and recently, it became a routine procedure for genotypically characterizing CIMMYT's International Septoria Observation Nurseries (ISEPTON) with linked SSRs (Table 16.9). Similar to the FHB markers, the ones for STB are also being constantly updated to incorporate new findings.

16.3.4 Other Diseases and Pests

Many other biotic stresses with global or regional importance exist in wheat. Some functional markers and many QTL associated with resistance genes for these diseases have been mapped and are applied in breeding. At CIMMYT, besides the major focus on rust, FHB, and STB, germplasm is additionally characterized for tan spot (caused by *Pyrenophora tritici-repentis* (Died.) Drechs.), stagonospora nodorum blotch (caused by *Parastagonospora nodorum* (Berk.) Quaedvlieg, Verkley and Crous), soilborne diseases such as the cereal cyst (*Heterodera avenae*) and root lesion nematode (*Pratylenchus thornei*), and Hessian fly (*Mayetiola destructor*).

16.4 Current Use of MAS for Quality Traits

New wheat varieties should meet specific grain quality requirements to satisfy the increasing demand for processed wheat-based foods. Grain quality is a variable concept, and its meaning depends on the type of flour to produce (whole meal flour, refined flour, semolina, etc.), the end product to manufacture (bread, biscuit, pasta, etc.), the process used to produce it (handmade, semi-mechanized, mechanized, etc.), and the consumer's preferences. In defining quality for any given end use, processing performance and end-product properties have to be considered.

Processing quality and end-product properties are determined by a set of complex traits, the most important being the endosperm texture or grain hardness, the content and composition of storage proteins (mainly glutenins), the composition of starch and non-starch polysaccharides, and, for some specific products, the color of the flour/semolina. The high variability in grain quality traits existing in wheat has led to the creation of thousands of varieties possessing many different grain composition combinations, allowing using wheat to manufacture many different types of foods.

Although the main quality traits are influenced by the environment and cropping practices, their expression is mainly controlled by qualitative genes and their allelic variations. The good association between genotype and phenotype for main grain quality parameters has made the use of these parameters possible to estimate the presence/absence of quality-related loci. Until now, functional markers for almost all important high-molecular-weight glutenin subunit (HMWGS) and low-molecular-weight glutenin subunit (LMWGS) associated with the gluten properties are available, as well as for genes related to grain hardness, protein content, flour/semolina color, and starch properties (Liu et al. 2012). Due to the complexity of analyzing some quantitative and qualitative traits using conventional non-molecular tools, these molecular markers have received attention and are being implemented in some breeding programs including CIMMYT.

16.4.1 Gluten Composition

When wheat flour is mixed with water to form dough, the storage proteins of the grain are aggregated in a viscoelastic protein network named gluten. The processing of most wheat-based products requires gluten strength and extensibility in a greater or lesser extent. Gluten elasticity or strength requirements depend on the processing conditions and the end product to be manufactured. Gluten is composed of a large number of proteins, mainly glutenins and gliadins. Glutenins contribute more to gluten strength, while gliadins do to extensibility and viscosity. It is impossible to understand dough/gluten viscoelastic properties without studying both kinds of proteins independently but coexisting together in the intricate gluten protein network and in the complex dough system. Among the glutenins, the HMWGSs are codified by the *Glu-A1*, *Glu-B1,* and *Glu-D1* loci (located at the long arm of chromosomes 1A, 1B, and 1D, respectively), and the LMWGSs are codified by the *Glu-A3*, *Glu-B3,* and *Glu-D3* loci (located at the short arms of chromosomes 1A, 1B, and 1D, respectively). Different alleles for each of the glutenin and gliadin genes have been detected and classified mainly by SDS-PAGE protein electrophoresis. Most of these alleles have been associated to high or poor quality, the clearest example being the association of the *Glu-D1d* allele (subunits 5+10) with higher elasticity and extensibility than that conferred by the allele *Glu-D1a* (subunits 2+12) (Payne et al. 1987).

The six *Glu-1* and *Glu-3* loci have been already characterized at a molecular level, and molecular markers are available for almost all the alleles (see Rasheed et al. 2013 for a review). Most of them are STS-type markers, although several SNP markers have been already developed and validated (http://www.cerealsdb.uk.net/). At CIMMYT up to now, SDS-PAGE electrophoresis has resulted still more time-efficient and has a lower cost than the use of several individual markers, as low-cost multiplex assays for all observed alleles do still not exist. The lines that carry SDS-PAGE subunit bands difficult to differentiate, mainly due to band overlapping, are additionally analyzed by the specific molecular markers. Those are usually the lines carrying the subunit 7 overexpressed ($Bx\text{-}7^{OE}$), which has been shown to increase the concentration of this subunit, which is difficult to detect in an SDS-PAGE gel. Other alleles difficult to identify by SDS-PAGE are *Glu-A3f* and *e*, as well as *Glu-B3f* and *g*. In those cases, the power resolution of the molecular markers (Table 16.10) is very useful to have a concluding result.

Table 16.10 Functional markers for grain quality traits

Gene	Allele	Marker name	Reference
Glu-B1	$Glu\text{-}B7^{OE}$	*TaBAC1215C06-F517/R964* *TaBAC1215C06-F24671/R25515*	Ragupathy et al. (2008)
Glu-A3	*Glu-A3f*	*LA1F/SA6R*	Wang et al. (2010)
	Glu-A3e	*LA1F, SA5R*	
Glu-B3	*Glu-B3f*	*SB6*	Wang et al. (2009)
	Glu-B3g	*SB7*	
Wx-A1	*Wx-A1b*	*MAG264*	Liu et al. (2005)
Wx-B1	*Wx-B1b*	*GBSS*	McLauchlan et al. (2001)
Wx-D1	*Wx-D1b*	*MAG269*	Liu et al. (2005)

16.4.2 Grain Hardness

Grain hardness could be considered the most important single factor determining the general end use of a wheat cultivar (Morris 2002). Hard wheat is for bread, while soft wheat is for biscuits, and the very hard, vitreous grain of durum wheat, is suitable for pasta. The importance of grain hardness resides in its influence on the level of damaged starch resulting during flour milling; the harder the grain, the higher the level of damaged starch in the flour, and the higher the water hydration capacity of the flour (Posner 2000).

Grain hardness is controlled in common wheat by *Pina-D1* and *Pinb-D1*, two small genes without introns located at the short arm of chromosome 5D (Morris 2002). These genes are codified for two proteins named puroindolines (PINA and PINB), associated with the membrane surrounding starch granules, and that have direct role in the definition of grain hardness (see Morris and Bhave 2008 for a review). When wild forms of both proteins are present, the grain texture is soft. However, if one of the proteins is missing or has a modified amino acid sequence, the texture will be hard or semi-hard. Up to date, four and fourteen different alleles leading to hard texture have been identified for *Pina-D1* and *Pinb-D1*, respectively (McIntosh et al. 2014). All these alleles, except *Pina-D1b*, are characterized by the presence of one SNP in the coding region that either changes the ORF leading to a premature stop codon or to the change of one amino acid in the protein sequence. The most common *Pin* alleles causing hard texture are *Pina-D1b* and *Pinb-D1b*. The first one, predominant in CIMMYT germplasm (Lillemo et al. 2006), is characterized by the almost complete deletion of the *Pina-D1* gene.

Although grain hardness is easily measured in the laboratory and shows only small environmental influence, the molecular markers related to the *Pin-D1* genes are useful to determine the different allele combinations associated with different hardness levels (Martin et al. 2001; Takata et al. 2010). For example, *Pina-D1b* is associated with a harder texture than *Pinb-D1b* and in some cases with higher flour yield. Knowing the *Pin* genotype can help to develop wheat with specific and desirable texture for a specific end product. At CIMMYT, parental lines are analyzed with respect to *Pina-D1* and *Pinb-D1* with developed SNP markers (http://www.cerealsdb.uk.net/). Markers are being used to identify lines carrying the uncommon *Pinb-D1b* allele to enhance variability. In addition, SNP markers for additional less frequently found alleles that could lead to a different texture such as *Pinb-D1c*, *Pinb-D1d*, or *Pina-D1m* are being developed to screen different wheat collections and introduce them into CIMMYT germplasm.

16.4.3 Starch Properties

Starch is mainly composed of amylose and amylopectin. The ratio of both macromolecules has a significant impact on starch and dough characteristics such as viscosity, extensibility, and expansion, particularly at the oven stage in bread-making, when hydration of macromolecules changes, mainly due to denaturation. In common wheat, the amount of amylose and amylopectin is 25–28 and 75–72 %, respectively. From the nutrition and health point of view, high amylose content is related to high-resistant starch concentration, which acts as pseudo-fiber in the human intestine during digestion and which is associated with a healthier diet due to its lower glycemic index and because it increases satiation with less ingest of food.

The amylose/amylopectin ratio is controlled by the enzymes responsible of their synthesis (Morell et al. 2001). The amylopectin synthesis is complex and is carried out by different starch synthases (SGP-1, SGP-2, and SGP-3), branching and debranching enzymes of the grain. Amylose is synthesized by the granule-bound starch synthase (GBSS) I commonly named waxy protein. In bread wheat, three different waxy proteins are present and controlled by the three *Wx* loci (*Wx-A1*, *Wx-B1,* and *Wx-D1*) located at chromosomes 7AS, 4AL, and 7DS, respectively (Yamamori et al. 1994). These proteins have shown polymorphism, denoting the existence of null alleles that lead to the reduction of the amylose content. The coding region of these genes is composed of eleven exons and ten introns of a total size between 2781 and 2862 bp. The null alleles are known as *Wx-A1b*, *Wx-B1b,* and *Wx-D1b,* and molecular markers have been validated (Liu et al. 2005; McLauchlan et al. 2001; Saito et al. 2009). At CIMMYT, there is an increasing interest in determining the variation in starch properties of modern germplasm, as well as diverse genetic resources. Although starch composition is not considered a major factor in defining processing quality, parental lines have been analyzed with molecular markers (Table 16.10) for the presence of different null alleles, finding significant presence of the *Wx-B1b* allele (11 % of the lines) but not for the others. The use of these markers, especially when breeding is oriented to improve very specific products such as biscuits, flat unleavened breads, or noodles, with very specific quality requirements, is useful because colorimetric assays to determine amylose content or electrophoretic separation or waxy proteins are time-consuming processes and interpretation of the results is not always easy.

16.4.4 Flour Color

Flour color is an important trait in the assessment of flour quality. The enzyme polyphenol oxidase (PPO) has been found to be involved in undesirable time-dependent browning of noodles, flat breads, and steam bread. PPO activity, although largely dependent on the environment, is variable among different genotypes. Six loci, two per each genome (*Ppo-1* and *Ppo-2*), have been characterized at a molecular level, and alleles associated to high or low PPO activity are available. The validation and implementation of markers for PPO activity at CIMMYThave been initiated as this physiological–biochemical trait that cannot be easily evaluated based on its phenotype. The use of these markers would allow the selection of wheat progenies in the early generations and would greatly improve selection efficiency for color.

16.4.5 Nutritional Quality

Plant breeding to develop biofortified crops with enhanced micronutrient concentrations has emerged as a sustainable solution to complement strategies such as supplementation or fortification, especially for micronutrient-deficient rural inhabitants with limited access to formal markets or health care and who rely heavily on locally grown staple food crops (Bouis et al. 2011). In particular, South Asia suffers from high population densities and alarming rates of malnutrition (Velu et al. 2012). With funding from the HarvestPlus Challenge Program and the CGIAR Research Program on Agriculture for Nutrition and Health, CIMMYT is leading a global effort to develop and disseminate to partners in South Asia high-yielding wheat varieties that contain high levels of grain Zn concentration. Identifying the QTL that regulate the accumulation of high Zn levels in the wheat grain would allow breeders to more efficiently develop biofortified cultivars by using closely linked molecular markers to screen and select the most favorable genotypes. Two novel QTLs of large effect for increasing GZnC on chromosomes 2Bc and 3AL were recently detected by Hao et al. (2014) from the cross between PBW343 and Kenya Swara. The two QTLs individually explained about 10–15 % of the total phenotypic variation. Closely linked markers, DArT markers, identified were recently converted to SNPs and MABC using best RILs as donor parents are underway for validation and introgression of the QTL in CIMMYT-advanced breeding lines.

16.5 Marker-Assisted Selection for Grain Yield and Developmental Traits

Grain yield is the most important trait plant breeders are interested in. It is reflecting the culmination of all the processes of vegetative and reproductive growth and development, and their interactions with the edaphic and aerial environments. Most of the improvements in grain yield have arisen through incremental genetic advances. For example, wheat varieties with reduced plant height were introduced to the global wheat industry during the Green Revolution. These varieties substantially improved grain yield through increased harvest index and straw strength (Borlaug 1968).

Grain yield is usually broken down into three components: number of spikes per area, grain number per spike, and grain size estimated as thousand grain weight (TGW). These yield components are sequentially fixed during the growth cycle, vary in terms of their heritability, and are not always positively correlated with yield. TGW usually shows stable heritability (Kuchel et al. 2007b) and can be further broken down into individual components including physical parameters (grain length, width, area) and grain-filling characteristics, which are also under independent genetic control. In the past decade, there have been significant advances in the understanding of the genetic control of grain size, shape, and grain-filling parameters in the diploid crop species especially in rice (Ikeda et al. 2013). Several genes with relatively large effects have been identified through map-based cloning and support the independent genetic control of grain length, width, and grain-filling parameters. In wheat, there is still a limited understanding of grain weight genetic control. Many studies have identified QTL for TGW, grain size, and shape (Gross et al. 2003; Kumar et al. 2006; Breseghello and Sorrells 2007; Tsilo et al. 2010; Rustgi et al. 2013; Sun et al. 2009; Zhang et al. 2010a, b), but no gene has yet been cloned. Many of the observed QTL are in addition in relatively wide genomic regions and have not been validated and fine-mapped and have therefore limited impact in breeding.

For some of the genes associated with TGW and grain shape cloned in rice, orthologs have been identified in wheat via comparative genetics. These genes play different roles in various stages of grain development and include (1) sucrose synthase genes (*TaSus1* and 2), which are correlated to dry matter accumulation (Hou et al. 2014); (2) cell wall invertase genes (*TaCwi-2A*, *-4A*, *-4B and* -5D) related to sink tissue development and carbon partitioning (Ma et al. 2012; Jiang et al. 2015); (3) *TaGW2* (*TaGw2-6A*, 6B, 6D), a orthologous gene to the rice gene *OsGW2* and associated with kernel width and weight by controlling endosperm cell number in both the cell division and late grain-filling phases (Su et al. 2011); (4) a cytokinin oxidase/dehydrogenase gene (*TaCKX6-D1*) that plays a principal role in controlling cytokinin levels and affects grain weight in wheat (Zhang et al. 2012); (5) *TaSAP1*, a member of the stress associated protein (SAP) gene family in wheat associated with grain weight, number of grains per spike, spike length, and peduncle length in multiple environments (Chang et al. 2013a, b); and (6) *TaGS-D1* and *TaGASR7*, two genes mainly related to grain length (Zhang et al. 2015; Dong et al. 2014).

The exact effect of these genes on TGW, grain size, or shape in wheat is still not well understood to date. For example, several studies have examined the role of *TaGW2* on grain size parameters and contradictory results have been reported. Two studies have described a SNP upstream of the putative start codon as significantly associated with wider grains and increased TGW in Chinese germplasm (Su et al. 2011; Zhang et al. 2013). However, each study found the positive association with the opposite SNP and a negative association between *TaGW2* expression levels and grain width. Yang et al. (2012) identified a *TaGW2* frameshift mutation in a large grain variety and associated this mutant allele with increased grain width and TGW in a large $F_{2:3}$-population. However, down-regulation of *TaGW2* through RNA interference (RNAi) resulted in decreased grain size and TGW in wheat, suggesting that *TaGW2* is a positive regulator of grain size. At CIMMYT, we have initially validated molecular markers related to some of the published genes in four different

Table 16.11 Allele frequencies of the favorable allele of diverse published genes related to TKW in four different data sets

Germplasm	Number of lines	*TaSus1*	*Tasus2-2B*	*TaCWi-2A*	*TaGW2-6A*
		% favorable allele			
Mexican landraces	33	0.58	0.09	0.85	0.2
Turkish landraces	153	1	0.37	0.86	0.01
Historical set of CIMMYT lines	54	0.39	0.00	0.83	0.13
Current CIMMYT elite lines	112	0.48	0.00	0.91	0.12

germplasm sets (Table 16.11). Frequencies of the published favorable alleles varied between the data sets, and the most frequent favorable allele was the allele for *TaCWi-2A*. Initial analyses evaluating the effect of each of the alleles on TGW were inconclusive and only positively consistent for *TaCWi-2A* over all germplasm sets and in different environments (data not shown).

Recently, Simmonds et al. (2014) positioned the *TaGW2-6A* gene within the 6A QTL interval of the population Rialto × Spark. This QTL on chromosome 6 has been consistent in different populations and showed significant effects over seasons and environments (Snape et al. 2007). The QTL has been introgressed into CIMMYT germplasm via Spring × Winter Wheat crosses using UK cultivars such as 'Premio' or 'Mercato'. The QTL has also been observed in a recent CIMMYT parent 'PFAU/SERI.1B//AMAD/3/WAXWING' which shows good yield stability. The QTL has most likely been introgressed via the line 'PFAU' which also has European lines in its pedigree. The effect of this QTL in CIMMYT germplasm has still to be determined. Overall, it is therefore too early to conclude if and how these genes affect TGW grain size and shape and additional experiments are required.

16.5.1 Plant Phenology

Variation in expression to phenology is the most essential physiological adaptation of wheat to its cropping system. Archiving the appropriate plant, phenology permits wheat varieties (1) to fit into the timeframe of the cropping cycle, (2) to avoid extreme weather events (e.g., frost, drought), and (3) to optimize the use of resources to maximize yield. The manipulation of plant phenology is therefore a common target. The genetic determination of plant phenology has demonstrated that it is a complex character which exhibits a continuous variation and is controlled by many genes scattered over the whole genome (Snape et al. 1996). In wheat, the genetic bases of flowering time have been well studied and related genes have been classified according to whether they respond to vernalization or to photoperiod or to earliness per se during the preanthesis developmental phases. Stelmakh (1998) estimated that the vernalization gene system accounts for about 70–75 %, the photoperiod gene system for about 20–25 %, and the earliness per se for about 5 % or the genetic variability in the flowering time of bread wheat. Cane et al. (2013) estimated the effects of frequent alleles of *Vrn-1*, *Ppd-D1, and Ppd-B1* genes on flowering time, which accounted for 53 % of the genotypic variance of the trait. Diagnostic markers to identify the allelic variation of the genes have been developed (for a recent review see Kamran et al. 2014) and can be used in breeding programs. A recent study by Chen et al. (2014) has shown that Phytochrome C and circadian clock output genes play an additionally in long-day induced flowering in wheat. Phytochrome C operates by light activation upstream of the vernalization and photoperiod pathways and on the regulation of the circadian clock.

16.5.1.1 Vernalization Requirement

Vernalization is the acquisition or acceleration of plant's ability to initiate the flowering process by exposure to cold (Chouard 1960). According to the vernalization requirements, wheat is classified to have winter or spring growth habit. Spring wheat may be insensitive or partly sensitive to vernalization, but winter wheat has a considerable vernalization requirement. Genetic differences are caused by allelic variation at *Vrn-1*, *Vrn-2*, *Vrn-3,* and *Vrn-4* loci (Distelfeld et al. 2009). Spring wheat and facultative wheat are manifested by the presence of one or more dominant alleles at *Vrn-1* which confer the insensitivity or partial sensitivity to vernalization. Winter wheat possesses dominant alleles at *Vrn-2*, a floral repressor which is considered to delay flowering until the plants are vernalized and recessive alleles at the other three loci (Trevaskis et al. 2007). Fu et al. (2005) sequenced the *Vrn-1* genes located on the homologues chromosomes 5 from diverse wheat accessions. Several *Vrn-1* alleles result from insertion and deletions in the promoter and intron-1 regions of the gene (Yan et al. 2004a, b; Fu et al. 2005; Diaz et al. 2012). The role in altering vernalization response of each allele differs. While some alleles have large effects on the growth habit, others are silent mutations and therefore unlikely to have any direct role in the vernalization response. The *Vrn-3* genes mapped on the homologues chromosomes 7 promote the transcription of *Vrn-1,* thereby accelerating flowering time further. Genetic variation has been observed in *Vrn-B3* and *Vrn-D3* (Yan et al. 2006, Chen et al. 2010). Rather limited information is available for *Vrn-4.* So far only one allele has been described which was designated *Vrn-D4* and assigned to chromosome 5D (Kato et al. 2003, Yoshida et al. 2010).

At CIMMYT, molecular markers linked to the *Vrn-1* and *Vrn-3* genes are utilized to evaluate CIMMYT-advanced wheat lines. Summarizing the results of 1041 recent advanced lines, the most common allele in CIMMYT wheat is the dominant spring allele *Vrn-D1a* (99 %), followed by *Vrn-B1a* (70 %). Stelmakh (1993) and Eagles et al. (2011) evaluated the genetic effect of the three *Vrn-1* genes and suggested a higher effect on heading date of *Vrn-D1* or *-A1* compared with *Vrn-B1.* The Japanese cultivar 'Akakomugi' is thought to be the donor parent of the *Vrn-*

D1a allele (Stelmakh 1990) which was later transferred into early Green Revolution cultivars such as 'Lerma Rojo' and 'Sonora 64.' These two cultivars are thought to be the potential source of the *Vrn-D1a* allele in South and Southeast Asian wheat (Stelmakh 1990; Van Beem et al. 2005). Stelmakh (1993) also concluded that the highest yield was predicted for varieties containing *Vrn-D1a*. *Vrn-A1a* is almost absent in CIMMYT wheat. Different recessive winter *vrn-A1* alleles (V and W) have been identified which are distinguished by a C/T SNP in the fourth exon of the gene and are also associated with copy number variation (Zhu et al. 2014). The 'W' allele is present in 80 % of CIMMYT wheat lines. The allele is characterized by a higher copy number variation, a greater vernalization requirement, and increased frost tolerance (Zhu et al. 2014). The allele was previously observed in CIMMYT 'Veery' lines and derivatives such as 'Attila' and 'Babax' (Eagles et al. 2011) and might have some adaptive advantage or is linked to another favorable allele. For *Vrn-3* genes, no variation was observed for *Vrn-B3* in CIMMYT wheat. However, the published allele of *Vrn-D3* from the cultivar 'Jagger' is present in 60 % of the CIMMYT lines evaluated. The *Vrn-D3* allele further promotes development and according to Chen et al. (2010) maximizes effects at physiological maturity. The effect of *Vrn-D3* on heading and maturity date could not be confirmed yet. Overall, the most common haplotype at *Vrn-1* and *Vrn-3* in CIMMYT wheat is *vrn-A1W*, *Vrn-B1a*, *Vrn-D1a*, and *Vrn-D3*.

16.5.1.2 Photoperiod-Sensitivity Genes

Photoperiod-sensitive wheat is stimulated to flower only in long days, and flowering is delayed under short days provided that any requirement for vernalization is met. In spring habit wheat, photoperiod-sensitive types cannot be grown as an overwinter crop in tropical or low latitude areas, since the day length requirement would not be satisfied in a short enough time frame to produce a commercially viable crop (Worland and Snape 2001). Photoperiod-insensitive wheat flowers independently of day length can be grown to maturity in long or short day environments. This is of particular advantage in warmer and dry climates as early flowering varieties are able to fill their grains prior to the onset of high temperatures and droughtstress occurring late in the season (Worland and Snape 2001). To date, three such genes have been identified, including *Ppd-A1*, *Ppd-B1,* and *Ppd-D1,* located on chromosomes 2A, 2B, and 2D, respectively. The primary influence of the genes is on ear growth and spikelet growth (Scrath et al. 1985). A novel photoperiod response gene designated as *Ppd-B2* has been mapped on wheat chromosome 7BS (Khlestkina et al. 2009). This gene accelerates flowering only under long photoperiods in contrast to the *Ppd-1* genes that induce earlier flowering irrespective of day length. *Ppd-D1* is the photoperiod-insensitive locus with the largest effect followed by *Ppd-B1* and *Ppd-A1* (Worland 1996). Photoperiod insensitivity is induced by In/Dels in the 5′ upstream region of the gene pseudo-response regulator genes; they do not exist in the photoperiod-sensitive alleles (Beales et al. 2007; Wilhelm et al. 2009; Nishida et al. 2013). Furthermore

Table 16.12 Allele frequency of reported *Vrn* and *Ppd* alleles in two sets of germplasm forming the international bread wheat screening nursery (IBWSN)

Trial	M45IWBSN			M46IBWSN		
Gene/Allele	No	No of lines	%	No	No of lines	%
Vrn-A1a	706	4	0.6	317	0	0.0
vrn-A1	706	702	99.4	317	317	100.0
Vrn-A1v	704	168	23.9	317	63	19.9
Vrn-A1w	704	536	76.1	317	254	80.1
Vrn-B1a	720	452	62.8	298	252	84.6
Vrn-B1b	720	20	2.8	298	16	5.4
vrn-B1	720	248	34.4	298	30	10.1
Vrn-D1a	706	702	99.4	319	319	100.0
vrn-D1	706	4	0.6	319	0	0.0
Vrn-D3a	698	452	64.8	321	187	58.3
vrn-D3	698	246	35.2	321	134	41.7
Ppd-A1a (GS105)	714	10	1.4	319	5	1.6
Ppd-A1b	713	5	0.7	298	2	0.7
Ppd-A1(Null)	701	145	20.7	318	133	41.8
Ppd-A1	714	554	77.6	319	181	56.7
Ppd-B1a (4x)	709	3	0.4	319	2	0.6
Ppd-B1a (3x)	720	497	69.0	308	207	67.2
Ppd-B1b (1x)	719	213	29.6	308	88	28.6
Ppd-B1b (1x)	720	6	0.8	308	10	3.2
Ppd-D1a	716	641	89.5	319	289	90.6
Ppd-D1b	705	76	10.8	319	30	9.4
Ppd-D1(null)	714	1	0.14	317	1	0.3

Diaz et al. (2012) showed that for *Ppd-B1*, alleles conferring altered flowering time had an increased copy number of the gene and altered gene expression.

In CIMMYT, wheat *Ppd-D1a* is predominant in CIMMYT wheat germplasm (Table 16.12). Across the same 1041 recent advanced lines described above, 95 % of the lines carry the *Ppd-D1a* allele. Since its beginning by Norman Borlaug and his colleges, the CIMMYT wheat program is based in Mexico and shuttles germplasm between two contrasting environments (NW Mexico, Ciudad Obregon [(27° N109°W) and central highlands in El Batan (20°N100°W)]. This shuttle breeding exposes wheat materials to diverse photoperiod and temperatures and to a range of important diseases led to the selection of photoperiod-insensitive lines. The *Ppd-A1a* alleles first described in durum wheat are present in 3–5 % CIMMYT bread wheat germplasm. The allele was transferred from durum wheat via synthetic hexaploid wheat derivative that has been incorporated with increasing number into the bread wheat breeding programs (Dreisigacker et al. 2008). The *Ppd-B1* alleles show the largest variation in CIMMYT wheat. Most of the reported alleles were observed, e.g., the four copy number variant initially identified in 'Chinese Spring.'

The most frequent allele is the three copy number variant first characterized in the Green Revolution line 'Sonora 64.'

16.5.1.3 Earliness

Earliness per se (*Eps*) genes are those that regulate flowering time independently of vernalization and photoperiod and are important for the fine-tuning of flowering time and for the wide adaptation of wheat to different environments. Among the contributing factors influencing time to flowering, earliness per se (*Eps*) has been least investigated. *Eps* loci have already been identified in wheat (Hoogendoorn 1985), and meta-QTL analysis of heading time in bread wheat revealed that numerous QTL colocated in chromosomal regions known to carry *Eps* loci (Hanocq et al. 2007; Griffiths et al. 2009). However, most of these *Eps* loci remained molecularly undefined, and only the *Eps-*A^m*1* locus in einkorn wheat has been fine-mapped and phenotypically characterized (Faricelli et al. 2010). Gawronski and Schnurbusch (2012) recently fine-mapped a second gene derived from einkorn wheat *Eps-*A^m*3*.

Under combined vernalization and photoperiod treatments, we identified an *Eps* QTL on chromosome 1DL using genome-wide association mapping in the Wheat Association Mapping Initiative (WAMI) population genotyped with the 90K Wheat Illumina SNP array (Sukumaran, personal communication). Subsequent BLAST searches indicated that the QTL region with sequence similarity identity higher than 96 % contained the *Mot1* and *ELF3* genes that were candidates for earliness per se from earlier studies in einkorn wheat, so is a likely orthologue of *Eps-*A^m*1*. A recent study using four independent pairs of NILs derived from a cross between Spark and Rialto winter wheat varieties identified the same region on 1DL for *Eps* in wheat suggesting that MAS of *Eps* effects is getting feasible (Zikhali et al. 2014).

16.5.2 Plant Height

Plant height is an important agronomic trait in cereal crops. It not only determines plant architecture but also contributes a lot to grain yield. The *Rht-B1b* (*Rht1*) and *Rht-D1b* (*Rht2*) semidwarfing genes were introduced into commercial wheat cultivars from the Japanese variety Norin10 in the 1960s as part of wheat improvement programs in the USA and at CIMMYT and lead to the first 'Green Revolution' wheat varieties. A reduction in plant height improved stem strength and thus lodging resistance and Harvest Index, the partitioning of assimilates to the developing grain (Borlaug 1968). The large increases in yield that followed the introduction of these dwarfing genes led to widespread adoption of the dwarfing genes throughout the world (Gale et al. 1985). Perfect STS and SNP markers were developed in wheat for these genes (Ellis et al. 2002, http://www.cerealsdb.uk.net/). *Rht1* and *Rht2* encode proteins involved in gibberellin signal transduction, but also have pleiotropic effects on plant growth, causing reductions in coleoptile length and seedling leaf area. These genes reduce the leaf elongation rate and coleoptile length. A number of alternative plant height genes have been observed that might be more suitable for final plant height without compromising early plant growth (Ellis et al. 2004). Examples are *Rht4*, *Rht5,* or *Rht8* that do not reduce the leaf elongation rate and coleoptile length and do not affect early growth. *Rht1* is predominant in CIMMYT germplasm due to the introgression of this gene during the 'green revolution.' Efforts are, however, underway to incorporate some of the alternative alleles, e.g., *Rht4*, *Rht5, or Rht13* into CIMMYT germplasm using MAS. The SNP assays are routinely used to evaluate the CIMMYT elite germplasm. For the alternative drawing genes, the markers reported in Ellis et al. (2005) are utilized.

16.6 Marker-Assisted Selection for Abiotic Stresses in Wheat

Abiotic stresses affect plant development, productivity, and grain quality in wheat. Research on plant responses to abiotic stresses and their impact on crop production continue to be a major focus in breeding, especially as in the current scenario of climate change, climate-resilient wheat is a necessity. Among the four main abiotic stresses (drought, heat, salinity, and metal toxicity) drought is the single most threat to food security.

16.6.1 Drought and Heat Tolerance

Plant responses to drought and heat stress are complex depending on the genotypes, environments, and the G × E. In addition, the difficulties to identify QTLs for traits under drought and heat are as follows: (1) the availability of mapping population with controlled height and phenology to avoid confounding effect of major genes and (2) the phenotyping procedure in a time frame in large populations that will avoid confounding masking effect of major genes on minor genes (Reynolds and Tuberosa 2008).

Drought tolerance phenotyping can be realized indirectly by measurements of morpho-physiological traits mainly water use, water-use efficiency, carbon partitioning to grain, carbon isotope discrimination to determine transpiration efficiency, canopy temperature, green leaf area, stay green, water soluble carbohydrates, above ground biomass, grain yield and root parameters, root biomass, rooting depth, and root development under drought conditions (Rashid et al. 1999; Foulkes et al. 2007). At present, these traits are followed in CIMMYT to perform trait-based crosses to combine the high-value alleles. Parents with contrasting desired physiological traits are selected from available evaluated germplasm, and crosses are made with recurrent parents or elite lines. The developed lines are then tested under different environmental conditions.

At the candidate gene level, among the most common gene networks and pathways related to drought tolerance, abscisic acid (ABA)-dependent and ABA-independent related pathways are the most studied. Abscisic acid is a plant growth regulator and stress hormone, which induces leaf stomata closure to reduce water loss through transpiration and decreases the photosynthetic rate in order to improve the water-use efficiency of plants. A major QTL affecting drought-induced ABA accumulation was located on chromosome 5A in wheat, and examples of source genotypes are the cultivars 'Ciano 67' and 'SQ1' (Quarrie et al. 1994). Furthermore, many families of transcription factors have been demonstrated to play a role in stress responses in plants. bZIP, DREB, WRKY, bHLH, MYB, and NAC transcription factors represent the major groups of regulatory genes of which some members are found to be involved in wheat stress tolerance. A very limited number of markers are developed for these genes. Wei et al. 2009 developed a functional marker for the *Dreb-B1* gene that is initially tested at CIMMYT and Chang et al. (2013a, b) identified linked markers to the gene *TaSAP-A1*.

Further international efforts using genome-wide mapping approaches to detect QTLs for grain yield under drought stress conditions have been made with some success. Fleury et al. (2010) summarized more than 20 QTL for drought in wheat. To give an example, a QTL on chromosome 3BL was detected under heat, drought, and high yield potential conditions that explained up to 22 % of the variance for grain yield and canopy temperature (Bennett et al. 2012). The same QTL on chromosome 3B was also associated with grain yield in the studies of Bonneau et al. (2013) and Sukumaran et al. (2015).

Heat stress mostly occurs in combination with drought, and the combined effect of drought and heat is severe than any one of the stresses (Prasad et al. 2011). Higher temperature above 30 °C at grain-filling period is detrimental to wheat crop yield. For heat-tolerant studies, several traits have shown promises, viz. light interception traits, rapid ground cover, canopy structure, radiation use efficiency, stay green, photosynthesis and reduced photorespiration, photoprotective metabolites, wax, membrane thermostability, spike fertility, water soluble carbohydrate, starch synthesis, and plant signaling (Cossani and Reynolds 2012).

While confronting high temperature stress and alleviation from damage of cellular protein structure essential for survival in stressed conditions, plant triggers a novel class of protein called HSPs. These HSPs serve as molecular chaperons to maintain conformational protein functions as well as cellular protein refolding, thereby protecting plants under HS conditions (Wang et al. 2004). So far only one attempt has been made to identify SNPs that differentiate heat-tolerant and heat-susceptible genotypes of wheat analyzing the heat-shock protein HSP16.9 as the target gene. DNA fragments covering a partial sequence of wheat HSP16.9 were amplified from the heat-tolerant genotype 'K7903' and heat-susceptible genotype 'RAJ4014' and subsequently analyzed for the presence of SNPs. One SNP was found between these genotypes, and the analysis of the corresponding amino acid sequence showed that the base transition (A/G) positioned at 31 amino acid resulted in a missense mutation from *aspartic acid* to *asparagine* residue (Garg et al. 2012). Allele-specific primers based on SNP explained 29 and 24 % phenotypic variation for grain weight and TEWs, respectively.

Despite the importance of heat tolerance, only a few studies have focused to identify QTL via genome wide scans. Yang et al. (2002) found QTL linked to grain-filling duration on the short arms of chromosomes 1B and 5A. Vijayalakshmi et al. (2010) reported QTL with significant effects on grain yield, grain weight, grain filling, stay green, and senescence-associated traits on 2A, 3A, 4A, 6A, 6B, and 7A under post-anthesis high temperature stress in wheat. Heat susceptibility index (HSI) calculated from agronomic traits is associated with heat stress tolerance, and QTLs were mapped on chromosomes 1A, 2A, 2B, and 3B for HSI calculated from the kernel characteristics under stress conditions applied during early grain-filling stage that explained up to 31 % of the variation in the traits (Mason et al. 2010). QTLs were mapped for heat stress using the Fischer susceptibility index on chromosomes 1B, 5B, and 7B that explained up to 44 % of the variation in the traits (Mohammadi and Zali 2010). QTL mapping for terminal heat stress has identified QTL in chromosome 2B, 7B, and 7D that colocalized for kernel weight, grain-filling duration, and canopy temperature difference. During flowering, higher temperature can cause pollen abortion and thereby low yield. QTLs for heat stress tolerance were identified in a cross on cultivar NW1014 (heat tolerant) and HUW468 (heat susceptible) using HSI of grain weight, grain-filling duration, grain yield, and canopy temperature depression on chromosomes 2B, 7B, and 7D. These explained up to 20 % of the phenotypic variation for the traits (Paliwal et al. 2012). Mondal et al. (2015) identified QTL for leaf wax content located on chromosomes 1B and 5A with the 5A QTL region showing localization with QTL for leaf and spike temperature depression, indicating a genetic link between these traits. Composite interval mapping by the study of Talukder et al. (2014) identified five QTL regions significantly associated with response to heat stress. Associations were identified for plasma membrane damage on chromosomes 7A, 2B, and 1D; SPAD chlorophyll content on 6A, 7A, 1B, and 1D; and thylakoid membrane damage on 6A, 7A, and 1D. The variability explained by these QTLs ranged from 11.9 to 30.6 % for thylakoid membrane damage, 11.4 to 30.8 % for SPAD chlorophyll content, and 10.5 to 33.5 % for plasma membrane damage.

The plant developmental genes in wheat for vernalization and photoperiod (*Vrn* and *Ppd*, respectively) are related to the performance of the lines under drought and heat stress (Bogard et al. 2014). Therefore, care must be taken to avoid these effects in gene discovery when developing mapping populations. In CIMMYT, to study the genetic basis of drought and heat tolerance and to make strategic crosses for trait integration and line development, mapping populations are available with restricted phenology and plant height (Sukumaran et al. 2013). The Seri × Babax recombinant inbred line (RIL) population was the first of this type of population developed with a phenology range of 10 days. Using this Seri × Babax population, QTLs were identified for drought and heat tolerance on chromosome 4A that colocalized with grain yield. Common QTLs for drought and heat tolerance were identified on 1B, 2B, 3B, 4A, and 7A using canopy temperature measurements (Pinto et al. 2010).

Canopy temperature measurements are surrogates for estimating stomatal conductance (Rebetzke et al. 2013).

The WAMI population was subsequently created and consists of 287 lines selected from a series of CIMMYT international nurseries that has a phenology range of 14 days. The CIMMYT WAMI population was studied through genome-wide association study (GWAS) at the population level and by a candidate gene approach (Edae et al. 2013, 2014; Lopes et al. 2014; Sukumaran et al. 2015). Edae et al. (2013) performed a candidate gene study and confirmed the effects of the dehydration-responsive element binding 1A (*DREB1A*) gene on NDVI, heading date, biomass, and spikelet number; the enhanced response to abscisic acid (*ERA1-A* and *ERA1-B*) genes on harvest index, flag leaf senescence, and flag leaf width; and the fructan 1-exohydrolase (*1-FEH-A and 1-FEH-B*) genes associated with grain yield and thousand kernel weight, respectively. A consistent QTL in chromosome 2DS for grain yield and yield components under contrasting moisture conditions in the USA and Ethiopia was additionally identified through GWAS (Edae et al. 2014). On chromosomes 5A and 6A, loci for grain yield, thousand kernel weight, grain number, and canopy temperature were detected in the WAMI grown in Mexico (Sukumaran et al. 2015).

A recently developed RIL population Synthetic × Weebil at CIMMYT has a phenology range of three days that was phenotyped under drought and heat conditions. Identifying QTL for heat and drought tolerance in these populations will be more independent of the confounding effects of phenology. This population was phenotyped though 90K Illumina Bead chip array (Wang et al. 2014), and the research is under progress to detect QTL.

The detected QTL in the phenology-controlled populations is recent, and the validation of the identified QTL is only underway. Based on the up-to-date obtained knowledge on the underlying mechanisms and architecture of heat and drought tolerance, the applicability of MAS for both traits has to been questioned to its inability to capture small-effect QTL. Genome-wide approaches such as GS attempt to avoid this deficiency by capturing both large- and small-effect QTL with dense molecular marker coverage to predict complex trait values (Meuwissen et al. 2001).

16.6.2 Metal Toxicity

Wheat is susceptible to excess amounts of aluminum (Al), boron (B), cadmium (Cd), and copper (Cu). Out of these, under low pH, Al is the most prevalent and most toxic to wheat plants (Delhaize and Ryan 1995; Kochian et al. 2005). A major gene (*TaALMT1*) on chromosome 4DL which encodes a malate transporter constitutively expressed on root apices has been identified in wheat cultivars 'BH1146,' 'Atlas 66,' and 'Chinese Spring' (Ma et al. 2005; Raman et al. 2005; Sasaki et al. 2004). Raman et al. (2006) developed SSR markers, *ALMT1-SSR3a* and *ALMT1-SSR3b,* and a CAPS marker from the repetitive InDels and substitution region of *the TaALMT1*gene which can be used in MAS, but have not been applied in CIMMYT yet. However, some studies demonstrated that more than one gene might be involved in Al tolerance in wheat. Tang et al. (2002) suggested that at least two genetic loci might contribute to Al resistance in 'Atlas 66.' Zhu et al. (2007) reported a minor QTL on chromosome 3BL of 'Atlas 66' in addition to the major QTL on chromosome 4DL. A Al tolerance QTL on chromosome 3BL was also contributed by 'Chinese Spring'(Navakode et al. 2009). QTLs on chromosome 3BL and 2A apart from the major effect gene on chromosome 4DL collectively explained 80 % of the phenotypic variation (Cai et al. 2008; Dai et al. 2013).

Boron toxicity occurs when plants are grown in alkaline or volcanic soils. Boron has the narrowest range between deficient and toxic soil solution concentration of all plant nutrients. Boron toxicity in wheat can cause poor root growth, low above ground biomass, low seed set and sterility, and low grain yield (Pallotta et al. 2014). *TaBot1L (Bo1)* and *Bo4* are the two major effect QTL forboron tolerance in wheat. The utilization of *Bo1* on the long arm of chromosome 7BL has been a long-term priority for marker-assisted selection in wheat breeding programs in Australia. At CIMMYT, the STS marker *AWW5L7* published by Schnurbusch et al. (2008) is used. Sources of resistance were, e.g., the Australian line 'Gladius.' *Bo4* is located on chromosome 4AL and was recently placed with the marker interval *Xabg390-4A–XksuG10-4A* (Pallotta et al. 2014).

16.7 Genomic Predictions in CIMMYT Wheat

For polygenic or quantitative traits (such as drought and heat tolerance), the difficulty of the large-scale validation and refinement of large-effect QTL limits the implementation of MAS strategies in plant breeding programs (Xu and Crouch 2008). With the development of modern genotyping and sequencing methods, the MAS theory has recently shifted to the use of genome-wide markers to predict the performance of both phenotyped and unphenotyped individuals (genomic selection (GS)) for polygenic or quantitative traits. Using genome-wide markers, every trait locus is likely to be in linkage disequilibrium (LD) with a minimum of one marker locus in the entire target population. In GS, a training population related to the breeding germplasm is genotyped with genome-wide markers and phenotyped in target environments. These data are used to derive a prediction equation that can then be applied to genotypes of unphenotyped individuals to predict breeding values that can be used to inform selection decisions (Meuwissen et al. 2001; Bernardo and Yu 2007; de los Campos et al. 2009). Genetically estimated breeding values (GEBV) open up several new routes for increasing genetic improvement rates in plant breeding programs. They offer opportunities to (1) increase the selection efficiency of preliminary and multi-location yield trials and (2) shorten the breeding cycle by repeated early generation selection, thus increasing the genetic gain per unit of time. Several GS studies, many of which were developed or used data generated by CIMMYT's wheat breeding program (see Crossa et al. 2014 for a recent review), have shown that GS can achieve reasonably high prediction accuracy. This has raised expectations about the prospects of implementing GS in wheat breeding programs. However, implementing GS in breeding also presents important challenges.

16.7.1 Prediction Results Using CIMMYT Wheat Data Sets

For inbreeding species such as wheat, their relatively small effective population size (N_e) is the main advantage for GS. The smaller the N_e, the smaller the number of independent chromosome segments in the genome that allow for higher GEBV accuracies (Lin et al. 2014). Most studies report prediction accuracy as the correlation between GEBVs and phenotypes. In CIMMYT wheat data sets, prediction accuracies for grain yield, for example, ranged from 0.3 (Poland et al. 2012) to 0.7 (Perez-Rodriguez et al. 2012); for days to heading, prediction accuracies between 0.4 (Poland et al. 2012) and 0.7 (Wimmer et al. 2013) were reported. Prediction accuracy depends on the prediction problem assessed and on several other factors, such as trait heritability, the relationship between the individuals to be predicted, and those used to train the models for prediction, sample size, number of markers, and G × E. A large number of prediction models have been developed or adopted from other fields to handle the high-dimensional marker data sets that are typical of GS. The various types of models respond differently because they vary in their assumptions when treating the variance of complex traits. In GS, the number of predictors (p) is usually far greater than the number of individuals (n). In such cases, estimates of ordinary least squares have poor prediction ability because marker effects are treated as fixed effects, which leads to multicollinearity and overfitting among predictors, thereby making the model unfeasible (for a review see Lorenz et al. 2011, or de los Campos et al. 2013). To further improve genome-wide predictions in wheat, the Biometrics Unit at CIMMYT has developed and recommended various prediction algorithms mainly for low heritable traits. The R (R Development Core Team 2010) package 'Bayesian Generalized Linear Regression (BGLR)' (de los Campos et al. 2013) is applied for most CIMMYT implementations.

16.7.1.1 Trait Heritability

Heritability allows comparing the relative importance of genes and environments to the variation of traits within and across populations. Heritability depends on the genetic properties of a trait, the range of typical environments in the studied population, as well as various interactions between genes and environmental factors. For traits with low heritability, genes contribute little to individual phenotypic differences; for highly heritable traits, genes are the main reason for individual differences. GS performs differently in traits with distinct genetic properties

16.7.1.2 Relationship Between the Individuals in Training and Selection Populations and Sample Size

One approach to implementing GS is to select individuals with the highest genetic merit in the early generations of a breeding cycle (e.g., selecting F_2-individuals). The selected individuals can be intercrossed, and the resulting progenies can be selected again a number of times before extracting inbred lines, also called rapid cycling GS. In a simulation study, we explored the relative importance of the relatedness between training and selection populations, sample size, and marker density for the accuracy of genomic prediction in an early generation selection approach (Hickey et al. 2014). For simulation, several biparental populations, each having 550 F_2-individuals, were created that were related to each other in different ways: biparental populations that one parent in common (BP-P), that have one grandparent in common (BP-G), or that are unrelated (BP-U). The accuracy of selection was evaluated on 50 unphenotyped F_2-individuals from a single biparental population using the correlation between the GEBVs and the true breeding values. For the phenotypes, a polygenic trait was simulated with 0.5 heritability, and a range of 50–10,000 SNP markers was tested.

The phenotypes and genotypes that were used to train the prediction equation were either generated inside the single biparental population or inside the other biparental populations (BP-P, BP-G, or BP-U) that were simulated to have relationships with the given single biparental population. Between 1 to 40 populations and 50 to 500 F_2-individuals per population were used to train the prediction equations. Figure 16.2 shows the accuracy of prediction inside a single biparental population. The accuracy of the breeding values increased as the size of the training population increased. Training with up to 50 phenotypes gave accuracies between 0.2 and 0.6, while training with 100 or more phenotypes gave accuracies of 0.8 or higher. The results when differently related populations were used for genomic prediction are displayed in Fig. 16.3. Prediction accuracies decrease with decreasing relatedness to the given single biparental population (BP-P → BP-U). Using information from unrelated populations generally gave low accuracies unless very large numbers of phenotypes (more populations and more individuals per population) were used. This means there is a trade-off between relationship and population size that affects prediction accuracy. When using information from close relatives, the marker associations are due to the linkage between markers and QTLs, whereas when using information from distant relatives, marker associations are due to LD. Closer relatives share longer chromosome segments or haplotypes; therefore, a training population with close relatives will have a smaller number of independent haplotypes and a larger sample size per haplotype, leading to more precise predications. Distant relatives share shorter haplotypes, and a training set will have a large number of independent haplotypes with different frequencies, leading to less accurate predictions.

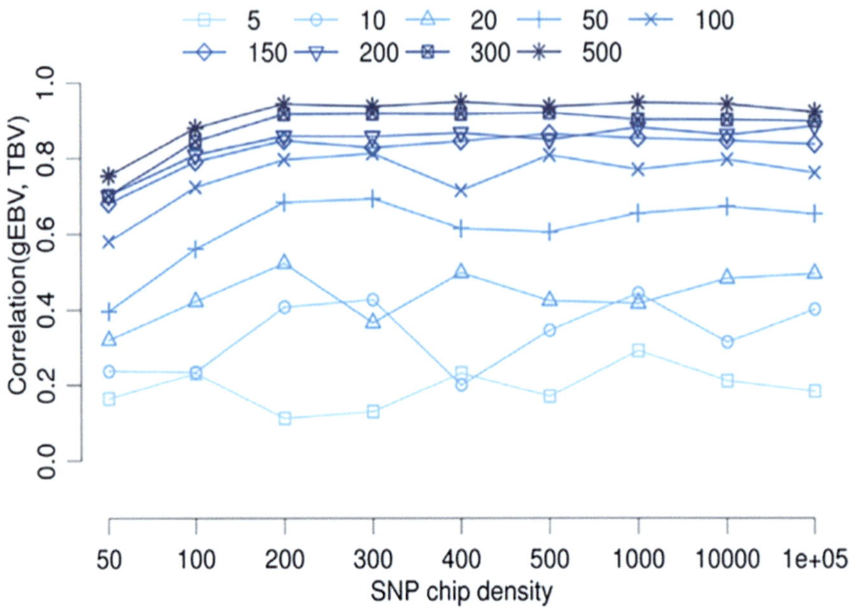

Fig. 16.2 Accuracy of breeding values inside a given biparental population when training in the same population with different numbers of markers and F_2s (from Hickey et al. 2014)

16.7.1.3 Marker Density

Genotypic data can still be expensive, especially when low-cost genotyping approaches such as genotyping by sequencing or genotyping strategies involving low-density SNP chips are not used. In the simulation study conducted by Hickey et al. (2014), marker density was considered as one important factor in genomic prediction. The simulation results in Fig. 16.3 show that the marker density required to obtain accurate genomic predictions depends on the degree of relatedness between the training and selection populations. With close relatives (e.g., BP-P), accurate predictions could be obtained with 200 markers. Increasing the marker density up to 10,000 markers did not improve prediction accuracy. A small number of markers are sufficient because the shared haplotypes and linkage blocks are large. When using distant relatives (e.g., BP-U in Fig. 16.3), more markers are required because of the lower LD between markers and QTLs. Similar results were also found by Solberg et al. (2008) and Meuwissen (2009).

16.7.1.4 Relationship Between Environments

Multi-environment trials are widely used by plant breeders to evaluate the relative performance of genotypes across environments. Multi-environment trials are

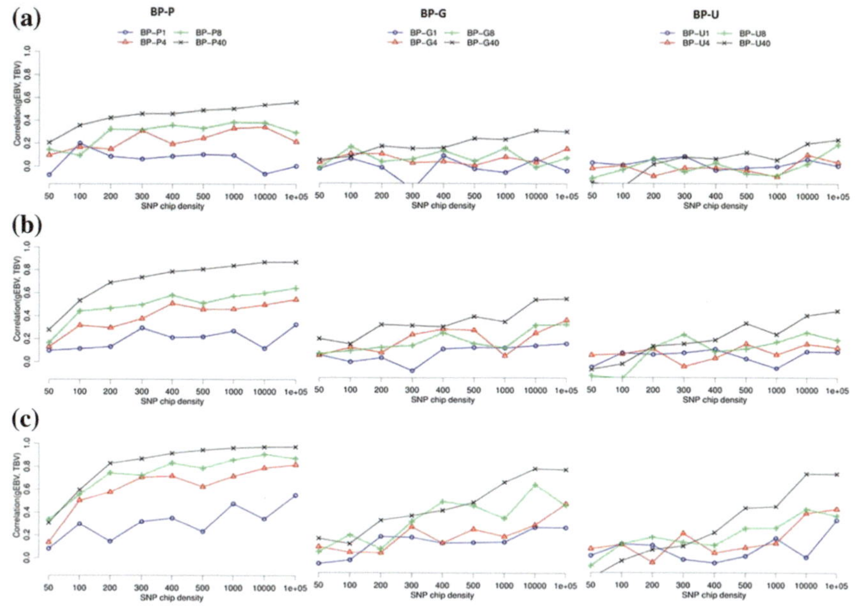

Fig. 16.3 Accuracy of breeding values inside a given biparental population when training in 1, 4, 8, or 40 biparental populations with one parent in common (BP-P1 to BP-P40), one grandparent in common (BP-G1 to BP-G40), or with no pedigree relationship (BP-U1 to BP-U40). A = 5 F_2s recorded in each population, B = 50 F_2s recorded in each population, C = 500 F_2s recorded in each population, *TBV*-true breeding value, and *Gebv*-genetically estimated breeding value (from Hickey et al. 2014)

conducted as GE introduces uncertainty into the measure of genotypic performance and complicates the selection of superior genotypes. Thus, accounting for G × E has always been a concern when analyzing agronomic data, and many different methods have been proposed for analyzing multi-environmental trials conducted by breeders.

Genome-wide markers provide a new tool that can be used in multi-environment trial analyses. Genome-wide markers do not change the fact of G × E; however, their use could enable better selection decisions. Burgueño et al. (2012) and Jarquin et al. (2013) used genome-wide markers and additional environmental factors for multi-environment trial analyses and showed that when gene and environmental interaction terms were introduced in the prediction equation, prediction accuracy increased, suggesting that the proportion of variance accounted for by the prediction model was higher. In Burgueño et al. (2012), the prediction accuracies showed the same pattern as the genetic correlations between environments. The interaction term allows borrowing information between environments, and for environments that are positively correlated, this increases prediction accuracy.

16.7.1.5 Modeling Selected Markers as Fixed Effects

A recent simulation study (Bernardo 2013) found that modeling a large-effect locus as fixed to be advantageous when trait heritability was greater than 0.5 and the proportion of genetic variance explained by the locus was greater than 0.25. Rutkoski et al. (2014) confirmed these results using GS as a potential tool to select for adult plant stem rust resistance. In a set of CIMMYT-advanced lines that were tested for adult plant stem rust resistance across environments and years, markers linked to the stem rust gene *Sr2* were applied and its results included as fixed effects in the prediction model that was more accurate than using genome-wide markers only. Overall, the levels of prediction accuracy found in this study indicate that GS can be effectively applied to improve stem rust APR in this germplasm.

16.7.2 *Implementation of GS*

GS has a great number of uses in a breeding program. Similar to MAS strategies, the introduction of GS is flexible and may vary for each breeding program, depending on the target traits and breeding scheme. The greatest potential use of GS is at points in the breeding program where selection using traditional methods is too expensive, time-consuming, or not biologically or logistically possible. Two main applications of GS are being studied in CIMMYT's global wheat breeding program: (1) to predict the genotypic value of individuals for potential release as cultivars and (2) to predict the breeding value of candidates in rapid-cycle populations.

16.7.2.1 Predicting the Genotypic Value of Individuals for Potential Release

The breeding methodology applied by the CIMMYT wheat breeding program includes modified bulk selection. After population advancement with selection for more heritable traits via shuttle breeding, inbred lines are extracted and tested in preliminary yield trials (PYTs) to identify superior entries which will then be evaluated in the following year in more extensive multi-environment yield trials and/or used as parents to begin another breeding cycle. CIMMYT PYTs usually include up to 10,000 genotypes, of which approximately 1000 are selected and evaluated in five to six different environments with two to three replications in the subsequent cycle.

CIMMYT PYTs are carried out in replicated yield trials in small plots and a single environment. GS could be useful to predict the GEBVs based on a large training population that includes previous breeding germplasm and amends the selection of lines. There is also a trade-off between the number of genotypes that enter the multi-environment trials and the number of plots per entry. A larger number of plots per entry allow a more accurate estimate of the performance of each genotype across environments, whereas a larger number of entries enhance the germplasm pool from which selections are made. If the number of plots is fixed, a larger number of entries can only be tested if they are divided across environments. Consequently, not all entries would be present in all environments, but the average genotypic performance across all environments could be determined using genomic prediction. Initial results testing this approach using diverse models that incorporate pedigree, marker, environment, and interaction terms into the prediction equation revealed relatively high prediction accuracies: an average 0.6 when 20 % of the entries were present in only one of five environments (unpublished data). These results indicate that not all entries have to be evaluated in all environments and that more entries could be tested. This approach can be optimized further by maximizing the relationship between the training and testing populations and by varying the number of genotypes and environments to be predicted. Several sister lines are usually present in each CIMMYT multi-environment trial. Dividing the sister lines across environments would additionally increase the relatedness between the training and testing populations and allow higher prediction accuracies.

The best performing lines in multi-environment trials are selected to form CIMMYT's international screening nurseries and yield trials, which are distributed globally via the International Wheat Improvement Network. Although trait heritabilities are high at this stage of the breeding program, genomic prediction could be useful for boosting the selection of lines to be included in each yield trial and that could potentially be released by national programs.

16.7.2.2 Predicting the Breeding Value of Candidates in Rapid-Cycle Populations

In a rapid-cycle GS breeding scheme, segregating populations can be genotyped at the seedling stage and then selected based on GEBVs derived from a related training population. The resulting F_2-candidates can be used to extract inbred lines or intercrossed. Applying GS rapid cycling in early generations (e.g., F_2) is a high-risk but high-turnover approach. In conventional breeding, early generation intercrossing is not practiced due to highly heterozygous and heterogeneous progenies and the unfeasibility of selecting for complex traits based on a single plant. Although genomic prediction accuracy may not be high, shortening the cycle time or generation interval is expected to increase genetic gains.

In a proof-of-concept experiment, CIMMYT has initiated a rapid-cycle GS scheme in wheat with grain yield as the target trait. Genomic prediction was applied in 40 F_2-populations, with two cycles of subsequent intercrossing within and between populations. In each generation, inbred lines were extracted based on genomic and conventional selection. Initial results in two seasons replicated trials of lines derived from single plants with a range of GEBVs have shown that higher GEBV individuals produce higher yielding derivative lines (Bonnett et al. unpublished). A yield increase of 7 % in derivatives of plants with the highest 13 % of GEBVs compared to the lowest 13 % of GEBVs was found following just one cycle of GS. This represents a 30 % realized gain from selection and is an encouraging initial result. A larger number of derivatives from simple cross F_2-derived bulks and 1st and 2nd cycle intercrosses compared to conventionally selected cohorts from the same initial crosses are currently being tested to extend the evaluation of rapid-cycle GS. These evaluations give the first indications of genetics gains from GS for a highly complex trait in an actual wheat breeding program.

16.8 Conclusions

To date, 30 different loci responsible for traits like resistance to various diseases, quality, and agronomy have been cloned in wheat, and 97 functional markers have been developed to categorize 93 alleles based on gene sequences (Liu et al. 2012). Many research institutes involved in wheat cultivar development and germplasm evaluation now have access to modern genotyping and sequencing technologies and have the expertise for gene discovery analysis; wherefore, the development of new functional markers is expected to grow. The development of low-cost SNP markers and related public databases such as GrainGenes (http://wheat.pw.usda.gov/GG2/index.shtml), CerealsDB (http://www.cerealsdb.uk.net/cerealgenomics/CerealsDB/indexNEW.php), and MAS wheat (www.maswheat.ucdavis.edu) will additionally encourage the widespread use of molecular markers for wheat improvement. The CIMMYT experience and examples show that marker-assisted breeding has gained importance among wheat breeders, pathologist, and physiologist. Application of MAS will be further promoted if several critical factors will be addressed in the future, including the following: (i) identified functional or linked markers need to be validated to show trait association with the desired genetic backgrounds grown under target environments, (ii) the DNA extraction and screening methodology will still need to be more cost-effective and time-saving, and (iii) improved databasing and bi-informatics pipelines will be needed to support rapid analyses.

A larger-scale implementation of GS in breeding programs will shift efforts from evaluating the whole plant to evaluating marker effects. We think that initial GS implementation should not significantly affect the way plant field trials are conducted in each breeding program as there are still many unanswered questions regarding how to achieve the optimal balance between genotyping and phenotyping and the best use of marker effect evaluations to maximize the overall genetic gain for single or multiple target traits in a particular breeding program. Significant challenges remain with respect to the successful implementation of GS. The cost of genotyping large numbers of SNPs is still an impediment, although technologies such as genotyping by sequencing (GBS) are reducing these costs significantly. Collecting large meaningful reference populations can also be costly due to extensive phenotyping. Furthermore, logistical limitations such as DNA extraction turnaround time, SNP genotyping, and biometric analysis have to be carefully considered. These challenges warrant further research and a progressive increase of

GS implementation in areas where studies inside and outside CIMMYT indicate is the greatest opportunities to accelerate genetic gain and transform marker-assistant plant breeding.

References

Abalo G, Tongoona P, Derera J, Edema R (2009) A comparative analysis of conventional and marker-assisted selection methods in breeding maize streak virus resistance in maize. Crop Sci 49:509–520

Anderson JA, Chao S, Liu S (2007). Molecular breeding using a major QTL for Fusarium head blight resistance in wheat. Crop Sci 47(Supplement_3):S-112

Arraiano LS, Worland AJ, Ellerbrook C, Brown JKM (2001) Chromosomal location of a gene for resistance to Septoria tritici blotch (*Mycosphaerella graminicola*) in the hexaploid wheat 'Synthetic 6x'. Theor Appl Genet 103:758–764

Bansal U, Bariana H, Wong D, Randhawa M, Wicker T, Hayden M, Keller B (2014) Molecular mapping of an adult plant stem rust resistance gene *Sr56* in winter wheat cultivar Arina. Theor Appl Genet 127:1441–1448

Basnet BR, Singh RP, Ibrahim AMH, Herrera-Foessel SA, Huerta-Espino J, Lan CX, Rudd JC (2014a) Characterization of Yr54 and other genes associated with adult plant resistance to yellow rust and leaf rust in common wheat Quaiu 3. Mol Breed 33:385–399

Basnet BR, Singh S, Lopez-Vera EE, Huerta-Espino J, Bhavani S, Jin Y, Rouse MN, Singh RP (2014b) Molecular mapping and validation of *SrND643*: A new wheat gene for resistance to stem rust race Ug99. Phytopathology (in press)

Beales J, Turner A, Griffiths S, Snape JW, Laurie DA (2007) A pseudo-response regulator is misexpressed in the photoperiod insensitive Ppd-D1a mutant of wheat (*Triticum aestivum* L.). Theor Appl Genet 115:721–733

Becker H (1993) Pflanzenzüchtung.Ulmerverlag, 327 S

Bennett D, Reynolds M, Mullan D, Izanloo A (2012) Detection of two major grain yield QTL in bread wheat (*Triticum aestivum* L.) under heat, drought and high yield potential environments. Theor Appl Genet 125:1473–1485

Bernardo R (2013) Genomewide selection when major genes are known. Crop Sci 54:68–75

Bernardo R, Yu J (2007) Prospects for genome-wide selection for quantitative traits in maize. Crop Sci 47:1082–1090

Bernardo AN, Ma HX, Zhang DD, Bai GH (2012) Single nucleotide polymorphism in wheat chromosome region harboring *Fhb1* for Fusarium head blight resistance. Mol Breed 29:477–488

Bjarko ME, Line RF (1988) Heritability and number of genes controlling leaf rust resistance in four cultivars of wheat. Phytopathology 78(4):457–461

Bogard M, Ravel C, Paux E, Bordes J, Balfourier F, Chapman S, Le Gouis J, Allard J (2014) Predictions of heading date in bread wheat (*Triticum aestivum* L.) using QTL-based parameters of an ecophysiological model. J Exp Bot 65:5849–5865

Bonneau J, Taylor J, Parent B, Bennett D, Reynolds M, Feuillet C, Langridge P, Mather D (2013) Multi-environment analysis and improved mapping of a yield-related QTL on chromosome 3B of wheat. Theor Appl Genet 126:747–761

Bonnett DG, Rebetzke GJ, Spielmeyer W (2005) Strategies for efficient implementation of molecular markers in wheat breeding. Mol Breed 15:75–85

Borlaug NE (1968) Wheat breeding and its impact on world food supply. Australian Academy of Science, Australia, pp 1–36

Bouis HE, Hotz C, McClafferty B, Meenakshi JV, Pfeiffer WH (2011) Biofortification: a new tool to reduce micronutrient malnutrition. Food Nutr Bull 32(Supplement 1):31S–40S

Breseghello F, Sorrells ME (2007) QTL analysis of kernel size and shape in two hexaploid wheat mapping populations. Field Crop Res 101(2):172–179

Buerstmayr H, Ban T, Anderson JA (2009) QTL mapping and marker-assisted selection for Fusarium head blight resistance in wheat: a review. Plant Breed 128:1–26

Burgueño J, de los Campos G, Weigel K, Crossa J (2012) Genomic prediction of breeding values when modeling genotype × environment interaction using pedigree and dense molecular markers. Crop Sci 52:707

Cai S, Bai GH, Zhang D (2008) Quantitative trait loci for aluminum resistance in Chinese wheat landrace FSW. Theor Appl Genet 117:49–56

Cane K, Eagles HA, Laurie DA, Trevaskis B, Vallance N, Eastwood RF, Gororo NN, Kuchel H, Martin PJ (2013) *Ppd-B1* and *Ppd-D1* and their effects in southern Australian wheat. Crop Pasture Sci 64:100–114

Chang J, Zhang J, Mao X (2013a) Polymorphism of TaSAP1-A1 and its association with agronomic traits in wheat. Planta 237:1495–1508

Chang J, Zhang J, Mao X, Li A, Jia J, Jing R (2013b) Polymorphism of *TaSAP1-A1* and its association with agronomic traits in wheat. Planta 237(6):1495–1508

Chen YH, Carver BF, Wang SW, Cao SH, Yan LL (2010) Genetic regulation of developmental phases in winter wheat. Mol Breed 26:573–582

Chen X, Coram T, Huang X, Wang M, Dolezal A (2013) Understanding molecular mechanisms of durable and non-durable resistance to stripe rust in wheat using a transcriptomics approach. Curr Genomics 14(2):111

Chen A, Li C, Hu W, Lau MY, Lin H, Rockwell NC, Martin SS, Jernstedt JA, Lagarias JC, Dubcovsky J (2014) PHYTOCHROME C plays a major role in the acceleration of wheat flowering under long-day photoperiod. Proc National Acad Sci USA 111:10037–10044

Cheng P, Chen XM (2010) Molecular mapping of a gene for stripe rust resistance in spring wheat cultivar IDO377s. Theor Appl Genet 121:195–204

Chouard P (1960) Vernalization and its relation to dormancy. Annu Rev Plant Physiol 11:191–237

Choulet F, Caccamo M, Wright J, Alaux M, Šimková H, Šafář J, Leroy P, Dolezel J, Rogers J, Eversole K, Feuillet C (2014) The wheat black jack: advances towards sequencing the 21 chromosomes of bread wheat. In: Tuberosa R et al (eds) Genomics of plant genetic resources. Springer Science, The Netherlands, pp 405–438

Cossani CM, Reynolds MP (2012) Physiological traits for improving heat tolerance in wheat. Plant Physiol 160:1710–1718

Cowger C, Hoffer ME, Mundt CC (2000) Specific adaptation by *Mycosphaerella graminicola* to a resistant wheat cultivar. Plant Pathol 49:445–451

Crossa J, Pérez P, Hickey J, Burgueño J, Ornella L, Cerón-Rojas J, Zhang X, Dreisigacker S, Babu R, Li Y, Bonnett D, Mathews K (2014) Genomic prediction in CIMMYT maize and wheat breeding programs. Heredity 112:48–60

Cuthbert PA, Somers DJ, Thomas J, Cloutier S, Brulé-Babel A (2006) Fine mapping *Fhb1*, a major gene controlling fusarium head blight resistance in bread wheat (*Triticum aestivum* L.). Theor Appl Genet 112:1465–1472

Cuthbert PA, Somers DJ, Brulé-Babel A (2007) Mapping of *Fhb2* on chromosome 6BS: a gene controlling Fusarium head blight field resistance in bread wheat (*Triticum aestivum* L.). Theor Appl Genet 114:429–437

Dai J, Bai G, Zhang D, Hong D (2013) Validation of quantitative trait loci for aluminum tolerance in Chinese wheat landrace FSW. Euphytica 192:171–179

Davies J, Berzonsky W, Leach G (2006) A comparison of marker-assisted and phenotypic selection for high grain protein content in spring wheat. Euphytica 152:117–134

de los Campos G, Naya H, Gianola D, Crossa J, Legarra A, Manfredi E, Weigel K, Cotes JM (2009) Predicting quantitative traits with regression models for dense molecular markers and pedigree. Genetics 182:375–385

de los Campos G, Pérez P, Vazquez AI, Crossa J (2013) Genome-enabled prediction using the BLR (Bayesian linear regression) R-package. Genome-wide association studies and genomic prediction. Springer, New York, pp 299–320

Delhaize E, Ryan PR (1995) Aluminum toxicity and tolerance in plants. Plant Physiol 107:315–321

Díaz A, Zikhali M, Turner AS, Isaac P, Laurie DA (2012) Copy number variation affecting the Photoperiod-B1 and Vernalization-A1 genes is associated with altered flowering time in wheat (*Triticum aestivum*). PLoSONE 7:e33234

Distelfeld A, Li C, Dubcovsky J (2009) Regulation of flowering in temperate cereals. Curr Opin Plant Biol 12:1–7

Dong L, Wang F, Liu T, Dong Z, Li A, Jing R, Mao L, Li Y, Liu X, Zhang K, Wang D (2014) Natural variation of TaGASR7-A1 affects grain length in common wheat under multiple cultivation conditions. Mol Breed 34(3):937–947

Dreisigacker S, Kishii M, Lage J, Warburton M (2008) Use of synthetic hexaploid wheat to increase diversity for CIMMYT bread wheat improvement. Aust J Agric Res 59:413–420

Dreisigacker S, Wang X, Cisneros BAM, Jing R, Singh, PK (2015). Adult-plant resistance to Septoria tritici blotch in hexaploid spring wheat. Theor Appl Genet 128(11): 2317–2329

Dubin HJ, Brennan JP (2009) Fighting a 'shifty enemy'. Int Collab Contain Wheat Rusts 2:19–24

Duveiller E, Mezzalama M, Murakami J, Lewis J, Ban T (2008) Global fusarium networking. Cereal Res Commun 36:11–19

Eagles HA, Cane K, Trevaskis B (2011) Veery wheats carry an allele of Vrn-A1 that has implications for freezing tolerance in winter wheats. Plant Breed 130:413–418

Eathington SR, Crosbie TM, Edwards MD, Reiter RS, Bull JK (2007) Molecular markers in a commercial breeding program. Crop Sci 47(Supplement 3):S154

Edae EA, Byrne PF, Manmathan H, Haley SD, Moragues M, Lopes MS, Reynolds MP (2013) Association mapping and nucleotide sequence variation in five drought tolerance candidate genes in spring wheat. Plant Gen 6:2

Edae EA, Byrne PF, Haley SD, Lopes MS, Reynolds MP (2014) Genome-wide association mapping of yield and yield components of spring wheat under contrasting moisture regimes. Theor Appl Genet 127:791–807

Ellis MH, Spielmeyer W, Gale K, Rebetzke G, Richards R (2002) Perfect markers for the Rht-B1b and Rht-D1b dwarfing genes in wheat. Theor Appl Genet 105(6–7):1038–1042

Ellis MH, Rebetzke GJ, Chandler P, Bonnett D, Spielmeyer W, Richards RA (2004) The effect of different height reducing genes on the early growth of wheat. Funct Plant Biol 31(6):583–589

Ellis MH, Rebetzke GJ, Azanza F, Richards RA, Spielmeyer W (2005) Molecular mapping of gibberellin-responsive dwarfing genes in bread wheat. Theor Appl Genet 111:423–430

Faricelli ME, Valarik M, Dubcovsky J (2010) Control of flowering time and spike development in cereals: the earliness per se Eps-1 region in wheat, rice, and Brachypodium. Funct Integr Genomics 10:293–306

Fleury D, Jefferies S, Kuchel H, Langridge P (2010) Genetic and genomic tools to improve drought tolerance in wheat. J Exp Bot 61(12):3211–3222

Flor HH (1942) Inheritance of pathogenicity in *Melampsora lini*. Phytopathology 32(653):e69

Forrest K, Pujol V, Bulli P, Pumphrey M, Wellings C, Herrera-Foessel S, Huerta-Espino J, Singh R, Lagudah E, Hayden M, Spielmeyer W (2014) Development of a SNP marker assay for the *Lr67* gene of wheat using a genotyping by sequencing approach. Mol Breed 34 (4):2109–2118

Foulkes MJ, Sylvester-Bradley R, Weightman R, Snape JW (2007) Identifying physiological traits associated with improved drought resistance in winter wheat. F Crop Res 103:11–24

Frisch M, Melchinger AE (2005) selection theory for marker-assisted backcrossing. Genetics 170:909–917

Frisch M, Bohn M, Melchinger AE (1999) Minimum sample size and optimal positioning of flanking markers in marker-assisted backcrossing for transfer of a target gene. Crop Sci 39:967–975

Fu D, Szucs P, Yan L, Helguera M, Skinner JS, Zitzewitz J, Hayes PM, Dubcovsky J (2005) Large deletions within the first intron in VRN-1 are associated with spring growth habit in barley and wheat. Mol Gen Genom 273:54–65

Gale MD, Youssefian S, Russell GE (1985) Dwarfing genes in wheat. In: GE Russell (ed) Progress in plant breeding. Butterworths, London, pp 1–35

Garg D, Sareen S, Dalal S, Tiwari R, Singh R (2012) Heat shock protein based SNP marker for terminal heat stress in wheat (‘*triticum aestivum*’ L.). Aust J Crop Sci 6:1516–1521

Gawronski P, Schnurbusch T (2012) High-density mapping of the earliness per se-3A^{m} (Eps-3A^{m}) locus in diploid einkorn wheat and its relation to the syntenic regions in rice and *Brachypodium distachyon* L. Mol Breed 30:1097–1108

Ghaffary SM, Robert O, Laurent V, Lonnet P, Margale E, van der Lee TA, Visser RG, Kema GH (2011) Genetic analysis of resistance to Septoria tritici blotch in the French winter wheat cultivars balance and Apache. Theor Appl Genet 123:741–754

Ghaffary SMT, Faris JD, Friesen TL, Visser RGF, van der Lee TAJ, Robert O, Kema GHJ (2012) New broad-spectrum resistance to Septoria tritici blotch derived from synthetic hexaploid wheat. Theor Appl Genet 124:125–142

Ghazvini H, Hiebert CW, Zegeye T, Liu S, Dilawari M, Tsilo T, Anderson JA, Rouse MN, Jin Y, Fetch T (2012) Inheritance of resistance to Ug99 stem rust in wheat cultivar Norin 40 and genetic mapping of *Sr42*. Theor Appl Genet 125:817–824

Gilbert J, Haber S (2013) Overview of some recent research developments in fusarium head blight of wheat. Can J Plant Pathol 35:149–174

Goodwin SB (2012) Resistance in wheat to septoria diseases caused by *Mycosphaerella graminicola* (*Septoria tritici*) and *Phaeosphaeria* (*Stagonospora*) *nodorum*. In: Sharma I (ed) Disease resistance in wheat. CABI, Wallingford, pp 151–159

Goodwin SB, Cavaletto JR, Hale IL, Thompson I, Xu SS, Adhikari TB, Dubcovsky J (2015) A new map location of gene Stb3 for resistance to Septoria tritici blotch in wheat. Crop Sci 55 (1):35–43

Griffiths S, Simmonds J, Leverington M, Wang Y, Fish L, Sayers L, Alibert L, Orford S, Wingen L, Herry L, Faure S, Laurie D, Bilham L, Snape J (2009) Meta-QTL analysis of, the genetic control of ear emergence in elite European, winter wheat germplasm. Theor Appl Genet 119:383–395

Groos C, Robert N, Bervas E, Charmet G (2003) Genetic analysis of grain protein-content, grain yield and thousand-kernel weight in bread wheat. Theor Appl Genet 106(6):1032–1040

Hanocq E, Laperche A, Jaminon O, Laine AL, Le Gouis J (2007) Most significant genome regions involved in the control of earliness traits in bread wheat, as revealed by QTL meta-analysis. Theor Appl Genet 114:569–584

Hanson WD (1959) Minimum family sizes for the planning of genetic experiments. Agron J 51:711–715

Hao Y, Velu G, Peña RJ, Singh S, Singh RP (2014) Genetic loci associated with high grain zinc concentration and pleiotropic effect on kernel weight in wheat (*Triticum aestivum* L.). Molecular Breed 34(4):1893–1902

He X, Singh PK, Duveiller E, Dreisigacker S, Singh RP (2013a) Development and characterization of International Maize and Wheat Improvement Center (CIMMYT) germplasm for fusarium head blight resistance. In: Alconada TM, Chulze SN (eds) Fusarium head blight in Latin America. Springer, Berlin, pp 241–262

He X, Singh PK, Duveiller E, Schlang N, Dreisigacker S, Singh RP (2013b) Identification and characterization of international Fusarium head blight screening nurseries of wheat at CIMMYT, Mexico. Eur J Plant Pathol 136:123–134

Helguera M, Khan IA, Dubcovsky J (2000) Development of PCR markers for the wheat leaf rust resistance gene *Lr47*. Theor Appl Genet 100:1137–1143

Helguera M, Khan IA, Kolmer J, Lijavetzky D, Zhong-qi L, Dubcovsky J (2003) PCR assays for the *Lr37-Yr17-Sr38* cluster of red resistance genes and their use to develop isogenic hard red spring wheat lines. Crop Sci 43:1839–1847

Helguera M, Vanzetti L, Soria M, Khan IA, Kolmer J, Dubcovsky J (2005) PCR markers for *Triticum speltoides* leaf rust resistance gene *Lr51* and their use to develop isogenic hard red spring wheat lines. Crop Sci 45:728–734

Herrera-Foessel SA, Singh RP, Huerta-Espino J, Rosewarne GM, Periyannan SK, Viccars L, Calvo-Salazar V, Lan CX, Lagudah ES (2012) *Lr68*: a new gene conferring slow rusting resistance to leaf rust in wheat. Theor Appl Genet 124:1475–1486

Hickey JM, Dreisigacker S, Crossa J, Hearne S, Babu R, Prasanna BM, Grondona M, Zambelli A, Windhausen VS, Mathews K, Gorjanc G (2014) Evaluation of genomic selection training population designs and genotyping strategies in plant breeding programs using simulation. Crop Sci 54:1476–1488

Holland J (2001) Epistasis and plant breeding. Plant Breed Rev 21:27–92

Hoogendoorn J (1985) A reciprocal F1 monosomic analysis of the genetic control of time of ear emergence, number of leaves and number of spikelets in wheat (*Triticum aestivum* L.). Euphytica 34:545–558

Hospital F, Chevalet C, Mulsant P (1992) Using markers in gene introgression breeding programs. Genetics 132:1199–1210

Hou J, Jiang QY, Hao CY, Wang YQ, Zhang HN, Zhang XY (2014) Global selection on sucrose synthase haplotypes during a century of wheat breeding. Plant Physiol 164:1918–1929

Ikeda M, Miura K, Aya K, Kitano H, Matsuoka M (2013) Genes offering the potential for designing yield-related traits in rice. Curr Opin Plant Biol 16(2):213–220

Jarquín D, Crossa J, Lacaze X, Du Cheyron P, Daucourt J, Lorgeou J, Piraux F, Guerreiro L, Pérez P, Calus M (2013) A reaction norm model for genomic selection using high-dimensional genomic and environmental data. Theor Appl Genet 1–13

Jiang Y, Jiang Q, Hao C, Hou J, Wang L, Zhang H, Zhang S, Chen X, Zhang X (2015) A yield-associated gene TaCWI, in wheat: its function, selection and evolution in global breeding revealed by haplotype analysis. Theor Appl Genet 128(1):131–143

Johnson R, Law CN (1973) Cytogenetic studies in the resistance of the wheat variety Bersée to *Puccinia striiformis*. Cereal Rusts Bull 1:38–43

Kamran A, Iqbal M, Spaner D (2014) Flowering time in wheat (*Triticum aestivum* L.): a key factor for global adaptability. Euphytica 197:1–26

Kato K, Yamashita M, Ishimoto K, Yoshino H, Fujita M (2003) Genetic analysis of two genes for vernalization response, the former Vrn2 and Vrn4, by using PCR based molecular markers. In: Pogna NE, Romano M, Pogna EA, Galterio G (eds) Proceedings of the 10th international wheat genetics symposium, vol 3. Instituto Sperimentale per la Cerealicoltura, Paestum, pp 971–973

Kelm C, Ghaffary SMT, Bruelheide H, Röder MS, Miersch S, Weber WE, Kema GHK, Saal B (2012) The genetic architecture of seedling resistance to Septoria tritici blotch in the winter wheat doubled-haploid population Solitär × Mazurka. Mol Breed 29:813–830

Khlestkina EK, Giura A, Roder MS, Borner A (2009) A new gene controlling the flowering response to photoperiod in wheat. Euphytica 165:579–585

Kochian V, Piñeros MA, Hoekenga OA (2005) The physiology, genetics and molecular biology of plant aluminum resistance and toxicity. Plant Soil 274:175–195

Kosellek C, Pillen K, Nelson JC, Weber EW, Saal B (2013) Inheritance of field resistance to Septoria tritici blotch in the wheat doubled-haploid population Solitär × Mazurka. Euphytica 194:161–176

Kuchel H, Warner P, Fox RL, Chalmers K, Howes N, Langridge P, Jefferies SP (2003) Whole genome based marker assisted selection strategies in wheat breeding. In: Pogna NE, Romano M, Pogna EA, Galterio G (eds) Proceedings of the 10th international wheat genet symposium pasteum, Italy, vol 1. Instituto Sperimentale per la Cerealicoltura, Rome, pp 144–147

Kuchel H, Fox R, Reinheimer J, Mosionek L, Willey N, Bariana H, Jefferies S (2007a) The successful application of a marker-assisted wheat breeding strategy. Mol Breed 20:295–308

Kuchel H, Williams KJ, Langridge P, Eagles HA, Jefferies SP (2007b) Genetic dissection of grain yield in bread wheat I. QTL analysis. Theor Appl Genet 115:1029–1041

Kumar N, Kuwal PL, Gaur A, Tyagi AK, Khurana JP, Khurana P, Balyan HS, Gupta PK (2006) QTL analyses for grain weight in common wheat. Euphytica 151:135–144

Lagudah ES, Krattinger SG, Herrera-Foessel S, Singh RP, Huerta-Espino J, Spielmeyer W, Brown-Guedira G, Selter LL, Keller B (2009) Gene-specific markers for the wheat gene *Lr34/Yr18/Pm28* which confers resistance to multiple fungal pathogens. Theor Appl Genet 119:889–898

Lan CX, Rosewarne GM, Singh RP, Herrera-Foessel SA, Huerta-Espino J, Basnet BR, Zhang YL, Yang EN (2014) Genetic analysis of resistance to leaf rust and stripe rust in wheat cultivar Francolin#1. Plant Dis 98(9):1227–1234

Li ZF, Lan CX, He ZH, Singh RP, Rosewarne GM, Chen XM, Xia XC (2014) Overview and application of QTL for adult plant resistance to leaf rust and powdery mildew in wheat. Crop Sci 54(5):1907–1925

Lillemo M, Chen F, Xia X, William M, Peña RJ, Trethowan R, He Z (2006) Puroindoline grain hardness alleles in CIMMYT bread wheat germplasm. J Cereal Sci 44:86–92

Lin F, Chen XM (2007) Genetics and molecular mapping of genes for race-specific all-stage resistance and non-race-specific high-temperature adult-plant resistance to stripe rust in spring wheat cultivar Alpowa. Theor Appl Genet 114:1277–1287

Lin Z, Hayes BJ, Daetwyler HD (2014) Genomic selection in crops, trees and forages: a review. Crop Pasture Sci 65(11):1177–1191

Liu L, He Z, Yan J, Zhang Y, Xia X, Peña RJ (2005) Allelic variation at the Glu-1 and Glu-3 loci, presence of the 1B.1R translocation, and their effects on mixographic properties in Chinese bread wheats. Euphytica 142:197–204

Liu SY, Hall MD, Griffey CA, McKendry AL (2009) Meta-analysis of QTL associated with fusarium head blight resistance in wheat. Crop Sci 49:1955–1968

Liu Y, He Z, Appels R, Xia X (2012) Functional markers in wheat: current status and future prospects. Theor Appl Genet 125:1–10

Liu J, Chang ZJ, Zhang XJ, Yang ZJ, Li X, Jia JQ, Zhan HX, Guo HJ, Wang JM (2013a) Putative *Thinopyrum intermedium*-derived stripe rust resistance gene *Yr50* maps on wheat chromosome arm 4BL. Theor Appl Genet 126:265–274

Liu Y, Zhang L, Thompson IA, Goodwin SB, Ohm HW (2013b) Molecular mapping re-locates the *Stb2* gene for resistance to Septoria tritici blotch derived from cultivar Veranopolis on wheat chromosome 1BS. Euphytica 190:145–156

Longin CF, Reif JC (2014) Redesigning the exploitation of wheat genetic resources. Trends Plant Sci 19(10):631–636

Lopes MS, Dreisigacker S, Peña RJ, Sukumaran S, Reynolds MP (2014) Genetic characterization of the wheat association mapping initiative (WAMI) panel for dissection of complex traits in spring wheat. Theor Appl Genet 1–12

Lopez-Vera EE, Nelson S, Singh RP, Basnet BR, Haley SD, Bhavani S, Huerta-Espino J, Xoconostle-Cazares B, Ruiz-Medrano R, Rouse MN, Singh S (2014a) Resistance to stem rust Ug99 in six bread wheat cultivars maps to chromosome 6DS. Theor Appl Genet 127:231–239

Lopez-Vera EE, Singh S, Bhavani S, Basnet BR, Singh RP, Huerta-Espino J, Rouse MN, Xoconostle-Cázares B, Ruiz-Medrano R (2014b) Identification of genes on wheat chromosome 2B conferring resistance to stem rust fungus race Ug99. In: Plant and animal genome XXII conference, 2014, 14–15 Jan, San Diego, California, USA

Lorenz AJ, Chao SM, Asoro FG, Heffner EL, Hayashi T, Iwata H, Smith KP, Sorrells ME, Jannink JL (2011) Genomic selection in plant breeding: knowledge and prospects. In: Sparks DL (ed) Advances in agronomy, vol 110. Elsevier Academic Press Inc., San Diego, CA, USA, pp 77–123

Luo PG, Hu XY, Ren ZL, Zhang HY, Shu K, Yang ZJ (2008) Allelic analysis of stripe rust resistance gene on wheat chromosome 2B. Genome 51:922–927

Ma H-X, Bai G-H, Carver BF, Zhou L-L (2005) Molecular mapping of a quantitative trait locus for aluminum tolerance in wheat cultivar Atlas 66. Theor Appl Genet 112:51–57

Ma DY, Yan J, He ZH, Wu L, Xia XC (2012) Characterization of a cell wall invertase gene TaCwi-A1 on common wheat chromosome 2A and development of functional markers. Mol Breed 29:43–52

Mago R, Bariana HS, Dundas IS, Spielmeyer W, Lawrence GJ, Pryor AJ, Ellis JG (2005) Development of PCR markers for the selection of wheat stem rust resistance genes *Sr24* and *Sr26* in diverse wheat germplasm. Theor Appl Genet 111:496–504

Mago R, Brown-Guedira G, Dreisigacker S, Breen J, Jin Y, Singh R, Appels R, Lagudah ES, Ellis J, Spielmeyer W (2011) An accurate DNA marker assay for stem rust resistance gene *Sr2* in wheat. Theor Appl Genet 122:735–744

Martin J, Frohberg R, Morris C (2001) Milling and bread baking traits associated with puroindoline sequence type in hard red spring wheat. Crop Sci 41:228–234

Mason RE, Mondal S, Beecher FW, Pacheco A (2010) QTL associated with heat susceptibility index in wheat (*Triticum aestivum* L.) under short-term reproductive stage heat stress. Euphytica 174:423–436

McCartney CA, Somers DJ, McCallum BD, Thomas J, Humphreys DG, Menzies JG, Brown PD (2005) Microsatellite tagging of the leaf rust resistance gene Lr16 on wheat chromosome 2BSc. Mol Breed 15:329–337

McIntosh RA, Yamazaki Y, Dubcovsky J, Rogers WJ, Morris C, Appels R, Xia XC (2014) Catalogue of gene symbols for wheat. http://www.shigen.nig.ac.jp/wheat/komugi/genes/macgene/2013/GeneSymbol.pdf. Accessed on 23 May 2014

McLauchlan A, Ogbonnaya FC, Hollingsworth B, Carter M, Gale KR, Henry RJ, Holton TA, Morell MK, Rampling LR, Sharp PJ, Shariflou MR, Jones MGK, Appels R (2001) Development of robust PCR-based DNA markers for each homoeo-allele of granule-bound starch synthase and their application in wheat breeding programs. Aust J Agric Res 52:1409–1416

Meuwissen THE (2009) Accuracy of breeding values of 'unrelated' individuals predicted by dense SNP genotyping. Genet Select Evol 41:35

Meuwissen THE, Hayes BJ, Goddard ME (2001) Prediction of total genetic value using genome-wide dense marker maps. Genetics 157:1819–1829

Miedaner T, Wilde F, Korzun V, Ebmeyer E, Schmolke M, Hartl L, Schoen CC (2009) Marker selection for fusarium head blight resistance based on quantitative trait loci (qtl) from two european sources compared to phenotypic selection in winter wheat. Euphytica 166:219–227

Mohammadi V, Zali AA (2010) Mapping QTLs for heat tolerance in wheat. J Agric Sci 10:261–267

Mondal S, Mason RE, Huggins T, Hays DB (2015) QTL on wheat (*Triticum aestivum* L.) chromosomes 1B, 3D and 5A are associated with constitutive production of leaf cuticular wax and may contribute to lower leaf temperatures under heat stress. Euphytica 201(1):123–130

Moreau L, Charcosset A, Gallais A (2004) Experimental evaluation of several cycles of marker-assisted selection in maize. Euphytica 137:111–118

Morell MK, Rahman S, Regina A, Appels R, Li Z (2001) Wheat starch biosynthesis. Euphytica 119:55–58

Morris CF (2002) Puroindolines: the molecular genetic basis of wheat grain hardness. Plant Mol Biol 48:633–647

Morris CF, Bhave M (2008) Reconciliation of D-genome puroindoline allele designationswith current DNA sequence data. J Cereal Sci 48:277–287

Navakode S, Weidner A, Lohwasser U, Röder MS, Börner A (2009) Molecular mapping of quantitative trait loci (QTLs) controlling aluminium tolerance in bread wheat. Euphytica 166 (2):283–290

Nishida H, Yoshida T, Kawakami K, Fujita M, Long B, Akashi Y, Laurie DA, Kato K (2013) Structural variation in the 5′ upstream region of photoperiod-insensitive alleles Ppd-A1a and Ppd-B1a identified in hexaploid wheat (*Triticum aestivum* L.), and their effect on heading time. Mol Breed 31:27–37

Olson EL, Brown-Guedira G, Marshall D, Stack E, Bowden RL, Jin Y, Rouse M, Pumphrey MO (2010) Development of wheat lines having a small introgressed segment carrying stem rust resistance gene *Sr22*. Crop Sci 50:1823–1830

Paliwal R, Röder MS, Kumar U, Srivastava JP, Joshi AK (2012) QTL mapping of terminal heat tolerance in hexaploid wheat (*T. aestivum* L.). Theor Appl Genet 125:561–575

Pallotta M, Schnurbusch T, Hayes J, Hay A, Baumann U, Paull J, Langridge P, Sutton T (2014) Molecular basis of adaptation to high soil boron in wheat landraces and elite cultivars. Nature 514:88–91

Payne PI, Nightingale MA, Krattiger AF, Holt LM (1987) The relationship between HMW glutenin composition and the bread-making quality of British-grown wheat varieties. J Sci Food Agric 40:51–65

Perez-Rodriguez P, Gianola D, González-Camacho JM, Crossa J, Manes Y, Dreisigacker S (2012) Comparison between linear and non-parametric regression models for genome-enabled prediction in wheat. G3 Genes Genomes Genet 2:1595–1605

Periyannan S, Moore J, Ayliffe M, Bansal U, Wang X, Huang L, Deal K, Luo M, Kong X, Bariana H, Mago R, McIntosh R, Dodds P, Dvorak J, Lagudah E (2013) The gene *Sr33*, an ortholog of barley *Mla* genes, encodes resistance to wheat stem rust race Ug99. Science 341:786–788

Pinto RS, Reynolds MP, Mathews KL, McIntyre CL, Olivares-Villegas JJ, Chapman SC (2010) Heat and drought adaptive QTL in a wheat population designed to minimize confounding agronomic effects. Theor Appl Genet 121:1001–1021

Poland J, Endelman J, Dawson J, Rutkoski J, Wu S, Manes Y, Dreisigacker S, Crossa J, Sanchez-Villeda H, Sorrells M, Jannink J-L (2012) Genomic selection in wheat breeding using genotyping-by-sequencing. Plant Genome 5(3):103–113

Posner ES (2000) Wheat. In: Kulp K, Ponte JGJ (eds) Handbook of cereal science and technology. Marcel Dekker, Inc, New York, pp 1–29

Prasad PVV, Pisipati SR, Momčilović I, Ristic Z (2011) Independent and combined effects of high temperature and drought stress during grain filling on plant yield and chloroplast EF-Tu expression in spring wheat. J Agron Crop Sci 197:430–441

Prigge V, Maurer HP, Mackhill DJ, Melchinger AE, Frisch M (2008) Comparison of the observed with the simulated distribution of the parental genome contribution in two marker-assisted backcross programs in rice. Theor Appl Genet 116:739–744

Prigge V, Melchinger AE, Dhillon BS, Frisch M (2009) Efficiency Gain of marker-assisted backcrossing by sequentially increasing marker density over generations. Theor Appl Genet 119:23–32

Qi LL, Pumphrey MO, Friebe B, Chen PD, Gill BS (2008) Molecular cytogenetic characterization of alien introgressions with gene *Fhb3* for resistance to *Fusarium* head blight disease of wheat. Theor Appl Genet 117:1155–1166

Quarrie SA, Gulli M, Calestani C, Steed A, Marmiroli N (1994) Location of a gene regulating drought-induced abscisic acid production on the long arm of chromosome 5A of wheat. Theor Appl Genet 89:794–800

Ragupathy R, Naeem HA, Reimer E, Lukow OM, Sapirstein HD, Cloutier S (2008) Evolutionary origin of the segmental duplication encompassing the wheat GLU-B1 locus encoding the overexpressed Bx7 (Bx7OE) high molecular weight glutenin subunit. Theor Appl Genet 116 (2):283–296

Raman H, Milgate A (2012) Molecular breeding for Septoria tritici blotch resistance in wheat. Cereal Res Commun 40:451–466

Raman H, Zhang K, Cakir M et al (2005) Molecular characterization and mapping of ALMT1, the aluminium-tolerance gene of bread wheat (*Triticum aestivum L.*). Genome 48:781–791

Raman H, Raman R, Wood R, Martin P (2006) Repetitive indel markers within the ALMT1 gene conditioning aluminium tolerance in wheat (*Triticum aestivum L.*). Mol Breed 18:171–183

Rasheed A, Mahmood T, Gul-Kazi A, Mujeeb-Kazi A (2013) An overview of omics for wheat grainquality improvement. In: Hakeem KR, Ahmad P, Ozturk M (eds) Crop improvement. Springer, New York, pp 307–344

Rashid A, Stark JC, Tanveer A, Mustafa T (1999) Use of canopy temperature measurements as a screening tool for drought tolerance in spring wheat. J Agron Crop Sci 182(4):231–238

Rebetzke GJ, Rattey AR, Farquhar GD (2013) Genomic regions for canopy temperature and their genetic association with stomatal conductance and grain yield in wheat. Funct Plant Biol 40:14–33

Ren RS, Wang MN, Chen XM, Zhang ZJ (2012) Characterization and molecular mapping of *Yr52* for high-temperature adult-plant resistance to stripe rust in spring wheat germplasm PI 183527. Theor Appl Genet 125:847–857

Reynolds M, Tuberosa R (2008) Translational research impacting on crop productivity in drought-prone environments. Curr Opin Plant Biol 11:171–179

Reynolds M, Foulkes MJ, Slafer GA, Berry P, Parry MAJ, Snape JW, Angus WJ (2009) Raising yield potential in wheat. J Exp Bot 60:1899–1918

Ribaut J-M, Hoisington D (1998) Marker-assisted selection: new tools and strategies. Trends Plant Sci 3:236–239

Risser P, Ebmeyer E, Korzun V, Hartl L, Miedaner T (2011) Quantitative trait loci for adult-plant resistance to *Mycosphaerella graminicola* in two winter wheat populations. Phytopathology 101:1209–1216

Rustgi S, Shafqat MN, Kumar N, Baenziger PS, Ali ML, Dweikat I, Campbell BT, Gill KS (2013) Genetic dissection of yield and its component traits using high-density composite map of wheat chromosome 3A: bridging gaps between QTLs and underlying genes. PLoS ONE 8(7):e70526

Rutkoski JE, Poland JA, Singh RP, Huerta-Espino J, Bhavani S, Barbier H, Rouse MR, Jannink J-L, Sorrells ME (2014) Genomic selection for quantitative adult plant stem rust resistance in wheat. The Plant Genome 7(3)

Saari EE, Prescott JM (1985) World distribution in relation to economic losses. In: Roelfs AP, Bushnell WR (eds) The cereal rusts vol II: diseases, distribution, epidemiology and control. Academic Press, Orlando, pp 259–298

Saito M, Vrinten P, Ishikawa G, Graybosch R, Nakamura T (2009) A novel codominant marker for selection of the null Wx-B1 allele in wheat breeding programs. Molecular Breed 23:209–217

Sambasivam PK, Bansal UK, Hayden MJ, Dvorak J, Lagudah ES, Bariana HS (2008) Identification of markers linked with stem rust resistance genes *Sr33* and *Sr45*. In: Appels R, Eastwood RF, Lagudah ES, Langridge P, Mackay M, Lynne (eds) Proceedings of the 11th international wheat genetics symposium. Sydney University Press, Sydney

Sasaki T, Yamamoto Y, Ezaki B, Katsuhara M, Ahn SJ, Ryan PR, Delhaize E, Matsumoto H (2004) A wheat gene encoding an aluminum-activated malate transporter. Plant J 37:645–653

Scarth R, Kirby EJM, Law CN (1985) Effects of the photoperiod genes *Ppd1* and *Ppd2* on growth and development of the shoot apex in wheat. Ann Bot 55(3):351–359

Simmonds J, Scott P, Leverington-Waite M, Turner AS, Brinton J, Korzun V, Snape J, Uauy C (2014) Identification and independent validation of a stable yield and thousand grain weight QTL on chromosome 6A of hexaploid wheat (*Triticum aestivum* L.). BMC Plant Biol 14:191

Simón MR, Cordo CA, Castillo NS, Struik PC, Börner A (2012) Population structure of *Mycosphaerella graminicola* and location of genes for resistance to the pathogen: recent advances in Argentina. Int J Agron 680275:7

Simons K, Abate Z, Chao S, Zhang W, Rouse M, Jin Y, Elias E, Dubcovsky J (2011) Genetic mapping of stem rust resistance gene *Sr13* in tetraploid wheat (*Triticum turgidum* ssp. *durum* L.). Theor Appl Genet 122:649–658

Snape JW, Quarrie SA, Laurie DA (1996) Comparative mapping and its use for the genetic analysis of agronomic characters in wheat. Euphytica 89:27–31

Snape JW, Foulkes MJ, Simmonds J, Leverington M, Fish LJ, Wang Y, Ciavarrella M (2007) Dissecting gene × environmental effects of wheat via QLT and physiological analysis. Euphytica 154:401–408

Solberg TR, Sonesson AK, Woolliams JA, Meuwissen THE (2008) Genomic selection using different marker types and densities. J Anim Sci 86:2447–2454

Spielmeyer W, Sharp PJ, Lagudah ES (2003) Identification and validation of markers linked to broad-spectrum stem rust resistance gene *Sr2* in wheat (*Triticum aestivum* L.). Crop Sci 43:333–336

Stelmakh AF (1990) Geographic distribution of Vrn genes in landraces and improved varieties of spring bread wheat. Euphytica 45:113–118

Stelmakh AF (1993) Genetic effect of Vrn genes on heading date and agronomic traits in bread wheat. Euphytica 65:53–60

Stelmakh AF (1998) Genetic systems regulating flowering response in wheat. Euphytica 100:359–369

Su Z, Hao C, Wang L, Dong Y, Zhang X (2011) Identification and development of a functional marker of TaGW2 associated with grain weight in bread wheat (*Triticum aestivum* L.). Theor Appl Genet 122(1):211–223

Sukumaran S, Ellis MH, Lopes MS, Pinto S, Bonnett D, Dreisigacker S, Mullan D, Reynolds MP (2013) Wheat mapping populations available at CIMMYT for yield potential research abstract the physiological breeding group in CIMMYT has assembled a number of wheat mapping populations for characterizing yield potential and adaptation of wheat lines to climate. In: 3rd international work. wheat yield consortium CENEB, CIMMYT, Cd. Obregon, Mexico, 5–7 March, pp 78–82

Sukumaran S, Dreisigacker S, Lopes M, Chavez P, Reynolds MP (2015) Genome-wide association study for grain yield and related traits in an elite spring wheat population grown in temperate irrigated environments. Theor Appl Genet 128:353–363

Sun X-Y, Wu K, Zhao Y, Kong F-M, Han G-Z, Jiang H-M, Huang X-J, Li R-J, Wang H-G, Li S-S (2009) QTL analysis of kernel shape and weight using recombinant inbred lines in wheat. Euphytica 165(3):615–624

Takata K, Ikeda TM, Yanaka M, Matsunaka H, Seki M, Ishikawa N, Yamauchi H (2010) Comparison of five puroindoline alleles on grain hardness and flour properties using near isogenic wheat lines. Breeding Science 60:228–232

Talbert LE, Blake NK, Chee PW, Blake TK, Magyar GM (1994) Evaluation of "sequence-tagged-site" PCR products as molecular markers in wheat. Theor Appl Genet 87:789–794

Talukder SK, Babar MA, Vijayalakshmi K, Poland J, Prasad PV, Bowden R, Fritz A (2014) Mapping QTL for the traits associated with heat tolerance in wheat (*Triticum aestivum L.*). BMC Genet 15(1):97

Tang Y, Garvin DF, Kochian LV, Sorrells ME, Carver BF (2002) Physiological genetics of aluminum tolerance in the wheat cultivar Atlas 66. Crop Sci 42:1541–1546

Tanksley SD, Young ND, Patterson AH, Bonierbale MW (1989) RFLP mapping in plant breeding. New tools for an old science. Bio/Technology 7:257–263

R Development Core Team (2010) R: a language and environment for statistical computing. Available at http://www.r-project.org/

Thapa R, Wise K, Goodwin S, Brown-Guedira G, Ohm H (2013) Determining the order of resistance genes *Qsng-3BS*, *Fhb1* and *Sr2* and combining them in coupling on wheat chromosome 3BS. In: Proceedings of the 12th international wheat genetics symposium, Pacifico Yokohama, Japan, p 84

Trevaskis B, Hemming MN, Dennis ES, Peacock WJ (2007) The molecular basis of vernalization-induced flowering in cereals. Trends Plant Sci 12:351–357

Tsilo TJ, Hareland GA, Simsek S, Chao S, Anderson JA (2010) Genome mapping of kernel characteristics in hard red spring wheat breeding lines. Theor Appl Genet 121:717–730

Uauy C, Brevis JC, Chen XM, Khan I, Jackson L, Chicaiza O, Distelfeld A, Fahima T, Dubcovsky J (2005) High-temperature adult-plant (HTAP) stripe rust resistance gene *Yr36* from *Triticum turgidum* ssp. *Dicoccoides* is closely linked to the grain protein content locus *Gpc-B1*. Theor Appl Genet 112:97–105

Uauy C, Brevis JC, Dubcovsky J (2006) The high grain protein content gene Gpc-B1 accelerates senescence and has pleiotropic effects on protein content in wheat. J Exp Bot 57(11):2785–2794

van Beem J, Mohler V, Lukman R, van Ginkel M, William M, Crossa J, Worland AJ (2005) Analysis of genetic factors influencing the developmental rate of globally important CIMMYT wheat cultivars. Crop Sci 45:2113–2119

Velu G, Singh RP, Huerta-Espino J, Peña RJ, Arun B, Mahendru-Singh A, Yaqub Mujahid M, SohufVS MavifGS, CrossaJ AlvaradoaG, Joshig AK, Pfeiffer WH (2012) Performance of biofortified spring wheat genotypes in target environments for grain zinc and iron concentrations. Field Crops Res 137:261–267

Vijayalakshmi K, Fritz AK, Paulsen GM, Bai G (2010) Modeling and mapping QTL for senescence-related traits in winter wheat under high temperature. Mol Breed 26:163–175

Wang W, Vinocur B, Shoseyov O, Altman A (2004) Role of plant heat-shock proteins and molecular chaperones in abiotic stress response. Trends Plant Sci 9:244–252

Wang J, Chapman SC, Bonnett DG, Rebetzke GJ, Crouch J (2007) Application of population genetic theory and simulation models to efficiently pyramid multiple gene via marker-assisted selection. Crop Sci 47:582–590

Wang C, Zhang Y, Han D, Kang Z, Li G, Cao A, Chen P (2008) SSR and STS markers for wheat stripe rust resistance gene Yr26. Euphytica 159(3):359–366

Wang LH, Zhao XL, He ZH, Ma W, AppelsR Peña RJ, Xia XC (2009) Characterization of low-molecular-weight glutenin subunit Glu-B3 genes and development of STS markers in common wheat (*Triticum aestivum* L.). Theor Appl Genet 118:525–539

Wang L, Li G, Peña RJ, Xia X, He Z (2010) Development of STS markers and establishment of multiplex PCR for Glu-A3 alleles in common wheat (*Triticum aestivum* L.). J Cereal Sci 51 (3):305–312

Wellings CR (2011). Global status of stripe rust: a review of historical and current threats. Euphytica 179(1): 129–141

Weng Y, Azhaguvel P, Devkota RN, Rudd JC (2007) PCR-based markers for detection of different sources of 1AL.1RS and 1BL.1RS wheat–rye translocations in wheat background. Plant Breeding 126: 482–486

Wilde F, Korzun V, Ebmeyer E, Geiger H, Miedaner T (2007) Comparison of phenotypic and marker-based selection for fusarium head blight resistance and don content in spring wheat. Mol Breed 19:357–370

Wilhelm EP, Turner AS, Laurie DA (2009) Photoperiod insensitive Ppd-A1a mutations in tetraploid wheat (*Triticum durum* Desf.). Theor Appl Genet 118(2):285–294

Wimmer V, Lehermeier C, Albrecht T, Auinger HJ, Wang Y, Schön CC (2013) Genome-wide prediction of traits with different genetic architecture through efficient variable selection. Genetics 195(2):573–587

Wang S, Wong, D, Forrest K, Allen A, Chao S, Huang BE, Maccaferri M, Salvi S, Milner S, Cattivelli L, Mastrangelo AM, Whan A, Stephen S, Barker G, Wieseke R, Plieske J, IWGSC, Lillemo M, Mather D, Appels R, Dolferus R, Brown-Guedira G, Korol A, Akhunova A, Feuillet C, Salse J, Morgante M, Pozniak C, Luo M, Dvorak J, Morell M,Dubcovsky J, Ganal M, Tuberosa R, Lawley C, Mikoulitch I, Cavanagh C, Edwards KJ, Hayden M, Akhunov E (2014) Characterization of polyploid wheat genomic diversity using a high-density 90,000 single nucleotide polymorphism array. Plant Biotechnol J 12(6):787–796

Worland AJ (1996) The influence of flowering time genes on environmental adaptability in European wheats. Euphytica 89:49–57

Worland T, Snape JW (2001) Genetic basis of worldwide wheat varietal improvement. In: Bonjean AP, Angus WJ (eds) The world wheat book: a history of wheat breeding. Intercept Ltd., London, pp 61–67

Xu Y, Crouch JH (2008) Marker-assisted selection in plant breeding: from publications to practice. Crop Sci 48:391–407

Xue S, Li GQ, Jia HY, Xu F, Lin F, Tang MZ, Wang Y, An X, Xu HB, Zhang LX, Kong ZX, Ma ZQ (2010) Fine mapping *Fhb4*, a major QTL conditioning resistance to Fusarium infection in bread wheat (*Triticum aestivum* L.). Theor Appl Genet 121:147–156

Xue S, Xu F, Tang M, Zhou Y, Li G, An X, Lin F, Xu H, Jia H, Zhang L, Kong Z, Ma Z (2011) Precise mapping *Fhb5*, a major QTL conditioning resistance to *Fusarium* infection in bread wheat (*Triticum aestivum* L.). Theor Appl Genet 123:1055–1063

Yamamori M, Nakamura T, Endo TR, Nagamine T (1994) Waxy protein deficiency and chromosomal location of coding genes in common wheat. TheorAppl Genet 89:179–184

Yan L, Loukoianov A, Blechl A, Tranquilli G, Ramakrishna W, SanMiguel P, Bennetzen JL, Echenique V, Dubcovsky J (2004a) The wheat VRN2 gene is a flowering repressor down-regulated by vernalization. Science 303:1640–1644

Yan L, Helgura M, Kato K, Fukuyama S, Sherman J, Dubcovsky J (2004b) Allelic variation at the Vrn-1 promoter region in polyploid wheat. Theor Appl Genet 109:1677–1686

Yan L, Fu D, Li C, Blechl A, Tranquilli G, Bonafede M, Sanchez A, Valarik M, Yasuda S, Dubkovsky J (2006) The wheat and barley vernalization gene *VRN3* is an orthologue of FT. Proc Natl Acad Sci USA 103:19581–19586

Yang J, Sears RG, Gill BS, Paulsen GM (2002) Quantitative and molecular characterization of heat tolerance in hexaploid wheat. Euphytica 126(2):275–282

Yang Z, Bai Z, Li X, Wang P, Wu Q, Yang L, Li L, Li X (2012) SNP identification and allelic-specific PCR markers development for TaGW2, a gene linked to wheat kernel weight. Theor Appl Genet 125(5):1057–1068

Yoshida T, Nishida H, Zhu J, Nitcher R, Distelfeld A, Akashi Y, Kato K, Dubcovsky J (2010) *Vrn-D4* is a vernalization gene located on the centromeric region of chromosome 5D in hexaploid wheat. Theor Appl Genet 120:543–552

Zhang W, Dubcovsky J (2008) Association between allelic variation at the Phytoene synthase 1 gene and yellow pigment content in the wheat grain. Theor Appl Genet 116:635–645

Zhang L-Y, Liu D-C, Guo X-L, Yang W-L, Sun J-Z, Wang D-W, Zhang A (2010a) Genomic distribution of quantitative trait loci for yield and yield-related traits in common wheat. J Integr Plant Biol 52(11):996–1007

Zhang W, Olson E, Saintenac C, Rouse M, Abate Z, Jin Y, Akhunov E, Pumphrey M, Dubcovsky J (2010b) Genetic maps of stem rust resistance gene *Sr35* in diploid and hexaploid wheat. Crop Sci 50:2464–2474

Zhang L, Zhao Y-L, Gao L-F, Zhao G-Y, Zhou R-H, Zhang B-S, Jia J-Z (2012) TaCKX6-D1, the ortholog of rice *OsCKX2*, is associated with grain weight in hexaploid wheat. New Phytol 195:574–584

Zhang X, Chen J, Shi C, Chen J, Zheng F, Tian J (2013) Function of TaGW2-6A and its effect on grain weight in wheat (*Triticum aestivum* L.). Euphytica 192:347–357

Zhang D, Wang B, Zhao J, Zhao X, Zhang L, Liu D, Dong L, Wang D, Mao L, Li A (2015) Divergence in homoeolog expression of the grain length-associated gene *GASR7* during wheat allo-hexaploidization. Crop J 3(1):1–9

Zhou LL, Bai GH, Carver B, Zhang DD (2007) Identification of new sources of aluminum resistance in wheat. Plant Soil 297(1–2):105–118

Zhu J, Pearce S, Burke A, See DR, Skinner DZ, Dubcovsky J, Garland-Campbell K (2014) Copy number and haplotype variation at the VRN-A1 and central FR-A2 loci are associated with frost tolerance in hexaploid wheat. Theor Appl Genet 127:1183–1197

Zikhali M, Leverington-Waite M, Fish L, Simmonds J, Orford S, Wingen LU, Goram R, Gosman N, Bentley A, Griffiths S (2014) Validation of a 1DL earliness *per se* (eps) flowering QTL in bread wheat (*Triticum aestivum*). Mol Breed 34:1023–1033

Erratum to: Use of Alien Genetic Variation for Wheat Improvement

P.K. Gupta

Erratum to:
Chapter 1 in: V.R. Rajpal et al. (eds.), *Molecular Breeding for Sustainable Crop Improvement*, Sustainable Development and Biodiversity 11, DOI 10.1007/978-3-319-27090-6_1

The original version of Chapter 1 was inadvertently published with an incorrect Figure 1.3. The correct figure should be as given below:

The online version of the original chapter can be found under
DOI 10.1007/978-3-319-27090-6_1

P.K. Gupta (✉)
Department of Genetics and Plant Breeding, Chaudhary Charan Singh University, Meerut, UP, India
e-mail: pkgupta36@gmail.com

V.R. Rajpal et al. (eds.), *Molecular Breeding for Sustainable Crop Improvement*, Sustainable Development and Biodiversity 11,
DOI 10.1007/978-3-319-27090-6_17

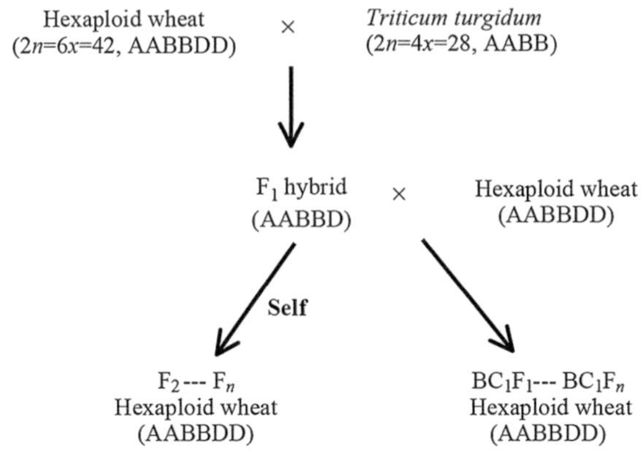

Fig. 1.3 Direct cross for transfer of one or more genes from tetraploid wheat into hexaploid wheat

Index

V.R. Rajpal et al. (eds.), *Molecular Breeding for Sustainable Crop Improvement*, Sustainable Development and Biodiversity 11,
DOI 10.1007/978-3-319-27090-6

MIX
Papier aus verantwortungsvollen Quellen
Paper from responsible sources
FSC® C105338

If you have any concerns about our products,
you can contact us on
ProductSafety@springernature.com

In case Publisher is established outside the EU,
the EU authorized representative is:
Springer Nature Customer Service Center GmbH
Europaplatz 3, 69115 Heidelberg, Germany

Printed by Libri Plureos GmbH
in Hamburg, Germany